Engineering Heat Transfer

Engineering Heat Transfer

Editor:

Harinirina Randrianarisoa, Ph.D.

ArclerPress

www.arclerpress.com

Engineering Heat Transfer

Editor: Harinirina Randrianarisoa, Ph.D.

© 2017 Arcler Press LLC
708 3rd Avenue, 6th Floor
New York
NY 10017
United States of America
www.arclerpress.com

ISBN: 978-1-68094-407-5

Library of Congress Control Number: 2016951052

About the Editor

Harinirina Randrianarisoa, Ph.D.

Hari Randrianarisoa obtained is Ph.D from Institute of Electronics, Microelectronics and Nanotechnology (IEMN) of the University of Science and Technology – Lille (France). His thesis focused for designing and manufacturing of thermal micro sensors. His interests span from on renewable energy especially solar energy. He has a strong knowledge in technology transfer field. He is currently working as a Business advisor for economic development organization in Québec City and as a consultant Scientist advisor.

Contents

List of Contributors

K. Kouloulias
Imperial College London, Mechanical Engineering Department, London SW7 2AZ, UK

A. Sergis
Imperial College London, Mechanical Engineering Department, London SW7 2AZ, UK

Y. Hardalupas
Imperial College London, Mechanical Engineering Department, London SW7 2AZ, UK

Michiel A.J. van Limbeek
Physics of Fluids, University of Twente, 7500 AE Enschede, The Netherlands

Minori Shirota
Physics of Fluids, University of Twente, 7500 AE Enschede, The Netherlands

Pascal Sleutel
Physics of Fluids, University of Twente, 7500 AE Enschede, The Netherlands

Chao Sun
Physics of Fluids, University of Twente, 7500 AE Enschede, The Netherlands

Andrea Prosperetti
Physics of Fluids, University of Twente, 7500 AE Enschede, The Netherlands
Department of Mechanical Engineering, Johns Hopkins University, Baltimore, MD 21218, USA

Detlef Lohse
Physics of Fluids, University of Twente, 7500 AE Enschede, The Netherlands

Christos N. Markides
Clean Energy Processes (CEP) Laboratory, Department of Chemical Engineering, Imperial College London, London SW7 2AZ, United Kingdom

Richard Mathie
Clean Energy Processes (CEP) Laboratory, Department of Chemical Engineering, Imperial College London, London SW7 2AZ, United Kingdom

Alexandros Charogiannis
Clean Energy Processes (CEP) Laboratory, Department of Chemical Engineering, Imperial College London, London SW7 2AZ, United Kingdom

Zhiwei Ma
Sir Joseph Swan Centre for Energy Research, Newcastle University, Newcastle upon Tyne NE1 7RU, United Kingdom

Huashan Bao
Sir Joseph Swan Centre for Energy Research, Newcastle University, Newcastle upon Tyne NE1 7RU, United Kingdom

Anthony Paul Roskilly
Sir Joseph Swan Centre for Energy Research, Newcastle University, Newcastle upon Tyne NE1 7RU, United Kingdom

Ilaria Guarracino
Clean Energy Processes (CEP) Laboratory, Department of Chemical Engineering, Imperial College London, South Kensington Campus, London SW7 2AZ, UK

Alexander Mellor
Blackett Laboratory, Department of Physics, Imperial College London, South Kensington Campus, London SW7 2AZ, UK

Nicholas J. Ekins-Daukes
Blackett Laboratory, Department of Physics, Imperial College London, South Kensington Campus, London SW7 2AZ, UK

Christos N. Markides
Clean Energy Processes (CEP) Laboratory, Department of Chemical Engineering, Imperial College London, South Kensington Campus, London SW7 2AZ, UK

Leonardo Micheli
Environment and Sustainability Institute, University of Exeter, Penryn Campus, Penryn, Cornwall TR10 9FE, UK

K.S. Reddy
Heat Transfer and Thermal Power Laboratory, Department of Mechanical Engineering, Indian Institute of Technology Madras, Chennai 600 036, India

Tapas K. Mallick
Environment and Sustainability Institute, University of Exeter, Penryn Campus, Penryn, Cornwall TR10 9FE, UK

Zhang Zhigang
School of Automobile Engineering, Chongqing University of Technology, Chongqing 400054, China
Key Laboratory of Safety Design and Reliability Technology for Engineering Vehicle, Changsha University of Science & Technology, Hunan 410114, China

Shi Xiaohui
School of Automobile Engineering, Chongqing University of Technology, Chongqing 400054, China

Guo Dong
School of Automobile Engineering, Chongqing University of Technology, Chongqing 400054, China

J. L. González-Santander
Universidad Católica de Valencia "San Vicente mártir", C/ Guillem de Castro 94, 46001 Valencia, Spain

Yuan Zhang
School of Mines, China University of Mining & Technology, No. 1 University Road, Xuzhou 221116, China
Key Laboratory of Deep Coal Resource Mining of Ministry of Education, China University of Mining & Technology

Zhijun Wan
School of Mines, China University of Mining & Technology, No. 1 University Road, Xuzhou 221116, China
Key Laboratory of Deep Coal Resource Mining of Ministry of Education, China University of Mining & Technology

Bin Gu
School of Mines, China University of Mining & Technology, No. 1 University Road, Xuzhou 221116, China

Hongwei Zhang
School of Mines, China University of Mining & Technology, No. 1 University Road, Xuzhou 221116, China

Peng Zhou
School of Mines, China University of Mining & Technology, No. 1 University Road, Xuzhou 221116, China

Gao Ma
College of Civil Engineering, Hunan University, Changsha 410082, China

Liang Huang
College of Civil Engineering, Hunan University, Changsha 410082, China

Libo Yan
Department of Organic and Wood-Based Construction Materials, Technical University of Braunschweig, Hopfengarten 20, 38102 Braunschweig, Germany
Centre for Light and Environmentally-Friendly Structures, Fraunhofer Wilhelm-Klauditz-Institut WKI, Bienroder

Hui Wang
College of Civil Engineering, Hunan University, Changsha 410082, China

Peng Yin
College of Civil Engineering, Hunan University, Changsha 410082, China

Zhaoyong Mao
School of Marine Science and Technology, Northwestern Polytechnical University, Xi'an 710072, China

Shaokun Yan
School of Marine Science and Technology, Northwestern Polytechnical University, Xi'an 710072, China

Ming Huang
Graduate School of Information Science, Nara Institute of Science and Technology, Ikoma 6300192, Japan

Toshiyo Tamura
Department of Biological Engineering, Osaka Electro-Communication University, Shijonawate, Japan

Zunyi Tang
Department of Biological Engineering, Osaka Electro-Communication University, Shijonawate, Japan

Wenxi Chen
Biomedical Information Technology Laboratory, The University of Aizu, Tsuruga, Aizuwakamatsu, Japan

Shigehiko Kanaya
Graduate School of Information Science, Nara Institute of Science and Technology, Ikoma 6300192, Japan

Yuan Zhang
School of Energy and Power Engineering, Jiangsu University, 301 Xuefu Road, Zhenjiang, Jiangsu 212013, China

Sunqi Zhuang
College of Urban Construction, Nanjing Tech University, Nanjing, China

Qian Wang
School of Energy and Power Engineering, Jiangsu University, 301 Xuefu Road, Zhenjiang, Jiangsu 212013, China

Jiapeng He
College of Urban Construction, Nanjing Tech University, Nanjing, China

Bin Zheng
School of Energy and Power Engineering, University of Shanghai for Science and Technology, Shanghai 200093, China
School of Transportation and Vehicle Engineering, Shandong University of Technology, Zibo 255000, China

Yongqi Liu
School of Transportation and Vehicle Engineering, Shandong University of Technology, Zibo 255000, China

Lichen Zou
School of Transportation and Vehicle Engineering, Shandong University of Technology, Zibo 255000, China

Ruiyang Li
School of Energy and Power Engineering, University of Shanghai for Science and Technology, Shanghai 200093, China

Odelu Ojjela
Department of Applied Mathematics, Defence Institute of Advanced Technology (Deemed University), Pune 411025, India

Pravin Kashyap Kambhatla
Department of Applied Mathematics, Defence Institute of Advanced Technology (Deemed University), Pune 411025, India

N. Naresh Kumar
Department of Applied Mathematics, Defence Institute of Advanced Technology (Deemed University), Pune 411025, India

Samir Kumar Das
Department of Applied Mathematics, Defence Institute of Advanced Technology (Deemed University), Pune 411025, India

Gautham Krishnamoorthy
Department of Chemical Engineering, University of North Dakota, Harrington Hall Room 323, 241 Centennial Drive, Grand Forks, ND 58202-7101, USA

Lauren Elizabeth Clarke
Department of Chemical Engineering, University of North Dakota, Harrington Hall Room 323, 241 Centennial Drive, Grand Forks, ND 58202-7101, USA

Petr Kracík
Institute of Power Engineering, Brno University of Technology, Faculty of Mechanical Engineering, Technická 2896/2, 616 69 Brno, The Czech Republic

Jiří Pospíšil
Institute of Power Engineering, Brno University of Technology, Faculty of Mechanical Engineering, Technická 2896/2, 616 69 Brno, The Czech Republic

Jamshid Khorshidi
Department of Engineering, Hormozgan University, Bandar Abbas, Iran

Salman Heidari
Department of Engineering, Hormozgan University, Bandar Abbas, Iran

Taasnim Ahmed Himika
Department of Mathematics & Physics, North South University, Dhaka 1229, Bangladesh

Md. Farhad Hasan
Department of Mathematics & Physics, North South University, Dhaka 1229, Bangladesh

Md. Mamun Molla
Department of Mathematics & Physics, North South University, Dhaka 1229, Bangladesh

Lei Yu
College of Population, Resources and Environment, Shandong Normal University, Jinan 250014, China

Jiemin Cheng
College of Population, Resources and Environment, Shandong Normal University, Jinan 250014, China

Zoubair Boulahia
Faculty of Sciences Aïn Chock, Laboratory of Mechanics, Hassan II University, BP 5366, Maarif, 20100 Casablanca, Morocco

Abderrahim Wakif
Faculty of Sciences Aïn Chock, Laboratory of Mechanics, Hassan II University, BP 5366, Maarif, 20100 Casablanca, Morocco

Rachid Sehaqui
Faculty of Sciences Aïn Chock, Laboratory of Mechanics, Hassan II University, BP 5366, Maarif, 20100 Casablanca, Morocco

Jyoti Sharma
Energy Research Centre, Panjab University, Chandigarh 160014, India

Urvashi Gupta
Dr. S. S. Bhatnagar University Institute of Chemical Engineering and Technology, Panjab University, Chandigarh 160014, India

R. K. Wanchoo
Dr. S. S. Bhatnagar University Institute of Chemical Engineering and Technology, Panjab University, Chandigarh 160014, India

Preface

Heat transfer is one of the most common physical phenomena. Most devices include heat sources and heat exchange with the environment. The problem of thermal management has become essential in many industrial sectors. It especially impacts the electronics industry for many years due to the increasing density of components. Thereby, the effects of significant overheating during thermal cycling in electronic components are numerous and affect both the reliability and durability of components and power circuits, compromising the operation and safety of complex systems. Furthermore, the sharing of a common thermal environment increasingly restricted by several heat sources makes the thermal coupling between critical and unpredictable components especially as cooling paths can be multiple. With regards to engineering heat transfer, this book discussed the applied thermal process, the work on the temperature control component. The research objects are addressed experimentally and numerically and lead to innovative design, also made in laboratory. The design of engineering project and development of model for heat transfer process are also carried out on this book to enable the study of systems and to optimize the design.

Editor
Harinirina Randrianarisoa, Ph.D.

CHAPTER 1

SEDIMENTATION IN NANOFLUIDS DURING A NATURAL CONVECTION EXPERIMENT

K. Kouloulias, A. Sergis, Y. Hardalupas

Imperial College London, Mechanical Engineering Department, London SW7 2AZ, UK

ABSTRACT

This study presents an experimental investigation of the thermophysical behavior of γ-Al_2O_3–deionized (DI) H_2O nanofluid under natural convection in the classical Rayleigh–Benard configuration, which consists of a cubic cell with conductive bottom and top plates, insulated sidewalls and optical access. The presence of nanoparticles either in stationary liquids or in flows affects the physical properties of the host fluids as well as the mechanisms and rate of heat and mass transfer. In the present work, measurements of heat transfer performance and thermophysical properties of Al_2O_3–H_2O nanofluids, with nanoparticle concentration within the range of 0.01–0.12 vol.%, are compared to those for pure DI water that serves as a benchmark. The natural convective chamber induces thermal instability in the vertical direction in the test medium by heating the medium from below and cooling it from above. Fixed heat flux at the bottom hot plate and constant temperature at the top cold plate are the imposed boundary conditions. The Al_2O_3–H_2O nanofluid is tested under different boundary conditions and various nanoparticle concentrations until steady state conditions are reached. It is found that while the Rayleigh number, *Ra*, increases with increasing nanoparticle concentration, the convective heat transfer coefficient and Nusselt number, *Nu*, decrease. This finding implies that the addition of Al_2O_3 nanoparticles deteriorates the heat transfer performance due to natural convection of the base fluid, mainly due to poor nanofluid stability. Also, as the nanoparticle concentration increases the temperature at the heating plate increases, suggesting fouling at the bottom surface; a stationary thin layer structure of nanoparticles and liquid seems to be formed close to the heating plate that is qualitatively observed to increase in thickness as the nanoparticle concentration increases. This layer structure imposes additional thermal insulation in the system and thus appears to be responsible in a big extend for the reported heat transfer degradation. Also, for relatively high nanoparticle concentrations of 0.06 and 0.12 vol.%, as the heating flux increases the rate of heat transfer deterioration increases. Specifically in the case of maximum

Citation: K. Kouloulias, A. Sergis, Y. Hardalupas, Sedimentation in nanofluids during a natural convection experiment, International Journal of Heat and Mass Transfer, Volume 101, October 2016, Pages 1193-1203, ISSN 0017-9310, http://dx.doi.org/10.1016/j.ijheatmasstransfer.2016.05.113.

nanoparticle concentration, 0.12 vol.%, when the turbulence intensity increases, by increasing the applied heat flux, the Nusselt number remains constant in comparison with lower nanoparticle concentrations. This behavior can be attributed mainly to the physical properties of the Al_2O_3 nanopowder used in this study and the resulting interactions between the heating plate and the nanoparticles.

INTRODUCTION

High performance cooling is one of the most vital needs of many industrial technologies to date, from micro/nanoelectronic mechanical systems (MEMS/NEMS) to nuclear reactors. Material science and technology has reached its limits regarding the increase of the solid-fluid interface area. In addition, downsizing appears to be the current trend. As a result, high heat fluxes have to be accommodated to ensure a safe, efficient and long-lasting operation of the heat exchangers and subsequently of the systems involved. As a consequence, there is a variety of engineering applications that demand the development of new coolants with superior thermophysical characteristics. One such application is fusion reactors in the nuclear energy sector, where extremely high heat fluxes (up to 20–30 MW/m^2) are present in both the current generation of experimental reactors, as well as future Fusion power plants [1].

Nomenclature

q''	applied heat flux [W/m^2]
A	surface area of the heating plate [m^2]
c_p	specific heat capacity [J/kg K]
d	distance [m]
F	power output from the heat exchanger [W]
Gr	Grashof number [–]
h	heat transfer coefficient [W/m^2 K]
k	thermal conductivity [W/mK]
L	characteristic length [m]
Nu	Nusselt number [–]
P	heat losses from the chamber [W]
Pr	Prandtl number [–]
Q	power input from the heating plate [W]
Ra	Rayleigh number [–]
T	temperature [K]
W	flow rate of cooling water [L/min]

Greek characters

α	thermal diffusivity [m^2/s]
β	thermal expansion coefficient [1/K]
μ	dynamic viscosity [N s/m^2]
ν	kinematic viscosity [m^2/s]
ρ	density [kg/m^3]
σ	uncertainty in the mean value of h [W/m^2 K]
φ	nanoparticle volume fraction [–]

Subscripts

c	cold
cl	lower surface of the cold plate
h	hot
hu	upper surface of the hot plate
$layer$	layer structure of nanoparticles and liquid
n	nanoparticles
nf	nanofluid
pl	aluminum plate
pth	thermocouple position
w	DI water

Recent advances in nanotechnology have allowed the development of a new category of coolants, namely nanofluids, as initially coined and proposed by Stephen U.S. Choi of Argonne National Laboratory in the early 1990s, while the first study was carried out in 1995 [2] and [3]. By definition, nanofluids are a new class of heat transfer fluids engineered by dispersing and stably suspending nanoparticles with typical size of the order of 1–100 nm in traditional heat transfer fluids [4]. Common materials used for nanoparticles include metals, metal oxides and carbon nanotubes of volumetric fractions usually between 0.001% and 10%, whereas popular carrier fluids include water, ethanol, ethylene glycol and engine oil [5] and [6]. During the last two decades, nanofluids have attracted great scientific interest due to their unexpectedly superior heat transfer properties over those of conventional fluids. More specifically, a very small amount of nanoparticles (<1 vol.%) can provide a dramatic improvement in the thermal properties of the host fluid. Based on a statistical analysis of available data in the literature, nanofluids offer an enhancement of 5–9% and 10–14% for the conductive and mixed conductive/convective heat transfer modes respectively, and 40–44% and 100–200% enhancement for pool boiling and critical heat flux respectively [7]. However, the insufficient understanding of the formulation and underlying physical mechanisms of heat transfer in nanofluids has led to an increased controversy and inconsistency among the numerous theoretical, numerical and experimental studies [8]. Therefore, before using nanofluids for commercial applications, an in-depth understanding of the physical mechanisms of flow and heat transfer in such fluids is needed [9].

Natural convection as a heat transfer mechanism is widely applied to a variety of applications, where the thermal management results only from the density difference of the working fluid involved. However, limited experimental studies have been carried out on nanofluid natural convection, compared to conduction and forced convection[10] and [11]. While substantial enhancement of heat transfer is reported for forced convection [12], [13] and [14], contradictory results are present between numerical and experimental studies for natural convection in the literature [15]. In most of the theoretical or numerical works, heat transfer enhancement is reported, whereas in experimental investigations unexpected deterioration is observed. Khanafer et al. [16] investigated numerically natural convection of copper-based nanofluids in a two-dimensional enclosure. Their results suggested substantial increase of the heat transfer rate for any given Grashof number. In addition, as the nanoparticle volume fraction increased, the heat transfer ratio also increased. Oztop and Abu-Nada [17] modeled natural convection of different nanofluids in a partially heated enclosure. They reported that the mean Nusselt number increased with the volume fraction of nanoparticles over the whole Rayleigh number range. Also, their results indicated the importance of both heater location and dimension, as well as the aspect ratio of the enclosure on heat transfer enhancement, when using nanofluids. Aminossadati and Ghasemi [18] performed a numerical study for natural convection of various nanofluids in a partially heated enclosure. They indicated that Al_2O_3, TiO_2 and even Cu and Ag nanoparticles provided high cooling performance by reducing the maximum temperature of the heat source significantly. Meng and Li [19] investigated numerically Al_2O_3–H_2O nanofluids under natural convection in a horizontal cylinder. Their work revealed that the nanofluid did not exhibit improved heat transfer performance than the base fluid. In addition, as the nanoparticle concentration increased, the heat transfer deterioration became more predominant.

Putra at al. [9] used Al_2O_3 and CuO nanoparticles suspended in water. Their experimental results indicated that the natural convective heat transfer of nanofluids was lower than that of pure water and that the deterioration increased with increased particle concentration. Also, the deterioration was greater for CuO than for Al_2O_3 nanofluids. This difference was attributed to the slip mechanism between the denser CuO particles, compared to Al_2O_3 particles, and the carrier fluid. Both the particle–fluid slip and the sedimentation mechanisms appeared to be of great importance for the heat transfer performance of the nanofluids. Wen and Ding [20] studied experimentally the performance of TiO_2–H_2O nanofluids under natural convection. They reconfirmed that deterioration of natural convective heat transfer increased with nanoparticle concentration. Possible explanations for this behavior were the convection induced by concentration difference, the particle–surface and particle–particle interactions and the modification of the dispersion properties. Nnanna [21] reported different experimentally measured convective heat transfer performance for small and large nanoparticle volumetric fractions, φ, for Al_2O_3–H_2O nanofluids under natural convection. For $0.2\% < \varphi > 2.0\%$, the natural convective heat transfer was enhanced, whereas for$\varphi >$ 2% heat transfer deterioration was reported. The augmentation was credited to the thermo-hydrodynamic effect that is caused by the changes in thermal conductivity and viscosity, where deterioration was attributed to the reduction of the Rayleigh number due to the increased kinematic viscosity. Li and Peterson [22] investigated experimentally the heat transfer characteristics of Al_2O_3–H_2O nanofluids. They reported a systematic deterioration of heat transfer with an increase of the volume fraction of nanoparticles. The observed performance was mainly attributed to the higher viscosity of nanofluids, the Brownian motion and thermophoretic motion of nanoparticles. Ni et al. [2] conducted fully turbulent natural convection experiments with water-based alumina nanofluids. Their results indicate that the convective heat transfer coefficient, the Nusselt and the Rayleigh numbers all decrease with increasing nanoparticle volumetric concentrations between 0% and 1.08%. The mass diffusion of the nanoparticles inside the nanofluid was suggested as a possible reason for the measured deterioration.

In cases where stationary or buoyancy-driven nanofluids are involved, the physical characteristics and the long-term stability of the nanoparticle suspension play key roles. The stability of an aqueous nanofluid is directly linked to its electrokinetic properties as high surface charge results in strong repulsive forces between the nanoparticles [23]. In the literature there are many effective methods for the preparation of stable suspensions. The most widely applied methods are the ultrasonic vibration, the electrostatic stabilisation and the steric stabilisation [23] and [24]. Numerous studies have been conducted on the analysis of nanofluid stability pointing out the critical effect of the pH of the carrier fluid [25], [26], [27] and [28]. The overall behavior of a nanofluid depends on whether the water is acidic or alkaline. The surface of Al_2O_3 nanoparticles in acidic and neutral DI water has a positive charge, whereas in alkali water it is negatively charged[26] and [29]. However, there is a certain pH value where a nanofluid reaches an equipotential point, the isoelectric point (IEP). At this point, the electric charge on the particles is zero [23] and [30]. As a result, the ideal pH corresponds to a high surface charge or zeta potential [27] and [31]. An increased zeta potential corresponds to stronger electrostatic repulsion forces among the charged nanoparticles due to the electric double layer that is formed around the particles. It is well studied that a suspension with zeta potential below 20 mV exhibits limited stability, whereas for 30 mV and above, it is characterized as physically stable [29]. However, the IEP depends not only on the ionic constitution of the base fluid, but also on the material, phase and purity of the nanoparticles. For γ-Al_2O_3 the IEP is near 7.7–7.9 [31]. Wen and Ding [28]concluded that for γ-Al_2O_3 a pH of 7 is appropriate for their experimental work, in the sense that it was reasonably away from the IEP and no damage would be caused to their heating surface. Pak and Cho reported that a pH value of 3 would ensure good homogeneous dispersion, whereas Adegbite [31] calculated the optimum pH value at 4. Mao et al [25] noted that when the pH was less than 7, the stability was improved, but for higher values sedimentation appeared in the system.

Therefore, the literature shows that the heat transfer behavior under natural convection conditions is unclear and both enhancement and deterioration have been observed, which suggests that additional understanding is required. Suggestions for the observed behavior have been provided in terms of electrostatic behavior of the nanoparticles in the base fluid and possible sedimentation. However, there is no conclusive experimental evidence for all the suggested sources influencing nanoparticle behavior and heat transfer rate and this forms the purpose of the currently reported research. The remaining paper presents the experimental arrangement and methodology in the next section, the results in terms of heat transfer performance – absolute and normalized heat transfer coefficient, Nusselt and Rayleigh numbers under different boundary conditions and for various nanoparticle volume fractions – and associated discussion in Section 3. Finally, a summary of the main conclusions is presented.

EXPERIMENTAL FACILITY AND PROCEDURE

Experimental Setup

The experimental setup is attempting to generate a classical Rayleigh–Benard system, in order to measure the heat transfer coefficient under natural convection. It comprises a cubic cell with a volumetric capacity of 1×10^{-3} m^3 with optical access through the side walls. A schematic diagram of the apparatus is shown in Fig. 1. The natural convection rig consists of many components that all together ensure a satisfactory operation, undisturbed flow fields and minimum heat losses. The bottom plate consists of an aluminum heating pan **A** of thickness 20 mm, which houses 5 parallel cartridge heaters of 300 W power output each. The heating pan **A** provides heat power to an insulating pan **B**that is placed on top and under the aluminum heating plate **C**. This configuration ensures that heat transfer from the heating plate in the downward direction is minimized. The heating plate **C**, which is 30 mm thick, consists of 4 parallel cartridge heaters of 80 W power output each. Around the heating plate, there is a Teflon plate **D** to eliminate any heat losses from the sides. The aluminum cubic cell **E** is the key component that houses the tested medium. It also provides perimetric optical access for possible visualization studies, by incorporating four quartz windows (2 large square [40 mm × 40mm] and 2 small rectangular [10 mm × 40 mm]). Around the cell, there are 4 specially designed insulating pans **F**, 34 mm thick, and also a transparent Plexiglas cover **G** of 10 mm thickness. This provides additional insulation and brings all the subcomponents of the chamber together. At the top of the test cell, there is an aluminum heat sink **H** connected with a cooling water supply. Finally, between the heat sink and the cubic cell, a Teflon plate **I** is placed to prevent their thermal connection.

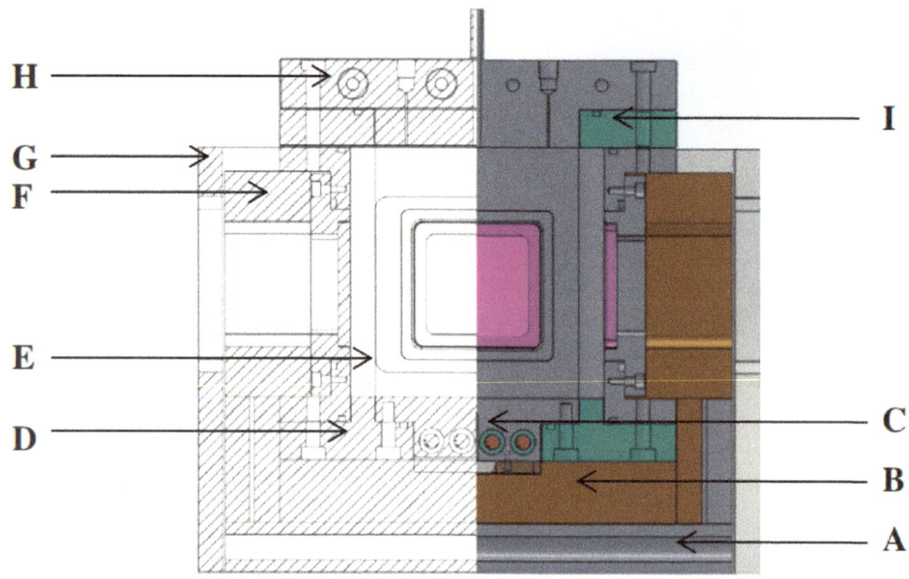

Figure 1. Schematic drawing of the natural convection chamber. The marked components are explained in the text.

Eleven thermocouples are placed at various locations inside and outside the chamber to monitor and evaluate heat transfer performance. From top to bottom, six thermocouples are placed at the heat exchanger. Two of them measure the temperature at the inlet and outlet of the cooling unit, whereas another four measure the temperature at the top, cold plate, 5 mm above the free surface of the cell. Two thermocouples are placed at the heating plate. The first one is vertically placed 6 mm below the cell's free surface to measure the temperature T_h of the bottom, hot plate. The second is placed on the bottom of the plate and measures the temperature there. One thermocouple is attached at the top surface of the heating pan and is used as a key indicator of the heat flux provided by the pan in order to match the temperature of the heating plate and thus eliminate the heat losses downwards. Finally, two thermocouples are embedded inside the heating plate and heating pan and measure the temperature there, so as to prevent any overheating or failure of the cartridge heaters.

Control System

The operation of the nanofluid chamber was controlled through LabVIEW software, coupled with National Instruments (NI) hardware and supplementary in-house electrical device. Through the software, the operation of the rig was monitored and controlled in real time, while the data, namely input power, energy losses and temperatures, were stored at a sampling rate of 1 Hz.

The NI hardware consists of a CompactDAQ chassis (cDAQ-9188), which integrates connectivity and signal conditioning into modular I/O for direct interfacing to any sensor or signal, and voltage input (NI 9213) and output (NI 9263) modules. The additional in-house electrical device, which is connected between the CompactDAQ chassis and the natural convection chamber, consists of two power controlling thyristors (United Automation PR1-DIN-2.5KW) and two temperature controllers (TC Direct 309100). Each thyristor corresponds to a heater assembly – one for the heating plate and the second for the heating pan. Both thyristors are connected to the NI voltage output module and the heaters (heating plate and pan). The temperature controllers are connected with two K-type thermocouples, located inside the heaters. Through an improved autotuning algorithm, they achieve fast and stable temperature control and prevent possible damages to the heaters and the rig in the unexpected case of over-heating.

The LabVIEW software incorporates different operating scenarios according to the selected boundary conditions. In this study, constant heat flux at the heating plate and constant temperature at the cooling unit were enabled. The main feature in this operating scenario is a virtual proportional–integrate–derivative (PID) controller that matches the temperature of the heating pan to the heating plate, hence eliminating any heat losses in the downward direction.

Nanofluid Preparation

In this study, no pH changers or surface activators were added during any stage of the nanoparticle production or the synthesis of the nanofluid as these could affect the thermophysical properties of the suspension [20]. For instance, Eastman et al. [32]reported that in samples where a small amount of thioglycolic acid (<1 vol.%) was added, the thermal conductivity was improved compared to non-acid-containing nanofluids. Also, Wen and Ding [28], studied the thermal behavior of carbon nanotube suspensions by using sodium dodecyl benzene sulfonate (SDBS). However, their results indicated that the added SDBS chemical structure was catastrophically altered at 69 °C causing severe nanofluid stability deterioration. Therefore, in the present study special care was paid to the nanofluid synthesis and the preparation process (optimum ultrasonication time, vortex mixing and high purity DI water were involved).

A two-step preparation method was followed, according to which the nanoparticles are dispersed as a dry powder into the base fluid. The γ-Al_2O_3 nanoparticles had a nominal particle size of less than 50 nm (quantified by TEM), a particle density of 4000 kg/m^3 and were supplied by Sigma Aldrich. Additional TEM measurements were performed in-house to identify the average particle size, shape and purity of the aluminum oxide nanopowder. For these measurements, a JEOL 2100F TEM operating at 200 kV was employed, according to the process described in Barrett et al. [6]. Fig. 2 presents TEM images for the aforementioned aluminum oxide powder. It can be observed that the nanoparticles are highly agglomerated and form large clusters. This was expected since no ultrasonication was performed and the nanopowder is initially in agglomerated form due to the strong Van der Waals forces. Also, this figure depicts both the size and shape of the nanoparticles. It can be seen that the identified ultra-fine particles look like nanoclay flakes and spheres with an average diameter of 10 nm. In addition, the material composition was measured and verified through TEM. In Fig. 3, spectral data for the nanopowder indicate a highly pure (~ 99%) aluminum oxide powder.

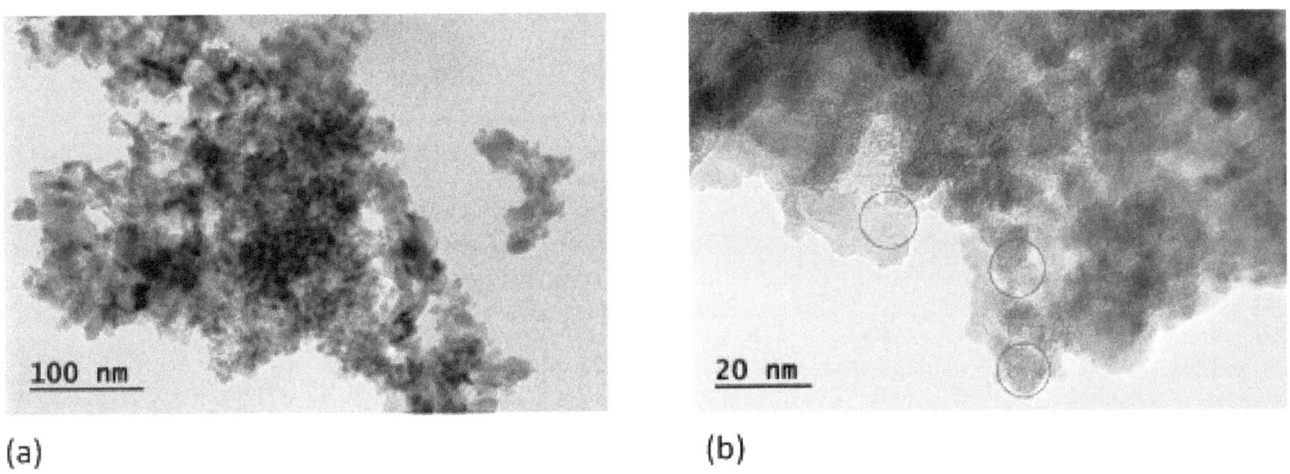

Figure 2. TEM images of γ-Al_2O_3 nanoparticles; (a) large particle agglomerate and (b) average particle size and shape in the sample.

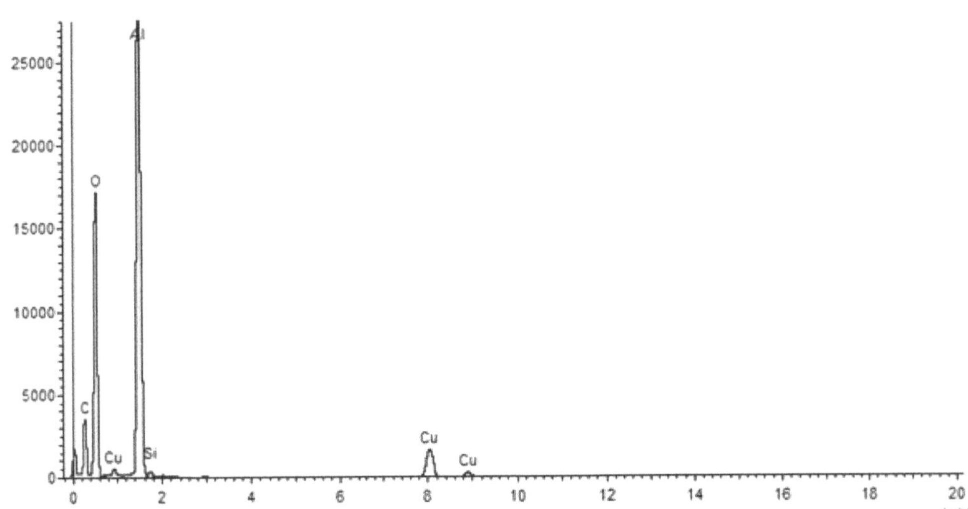

Figure 3. TEM spectral data revealing highly pure aluminum oxide powder with minor impurities in the sample.

High purity DI water, supplied by VWR, was used as the carrier fluid. At the beginning of every experiment, the required amount of nanopowder was measured by weight, using a high accuracy electronic balance (Sartorius semi-microbalance R 200 D). Then, the powder was added to a 15 ml glass test tube and filled up with DI water. The next step was the ultrasonication process that is followed in order to break the agglomerates into their primary particle size. The reason is that the commercially available dry nanoparticles are initially in an agglomerated form due to the strong Van der Waals attractive forces among the particles [28], [30] and [33]. An ultrasonic bath (Pulsatron KC2 by Guyson International Ltd) was used in combination with an analogue vortex mixer (mini vortex mixer by VWR) for a period of 5.0 h [6] and [29] to effectively break and disperse the weakly-bonded agglomerates by high shearing.

Methodology

The first step of the current study was a careful evaluation and characterization of the apparatus used, in order to select appropriate boundary conditions for the needs of this work. Three different power inputs at the heating plate were used to test the nanofluid under varying induced flow regimes. The minimum power input was set at 125 W, which ensures that the maximum heat losses from the chamber are less than 5% of the input heat flux. The intermediate power input was set at 150 W and the maximum at 175 W, since beyond this power boiling is initiated. γ-Al_2O_3–DI H_2O nanofluids of various concentrations, ranging from 0.01 to 0.12 vol.%, were synthesized and tested for all three power inputs. The heat transfer results of this study were considered and compared under steady state conditions. In the absence of pH changers and dispersants, there was a need to perform the nanofluid experiment fast enough to eliminate any issues related to the stability of the suspension. For this reason, the nanofluid experiments were performed when steady state conditions for pure DI water had been reached. The experimental procedure involved the deactivation of the heating elements for a short period of time (~5 min) and the removal of the top, cold plate. Then, the sonicated nanofluid samples were poured into the chamber and stirred well with the rest of the base fluid to ensure good dispersion. Finally, the heaters were switched on again and the rig was operated until steady state conditions were reached. At steady state (about 3.0 h after switching on the heaters), the equivalent temperatures of the heating and cooling plates were calculated by averaging a sample record of 1000 sets of temperature measurements recorded at a sampling frequency of 1 Hz via the corresponding hot and cold plate thermocouples. This averaging process is followed in order to eliminate the temporal and spatial temperature fluctuations on the plates arising due to turbulence and buoyancy effects. Subsequently, the heat transfer properties, namely heat transfer coefficient, Nusselt and Rayleigh numbers were calculated.

RESULTS AND DISCUSSION

In this part, results of the heat transfer performance under natural convection of nanofluid are presented and compared with DI water. More specifically, the absolute and normalized (with the value for DI water) heat transfer coefficient as a function of nanoparticle volume fraction and power input is depicted. In addition, Rayleigh number in terms of power input and Nusselt as a function of Rayleigh number are illustrated to identify their dependency and evaluate the resulting heat transfer performance.

The section is divided into six subsections; at first, the quality of the nanofluid is evaluated in terms of stability, by performing a qualitative visualization study of a 'cold' nanofluid placed in a glass flask. Following, the thermophysical properties of the test fluids are considered by including some widely available correlations used in the present study. Then, the temporal evolution of the temperature difference between the heating and cooling plate until steady state conditions is presented for both water and nanofluid. Finally, the influence of the nanoparticle volume fraction as well as the heat flux on the heat transfer properties of the nanofluid is investigated and finally, an extensive discussion on the findings is included.

Nanofluid Stability

For the study of the thermophysical properties of nanofluids, pure nanosuspensions were used without any surface additives or pH controllers. Thus, any mechanism responsible for the altered thermophysical properties of nanofluids can be better identified and studied. Therefore, in such cases, the purity of DI water as well as its initial pH value play dominant roles. The quality of water is strongly associated with the followed purification process. Therefore, technical standards on water quality have been established by many national/international organizations. One such is the international organization for standardization (ISO), according to which there are certain limits to categorize water in terms of purity. In this study, high purity DI water (grade 3 according to ISO 3696 standard) that is produced by distillation and capacitive deionisation was used with an initial pH value of 6.9. However, both the introduction of Al_2O_3 nanoparticles in the base fluid and the CO_2 of the atmosphere affect the final pH value of the nanofluid.

A visualization study, Fig. 4, was conducted to qualitatively assess the stability of a freshly prepared nanofluid and record any possible sedimentation or separation in terms of time. The study was performed for a 'cold' nanofluid (no heat flux was induced), placed in a 1.0 L rectangular glass flask that approaches the shape of the chamber's main body. It can be observed that after 24 h, no visible change in the nanofluid stability took place. However, in the middle of the first week after the preparation of the nanofluid mixture, the suspension became less bright, indicating significant nanoparticle deposition. During the period between the first half week and up to the second week, the nanofluid appeared qualitatively unchanged in terms of stability, while in the third week concentration stratification layers were visible. In addition, deposition of aluminum nanoparticles or agglomerates was noticed at the bottom of the flask even during the first few hours.

Figure 4. Stability of Al_2O_3–water nanofluid (φ = 0.06 vol.%) as a function of time.

Thermophysical Properties of Nanofluids

The Al_2O_3–H_2O nanofluid is a well-studied suspension concerning its physical properties. The water properties, namely dynamic viscosity, μ_w, the thermal conductivity, k_w, the heat capacity, c_p and the thermal expansion coefficient, β, were obtained from Rohsenow et al. [34]. As for the nanofluid, they were calculated by widely used empirical correlations and formulas from the literature. Also, both the properties of water and nanofluid were considered at the mean temperature of the hot and cold plates.

The density of a nanofluid, ρ_{nf}, can be obtained from the following widely accepted formula [2], [5], [19], [35], [36], [37], [38] and [39] which relates the density of the base fluid, ρ_w, with the density and the volume fraction of the nanoparticles, ρ_n and φ, respectively

$$\rho_{nf} = (1-\varphi)\rho_w + \varphi\rho_n \quad (1)$$

In a similar way, the heat capacity of the nanosuspension, $c_{p,nf}$, can be expressed by the extensively used correlation [2], [35] and [40]:

$$c_{p,nf} = \frac{\varphi\rho_n c_{p,n} + (1-\varphi)\rho_w c_{p,w}}{(1-\varphi)\rho_w + \varphi\rho_n} \quad (2)$$

For the nanofluid viscosity, μ_{nf}, we adapt an empirical formula that was initially obtained by Maiga et al. [41] but also verified and used by Ni et al. [2]

$$\mu_{nf} = \mu_w(1 + 7.3\varphi + 123\varphi^2) \quad (2)$$

It is based on experimental data using aluminum based nanofluid with an average particle diameter of 20 nm, which was very close to ours, for room temperature. Nevertheless, it was experimentally observed that the measured viscosity did not have a strong temperature dependency and thus gave good representation over a wide range of temperatures.

Finally, for the conductivity, k_{nf}, a fitting formula used by Ni et al. [2], which is based on an empirical relationship [41], was involved. This has been evaluated with different Al_2O_3–H_2O nanofluid data sets over comparable ranges of temperature and concentration with the present work:

$$k_{nf} = k_w[1 + (-49.796 + 0.178T)\varphi + (535.576 - 1.840T)\varphi^2] \quad (3)$$

In the current investigation the particle size involved was smaller than that of Ni et al. [2], but for relatively low nanoparticle concentrations – such as those used in this study – Eq.(4) can be approximately reduced to the theoretical correlation from the effective medium theory [2] and [42]. Therefore, no significant augmentation of the nanofluid thermal properties was expected.

In free convection applications, the heat transfer correlations are of the form of Nu=f(Gr,Pr), where the Nusselt number (Nu) is a function of Grashof (Gr) and Prandtl (Pr) numbers. The Nu number relates the convective with the conductive heat transfer mode and is given by,

$$Nu = \frac{hL}{k} \quad (5)$$

On the other hand, Gr governs the transition in the free convection boundary layer and Prcontrols the relative thickness of the momentum and thermal boundary layers. These are given by

$$Gr = \frac{g\beta\Delta TL^3}{v^2} \quad (6)$$

$$Pr = \frac{v}{\alpha} \quad (7)$$

The product of the Gr and Pr numbers is known as Rayleigh number, Ra and is commonly used for free convection applications. This is given by,

$$Ra = \frac{g\beta\Delta TL^3}{v\alpha} \quad (8)$$

At steady state conditions, it is assumed that the temperature at the heating and cooling plates is uniform due to the highly conductive material of the aluminum plates and the insulation Teflon plates that surround them. Thus, the heat diffusion equation was adopted to calculate the surface temperatures T_{hu} and T_{cl}:

$$T_{hu} = T_h - \frac{q''d_h}{k_{pl}} \quad (9)$$

$$T_{cl} = T_c + \frac{q''d_c}{k_{pl}} \tag{10}$$

As a result, the heat transfer coefficient is given by,

$$h = \frac{Q}{A^2(T_{hu} - T_{cl})} \tag{11}$$

Transient and Steady State of the Chamber Wall Temperature

The experiments were conducted until steady state conditions were reached and temperature signals were acquired and recorded every second. Due to the methodology followed for the nanofluid, no direct comparison can be made concerning the time required to achieve steady state between pure water and nanofluid. Steady state condition is assumed to be attained when the temperature difference between the hot and the cold plates becomes constant and the energy rate absorbed by the cooling unit reaches a constant value. In Fig. 5, for DI water, a nearly constant temperature difference is reached after an hour for all three input power inputs. However, steady state conditions are reached after 3.0 h where the power absorbed by the cooling stabilises at a particular level, as depicted on Fig. 6.

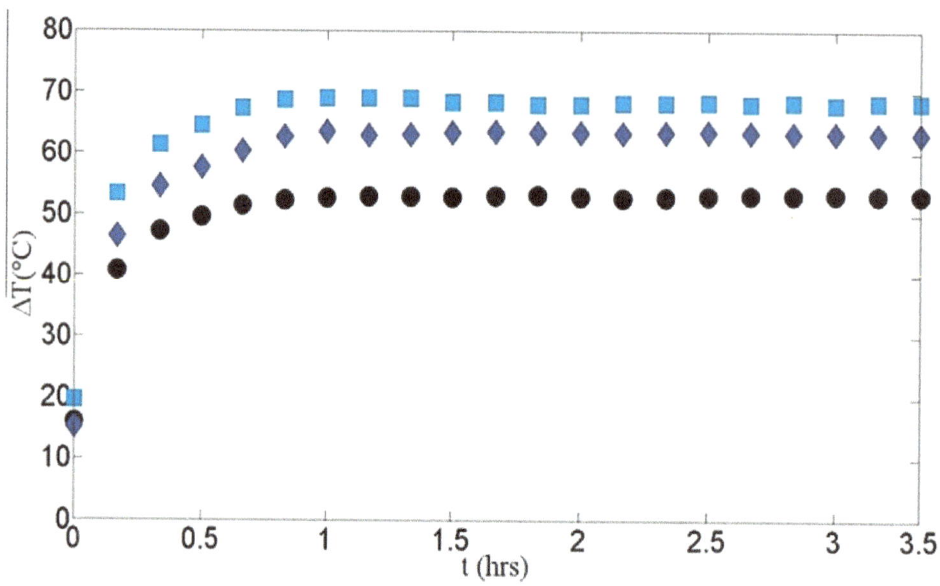

Figure 5. Temporal evolution of the time-averaged temperature difference between the heating and cooling plate for power input of 125 (circles), 150 (rhombus) and 175 W (squares), indicating the transient behavior and the steady state conditions, for pure DI water.

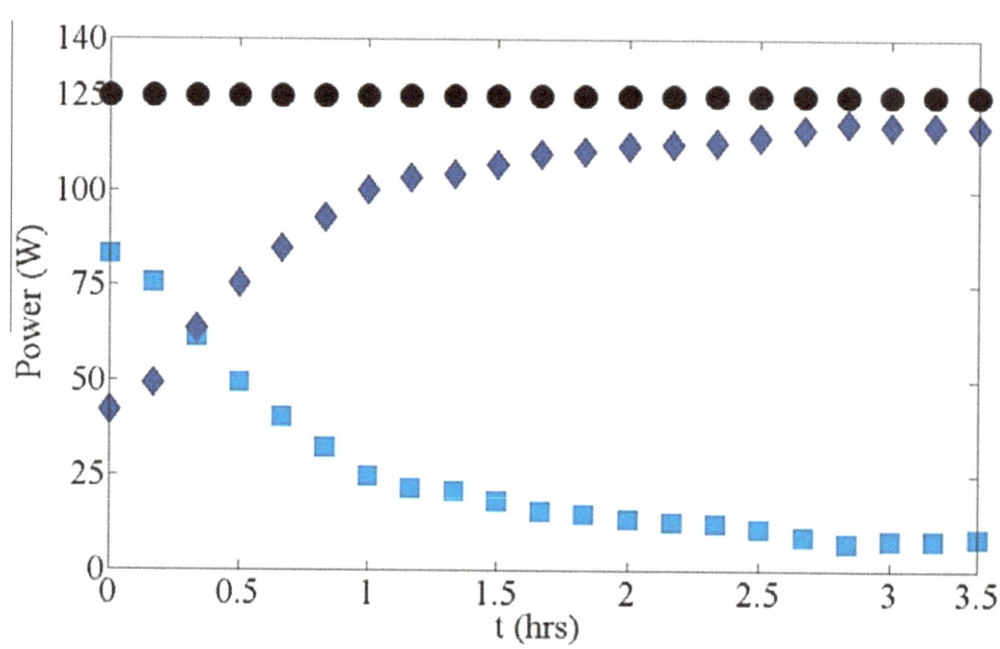

Figure 6. Temporal evolution of the time-averaged power input (circles), power output at the heat exchanger (rhombus) and heat losses (squares) from the chamber for pure DI water under 125 W.

Influence of Nanoparticle Volume Fraction on Heat Transfer Properties

The natural convective heat transfer coefficient strongly depends on the geometry of the convective chamber as well as on the thermophysical properties of the testing fluids. Some additional factors are the configuration and orientation of the heating elements, the method of heating and the roughness of the chamber's walls [30]. In this study, pure DI water and nanofluid experiments were conducted for three different boundary conditions and various nanoparticle volumetric concentrations, namely 0.01%, 0.03%, 0.06% and 0.12%. The main purpose was to investigate the convective heat transfer performance of nanofluids and to study the influence of the nanoparticle concentration in the suspension. As a first step, constant heat flux, $q'' = 19.5313$ kW/m² (corresponding to input power of 125 W) was applied to the bottom plate and constant temperature, $T_c = 21$ °C (average value) was imposed to the cold plate to study the heat transfer performance and the flow dynamics of the nanofluid as a function of the nanoparticle volumetric concentration. InFig 7, the normalized heat transfer coefficient of nanofluid to that of water for various nanoparticle concentrations is presented at constant power input of 125 W (circles). It can be observed in Fig. 7 that the addition of aluminum nanoparticles to the aqueous base fluid deteriorates the heat transfer coefficient. As the nanoparticle concentration increases, the heat transfer coefficient decreases. Also, even though the flow becomes more turbulent due to the addition of nanoparticles, the convective heat transfer mode is reduced. The next step involved the examination of the impact of more intense turbulent flows on the heat transfer performance of the nanofluid as the nanoparticle concentration was increased. For this case, heat flux, $q'' = 23.4375$ and 27.3438 kW/m² (corresponding to input power of 125 and 175 W, respectively) at the bottom plate was imposed, while the temperature of the cold plate was maintained constant, $T_c = 21$ °C (average value). In Fig. 7, the normalized heat transfer coefficient of nanofluid to that of water for various nanoparticle concentrations is presented at constant power, $Q = 150$ W (rhombus) and 175 W (squares). Even under a more intense turbulent flow, the addition of nanoparticles results in deterioration of the heat transfer performance relative to the base fluid. For all three power inputs, the non-dimensional heat transfer parameters are presented in Table 1.

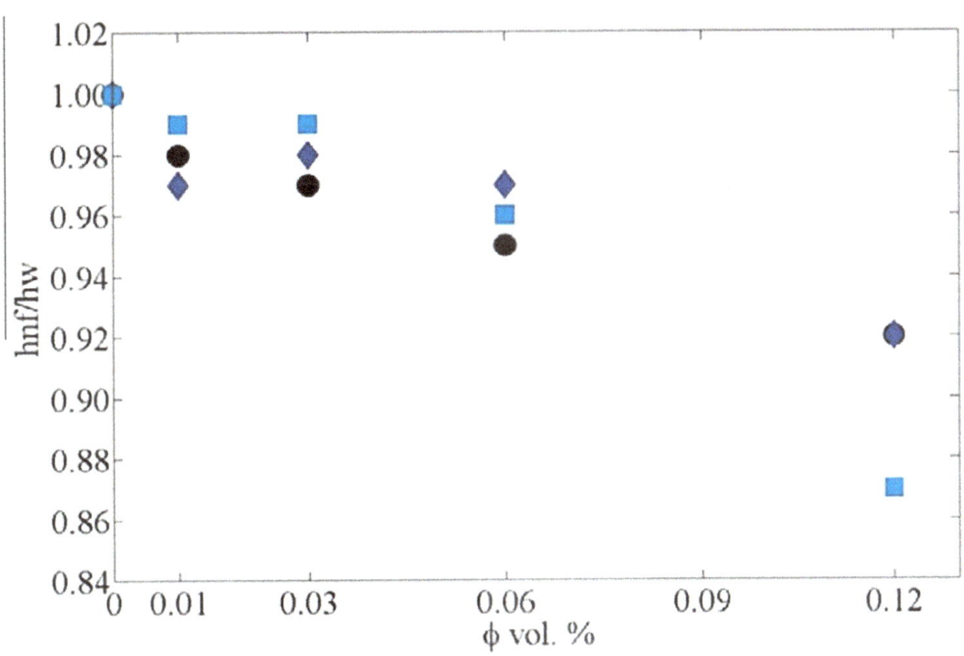

Figure7. Convective heat transfer coefficient of the nanofluid normalized by that of pure DI water, h_{nf}/h_w for power input of 125 (circles), 150 (rhombus) and 175 W (squares).

Table 1. Experimental conditions and thermophysical properties for power input of 125, 150 and 175 W for DI water and different nanoparticle volume fractions.

Oper. scenario	Q (W)	q'' (kW/m²)	φ (vol.%)	ΔT (°C)	h (W/m² K)	Nu	Pr	Ra (×10⁹)
A	125	19.5313	0.00	52.53	372	58.0	3.7	2.5
			0.01	53.70	364	56.9	3.7	2.6
			0.03	54.15	361	56.3	3.7	2.6
			0.06	55.26	353	54.9	3.7	2.7
			0.12	57.05	342	53.3	3.7	2.7

B	150	23.4375	0.00	61.63	380	59.1	3.5	3.2
			0.01	63.77	368	57.0	3.4	3.6
			0.03	63.05	372	57.5	3.4	3.5
			0.06	63.31	370	57.1	3.4	3.5
			0.12	66.70	351	53.7	3.1	4.1
C	175	27.3438	0.00	67.74	404	62.4	3.1	4.2
			0.01	68.60	399	61.3	3.0	4.6
			0.03	68.27	401	61.5	3.0	4.5
			0.06	70.79	386	59.0	3.0	4.7
			0.12	77.54	353	53.5	3.0	5.2

Influence of the Input Heat Flux on the Thermophysical Properties

In this section, water and nanofluid data are compared for the nanoparticle volumetric concentrations, φ = 0.03%, 0.06% and 0.12% for all power inputs, Q, of 125, 150 and 175 W. The purpose is to identify any changes in the absolute value of the heat transfer coefficient trendline as the imposed heat flux increases for pure water and nanofluid. In Fig. 8, the heat transfer coefficient of pure DI water and nanofluid in terms of the power input level is presented. It can be seen that as the power input increases, the heat transfer coefficient increases for all the cases. However, for the maximum nanoparticle concentration the rate of enhancement is notably less.

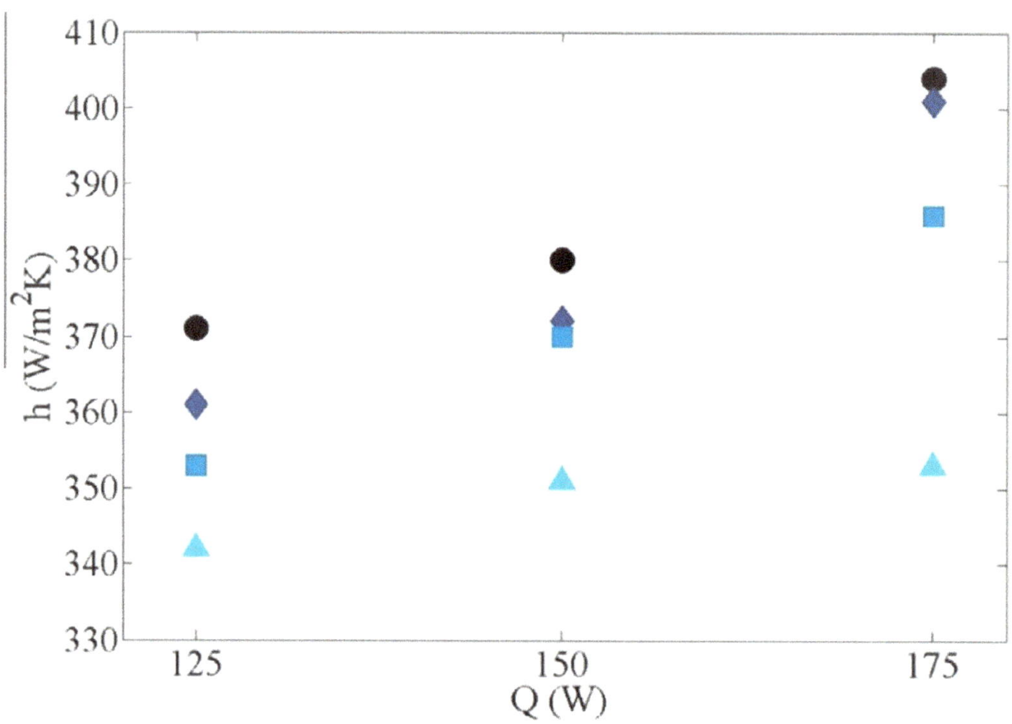

Heat transfer coefficient of pure DI water (circles), nanofluid of φ = 0.03 (rhombus), 0.06 (squares) and 0.12 vol.% (triangles) as a function of power input.

In Fig. 9, the Rayleigh number, *Ra*, is presented as a function of the power input level for water and nanofluid. As the nanoparticle concentration and the power input level are raised, the flow becomes more and more turbulent, as revealed by the increase of the*Ra*. Nonetheless, the increment is near 20% for the maximum nanoparticle concentration. In Fig. 10, *Nu* for water and nanofluid as a function of *Ra* is depicted. In most of the cases, as *Ra* increases, *Nu* increases. However, for the maximum nanoparticle concentration, *Nu* becomes nearly constant. The observed *Ra–Nu*dependency indicates a relationship of the type,

$$Nu = cRa^{n} \tag{12}$$

which has also been presented by other numerical and experimental studies [2],[43] and [44]. In this work, the constant "*c*" and the index "*n*" appear to be strongly related to the nanoparticle concentration and the stability characteristics of the suspension. For the nanofluid with maximum nanoparticle concentration, the relationship of Eq. (12) is no longer valid, due to the significant nanoparticle deposition on the hot, bottom plate.

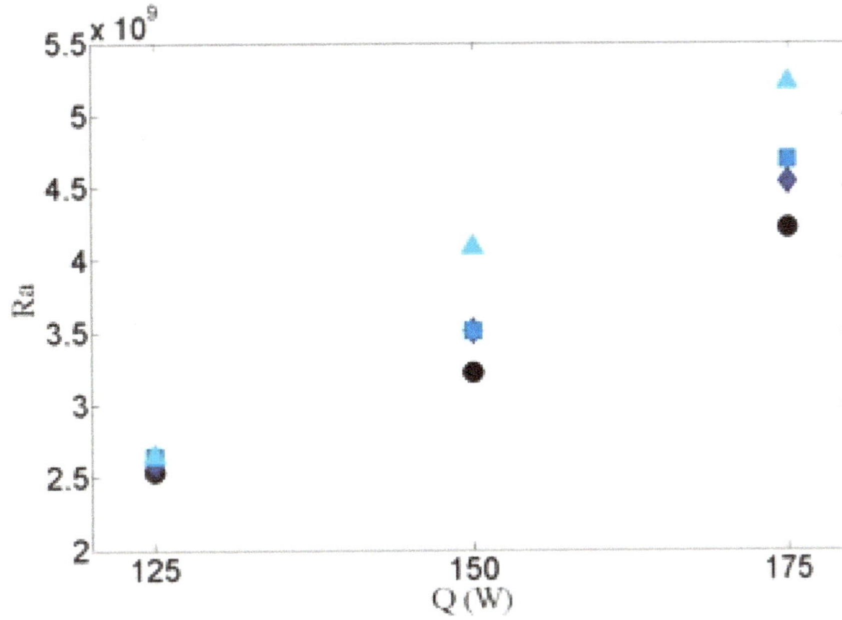

Figure 9. Rayleigh number for pure DI water (circles), nanofluid of 0.03 (rhombus), 0.06 (squares) and 0.12 vol.% (triangles) as a function of power input.

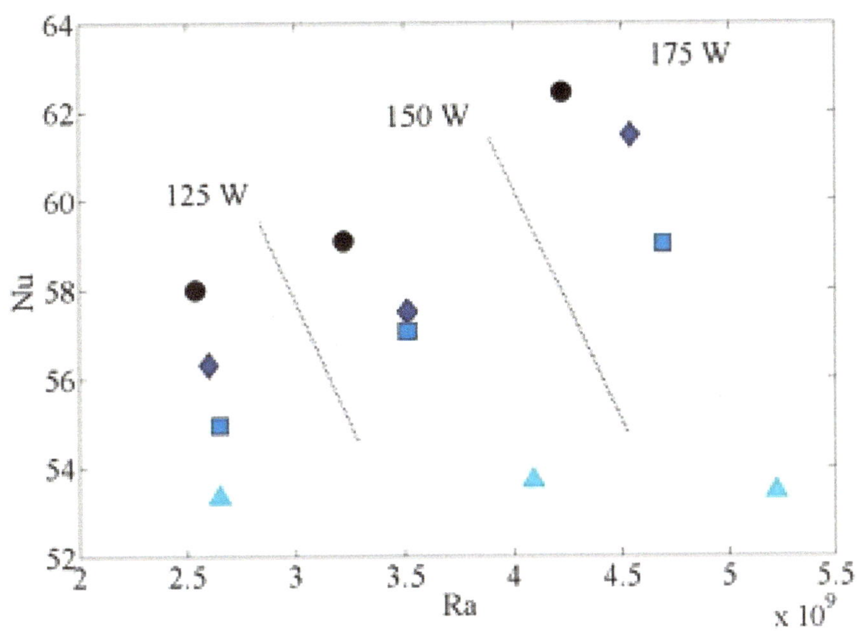

Figure 10. Nusselt number for pure DI water (circles), nanofluid of $\varphi = 0.03$ (rhombus), 0.06 (squares) and 0.12 vol.% (triangles) as a function of *Ra*, while the power input increases from 125 to 150 and 175 W.

Repeatability Test and Error Analysis

To ensure reliability of the averaged results for DI water, the uncertainty in the mean value of the heat transfer coefficient, σ_w was calculated according to Kirkup [45] and found to be 2 W/m^2 K for all three power inputs. For the nanofluid, due to time and cost limitations, it was not possible to perform such an extended analysis for every nanoparticle volume fraction, for all three power inputs. Therefore, a repeatability test was performed for an arbitrary selected nanofluid of a specific nanoparticle volume fraction. The purpose was to ensure repeatability of the nanofluid results and to identify whether any systematic uncertainties are induced either by the nanofluid preparation or the experimental process followed. For the 0.03 vol.% nanofluid under 125 W, a set of five individual measurements was considered and subsequently, a comparison of the temporal evolution of the temperature difference between the heating and cooling plate for the whole set was performed. As illustrated in Fig. 11, the temperature differences at steady state conditions are very close to each other, with the standard deviation being 0.59 °C. Concerning the uncertainty in the mean, σ_{nf} for the heat transfer coefficient, it was found to be the same with that of DI water, 2 W/m^2 K. Through the repeatability tests, the uncertainty of the measured value of the heat transfer coefficient was found to be very small and as a consequence, the individual calculated values are considered highly precise.

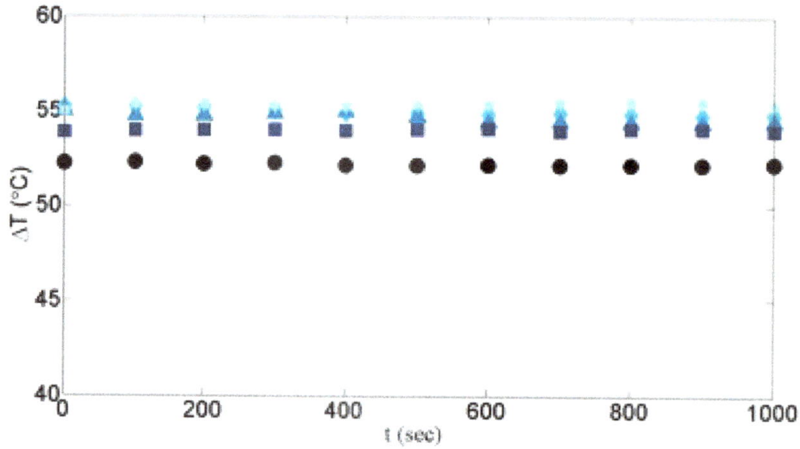

Figure 11. Temporal evolution of the time-averaged temperature difference between the heating and cooling plate at steady state conditions, for a set of five individual experiments with DI water.

The calculation of the uncertainty in both the values and numbers that characterize the thermal performance of liquids requires the determination of the uncertainties in the measured quantities through error propagation. Main sources of uncertainty in the calculations were the resolution of the flow meter employed at the cold plate, ±0.055 L/min used to calculate the heat losses from the sides of the rig, the tolerance of the thermocouples, ±1.1 °C of the reading value, and the resolution and measurement error of the control instrumentation, ±0.9 °C for the temperature range of interest. In this study, where simple arithmetic calculations were required, the mean uncertainty in the heat losses from the sides of the rig, for instance, were calculated by,

$$\Delta(P) = \Delta(F) = \left|\frac{\partial F}{\partial W}\right|\Delta(W) + \left|\frac{\partial F}{\partial T}\right|\Delta(T) \qquad (13)$$

where it was found to be 7.7 W. In a similar way, the mean and the maximum fractional uncertainty (%) of the heat transfer coefficient were 6.7% and 7.9% respectively.

DISCUSSION

An important finding of this study is that the addition of γ-Al_2O_3 nanoparticles to pure DI water deteriorates the natural convective heat transfer performance under the specific experimental conditions. This result is consistent with other experimental observations in the literature, where the heat transfer coefficient, *h* and *Nu* decrease with the addition of nanoparticles, even though *Ra* increases. However, natural convective heat transfer deterioration has been reported not only for Al_2O_3–H_2O nanofluids [2], [9], [21] and [22], but also for CuO [9] and TiO_2–H_2O nanofluids [20]. Such a finding is contradictory to the general expectation that the heat transfer in nanofluids increases. In the vast majority of nanofluid studies for conduction [29], [46], [47], [48], [49], [50], [51], [52], [53], [54], [55],[56], [57] and [58], forced convection [12], [59] and [60], boiling [28] and [61] and critical heat flux [62], [63] and [64], heat transfer enhancement was reported. Even more, heat transfer increase was observed in many numerical natural convection studies, making the overall behavior of nanofluids even more questionable and unclear. Up to date, the reported heat transfer deterioration under natural convective experimental studies is attributed to some widely addressed, but not systematically verified mechanisms, suggested in the literature. Savithiri et al. [65] performed a scaling analysis for the investigation of slip mechanisms in nanofluids and they concluded that properties like the shape, size, concentration and temperature of the nanoparticles have major effects on the slip forces. Some widely reported mechanisms, responsible for the heat transfer deterioration in natural convective studies, include: the particle–fluid slip, the particle–surface and particle–particle interactions, the convection induced by particle concentration differences, the reduction of *Ra* due to the increased kinematic viscosity, the Brownian motion, the thermophoretic motion of nanoparticles and the sedimentation. Some speculated proposed mechanisms, attempting to explain the alteration (enhancement and/or reduction) of the general heat transfer performance of nanofluids, can be found in the extended literature review by Sergis and Hardalupas [7]. However, there is not yet conclusive experimental evidence to support them.

In the present experimental investigation, the fouling of the bottom surface appears to play dominant role for the deterioration observed. This is mainly due to the inadequacy of the system forces to ensure both long and short-term nanofluid stability, in the absence of any surfactant or pH changer. Therefore, strong attractive Van der Waals forces between the particles present in the suspension result in the formation of nanoparticle agglomerates that could lead or accelerate nanoparticle settling at the bottom of the rig. Also, as the nanoparticle volumetric concentration increases, the temperature at the heating plate increases. This can be credited to the increased fouling of the heating surface due to the formation of a thin stationary layer structure near the bottom that consists of nanoparticles, agglomerates and water molecules and impose additional thermal resistance in the system that obstructs the efficient heat transfer. This observation could be qualitatively verified for the first time, since the experiments were performed within a chamber that had optical access. Fig. 12 illustrates severe nanoparticle deposition at the bottom of the chamber for a highly unstable nanofluid, γ-Al_2O_3 nanoparticles suspended in low-quality DI water, with a pH value that approaches the IEP of aluminum oxide. It is observed that a thick layer of nanoparticles and water molecules has been formed at the bottom of the chamber that appears to be responsible for the increased temperature at the heating plate. Fig. 13 provides images that demonstrate the layer formation within the chamber for a nanofluid of maximum nanoparticle concentration, 0.12 vol.%. More specifically, Fig. 13(a) shows the deposition of nanoparticles at the bottom plate of the rig, verified when pumping the nanofluid out of the chamber, at the end of the experiment, where Fig. 13(b) depicts the fouling of the heating surface with nanoparticles and agglomerates that remain adhered to the surface of the heating plate. It was also noticed that when the imposed heat flux was increased, the adhesion of nanoparticles at the heating plate was more intensive. This can be mainly credited to the thermophysical properties of the γ-Al_2O_3 nanopowder used in this study and to the resulting interactions between the heating surface and the deposited nanoparticles and agglomerates at the bottom, caused by the reduction of surface's wettability. To summarise, particle–particle and particle–fluid interactions that led to significant sedimentation as well as particle–surface interactions between the nanoparticles and the heating surface appear to be the major mechanisms that affected the heat transfer behavior under natural convection in the current investigation. In similar studies, such as the one conducted by Ni et al. [2], where a very small amount of dispersant was added in the suspension to minimise nanoparticle agglomeration the decrease of the heat transfer coefficient in nanofluids was not that intense, even for higher nanoparticle concentrations. In their experimental investigation, sedimentation seems not to be a major deteriorating mechanism, as the *Ra* number decreases with the increasing volume fraction of nanoparticle in the base fluid.

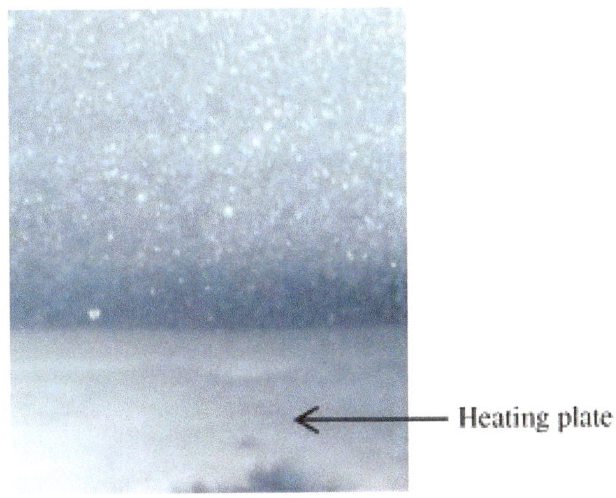

Figure 12. Severe nanoparticle deposition at the bottom of the chamber for a highly unstable nanofluid, pH ~ 8 as observed from the optical access of the chamber.

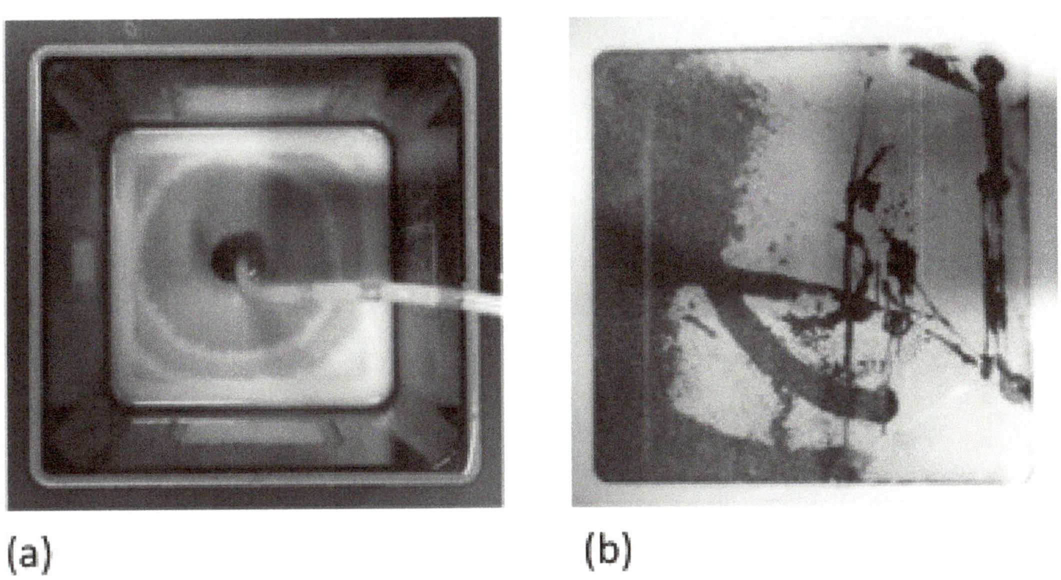

Figure 13. (a) Fouling of the heating surface observed when pumping the nanofluid out of the rig and (b) increased adhesion of nanoparticles on the surface for maximum nanoparticle volume fraction, 0.12 vol.% under maximum power input of 175 W, noticed at the end of the experiment. The photos of the heating plate were obtained from the top of the vessel after removing the cooling plate.

The effect of fouling of the heating surface on the heat transfer coefficient of nanofluid is modeled numerically, in order to predict the temperature gradient between the hot and cold plates. The significance of the fouling effect is evaluated qualitatively to verify the experimental results. To start with, a nanofluid of maximum nanoparticle concentration 0.12 vol.% is considered under the operating conditions A (Table 1). At steady state conditions, a thin film that consists entirely of Al_2O_3 is assumed to have uniformly covered the heating plate of the rig. Therefore, an additional thermal resistance should be also considered in the heat transfer calculations for nanofluid, according to Eq. (14)

$$Q = \frac{\Delta T_w}{\frac{1}{h_w A}} = \frac{\Delta T_{layer}}{\frac{L}{k_{layer} A}} = \frac{\Delta T_{nf}}{\frac{L}{k_{layer} A} + \frac{1}{h_{nf} A}} \tag{14}$$

By substituting the values from Table 1 and assuming a small layer thickness$L < 0.200$ mm and $k_{layer} = 40$ W/mK, it is found that $\Delta T_{layer} < 0.10$ °C. Therefore, nanoparticle deposition on its own cannot explain the increased temperature of the heating surface and thus the reported convective heat transfer deterioration.

In a different approach for the fouling effect of the heating plate, a thin layer that contains nanoparticles, agglomerates and water molecules is considered. Like the previous analysis, nanofluid of maximum nanoparticle concentration, 0.12 vol.% is considered under the operating conditions A (Table 1) at steady state. Suppose that 25% of the nanoparticles, for instance, have settled out of the suspension and deposited close to the heating plate in a volume of 100 × 100 × 0.2 mm³ that corresponds to a nanoparticle concentration of $\varphi = 15$ vol.% in that layer. Therefore, by substituting the values fromTable 1 and taking into account that $L = 0.200$ mm and $k_f = 0.643$ W/mK at 20 °C and$k_{layer} = 0.93$ W/mK (calculated from the effective medium theory, where $\frac{k_{layer}}{k_w} = 1 + 3\varphi$), it is found that $\Delta T_{layer} = 4.20$ °C and $\Delta T_{nf} = 56.73$ °C. By comparing this value with the experimentally measured, 57.05 °C, excellent agreement is noticed. By following the same procedure, but under the operating conditions B (Table 1), it is reported that that $\Delta T_{layer} = 5.04$ °C and $\Delta T_{nf} = 66.67$ °C, which is very close to the experimentally measured, 66.70 °C. Finally, the same process is performed for operating conditions C (Table 1), where $\Delta T_{layer} = 5.88$ °C and $\Delta T_{nf} = 73.62$ °C, while the experimentally observed is 77.54 °C. Therefore, the experimentally reported effects of fouling near the heating plate at the bottom of the rig can be modeled successfully, regardless of the assumptions considered for the thickness or the concentration of that layer. As a consequence, in stationary liquids and low-velocity applications, special attention should be paid to the nanofluid stability as the forces acting in the system, namely buoyant and gravitational appears to be inadequate to ensure that. Hence, additional stabilization methods such as the electrokinetic stabilization and careful selection of nanopowder, in terms of formulation and physical characteristics, may provide a solution.

CONCLUSIONS

This study examines experimentally the heat transfer characteristics of γ-Al_2O_3–deionized (DI) H_2O nanofluids under natural convection. Heat transfer measurements were obtained in a classical Rayleigh–Benard chamber and compared with pure DI water results. According to our findings the addition of aluminum nanoparticles in the aqueous base fluid and in the absence of any pH changer or surface activator deteriorates the heat transfer performance. Furthermore, the deterioration increases as the nanoparticle concentration increases. In the case of nanofluid, the increased temperature gradient in the chamber is caused by the increased temperature at the heating plate, suggesting intense fouling of that surface. However, this cannot be fully explained by pure nanoparticle deposition. One possible reason could be the formation of a thin stationary layer structure that consists of nanoparticles, agglomerates and water molecules close to the heating surface that has been qualitatively observed. This behavior is attributed to the poor short-term stability of the nanofluid that could be disrupted by the nanoparticle agglomerates that are being formed in the suspension. In the absence of any surfactant or pH changer, the nanoparticles exhibit low surface charge and thus the electrostatic repulsion forces among them are weak. Also, due to the lack of significant flow momentum and agitation, specifically for the investigated natural convection application (versus e.g. forced convection cases), the agglomeration process seems to be accelerated further. In the present study, as the applied heat flux increases, the induced flow becomes more turbulent and as expected the convective heat transfer is enhanced. However, for maximum nanoparticle concentration of 0.12 vol.%, this is not the case and*Nu* remains constant. This can be credited to the presence of a denser nanoparticle layer structure that results in confined liquid contact with the surface and thus limited surface wettability. Therefore, when low-velocity applications, such as natural convection ones, are involved special attention should be paid on ways to prevent or eliminate the formation of such layer structure. This can be done primarily by improving the attained stability of the suspension, through the employment of stabilization methods, such as the electrostatic. In this way, the surface charge of the nanoparticles can be increased, strong repulsive forces among them will be established and, as a consequence, any possible agglomeration will be eliminated. To summarise, a combination of a proper selection of nanopowder, in terms of its physical characteristics, along with the addition of a pH changer could provide a first solution towards this direction.

ACKNOWLEDGEMENTS

This work was funded by the RCUK Energy Program and EURATOM. The views and opinions expressed herein do not necessarily reflect those of the European Commission.

We would like to thank CCFE for the financial support through the CASE EPSRC studentship that allowed this research and especially Dr. Barrett Thomas for his input to this project. Finally, we would like to thank Dr. Byon Chan for his valuable help in designing and manufacturing the thermal rig. Data supporting this publication can be obtained by request from MultiphaseFlows@imperial.ac.uk.

REFERENCES

1. J. Milnes, A. Burns, D. Drikakis, Computational modelling of the HyperVapotron on cooling technique, Fusion Eng. Des. 87 (9) (2012) 1647–1661.
2. R. Ni, S.-Q. Zhou, K.-Q. Xia, An experimental investigation of turbulent thermal convection in water-based alumina nanofluid, Phys. Fluids 23 (2) (2011) 022005.
3. S.U.S. Choi, J.A. Eastman, Enhancing thermal conductivity of fluids with nanoparticles, Dev. Appl. Non Newtonian Flows 66 (99–105) (1995).
4. S.U.S. Choi, Nanofluids: from vision to reality through research, J. Heat Transfer 131 (3) (2009) 1–9.
5. J. Buongiorno, Convective transport in nanofluids, J. Heat Transfer 128 (3) (2006) 240–250.

6. T.R. Barrett, S. Robinson, K. Flinders, A. Sergis, Y. Hardalupas, Investigating the use of nanofluids to improve high heat flux cooling systems, Fusion Eng. Des. 88 (9–10) (2013) 2594–2597.

7. A. Sergis, Y. Hardalupas, Anomalous heat transfer modes of nanofluids: a review based on statistical analysis, Nanoscale Res. Lett. 6 (1) (2011) 391.

8. D. Wen, G. Lin, S. Vafaei, K. Zhang, Review of nanofluids for heat transfer applications, Particuology 7 (2) (2009) 141–150.

9. N. Putra, W. Roetzel, S.K. Das, Natural convection of nano-fluids, Heat Mass Transfer 39 (2003) 775–784.

10. G. Puliti, S. Paolucci, M. Sen, Nanofluids and their properties, Appl. Mech. Rev. 64 (3) (2011) 030803-1–030803-23.

11. S.K. Das, S.U.S. Choi, W. Yu, T. Pradeep, Nanofluids, Science and Technology, Wiley, New Jersey, 2007.

12. Y. Xuan, Q. Li, Investigation on convective heat transfer and flow features of nanofluids, J. Heat Transfer 125 (1) (2003) 151–155.

13. A.K. Nayak, M.R. Gartia, P.K. Vijayan, An experimental investigation of singlephase natural circulation behavior in a rectangular loop with Al2O3 nanofluids, Exp. Therm. Fluid Sci. 33 (1) (2008) 184–189.

14. S. Kakaç, A. Pramuanjaroenkij, Review of convective heat transfer enhancement with nanofluids, Int. J. Heat Mass Transfer 52 (13–14) (2009) 3187–3196.

15. S. Savithiri, A. Pattamatta, S.K. Das, In Al2O3/water nanofluid, Numer. Heat Transfer Part A Appl. 68 (10) (2015) 1106–1124.

16. K. Khanafer, K. Vafai, M. Lightstone, Buoyancy-driven heat transfer enhancement in a two-dimensional enclosure utilizing nanofluids, Int. J. Heat Mass Transfer 46 (2003) 3639–3653.

17. H.F. Oztop, E. Abu-Nada, Numerical study of natural convection in partially heated rectangular enclosures filled with nanofluids, Int. J. Heat Fluid Flow 29 (5) (2008) 1326–1336.

18. S.M. Aminossadati, B. Ghasemi, Natural convection cooling of a localised heat source at the bottom of a nanofluid-filled enclosure, Eur. J. Mech. B/Fluids 28 (5) (2009) 630–640.

19. X. Meng, Y. Li, Numerical study of natural convection in a horizontal cylinder filled with water-based alumina nanofluid, Nanoscale Res. Lett. 10 (1) (2015) 1–10.

20. D. Wen, Y. Ding, Formulation of nanofluids for natural convective heat transfer applications, Int. J. Heat Fluid Flow 26 (6) (2005) 855–864.

21. A.G.A. Nnanna, Experimental model of temperature-driven nanofluid, J. Heat Transfer 129 (2007) 697–704.

22. C.H. Li, G.P. Peterson, Experimental studies of natural convection heat transfer of Al2O3/DI water nanoparticle suspensions (nanofluids), Adv. Mech. Eng. 2 (2010) 742739.

23. A. Ghadimi, R. Saidur, H.S.C. Metselaar, A review of nanofluid stability properties and characterization in stationary conditions, Int. J. Heat Mass Transfer 54 (17–18) (2011) 4051–4068.

24. V. Trisaksri, S. Wongwises, Nucleate pool boiling heat transfer of TiO2–R141b nanofluids, Int. J. Heat Mass Transfer 52 (5–6) (2009) 1582–1588.

25. C. Mao, H. Zou, X. Zhou, Y. Huang, H. Gan, Z. Zhou, Analysis of suspension stability for nanofluid applied in minimum quantity lubricant grinding, Int. J. Adv. Manuf. Technol. 71 (9–12) (2014) 2073–2081.

26. B.C. Pak, Y. Cho, Hydrodynamic and heat transfer study of dispersed fluids with submicron metallic oxide particles, Exp. Heat Transfer 11 (2) (1998) 151–170.

27. B.P. Singh, R. Menchavez, C. Takai, M. Fuji, M. Takahashi, Stability of dispersions of colloidal alumina particles in aqueous suspensions, J. Colloid Interface Sci. 291 (1) (2005) 181–186.

28. D. Wen, Y. Ding, Experimental investigation into the pool boiling heat transfer of aqueous based c-alumina nanofluids, J. Nanopart. Res. 7 (2005) 265–274.

29. J.-H. Lee, K.S. Hwang, S.P. Jang, B.H. Lee, J.H. Kim, S.U.S. Choi, C.J. Choi, Effective viscosities and thermal conductivities of aqueous nanofluids containing low volume concentrations of Al2O3 nanoparticles, Int. J. Heat Mass Transfer 51 (11–12) (2008) 2651–2656.

30. D. Wen, Y. Ding, Natural convective heat transfer of suspensions of titanium dioxide nanoparticles (nanofluids), IEEE Trans. Nanotechnol. 5 (3) (2006) 220– 227.

31. S. Adegbite, Particle characterisation and grinding behaviour of gammaalumina slurries prepared in a stirred media mill, J. Mater. Sci. Res. 2 (1) (2012) 135–147.

32. J.A. Eastman, S.U.S. Choi, S. Li, W. Yu, L.J. Thompson, Anomalously increased effective thermal conductivities of ethylene glycol-based nanofluids containing copper nanoparticles, Appl. Phys. Lett. 78 (6) (2001) 718.

33. H. Zhu, C. Li, D. Wu, C. Zhang, Y. Yin, Preparation, characterization, viscosity and thermal conductivity of CaCO3 aqueous nanofluids, Sci. Chin. Technol. Sci. 53 (2) (2010) 360–368.

34. W.M. Rohsenow, J.P. Hartnett, Y.I. Cho, Handbook of HEAT TRANSFER, third ed., McGraw-Hill, New York, 1998.

35. J. Bi, K. Vafai, D.M. Christopher, Heat transfer characteristics and CHF prediction in nanofluid boiling, Int. J. Heat Mass Transfer 80 (2015) 256–265.

36. E. Büyük, Natural convection of water-based nanofluids in an inclined enclosure with a heat source, Int. J. Therm. Sci. 48 (11) (2009) 2063–2073.

37. G.A. Sheikhzadeh, Numerical study of natural convection in an inclined cavity with partially active side walls filled with Cu–water nanofluid, Int. J. Eng. 24 (3) (2011) 279–291.

38. I. Rashidi, O. Mahian, G. Lorenzini, C. Biserni, S. Wongwises, Natural convection of Al2O3/water nanofluid in a square cavity: effects of heterogeneous heating, Int. J. Heat Mass Transfer 74 (2014) 391–402.

39. Z. Haddad, E. Abu-Nada, H.F. Oztop, A. Mataoui, Natural convection in nanofluids: are the thermophoresis

and Brownian motion effects significant in nanofluid heat transfer enhancement?, Int J. Therm. Sci. 57 (2012) 152–162.

40. E.V. Timofeeva, W. Yu, D.M. France, D. Singh, J.L. Routbort, Nanofluids for heat transfer: an engineering approach, Nanoscale Res. Lett. 6 (1) (2011) 182.

41. S.E.B. Maïga, C.T. Nguyen, N. Galanis, G. Roy, Heat transfer behaviours of nanofluids in a uniformly heated tube, Superlattices Microstruct. 35 (3–6) (2004) 543–557.

42. E.V. Timofeeva, A.N. Gavrilov, J.M. McCloskey, Y.V. Tolmachev, S. Sprunt, L.M. Lopatina, J.V. Selinger, Thermal conductivity and particle agglomeration in alumina nanofluids: experiment and theory, Phys. Rev. E 76 (6) (2007) 061203.

43. G. Barakos, E. Mitsoulis, D. Assimacopoulos, Natural convection flow in a square cavity revisited: laminar and turbulent models with wall functions, Int. J. Numer. Methods Fluids 18 (1994) 695–719.

44. R.K. MacGregor, A.F. Emery, Free convection through vertical plane layers – moderate and high Prandtl number fluids, J. Heat Transfer 91 (3) (1969) 391– 401.

45. L. Kirkup, Experimental Methods; an Introduction to the Analysis and Presentation of Data, John Wiley & Sons, 1994.

46. J.-H. Lee, S.-H. Lee, S.P. Jang, Do temperature and nanoparticle size affect the thermal conductivity of alumina nanofluids?, Appl Phys. Lett. 104 (16) (2014) 161908.

47. X. Wang, X. Xu, S.U.S. Choi, Thermal conductivity of nanoparticle - fluid mixture, J. Thermophys. Heat Transfer 13 (4) (1999) 474–480.

48. C.H. Li, G.P. Peterson, Experimental investigation of temperature and volume fraction variations on the effective thermal conductivity of nanoparticle suspensions (nanofluids), J. Appl. Phys. 99 (2006).

49. X. Zhang, H. Gu, M. Fujii, Experimental study on the effective thermal conductivity and thermal diffusivity of nanofluids, Int. J. Thermophys. 27 (2) (2006) 569–580.

50. S. Shaikh, K. Lafdi, R. Ponnappan, Thermal conductivity improvement in carbon nanoparticle doped PAO oil: an experimental study, J. Appl. Phys. 101 (6) (2007) 064302.

51. H. Xie, J. Wang, T. Xi, Y. Liu, F. Ai, Q. Wu, Thermal conductivity enhancement of suspensions containing nanosized alumina particles, J. Appl. Phys. 91 (2002) 4568–4572.

52. S.M.S. Murshed, K.C. Leong, C. Yang, Enhanced thermal conductivity of TiO2 water based nanofluids, Int. J. Therm. Sci. 44 (2005) 367–373.

53. K.C. Leong, C. Yang, Investigations of thermal conductivity and viscosity of nanofluids, Int. J. Therm. Sci. 47 (2008) 560–568.

54. R.S. Vajjha, D.K. Das, Experimental determination of thermal conductivity of three nanofluids and development of new correlations, Int. J. Heat Mass Transfer 52 (21–22) (2009) 4675–4682.

55. X. Wei, H. Zhu, T. Kong, L. Wang, Synthesis and thermal conductivity of Cu2O nanofluids, Int. J. Heat Mass Transfer 52 (2009) 4371–4374.

56. S.K. Das, N. Putra, P. Thiesen, W. Roetzel, Temperature dependence of thermal conductivity enhancement for nanofluids, J. Heat Transfer 125 (2003) 567– 574.

57. S.U.S. Choi, Z.G. Zhang, W. Yu, F.E. Lockwood, E.A. Grulke, Anomalous thermal conductivity enhancement in nanotube suspensions, Appl. Phys. Lett. 2252 (2001) 14–17.

58. J.A. Eastman, S.U.S. Choi, S. Li, W. Yu, L.J. Thompson, Anomalously increased effective thermal conductivities of ethylene glycol-based nanofluids containing copper nanoparticles, Appl. Phys. Lett. 78 (6) (2001) 718–720.

59. U. Rea, T. McKrell, L.-W. Hu, J. Buongiorno, Experimental study of laminar convective heat transfer and viscous pressure loss of alumina-water nanofluid, in: Micro/Nanoscale Heat Transfer International Conference, 2008, pp. 1–6.

60. K. Sik, S. Pil, S.U.S. Choi, Flow and convective heat transfer characteristics of water-based Al2O3 nanofluids in fully developed laminar flow regime, Int. J. Heat Mass Transfer 52 (1–2) (2009) 193–199.

61. G.P. Narayan, K.B. Anoop, S.K. Das, Mechanism of enhancement/deterioration of boiling heat transfer using stable nanoparticle suspensions over vertical tubes, J. Appl. Phys. 102 (7) (2007) 074317.

62. I.C. Bang, S.H. Chang, Boiling heat transfer performance and phenomena of Al2O3–water nano-fluids from a plain surface in a pool, Int. J. Heat Mass Transfer 48 (12) (2005) 2407–2419.

63. H. Seon, H. Kim, H. Jo, S. Kang, W. Chang, M. Hwan, Experimental study of critical heat flux enhancement during forced convective flow boiling of nanofluid on a short heated surface, Int. J. Multiphase Flow 36 (5) (2010) 375– 384.

64. H. Kim, M. Kim, Experimental study of the characteristics and mechanism of pool boiling CHF enhancement using nanofluids, Heat Mass Transfer 45 (7) (2007) 991–998.

65. S. Savithiri, A. Pattamatta, S.K. Das, Scaling analysis for the investigation of slip mechanisms in nanofluids, Nanoscale Res. Lett. 6 (1) (2011) 471.

CHAPTER 2

VAPOUR COOLING OF POORLY CONDUCTING HOT SUBSTRATES INCREASES THE DYNAMIC LEIDENFROST TEMPERATURE

Michiel A.J. van Limbeek [a], Minori Shirota [a], Pascal Sleutel [a], Chao Sun [a], Andrea Prosperetti [a,b], Detlef Lohse [a]

[a] Physics of Fluids, University of Twente, 7500 AE Enschede, The Netherlands

[b] Department of Mechanical Engineering, Johns Hopkins University, Baltimore, MD 21218, USA

ABSTRACT

A drop impacting a smooth solid surface heated above the saturation temperature can either touch it (contact boiling) or not (film boiling), depending on the surface temperature. The heat transfer is greatly reduced in the latter case by the insulating vapour layer under the drop. In contrast to previous studies, here we use a relatively poor thermally conducting glass surface. Using a total internal reflection method, we visualise the wetting dynamics of the drop on the surface. We discover a new touch-down process, in which liquid–solid contact occurs a few hundred microseconds after the initial impact. This phenomenon is due to the cooling of the solid surface by the generation of vapour. We propose a model to account for this cooling effect, and validate it experimentally with our observations. The model leads to the determination of a thermal time scale (about 0.3 ms for glass) for the cooling of the solid. We conclude that when the impact time scale of the drop on the substrate (drop diameter/impact velocity) is of the order of the thermal time scale or larger, the cooling effect cannot be neglected and the drop will make contact in this manner. If the impact time scale however is much smaller than the thermal time scale, the surface remains essentially isothermal and the impact dynamics is not affected.

INTRODUCTION

In spray cooling applications and others, it is necessary to maintain contact between the droplets and the heated surface. As has been known for centuries [1] and [2], contact is abruptly lost once the surface of the hot solid exceeds a rather well defined value, called the Leidenfrost temperature T_L. This temperature depends on the thermo-physical properties of the liquid and the solid [3] and [4], the micro-structure of the surface[5] and [6] and the impact velocity U [7]. In the static Leidenfrost case (U=0), in which no impact dynamics are present, the lifetime of the drop is much longer than a thermal time scale $\tau_{th}=k_s\rho_s C_{p,s}h^{-2}$ [3], [8] and [9]. Here, ρ_s is the solid density, k_s the thermal conductivity [W/(km)], $C_{p,s}$ the specific heat and h the heat transfer coefficient from the solid to the drop. In general, the local cooling of the solid produced by the drop causes the Leidenfrost temperature T_L to depend on the solid's thermal properties [3] and [8]. However, if the

Citation: Michiel A.J. van Limbeek, Minori Shirota, Pascal Sleutel, Chao Sun, Andrea Prosperetti, Detlef Lohse, Vapour cooling of poorly conducting hot substrates increases the dynamic Leidenfrost temperature, International Journal of Heat and Mass Transfer, Volume 97, June 2016, Pages 101-109, ISSN 0017-9310, http://dx.doi.org/10.1016/j.ijheatmasstransfer.2016.01.080.

drop lifetime is shorter than τ_{th}, the solid remains essentially isothermal and T_L reaches a lower limiting value.

When the impacting drop has an initial downward velocity $U \neq 0$, the phenomenon is referred to as dynamic Leidenfrost effect. Due to its widespread applications, this dynamic case has drawn much interest [6], [10], [11], [12], [13], [14], [15] and [16]. By the dynamic pressure of the impact, the drop partly overcomes the cushioning of the vapour layer and only touches down at plate temperatures *higher* than the static T_L. The *dynamic* Leidenfrost temperature is found to depend on the drop size, impact velocity, liquid properties, surface roughness [11] and [13] and solid vibrations [12]. Some studies investigate scaling laws for the maximum radius of the deforming drop [11], [13] and [17] or models for the time evolution of the impact dynamics [14], [18] and [19].

The focus of this paper is the study of impacts on the smooth surface of glass, which has a relatively poor thermal conductivity and diffusivity, as little is known about how these features affect T_L. We experimentally examine the liquid–solid touch-down mechanism and how it is affected by the limited heat transfer supplied by the glass. The impacting liquid drop consists of ethanol and has an initial size D_0 and velocity U . We measure the thermal time scale τ_{th} and compare it with the impact time scale $\tau_{imp}=D_0/U$. At times larger than τ_{imp} all downward momentum is transferred into the radial direction and the drop is no longer forced into contact with the substrate. Our data shows that when τ_{th} is larger than τ_{imp}, the substrate remains essentially isothermal. In the opposite case, the surface is appreciably cooled down by the (partial) evaporation of the drop and the cooling caused by the drop cannot be neglected. By using microdroplets instead of millimetre sized drops, we recover an isothermal impact, as $\tau_{imp}=D_0/U$ is then decreased by the decrease of the drop size.

EXPERIMENTAL SETUP AND PROCEDURE

As sketched in Fig. 1, the experimental setup consists of a drop generator, a smooth and dry glass surface on which the drops impact, and two cameras to study the phenomenon.

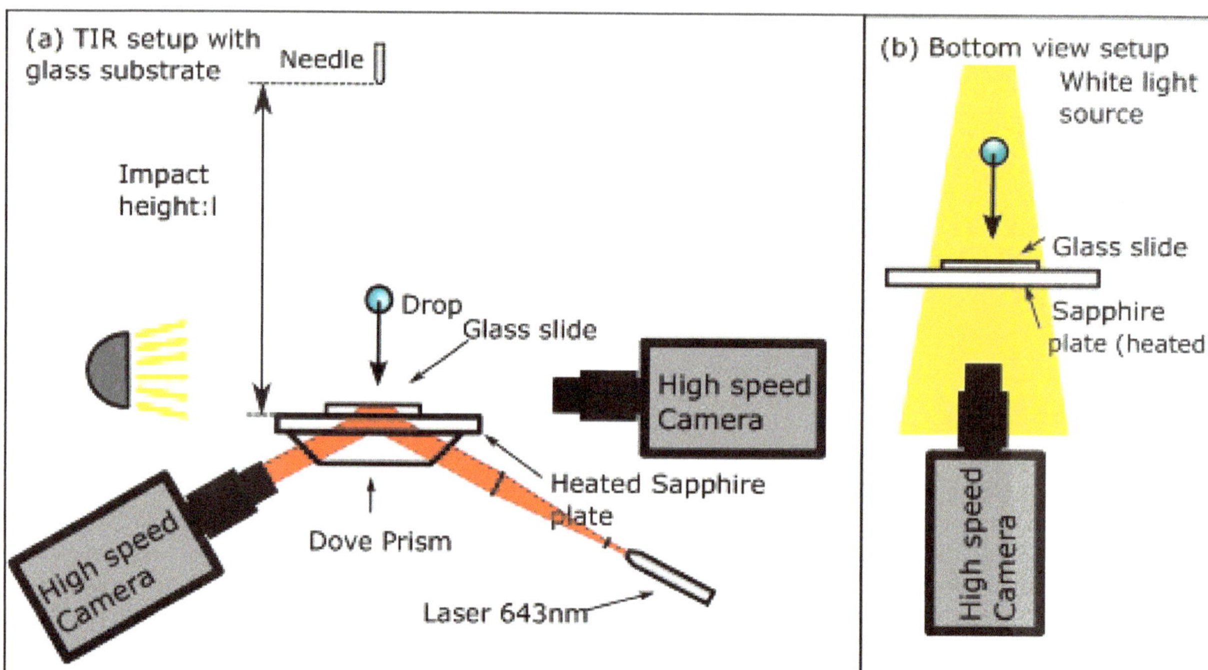

Figure 1. Schematic of the experimental setups used: the total internal refraction (TIR) setup (a) and the bottom view setup (b).

Drops with a diameter between 2.1 and 2.6 mm are released from a needle from a variable height. The resulting impact velocity ranges between 0.4 m/s and 3.9 m/s, with corresponding Weber numbers $We=\rho_l U^2 D_0/\sigma_l$ between 16 and 1480; here ρ_l and σ_l are the liquid density and surface tension coefficient. We also study the impact of 300 µm-diameter microdroplets, which are generated as a continuous stream by an inkjet nozzle, with an impact velocity of 8.9 m/s. The stream is blocked by mechanical shutters, which can open briefly to allow a few drops to reach the surface. This arrangement also scatters the drops in space, so that they do not interact with each other during impact. In all experiments we measured both the diameter and impact velocity of each drop with a high speed camera (Photron Fastcam SA1.1) at 10 000 fps.

The ethanol drops impact a smooth glass microscope slide, the properties of which are listed in Table 1. In order to ensure constant temperature on the opposite side of the glass slide, we place a sapphire disc, which has a much higher thermal conductivity than glass, between the slide and the heater. The thermal penetration depth in the glass during the entire impact process can be estimated to be $\sqrt{\alpha_s \tau_{imp}} \approx 10\,\mu m$, well below the 1 mm thickness of the slide. Before the impact, the slide surface is set at a temperature $Tsur_{,0}$ from 80 °C to 590 °C by an electric heater; $Tsur_{,0}$ was measured by a surface probe before each experiment. In view of the importance of the surface roughness for the Leidenfrost temperature (and the splashing threshold [20]), we measured by atomic force microscopy the RMS roughness of the glass slide, finding it smaller than 10 nm.

Table 1. Relevant properties of ethanol and glass, and for comparison sapphire.

Property	Units	Ethanol	Glass	Sap-phire
Density ρ	[kg/m3]	789	2520	3970
Latent heat L	[kJ/kg]	837		
Specific heat Cp	[kJ/kgK]	2.4	816	776
Surface tension σ	[mN/m]	23		
Thermal conductivity k	[W/Km]		1	32
Thermal diffusivity α	[m2/s]		4×10-7	1×10-5
kρCp	[J2/K2sm4]		2×106	1×108

Next to the aforementioned side-view camera, we use two different imaging techniques to view the phenomena occurring during impact from below. In both techniques, the bottom view cameras (Photron SAX-2) were synchronised with the side view camera. The transparency of the setup enables us to use back light imaging through the substrate to obtain the bottom view (Fig. 1b). To measure the wetted area we use total internal reflection imaging (TIR), which was used previously in literature in both impact[20] and [21] and boiling studies [16] and [22]. Here, the light is coupled into the glass slide by a dove prism and, when the surface is dry, reflects from the top surface, so that all incident light enters the camera. However, if the surface is wetted by the drop, the refractive index changes from 1 (air) to 1.4 (ethanol) and the light can exit the slide through the drop, resulting in a dark spot on the camera. Just before the drop contacts the slide, it traverses the thickness of the evanescent wave, resulting in loss of intensity on the camera. As a consequence of the Fresnel equations, the intensity decreases exponentially with the distance from the slide surface, I(y)/I(∞)=1-exp(-βy), where $\beta = 4\pi/(\lambda n)\sqrt{\sin^2\theta - n^2}$ [23]. In our case, $\beta = 80\,\mathrm{nm}^{-1}$ depending on wavelength λ, refractive index ratio $n=n_t/n_i$ (with *t* and *i* indicate transmitted and incident) and the incident angle θ of the laser beam with the normal of the surface. This circumstance permits us to map the local height of the drop surface over the slide from the light intensity measured by each pixel within 12.5% uncertainty (see Appendix B). We correct for light reflections from the droplet interface, according to Court and von Willisen [24]. By this TIR technique we can clearly distinguish whether or not portions of the drop under-surface are in direct contact with the slide. Using a long-distance microscope (Navitar 12× Telecentric variable zoom system) we obtained a field of view of 15 mm and 1 mm for the study of millimetric and microdrops respectively. The side view and bottom view are illuminated by a white light source, while for the TIR recordings a continuous red diode laser (λ=643 nm) is used. The glass slide is optically coupled to the prism and sapphire disc by silicon oil to allow the laser light to propagate through all materials. The frame rate for the TIR imaging was 40 000 fps and we used a shutter speed of 300 ns to prevent motion blurring. In the microdroplet case, the TIR frame rate was increased to 100 000 fps for higher temporal resolution.

RESULTS AND DISCUSSION

Observations

Fig. 2 shows four examples of 2.6 mm-diameter drops impacting the glass slide at various temperatures with a velocity of U=1.3 m/s (*We* = 1293). For each example, the sequence on the left shows the side view of the impact. The sequence on the right is a composite of the TIR recording (left half), which shows the wetted area, and the bottom view (right half). These images are recorded in different experiments. However, we are confident that both can be considered as showing essentially the same event due to the repeatability of the experiment and to the fact that both recordings are synchronised with the side view.

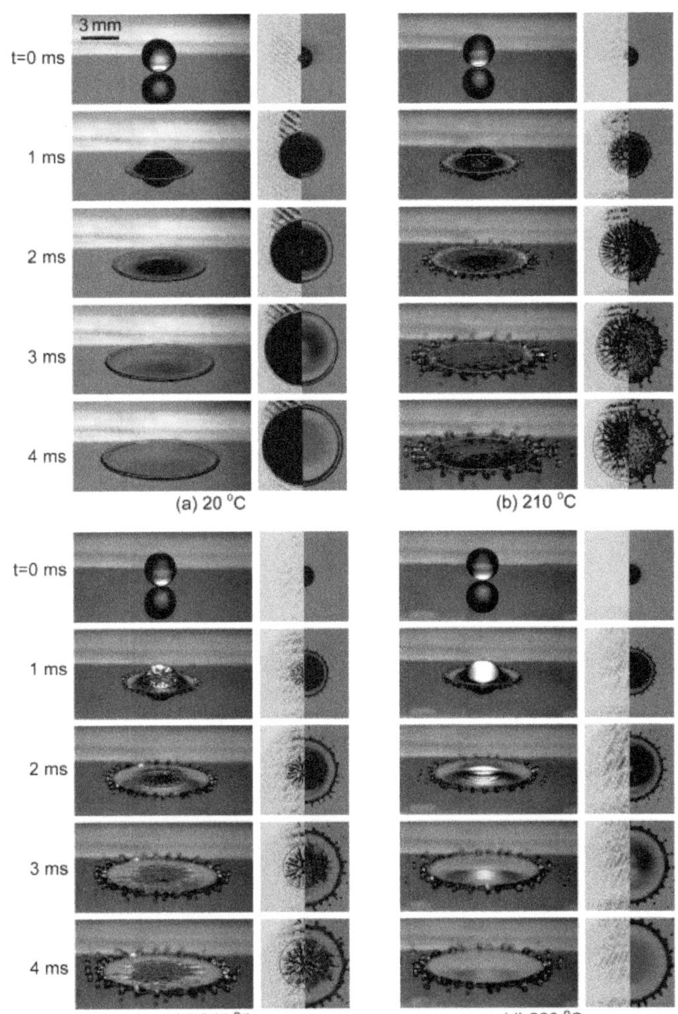

Figure 2. Sequences of ethanol drops impacting on a glass substrate with velocity of 1.3 m/s. The substrate is at an initial temperature of $Tsur_0$=20 (a), 210 (b), 245 (c) and 280 °C (d). In each sequence, the left images in each pair show the side-view, while the right ones are a composite of the bottom view (right part) and the total internal

reflection TIR (left part) recordings. The red circles in the side-view represent the measured wetted area from the TIR recordings. Sequence (a) shows a typical cold case, where all three recordings reveal the same wetted area. Impacts in the transition regime are shown in sequences (b) and (c), where the contact line spreading is suppressed (TIR), while a lamella spreads and fragments as visible in the bottom view. Sequence (d) shows an impact in the Leidenfrost regime, with no touch-down, which cannot be deduced from the bottom view but is revealed by the TIR. Some 'shadows' in the TIR images are artefacts of the optical system. The movies can be found online [25]. (For interpretation of the references to colour in this figure caption, the reader is referred to the web version of this article.)

For impacts from room temperature (Fig. 2a) up to 186 °C we observe the spreading of the drops on the solid surface to proceed at velocities similar to those reported in previous studies of the impact on cold surfaces [26], [27] and [28]. The radius from the TIR data corresponds to that of the moving contact line, represented by a red circle, in both the bottom and side views. Heating the substrate further changes this correspondence: at 210 °C we see that the growth of the wetted area is suppressed: the drop liquid is now partially levitating above the solid surface, forming a lamella which hovers over its own vapour (Fig. 2b). As the drop is now partly levitating and partly touching the solid, it is in a transitional state between contact and Leidenfrost boiling. With increasing temperature, the wetted area shrinks further (Fig. 2c) until at 280 °C the surface is never wetted by the drop, which is thus in the Leidenfrost state (Fig. 2d). This sequence of behaviours with increasing temperature is sketched in Fig. 3a.

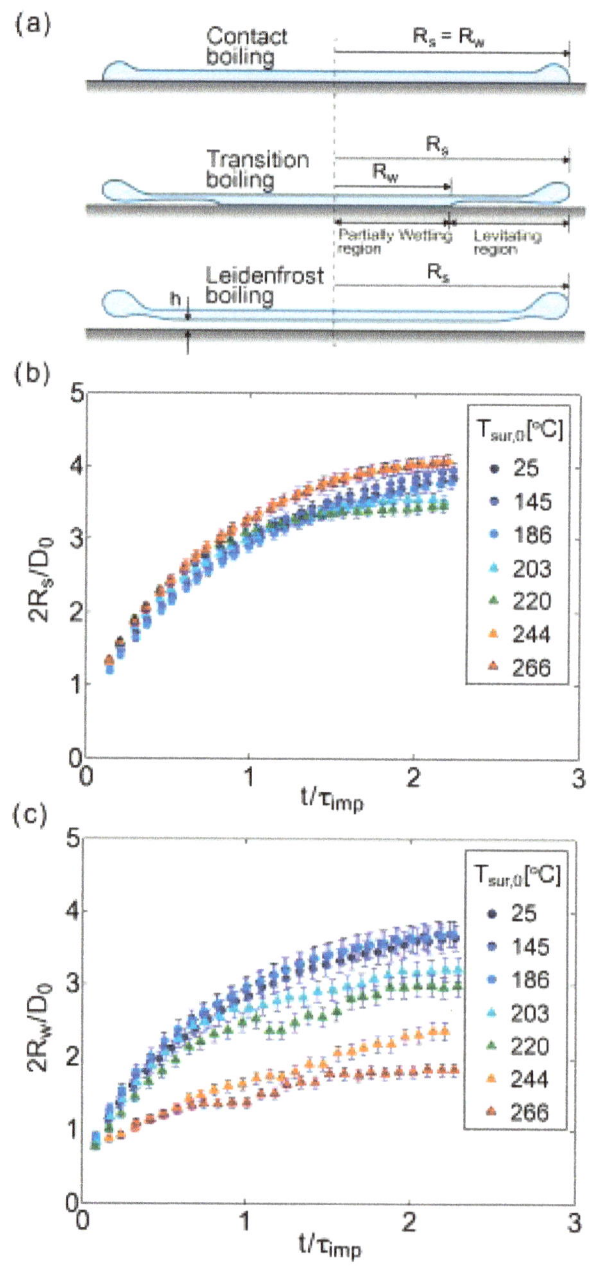

Figure 3. (a) Schematic of the three different spreading states: contact, transition and Leidenfrost boiling, indicating the difference between the radius of the wetted area R_w and spreading front R_s (within 5 and 10% uncertainty, see Appendix A). Observed time evolution of R_s (b) and R_w (c) of ethanol drops impacting at $U = 1.3\ m/s$ for increasing surface temperatures. Data for cold and contact boiling is indicated by °. For contact boiling R_w=R_s, while the lamella speed increases in the transition regime △; the wetting speed instead decreases, as compared to °. Later ($t = 2\ ms$), the lamella spreading stalls due to fragmentation (see Fig. 2b). The absence of wetting above 266 °C shows the drops to be in the Leidenfrost state. Details on errorbars can be found in Appendix A.

In the transition boiling regime, the TIR recordings show a fingering pattern (Fig. 2b and c), as reported for earlier experiments on glass [16]. However, preliminary results (not shown here) for impacts on a sapphire plate with a comparable smoothness do not exhibit this fingering pattern [14]. As sapphire is a better thermal conductor than glass, these observations suggest that the thermal properties of the substrate play a role in the appearance of this pattern.

As exemplified in Fig. 2, for each surface temperature $Tsur_{,0}$, the bottom views (rightmost images) provide information on the spreading front of the drop and on the wetted area. We define the radius $R_s(t)$ of the spreading front as the radius of the dark circle marking the edge of the drop in the bottom view images. The radius $R_w(t)$ of the wetted area is defined as the radius of the smallest circle encompassing the wetted area, as shown by the red circle in the figure. The data obtained in this way are shown in Fig. 3for various surface temperatures up to the incipience of the Leidenfrost state. We find that $R_s(t)$ (Fig. 3b) increases only slightly with increasing $Tsur_{,0}$. At the lower temperatures, $R_w(t)$ (Fig. 3c) keeps pace with $R_s(t)$. However, from the start of the transition regime at 203 °C , this correspondence between the two quantities is lost and $R_w(t)$ grows much less than $R_s(t)$. At the end of the transition regime, R_w settles at about 0.75 D_0. Above 266 °C, no wetting is found and the drops are in the Leidenfrost state. It is also noteworthy that a non-monotonic increase of R_s with $Tsur_{,0}$ is found in Fig. 3b for $t/\tau_{imp}>1$. This is caused by the lamella breakup which is strongest at the beginning of the transition boiling as shown in Fig. 2.

We performed experiments for different impact velocities as well. The observed boiling behaviours are plotted in the phase space ($U, Tsur_{,0}$) in Fig. 4. Three main regions can be identified: for low substrate temperatures we find stable contact boiling, represented by blue circles. Here, the growth of the wetted area keeps pace with the spreading as for impact on unheated surfaces. In orange we indicate the aforementioned transition region where a lamella is ejected. Here, the wetting velocity of the contact line lags behind the expansion rate of the lamella. This regime is characterised by two different behaviours, indicated by △ and ◊, to be discussed later. For high temperatures we find no wetted area and the impacts are in the Leidenfrost state, shown as red squares.

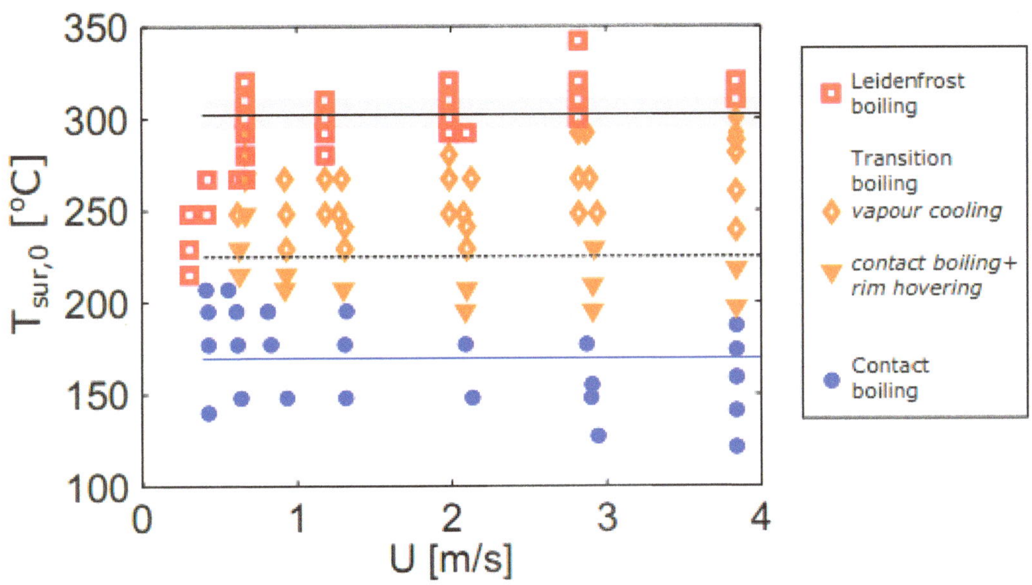

Figure 4. Phase space with an indication of the different boiling behaviours. The dashed line is a guide to the eye, indicating the boundary between *contact boiling + rim hovering* and *vapour cooling*. The solid black line denotes the transition to the Leidenfrost state, according to Eq. (2); the shaded area indicates the estimated error. The Leidenfrost transition for isothermal impacts on sapphire is shown for comparison by the blue line at $T_{sur,0} = 170\,^{\circ}C$ (data from Staat et al. [15]). (For interpretation of the references to colour in this figure caption, the reader is referred to the web version of this article.)

Subcooling Effect Causes Touch-Down

The points denoted by △ in the phase space of Fig. 4 indicate drops that, even though partially levitated, immediately exhibit stable boiling on the wetted area; this regime is indicated as *contact boiling + rim hovering* in the legend of the figure. This behaviour is strikingly different from that of drops that levitate at the very beginning of the impact, but then touch down at a later time. This regime, to which we refer as *vapour cooling* , is marked by ◊ in Fig. 4. A detailed example of the drop behaviour in this case is shown in Fig. 5a; the drop diameter is 2.3 mm, the impact velocity U=3.84 m/s and the glass slide temperature $Tsur_{,0}$ is 292 °C. For each frame in this figure, the upper half shows the TIR image, while the lower half is the reconstructed height profile of the underside of the drop in tens of nm resolution. Here the drop first remains levitated (i–iv), but in frames (v) and (vi) it starts to randomly touch the surface, exhibiting what may be called unstable boiling. This behaviour is different from what we earlier referred to as stable boiling in which the liquid–solid contact does not fluctuate appreciably. Later, the drop wets the surface and a stable fingering pattern is observed, as discussed before. It is also interesting to note that no drop is visible in frame (ii), although it reappears in frame (iii). This behaviour implies that the pressure under the drop is initially so large as to briefly push it back. We have encountered this phenomenon only in the upper range of the temperatures investigated.

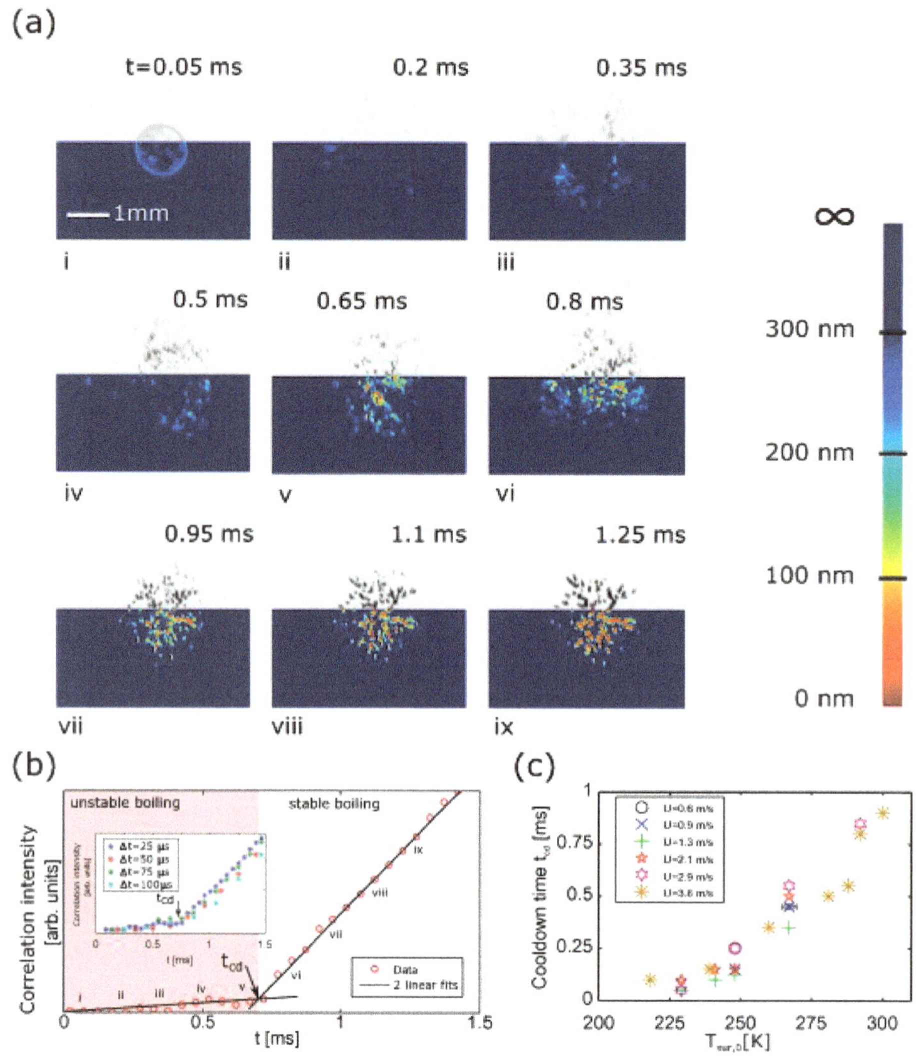

Figure 5. (a) TIR data of an ethanol drop, with a diameter of 2.3 mm, impacting at $U = 3.8$ m/s a heated glass surface of $T_{sur,0} = 292$ °C. The upper half of each frame shows the measured light intensity. As explained in the text, from these data we deduce the local height of the drop surface above the solid, i.e., the local thickness of the vapour layer, which is shown in the lower half (with 9% uncertainty, see Appendix B). The sequence shows the change from levitation (i–iv) to unstable contact (v,vi) to eventually stable boiling (vii–ix). The absence of the signal in (ii) results from a brief upward motion of the drop surface, only observed for high surface temperatures (b): Time series of the correlation coefficient for the data shown in (a) with matching roman numerals. The inset shows the effect of correlating pairs of non-consecutive images separated by 1–3 frames. (c): Measured cool-down times (defined in the text) for various impact velocities as a function of initial substrate temperature. For clarity, only one set of error bars is shown.

This delayed touch-down mechanism is due to the cooling that the vapour generation induces in the solid. Due to the small thermal conductivity of glass, $T_{sur}(t)$ drops significantly during the early stages of the impact to such an extent that the drop levitation cannot be maintained. Such effects of the solid thermal properties are frequently encountered in heat transfer phenomena. Examples are the influence of the thermal conductivity of a hot metal being wetted by the advancing front of a cold liquid[29] and [30], the similar effect of the substrate influencing the waiting time between two successive bubbles in boiling [31] and the reduction of the Nusselt number in turbulent thermal convection [32].

The situation that is established during the drop impact is characterised by the very short distance (less than 1 μm) that separates the liquid surface from the solid and a much longer scale parallel to the solid surface. This large scale separation justifies the use of a one-dimensional conduction model in the solid. We assume the drop surface to remain at the saturation temperature T_b and model the heat flux out of the solid as $h(T_{sur}(t)-T_b)$, where h is a heat transfer coefficient. We may expect h to be enhanced beyond the pure conduction value (k_v/δ_v, with k_v the thermal conductivity of the vapour and δ_v the thickness of the vapour layer) by the vapour flow between the drop and the surface, although the order of magnitude should not be altered. This expectation is based on the well-known relation of the Nusselt number for fully developed laminar flow in ducts [33]. The solution of the conduction problem thus posed is well known [3], [34], [35] and [36]. The temperature of the solid surface $T_{sur}(t)$ is found to be given by:

$$\frac{T_{sur}(t) - T_b}{T_{sur,0} - T_b} = \exp\left(\frac{t}{\tau_{th}}\right)\operatorname{erfc}\left(\sqrt{\frac{t}{\tau_{th}}}\right), \tag{1}$$

in which the characteristic thermal time scale is given by $\tau_{th}=k_s\rho_s C_{p,s}h^{-2}$.

In order to determine τ_{th}, as the heat transfer coefficient h is unknown for our system, for each experiment we measure the cool-down time t_{cd}, i.e., the time before stable contact boiling is observed. Then we apply Eq. (1) to fit our data and obtain τ_{th} from which we calculate h. The cool-down time t_{cd} is obtained by studying the temporal correlation coefficient between image pairs. This coefficient is a measure of the degree of similarity between two successive frames. For $t<t_{cd}$ the drop is either still levitating or randomly touching the substrate. The wetted area is then rapidly varying and the resulting correlation between successive images is poor. When the surface is cooled down by the evaporation, on the other hand, the wetted patches become more stable, resulting in a high correlation coefficient. We used sequences such as that of Fig. 5a to create a time series of correlation coefficients, an example of which is shown in Fig. 5b. Here, the sudden increase in correlation is readily identifiable as occurring for $t \approx 0.7$ ms, where two straight lines fitted to the correlation data intersect. Skipping frames (inset) reveals that the technique is fairly insensitive to the frame rate (provided, of course, it is not too short). We apply this technique to measure t_{cd} for various impact velocities and initial plate temperatures $Tsur_{,0}$. The results are presented in Fig. 5c and show that, for higher temperatures, it takes more time for the surface to cool down to the temperature level required for stable boiling. Furthermore, the data for different impact velocities exhibit a near-collapse. This is in agreement with the dashed line in Fig. 4 being independent of velocity as well. Both observations indicate that the cooldown time is at most weakly dependent on the impact velocity. However we do observe for high initial surface temperatures only stable boiling for high impact velocities. This is due to the higher dynamic pressure, forcing the drop more strongly towards the surface[28] and [37].

Once t_{cd} has been determined, we can proceed to use the information to estimate the thermal time scale τ_{th} appearing in Eq. (1). Since for $t=t_{cd}$ the surface temperature becomes too low to support the drop we deduce that, at this instant, the surface temperature equals the static Leidenfrost temperature for isothermal surfaces, i.e., $T_{sur}(t=t_{cd}) = 160\,^{\circ}\mathrm{C}$[3]. We plot in Fig. 6 for each experiment $(160\,^{\circ}\mathrm{C} - T_b)/(T_{sur,0} - T_b)$ against tcd. The black lines correspond to various solutions of Eq. (1) and the best fit of the data is the blue line with τ_{th}=0.3 ms. The fit is not very sensitive in this range either to the precise value of t_{cd} or to the value assumed for the static isothermal Leidenfrost temperature. Although the fit is reasonably close to the data, there seems to be some difference between large and small values of t_{cd}, which corresponds to a slight increase of the thermal time scale with t_{cd} and, therefore, with $Tsur_{,0}$. This results implies that the average heat transfer coefficient decreases somewhat with increasing surface temperature. Possible causes can be a larger initial vapour layer thickness as would be implied by the behaviour described before in connection with frame (ii) of Fig. 5a and increased convection cooling by the vapour.

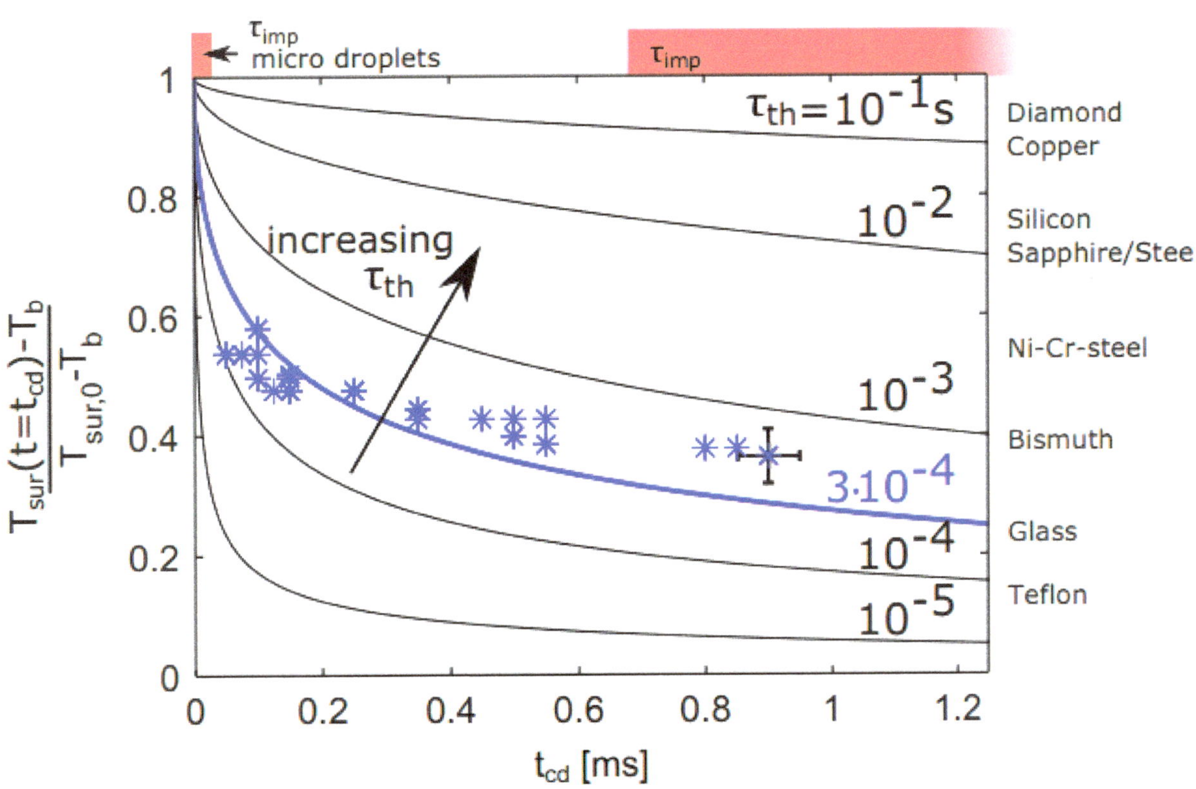

Figure 6. As argued in the text, for $t=t_{cd}$, the surface temperature $T_{sur}(t_{cd})$ should equal $160\,^{\circ}\mathrm{C}$. The asterisks show the value $(T_{sur}(t_{cd})-T_b)/(Tsur_{,0}-T_b)$ for various values of $Tsur_{,0}$, in correspondence of the t_{cd} measured in each experiment, as presented in Fig. 5c. For clarity, only one set of error bars is shown. The best fit of the data by Eq. (1) is presented by the thick blue line; the remaining black lines present other solution corresponding with different thermal time scales. These solutions match various materials listed on the right, assuming that for impacts on smooth materials the heat transfer coefficient through the vapour layer has the constant value measured in the present study (here 8×10^4 W/(km^2)). The red areas at the top of the figure indicate the impact time scales examined here for micro- and millimetric drops. (For interpretation of the references to colour in this figure caption, the reader is referred to the web version of this article.)

It is interesting to compare the thermal time scale τ_{th} with the time $\tau_{imp}=D_0/U$characteristic of the impact, by which all the downward momentum is converted into the radial direction and the dynamic pressure vanishes [38]. This is also the longest time the residual momentum can maintain the drop against the hot surface. Thus, if τ_{th} is equal or smaller than τ_{imp}, the surface will cool down significantly. This is the case for all the impacts of millimetric drops studied here, for instance: in Fig. 5 $\tau_{imp} = 0.6$ ms which is approximately τ_{th}.

From the thermal time scale τ_{th} we can deduce an estimate for the heat transfer coefficient $h = \sqrt{(k_s \rho_s C_{p,s}/\tau_{th})}$, which is therefore found to be 8×10^4 W/(m^{2}K). For pure conduction through the vapour layer, this result corresponds to a thickness $\delta_v \approx 100\,\mathrm{nm}$. This estimate is in agreement with our measurements of the minimum vapour thickness (for example Fig. 5) obtained from the evanescence wave attenuation length. Furthermore, assuming that this thickness is typical for impacts on smooth surfaces, one can now predict whether or not an impact can be considered as isothermal: the impact conditions yield τ_{imp} and one estimates $\tau_{th} = (100\,\mathrm{nm}/k_v)^2 k_s \rho_s C_{p,s}$ from the surface properties. The surface cooling for various materials based on their estimatedτth is shown in Fig. 6.

Focussing in more detail on the phase space of Fig. 4, we now aim to predict the boundary between the transition region (orange) and the Leidenfrost region (red). For all impact velocities we select the lowest surface temperature resulting in the Leidenfrost state and measure the time that the drop is within 100 nm distance of the substrate, seeTable 2. The average time observed is 0.51 ms without any significant velocity dependence, in accordance with our previous observations (Fig. 5c). This 'residence time' is typically shorter than τ_{imp}. Using these observations, we can calculate $Tsur_{,0}$by solving Eq. (1) with $T_{sur}(t = 0.51\ \mathrm{ms}) = 160\,^\circ\mathrm{C}$, the static isothermal T_L for ethanol:

$$T_{sur,0} = \frac{160\,^\circ\mathrm{C} - 80\,^\circ\mathrm{C}}{\exp(0.51/0.3)\mathrm{erfc}\left(\sqrt{0.51/0.3}\right)} + 80\,^\circ\mathrm{C} = 304\,^\circ\mathrm{C}. \tag{2}$$

The obtained prediction is plotted in Fig. 4 by a black solid line; the shaded area indicates the estimated error. We see that including this correction for the non-isothermal behaviour of the solid yields a good prediction for the dynamic Leidenfrost temperature.

Table 2. Observed residence time of the drop within 100 nm from the solid surface for impact on a glass slide. The last column is the estimated Leidenfrost temperature obtained similarly to (2), using for the residence time the value listed in column four in place of 0.51 ms.

U [m/s]	$We=\rho_l U^2 D_0/(2\sigma_l)$	Lowest T_L [°C]	Residence time [ms]	T_L by Eq. (2) [°C]
0.4	8	267	0.4	285±10
0.6	27	280	0.45	294±10
1.3	111	280	0.6	318±10
2.1	279	290	0.55	310±10
2.9	538	300	0.5	303±10
3.8	740	310	0.6	318±10

Micro Droplets show Isothermal Behaviour

The previous analysis assumed the elevated Leidenfrost temperature found for impacts on a glass surface to originate from the thermal properties of the substrate. In order to strengthen this conclusion, we also studied the case in which $\tau_{imp} \ll \tau_{th}$ by considerably shortening the impact time scale τ_{imp}. This objective is achieved by using microdroplets, as the smaller diameter $D_0 = 300\ \mu\mathrm{m}$ lowers $\tau_{imp}=D_0/U$ to 30 μs, satisfying $\tau_{imp} \ll \tau_{th}$. In this case we expect even a glass substrate to behave as isothermal with no observable cool-down effect. A sequence showing the full impact and spreading phases of a microdrop impacting at $U = 8.9\ \mathrm{m/s}$ a glass surface with $T_{sur,0} = 265\,^\circ\mathrm{C}$ (just below T_L) is shown in Fig. 7.

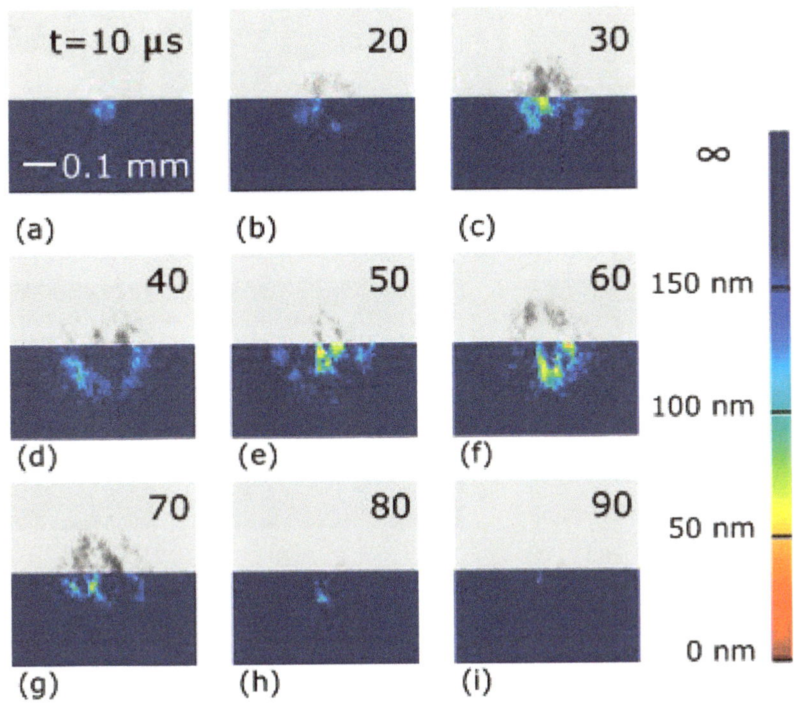

Figure 7. Similarly to Fig. 5, each frame is a composite of the TIR-measured light intensity and the corresponding thickness (with 12.5% uncertainty, see Appendix B) of the vapour film for an 300 μm-diameter ethanol droplet impacting a glass substrate just below the Leidenfrost temperature $T_{sur,0} = 265\,^{\circ}\mathrm{C}$. The impact velocity is 8.9 m/s (*We*=1874). No stable boiling is observed as $\tau_{imp} \approx 30\,\mu s$, which is much shorter than τ_{th}.

Comparing the results with the millimetric drops shown in Fig. 5, we first observe a levitating droplet as well. From $t = 50\,\mu s$ on, the drop touches the surface randomly. However, in contrast with the millimetric case, no stable boiling is observed for the microdroplet impact. Varying the impact velocity between 2 and 8.9 m/s, we did not observe any cool-down effect. In this case, the touch-down mechanism then is purely governed by the impact dynamics, as expected from the estimate of the time scales. We conclude therefore that the same surface that exhibits cooling with large drops, behaves as isothermal when the time scale of the impact dynamics is shorter than the thermal time scale.

CONCLUSIONS

We studied the impact of ethanol drops on superheated smooth glass substrates. We used high speed total internal reflection imaging to study the dynamics of the wetted area of the surface. We observed a new touch-down mechanism for millimetric drops caused by the cooling of the solid surface. By comparing the time scale τ_{th} for the cooling of the glass surface with the impact time scale $\tau_{imp}=D_0/U$ we concluded that a millimetric drop causes a significant cooling the glass surface. While the drop remains levitated during the early stages of the impact, this cooling causes it to eventually touch the solid. By means of a one-dimensional conduction model, we developed an estimate of the dynamic Leidenfrost temperature for a poorly conducting solid. We calculated τth for glass to be 0.3 ms, which is well below τ_{imp} for the impact events studied here. For impact velocities in the range of 0.6–3.8 m/s, investigated in this study, the vapour thickness calculated from τth is approximately 100 nm, consistent with our measurements. Using this information, together with the substrate properties one can thus a priori estimate the thermal time scale.

The previous conclusions are consistent with the results of another set of experiments on the impact of micro-droplets for which τ_{imp} is much shorter than τ_{th}. In this case, indeed, we do not observe any cooling effect and the surface remains essentially isothermal.

To summarise, therefore, for $\tau_{imp} \ll \tau_{th}$, the solid remains essentially isothermal and the touch-down of the drop is governed by the dynamic pressure of the impact. For $\tau_{imp} > \tau_{th}$, on the other hand, the substrate is cooled during the spreading of the drop, which eventually touches the solid surface.

ACKNOWLEDGEMENTS

This work was partially supported by Fundamenteel Onderzoek der Materie, NanoNextNL and by an ERC Advanced Grant.

Appendix A. Image processing

The three types of recordings we obtain in our experiments are processed in different ways, using Matlab programming. The side-view images are used for the measurements of the diameter and impact velocity by a simple edge detection technique. The classification of the boiling regimes are done manually by comparing all three image types by characteristics as explained in details in Section 3.1.

The image analysis of the TIR data consists mainly of two processes: firstly, the image transformation and secondly the detection of the wetted area, as shown in Fig. A.8(a–f). Even though a droplet wets the substrate in an axisymmetric manner, the TIR image gives an ellipsoid as shown in (a), due to the optical transformations in the setup. The original image is first binarized after background subtraction as shown in (b). From the binary image, the angle of rotation and aspect ratio are measured; for this measurement, the frame corresponding to $t/t_{imp}=0.25$ of room temperature experiment was typically used. Using the linear transformation, the circular droplet, (c), is reconstructed.

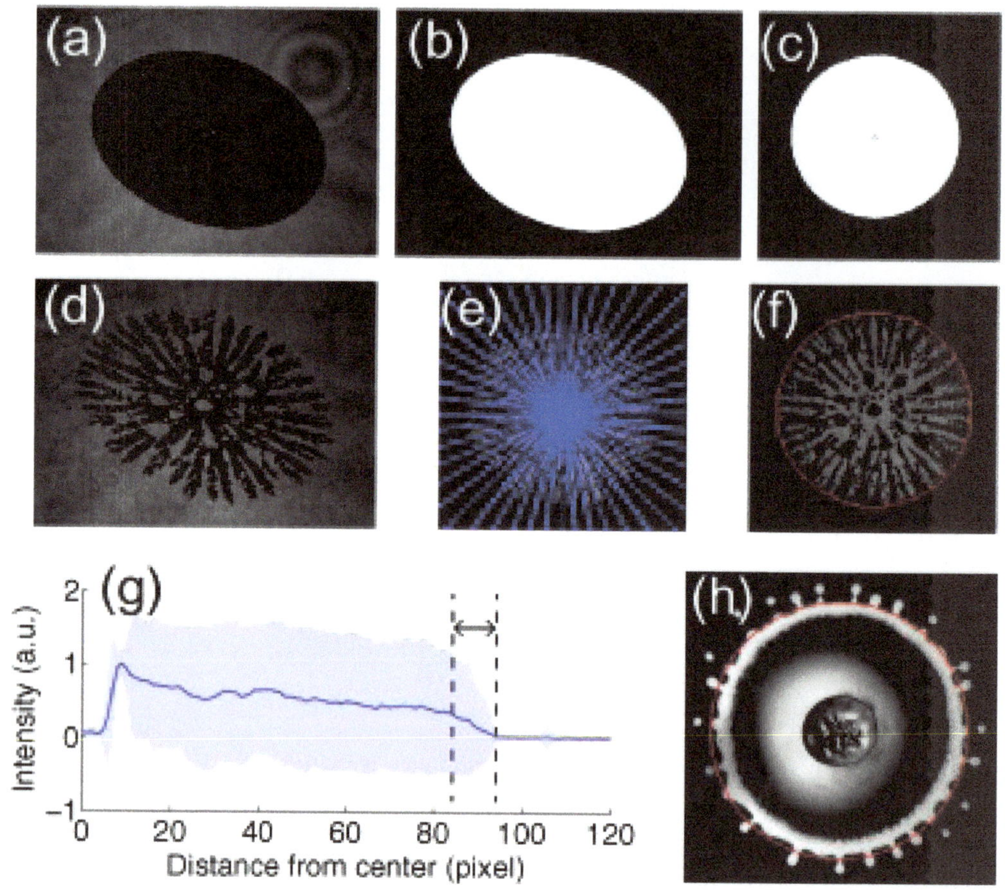

Figure A.8. Typical examples of (a–c) images used for calibration of TIR image transform, (d–f) TIR images in transition boiling regime showing the transformation with the rate given in (a–c), (g) average intensity (solid line) together with standard deviation (shaded area) along the lines radially spread outward from the centre of wetted area (blue lines in (e)), and (h) an image of bottom view with backlighting after the background subtraction. The red circles in (f) and (h) show the detected wetting and spreading front, respectively. The uncertainty caused by the non-uniform growth rate of the fingers is shown by the arrow in (g). (For interpretation of the references to colour in this figure caption, the reader is referred to the web version of this article.)

The linear transformation obtained from the room temperature experiment was also used in the analysis of different temperature cases. A typical example of impact on a heated surface is shown in Fig. A.8(d–f), where the initially ellipsoidal fingering pattern in (d) is well transformed back to the circular one in (f). Note here that the envelope circle shown as the red line in (f) was determined from the average-intensity profile, such as shown in (g), as the point of abrupt decrease in mean value (solid line) and standard deviation (shaded area); the intensity profile was calculated from intensities along 50–100 lines spreading radially outward from the droplet centre, see (e).

The radius of spreading front R_s was also determined from the bottom-view image as in the same manner for TIR analysis except for the linear transformation. We set the thresholding condition in such a way that the foot of the fingers at the rim of the droplet was detected as shown by the red circle in Fig. A.8(h).

The main cause of the uncertainty of the radius of the spreading front R_s arises from the non-uniform growth rate of the fingers pinched off from the peripheral rim; such non-uniform behaviour result in an uncertainty for defining the spreading front from the image intensity profile, a typical example of which is shown by the arrow in Fig. A.8(g). Similarly, for the radius of the wetting area R_w, the non-uniform wetting rate of the fingering pattern causes the largest uncertainty. In summary, the uncertainty is, at most, 5% for R_s and 10% for R_w.

Appendix B. Error estimate of the measured height from TIR measurement

Starting from the equation for the height h and the normalised intensity [24]: $I' = \frac{I}{I_\infty} = EE^*$, with E being the electric field and E^* its complex conjugate.

$$h = -\beta \ln \left(\frac{r_{12} - E}{r_{32} - r r_{12} r_{32}} \right), \quad \text{(B.1)}$$

$$\text{where } \beta = \frac{\lambda n}{4\pi} \frac{1}{\sqrt{\sin(\theta_i)^2 - n^2}} \quad \text{(B.2)}$$

Here λ is 643 nm, the wavelength of the light source, n=1/1.51 the ratio of refractive indices of the air and glass and θ_i is the angle between the laser light and the normal of the glass surface, resulting in β=81 and 59 nm^{-1}, with θ_i=52° and 61°, for the millimetric drops and micro droplet measurements respectively. The functions r_{12} and r_{32} are the reflectivity coefficients as given by the Fresnel equations for the solid–air and air–liquid interface for p -polarised light respectively. These functions depend on n, θ_i and n_d, the refractive index of the droplet liquid. We then calculate numerically the error which is presented in Fig. B.9.

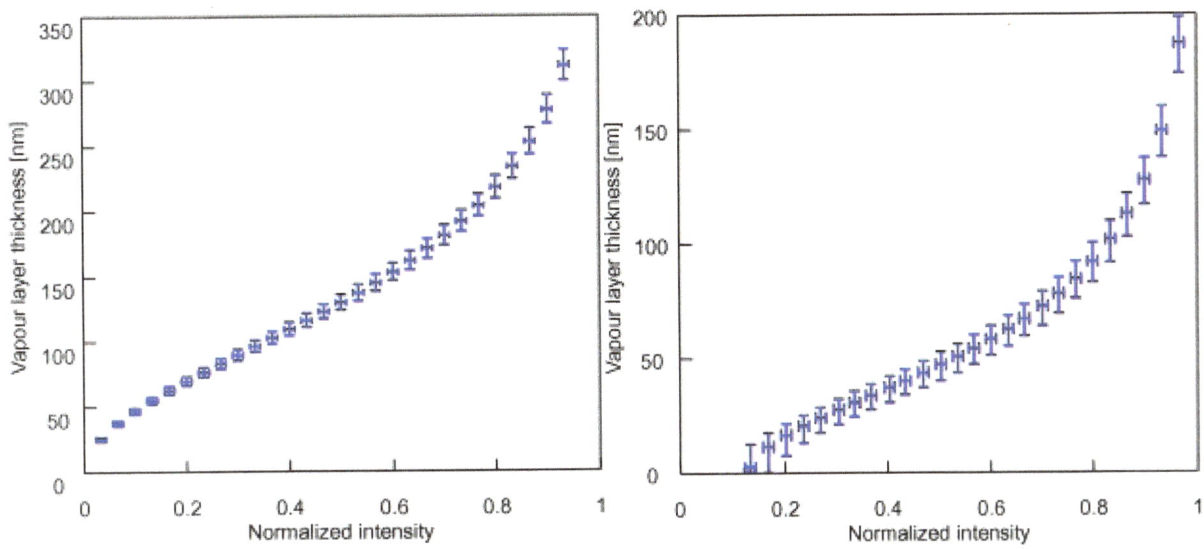

Figure. B.9. Error estimate for the millimetric (left) and micro droplet (right) measurements. The vertical errorbars the combined uncertainties in refractive indices, laser wavelength and angle of incidence of the laser.

Here the following uncertainties are used to estimate the total error: $\Delta\lambda$=0.5 nm,Δ_{nti}=0.001 where the refractive index was evaluated from room temperature to 400 °C [39], and $\Delta\theta_i$=0.5° from our analysis uncertainty. We estimate the error in the measured intensity to be 2% originating from fluctuations in the laser intensity. From this analysis, we can approximate the error Δh to increases linearly with height. The average relative error Δh/h is 9% for millimetric drops and 12.5% for the micro droplet measurements.

REFERENCES

1. J.G. Leidenfrost, De aquae communis nonnullis qualitatibus tractatus, Ovenius, Duisburg, 1756.
2. D. Quéré, Leidenfrost dynamics, Annu. Rev. Fluid Mech. 45 (2013) 197–215.
3. K. Baumeister, F. Simon, Leidenfrost temperature—its correlation for liquid metals, cryogens, hydrocarbons, and water, J. Heat Transfer 95 (1973) 166– 173.
4. J.D. Bernardin, I. Mudawar, The Leidenfrost point: experimental study and assessment of existing models, J. Heat Transfer 121 (1999) 894–903.
5. H. Kim, B. Truong, J. Buongiorno, L.-W. Hu, On the effect of surface roughness height, wettability, and nanoporosity on Leidenfrost phenomena, Appl. Phys. Lett. 98 (2011) 083121.
6. H.-M. Kwon, J.C. Bird, K.K. Varanasi, Increasing Leidenfrost point using micro– nano hierarchical surface structures, Appl. Phys. Lett. 103 (2013) 201601.
7. T. Tran, H.J.J. Staat, A. Prosperetti, C. Sun, D. Lohse, Drop impact on superheated surfaces, Phys. Rev. Let. 108 (2012) 036101.
8. L.H.J. Wachters, H. Bonne, H.J. van Nouhuis, The heat transfer from a hot horizontal plate to sessile water drops in the spherodial state, Chem. Eng. Sci. 21 (1966) 923–936.
9. A.-L. Biance, C. Clanet, D. Quéré, Leidenfrost drops, Phys. Fluids 15 (2003) 1632–1637.
10. H. Fujimoto, Y. Oku, T. Ogihara, H. Takuda, Hydrodynamics and boiling phenomena of water droplets impinging on hot solid, Int. J. Multiphase Flow 36 (2010) 620–642.
11. H. Nair, H.J.J. Staat, T. Tran, A. van Houselt, A. Prosperetti, D. Lohse, C. Sun, The Leidenfrost temperature increase for impacting droplets on carbon-nanofiber surfaces, Soft Matter 10 (2014) 2102–2109.
12. B.T. Ng, Y.M. Hung, M.K. Tan, Suppression of the Leidenfrost effect via low frequency vibrations, Soft Matter 11 (2015) 775–784.
13. T. Tran, H.J.J. Staat, A. Susarrey-Arce, T.C. Foertsch, A. van Houselt, H.J.G.E. Gardeniers, A. Prosperetti, D. Lohse, C. Sun, Droplet impact on superheated micro-structured surfaces, Soft Matter 9 (2013) 3272–3282.
14. M. Shirota, M.A.J. van Limbeek, C. Sun, A. Prosperetti, D. Lohse, Dynamic Leidenfrost effect: relevant time and length scales, Phys. Rev. Let. 116 (2016) 064501.
15. H.J.J. Staat et al., Phase diagram for droplet impact on superheated surfaces, J. Fluid Mech. 779 (2015) R3.
16. M. Khavari, C. Sun, D. Lohse, T. Tran, Fingering patterns during droplet impact on heated surfaces, Soft Matter 11 (2015) 3298–3303.
17. A.-L. Biance, F. Chevy, C. Clanet, G. Lagubeau, D. Quéré, On the elasticity of an inertial liquid shock, J. Fluid Mech. 554 (2006) 47–66.
18. H. Lastakowski, F. Boyer, A.-L. Biance, C. Pirat, C. Ybert, Bridging local to global dynamics of drop impact onto solid substrates, J. Fluid Mech. 747 (2014) 103–118.
19. E. Villermaux, B. Bossa, Drop fragmentation on impact, J. Fluid Mech. 668 (2011) 412–435.
20. J.M. Kolinski, L. Mahadevan, S.M. Rubinstein, Drops can bounce from perfectly hydrophilic surfaces, Europhys. Lett. 108 (2014) 24001.
21. J.M. Kolinski, S.M. Rubinstein, S. Mandre, M.P. Brenner, D.A. Weitz, L. Mahadevan, Skating on a film

of air: drops impacting on a surface, Phys. Rev. Lett. 108 (2012) 074503.

22. J. Kim, Spray cooling heat transfer: the state of the art, Int. J. Heat Fluid Flow 28 (2007) 753–767.
23. E. Hecht, Optics, fourth ed., Addison Wesley Longman Inc, Boston, 1998.
24. I.N. Court, F.K. von Willisen, Frustrated total internal reflection and application of its principle to laser cavity design, Appl. Opt. 3 (6) (1964) 719–726.
25. M.A.J. van Limbeek, M. Shirota, P. Sleutel, C. Sun, A. Prosperetti, D. Lohse, Vapour cooling of poorly conducting hot substrates increases the dynamic Leidenfrost temperature, Int. J. Heat Mass Transfer 97 (2016) 101–109.
26. S. Mandre, M. Mani, M.P. Brenner, Precursors to splashing of liquid droplets on a solid surface, Phys. Rev. Lett. 102 (2009) 134502.
27. S. Mandre, M.P. Brenner, The mechanism of a splash on a dry solid surface, J. Fluid Mech. 690 (2012) 148–172.
28. G. Riboux, J.M. Gordillo, Experiments of drops impacting a smooth solid surface: a model of the critical impact speed for drop splashing, Phys. Rev. Lett. 113 (2014) 024507.
29. R.B. Duffey, D.T.C. Porthouse, The physics of rewetting in water reactor emergency core cooling, Nucl. Eng. Des. 25 (1973) 379–394.
30. M.K. Agrawal, S.K. Sahu, Analysis of conduction-controlled rewetting of a hot surface by variational method, Heat Mass Transfer 49 (2013) 963–971.
31. V.K. Dhir, Mechanistic prediction of nucleate boiling heat transfer–achievable or a hopeless task?, J Heat Transfer 128 (2006) 1–12.
32. R. Verzicco, Effects of nonperfect thermal sources in turbulent thermal convection, Phys. Fluids 16 (2004) 1965–1979.
33. A. Bejan, Heat Transfer, first ed., John Wiley @ Sons Inc, New York, 1993.
34. S. Chandra, C. Avedisian, On the collision of a droplet with a solid surface, Proc. R. Soc. Lond. A: Math. Phys. Eng. Sci. 432 (1991) 13–41.
35. H. Carslaw, J. Jaeger, Conduction of Heat in Solids, first ed., Clarendon Press, Oxford, 1959.
36. Y. Qiao, S. Chandra, Boiling of droplets on a hot surface in low gravity, Int. J. Heat Mass Transfer 39 (1996) 1379–1393.
37. J. Philippi, P.-Y. Lagrée, A. Antkowiak, Drop impact on a solid surface: short time self-similarity, 2015. .
38. M. Mani, S. Mandre, M.P. Brenner, Events before droplet splashing on a solid surface, J. Fluid Mech. 647 (2010) 163–185.
39. T. Toyoda, M. Yabe, The temperature dependence of the refractive indices of fused silica and crystal quartz, J. Phys. D 16 (1983) L97L100.

CHAPTER 3

AN EXPERIMENTAL STUDY OF SPATIOTEMPORALLY RESOLVED HEAT TRANSFER IN THIN LIQUID-FILM FLOWS FALLING OVER AN INCLINED HEATED FOIL

Christos N. Markides , Richard Mathie, Alexandros Charogiannis

Clean Energy Processes (CEP) Laboratory, Department of Chemical Engineering, Imperial College London, London SW7 2AZ, United Kingdom

ABSTRACT

This paper describes the development of an experimental technique that combines simultaneous planar laser-induced fluorescence (PLIF) and infrared (IR) thermography imaging, and its application to the measurement of unsteady and conjugate heat-transfer in harmonically forced, thin liquid-film flows falling under the action of gravity over an inclined electrically heated-foil substrate. Quantitative, spatiotemporally resolved and simultaneously conducted measurements are reported of the film thickness, film free-surface temperature, solid–liquid substrate interface temperature, and local/instantaneous heat flux exchanged with the heated substrate. Based on this information, local and instantaneous heat-transfer coefficients (HTCs) are recovered. Results concerning the local and instantaneous HTC and how this is correlated with the local and instantaneous film thickness suggest considerable heat-transfer enhancement relative to steady-flow predictions in the thinner film regions. This behaviour is attributed to a number of unsteady/mixing transport processes within the wavy films that are not captured by laminar, steady-flow analysis. The Nusselt number Nu increases with the Reynolds number Re; at low Re values the mean Nu number corresponds to 2.5, in agreement with the steady-flow theory, while at higher Re, both the Nu number and the HTC exhibit significantly enhanced values. Evidence that the HTC becomes decoupled from the film thickness for the upper range of observed film thicknesses is also presented. Finally, smaller film thickness fluctuation intensities were associated with higher HTC fluctuation intensities, while the amplitude of the wall temperature fluctuations was almost proportional to the amplitude of the HTC fluctuations.

Citation: Christos N. Markides, Richard Mathie, Alexandros Charogiannis, An experimental study of spatiotemporally resolved heat transfer in thin liquid-film flows falling over an inclined heated foil, International Journal of Heat and Mass Transfer, Volume 93, February 2016, Pages 872-888, ISSN 0017-9310, http://dx.doi.org/10.1016/j.ijheatmasstransfer.2015.10.062.

INTRODUCTION

Liquid films falling under the action of gravity are classical free-surface flows that are either directly encountered in or representative of a wide range of industrial applications, including wetted-wall absorbers, heat exchangers, condensers, evaporators and reactors. The practical interest in these flows arises primarily from their high surface-to-volume ratios and excellent heat and mass transfer characteristics. It therefore comes as no surprise that extensive theoretical [1], [2], [3] and [4], numerical [5], [6], [7], [8] and [9] and experimental [10], [11], [12] and [13] efforts have been devoted to their study over the past decades. The extraction of reliable *detailed* experimental data has, however, proven particularly challenging owing to a number of challenges inherent to these flow systems, such as the restricted fluid domains under observation (often sub-mm) and the intermittent nature of the moving and wavy interface. The present investigation is motivated by the strong demand for detailed, spatiotemporally resolved hydrodynamic and heat-transfer data in planar falling-film flows. This information is necessary for the purpose of improving our fundamental understanding of these important flows, and also for furthering the development and validation of advanced analytical tools and numerical modelling codes, by providing sophisticated closures for accurate and reliable predictions. Beyond this fundamental interest, the availability of the information can also prove valuable in our efforts to appreciate the transport capabilities of these flows and our in-depth understanding of the associated underlying heat-transfer mechanisms, which can then be harnessed in practical situations to improve performance and reduce the size and cost of equipment.

This paper describes a simultaneous application of optical-diagnostic techniques for the provision of the aforementioned spatiotemporally resolved heat-transfer information in thin liquid-films flowing down an inclined heated foil. Optical techniques are recognized as advanced measurement tools, capable of providing detailed, non-intrusive multi-dimensional information on concentration, phase, temperature, velocity and other scalar and/or vector fields with high spatial and temporal resolution. As a result, they are becoming increasingly prevalent in scientific and engineering research and development fields. In particular, the heat-transfer community is moving gradually towards experiments that can combine heat-transfer measurements such as of temperature, heat flux, heat-transfer coefficient (HTC) and in a few cases the velocity field within the fluid domains, thus moving towards a comprehensive understanding of their associated interactions leading to a particular heat-exchange performance.

Specifically, the experimental technique reported here, namely PLIF-IR, involves a combination of laser-induced florescence (PLIF) and infrared (IR) thermography, and was developed in order to enable simultaneous measurements of the film free-surface height relative to the solid substrate (film thickness), the film free-surface and solid–liquid substrate interface temperatures, and the instantaneous heat flux exchanged between the liquid film and the heated substrate. By extension, the acquired data allow the estimation of local and instantaneous HTCs, and an investigation of their variation as a function of important flow parameters. Another important aspect of this effort is our specific interest in developing an experimental methodology that allows the explicit study of unsteady and conjugate heat-transfer [14] and [15]. In these problems the temperature*and* heat flux at the solid–liquid interface (wetted substrate side) both time-vary, and it is necessary to measure directly these variations. To address this challenge, the employment of a thin metal foil as a substrate enables IR measurements at its underside (un-wetted substrate side) to provide temperature and heat flux information on the wetted solid–liquid surface.

A considerable body of previous research has been dedicated to the development of a variety of experimental measurement techniques and their adaptation to the study of isothermal film-flows. Relevant experimental methodologies can broadly be classified as conventional or advanced, depending on the practice employed towards retrieving the desired data, primarily focusing on recovering the interface shape (film thickness and wave dynamics) and mass-transfer characteristics (entrainment, reposition, etc.). Examples of the former, (i.e. conventional, non-optical methods) are hot-wire anemometry (HWA) [16] and [17], electrical conductivity [18] and [19] and capacitance probes [20] and [21]. Such measurement approaches have played an important role in improving our understanding these flows and have contributed to many advancements made in this field, yet they are associated with certain challenges; in some cases they are intrusive, can suffer from spatial or temporal resolution limitations, are by and large point-measurements, and often involve complex and cumbersome calibration procedures, thus progressively favouring the implementation of non-intrusive optical methods.

Amongst a plethora of relevant optical-diagnostic techniques, fluorescence-based methods have proven to be very effective and are therefore popular amongst researchers engaging in film-flow investigations. These methods employ a tracer (either occurring naturally in the flow or an added chemical substance) that is excited using a laser source. Fluorescence-based methods can be further classified as laser-induced fluorescence (LIF) imaging or planar laser-induced fluorescence (PLIF) imaging. The primary difference between the two is that LIF utilizes the fluorescence intensity, measured at or near the incident light direction, to quantify the film thickness along a line, over a one-dimensional (1-D) or two-dimensional (2-D) domain. PLIF on the other hand relies on planar illumination across the flow field and imaging from the side, and can be used to directly identify the extent of the liquid domain, and to resolve with a higher level of detail, resolution and accuracy 1-D spatiotemporal film thickness variations along the illuminated plane. Consequently, one or both film boundaries need to be identified in raw PLIF images in order to measure the spatial extent of the film. Another advantage of the latter approach (PLIF) is that the fluid-flow domain is visualized directly, allowing for spatiotemporally resolved velocity or temperature data to be obtained simultaneously with the film thickness. Examples of experimental investigations employing LIF imaging are reported in Refs. [22] and [23], and more recently by Alekseenko et al. [24]. PLIF-based studies are presented in Refs. [14], [25], [26] and [27] in various multiphase/interfacial flows.

Beyond isothermal film-flow investigations, fluorescence-based methods using temperature-sensitive markers have been employed when evaluating the thermal performance of microchannel devices and other microfluidic systems [28] and [29], as well as plate heat-exchangers in the presence of oscillating flows [30], [31] and [9]. Temperature measurements with an absolute accuracy of 0.5 K, or better, have been reported [28], [32] and [33]. Compared to isothermal film-flows, the study of diabatic (heated/cooled) film-flows which entail liquid-temperature measurements is more demanding, owing to the fact that the liquid is often in physical and thermal contact with a hot solid-substrate which is typically metallic and opaque. Illumination of the liquid domain from the solid side is unfeasible in such a configuration, with the (intermittent and disturbed) liquid free-surface providing the only optical access to the liquid domain. In this case, refraction at the wavy interface results in strong beam-steering that must be accounted for by post-processing correction algorithms. Such measurements are also subjected to the restrictions noted earlier, and any available studies are consequently limited to single-point measurements.

Amongst others, the excellent efforts at RWTH Aachen University [34] and [35] to conduct simultaneous film-thickness and temperature measurements by employing diacetyl fluorescence and phosphorescence imaging are worthy of mention. In more detail, time-varying local HTCs were reported at the measurement position (point), suggesting a strong correlation between the film thickness and heat transfer (+80% enhancement ahead of the wave and -40% deterioration below the wave crest). Unfortunately, the particular experimental effort was restricted to a narrow flow parameter range (*Re*=126), and the role of important experimental parameters on the wave regime and HTC performance was not within its scope. Furthermore, the results were recovered on the assumption that the velocity within the film is well-described by the theoretical parabolic Nusselt profile (for 1-D, laminar and fully developed steady-flow dominated by the balance of gravity and viscosity only). Deviations from this profile on account of two-dimensional (2-D) and three-dimensional (3-D) unsteadiness have, however, been confirmed [10] and [27], which suggests that additional experiments to confirm these findings are necessary.

Further improvements to our insight into the complex and unsteady flow phenomena associated with thin films and their link to heat-transfer performance can be attributed to the extensive body of experimental research carried out at the Kutateladze Institute of Thermophysics of the Siberian Branch of the Russian Academy of Sciences [36], [12],[37], [38] and [39] and at the Institute of Thermophysics of the Technical University of Darmstadt [40], [41], [42] and [43]. The experimetnal efforts at Kutateladze aimed at examining the process of rivulet formation due to thermocapillary convection, employing a wide range of optical techniques including IR thermography, particle velocimetry, shadowgraphy and LIF. In Ref. [12], a range of locally (non-uniformly) heated falling film flows between *Re*=300 and *Re*=500. Even when low heat fluxes were imparted onto the vertically falling water films, significant thermocapillary effects were observed. Non-uniform heating was also employed in Ref. [36] to induce the formation of a horizontal bump-shaped free-surface deformation across the film, which was found to transition to a rivulet pattern above a critical heat input. By depositing Aluminium particles onto the liquid surface, a highly non-linear 3-D flow was revealed, while a stagnation point was identified at the top of "horseshoe-like" structures. The separation between adjacent rivulets in experiments carried out with uniform heating was shown to depend weakly on the film inclination angle above 15° [37], while positive temperature gradients in the flow direction resulted in an increase in the wave amplitude. The evolution of 3-D waves into rivulet structures along with the dependence of the rivulet wavelength on the *Re* and applied heat flux were studied in Refs. [38] and [39]. Shorter wavelengths ensued when increasing the heat flux, while longer wavelengths were observed when increasing the*Re*. The work at Darmstadt has been dedicated to the characterization of the thermal performance of a variety of film flows. For example, experimental studies of the heat transfer (including by evaporation) of laminar water falling-films over smooth, as well as patterned plates, are reported in Refs. [40] and [41], using IR thermography and confocal chromatic imaging (CCI). Significant heat transfer enhancements were noted compared to a smooth-plate arrangement when grooved plates were employed. Furthermore, different mechanisms of film rupture were also investigated. Other heat transfer systems that have been extensively investigated (in a similar fashion) by the same research group include micro-channel and micro-tube flows (see, for example, Refs. [42] and [43]). It is noted that this work is mostly dedicated to reporting time-averaged HTCs and characterizing the overall heat transfer performance of the examined systems, while the aforementioned work at Kutateladze focuses mainly on the characterization of hydrodynamic flow-regimes.

The experimental work presented here is specifically associated with the application of the PLIF-IR technique, which employs fluorescence in conjunction with IR thermography. As noted earlier, the free-surface temperature of liquid films has been measured extensively by various researchers with IR imagers, sometimes with filter-equipped lenses that block the wavelengths over which the working fluid is transparent. The key aspect of this measurement tactic is the absorptivity of the fluid, which is required to be sufficiently high at the wavelengths used for imaging. This ensures the observation of a thin layer of fluid near the interface (i.e. of the free-surface), without contamination from the liquid bulk deeper within the film, or from the solid substrate. For example, Refs. [44] and [45] used an IR camera to measure the interfacial temperature of silicone-oil films; in this case the absorption coefficient allowed 99% of the radiation to be absorbed within a thin (10–15 μm) layer at the surface. IR thermography has also been extensively employed in water film-flows (e.g. Refs. [46] and [47]), as well as many others.

The present paper is an extension of the experimental investigation reported in Mathie and Markides [15]. Contrary to that work, which focused exclusively on the recovery of the so-called 'augmentation ratio' in unsteady flows including film flows in the same apparatus with the same PLIF-IR technique, and only in select experimental conditions, the focus of the present work is on the full, detailed characterization of unsteady heat-transfer over a range of thin liquid-film flow-conditions in this apparatus, thus exploring the effects of important variables and parameters on the obtained heat transfer. The paper is structured as follows: in the next sections (Sections 2 and 3) the experimental apparatus is presented briefly, along with the employed optical measurement techniques, namely PLIF imaging and IR thermography. Also included is a description of the relevant data processing methodologies. In Section 4 the experimental results are presented. The results are organized in two sections, one dedicated to noteworthy qualitative observations regarding any interfacial thermal features and the phenomenology and dynamics of the examined film flows, and the second pertinent to the quantitative heat-transfer characterization. Finally, the paper closes with a summary of the main conclusions (Section 5) derived from the present study.

EXPERIMENTAL METHODS

Flow Apparatus

A series of experiments were conducted in order to measure the instantaneous temperature and heat flux across wavy films developing over a main test-section that comprises an inclined, electrically heated Titanium foil of thickness a=50 μm [15]. Heating of the test-section substrate was achieved by holding the foil taught between two 10 mm-thick copper electrodes and applying a current using a DC power supply rated to 10 V and 1000 A (i.e. 10 kW), such that a maximum heat-flux of $\bar{\dot{q}} = 4.2$ W cm^{-2} over the 1200 cm^2 surface area of the foil was possible. Beyond the part of the apparatus that was employed for generating the heated substrate over which the flows were considered, the experimental arrangement also comprised a preparation section where the film flows were initiated, developed and delivered to the substrate and a closed flow-loop via which the liquid circulated to the test section (Fig. 1). This was preferred to an open-loop arrangement, as it allowed for the fluorescent dye concentration to be preserved throughout the duration of the experiments. When in operation, the liquid initially passed through a counterflow heat exchanger supplied by Alfa Laval, in order to modulate its temperature before being controlled, measured and entering the preparation and test sections. The cooling stream used for the heat exchanger was supplied by the water mains, with the rate of heat extraction regulated by using Valve V1 and temperature indications on both sides of the heat exchanger from 1 mm K-type thermocouples. A second valve, Valve V2, was used in conjunction with a flow-recirculation loop (as illustrated in Fig. 1) for bypassing part of the flow back to the tank, thus allowing a lower flow-rates to be achieved in the test section with the same pump; this maintained higher flow-rates through the pump and the heat exchanger, which were operated more efficiently and closer to their design specifications.

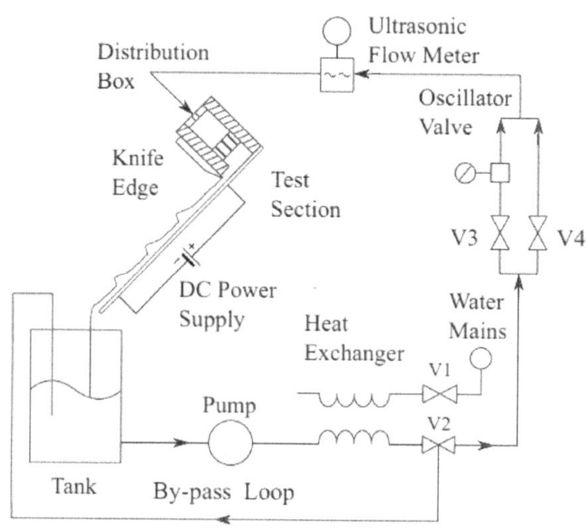

Figure 1. Apparatus schematic showing the distribution box, test section, pump and oscillator valve.

In the present work, pure water and water–ethanol solutions were used as the liquid phase, the former during the development and testing of the experimental setup, and the later during the main heat transfer experiments reported in this paper. Water has a high surface tension, resulting to partial dewetting even when modest heating is applied to the foil. This phenomenon not only hampers the reliability and reproducibility of the experimental results, but also presents a threat to the foil as unwetted regions will heat up disproportionately and deform. In order to prevent this outcome, an ethanol–water solution (20% ethanol and 80% water by volume) was preferred, which compared to pure water displays substantially higher wettability, allows for higher heat inputs to be applied, and reduces the risk of damaging the setup. The density ρ_f, kinematic viscosity ν_f and surface tension σ_f of the employed solution at 20 °C correspond to 978 kg m^{-3}, 1.53 × 10^{-6} m^2 s^{-1} and 4.2 × 10^{-2} N m^{-1} respectively.

In order to selectively trigger the rapid growth of waves and obtain developed wave-regimes within a relatively short test-section, the flow was split downstream of the bypass valve (V2) by diverting a portion of it through a rotating throttle valve (oscillator valve, V3 in Fig. 1). Torque to drive the oscillator valve was provided by a 5 V, 1 A, 1.8° step-angle stepper-motor controlled by an in-house developed controller, thus achieving a maximum rotational (i.e. wave) frequency of approximately 11 Hz. After the oscillator valve, the pulsating supply was recombined with the steady supply that was also controlled by an independent valve (V4). The total flow-rate was then measured by using an ultrasonic flowmeter (UF25B by Cynergy3 Components). In more detail, the recorded pulse-sequence (one TTL pulse generated for each mL of flow) was used to recover instantaneous and cumulative flow-rate time-traces, from which the mean flow-rates and intensity of flow-rate pulsations were calculated. The rotational speed of the pump, and therefore the film flow-rate, was modulated by using a 1200 V A variable autotransformer (Carroll & Meynell). Finally, the flow was distributed uniformly across the heated area of the foil by a flow-preparation section of width w=0.3 m, that featured a series of grids, meshes and a honeycomb, as well as a knife-edge slot along with a set of micrometer stages that were used to adjust and set the initial height of the liquid films as these were introduced over the heated-foil test-section.

A complete view of the apparatus is given in Fig. 2. Measurements were performed at two streamwise positions; at an 'upstream' location a distance x=185 mm from the distribution box knife-edge at the flow inlet, and at a 'downstream' location x=285 mm from the inlet. In the arrangement shown in Fig. 2(a) the IR camera is positioned underneath the test section with the laser sheet-optics illuminating a vertical slice along the direction of the flow, whereas in a second arrangement (Fig. 2(b)), the two PLIF cameras are positioned to observe a horizontal slice across the film, excited by the laser sheet-optics that were positioned on top of the test section (at the top of the image). Experiments with the IR camera in this position were also performed. A fluorescent target used to correct for pulse-to-pulse laser-intensity variations can also be seen, located just above the foil. The Titanium foil and distribution box were installed on an aluminium support-frame with the inclination angle fixed to β=40° to the horizontal. The distribution box, which comprised grids, meshes and a honeycomb, functioned as a settling chamber, breaking down and dissipating secondary flows, large-scale eddies and turbulence, and creating a uniform flow at the test-section inlet. Finally, a type-K thermocouple was installed inside the chamber allowing for the liquid temperature to be monitored. Using the mean flow-rate measurement from the ultrasonic flowmeter, a Reynolds number can be defined at the flow inlet, as:

$$Re = \frac{DU}{\nu_f} = \frac{\Gamma}{\nu_f}. \tag{1}$$

In this expression, U stands for the local bulk-fluid velocity, D for the channel depth at the inlet (20 mm), Γ for the flow rate per unit channel width w , and ν_f for the kinematic viscosity of the liquid.

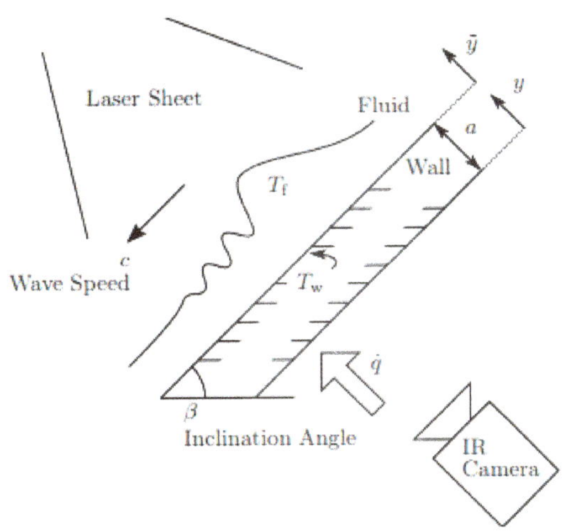

(a) Test-section schematic showing the position of the laser sheet and IR camera relative to the heated foil when observing a thin slice along the direction of the flow (reproduced from Ref. [15]).

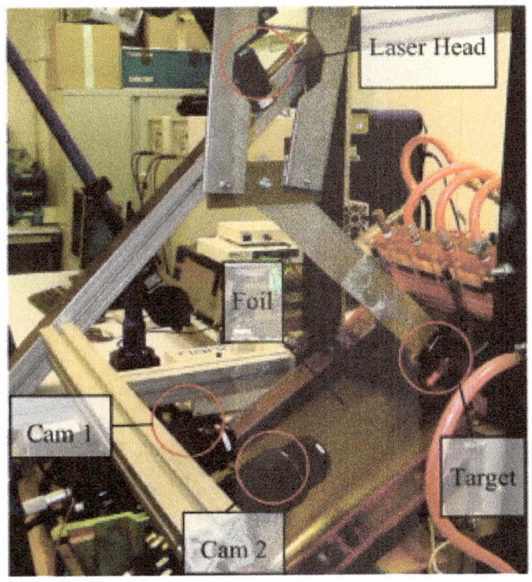

(b) Test-section arrangement showing the position of the laser sheet-optics and PLIF cameras relative to the heated foil when observing a horizontal slice across the direction of the flow.

Figure 2. Apparatus including the heated-foil substrate/test-section, flow-preparation/distribution setup and optical arrangement. Showing the position of the laser sheet-optics and PLIF cameras relative to the heated foil when observing a horizontal slice across the direction of the flow.

Measurement Techniques: PLIF imaging

A 2-D PLIF imaging technique was used to characterize the local and instantaneous film thickness and film free-surface temperature simultaneously. To that end, a temperature-sensitive fluorescent dye (Rhodamine B) was dissolved in the liquid phase (here, pure water or a water–ethanol solution), forming a low concentration (~1 g/L), perfectly mixed aqueous-dye solution. Amongst others, the well-characterized strong dependence of Rhodamine B fluorescence to temperature variations was a primary reason for employing this particular tracer in these present experiments.

The aqueous liquid-phase was excited by a 0.1 mm-thick laser-sheet, generated by a dedicated sheet optics and lens combination manufactured by LaVision that was positioned over the wavy interface and pointing downwards. Furthermore, a linear transition-stage was employed, so that the location of the excitation plane could be adjusted with a high degree of accuracy. The sheet optics were connected via a guiding arm to a dual-cavity, frequency-doubled Nd:YAG laser (Nano-L-50-100PV by Litron Lasers), capable of approximately 50 mJ/pulse at the selected excitation wavelength (532 nm) and a frequency of 100 Hz per head. Each head could be fired separately by the software, allowing a maximum continuous sampling of 200 Hz, and therefore, a temporal resolution down to 5 ms.

The resulting fluorescent light was collected by a pair of synchronized LaVision VC-Imager Pro HS 500 cameras equipped with 1280×1024 pixel CMOS sensors and Sigma 105 mm f/2.8 Macro lenses. Two cameras were used, positioned next to each other, in order to expand the field-of-view (FoV) of the measurement (by employing image stitching), while maintaining a high spatial-resolution (approximately 20–30 μm/pixel). In addition to the employment of extension rings in achieving the desired magnification, a Scheimpflug adaptor was installed on one camera (the one positioned at an angle to the excitation plane) to focus the entire film domain accurately and sharply onto the image. Finally, optical long-pass 540 nm dichroic/interference filters were installed on both camera lenses, in order to block any scattered laser-light while allowing the red-shifted fluorescence emission to be collected by the detectors, thus improving the signal-to-noise ratio (SNR). Both cameras and lasers were synchronized by a LaVision High-Speed Controller (HSC), operated by using the LaVision DaVis software.

Measurement Techniques: IR Thermography

Along with the liquid-film thickness and free-surface temperature measurements extracted by PLIF imaging, the foil (un-wetted) underside temperature was measured simultaneously by an IR camera. The corrected measurement was then used to recover the (wetted) solid–liquid substrate temperature and the heat flux exchanged between the liquid film and the heated substrate, based on the knowledge of the imposed volumetric (Joule) heating-rate. Two different IR cameras were employed, a Cedip Titanium and a FLIR SC3000. With each laser pulse, the laser (PLIF) system provided the triggering signal for the IR camera, as well as the data acquisition (DAQ) system used to record the flowmeter and thermocouple output signals. The recorded thermocouples were those positioned inside the flow preparation/distribution box, at the film inlet and in thermal contact with the underside of the foil, the latter in a series of separate experiments conducted for IR calibration and validation purposes.

In reference to the setup schematic presented in Fig. 2, the total incident radiation-flux seen by the IR camera, I_{cam}, is a linear superposition of the radiation emitted by the foil underside, which is at a temperature T_u, and any direct or reflected background radiationI_{bg}:

$$I_{\mathrm{cam}} = \sigma\epsilon T_{\mathrm{u}}^4 + I_{\mathrm{bg}}, \qquad (2)$$

where σ=5.67×10^{-8} W m^{-1} K^{-4} is the Stefan–Boltzmann constant, and ϵ is the emissivity of the Titanium foil. Bare, polished Titanium has a very low emissivity, of approximately 0.1, whereas the thin oxide-layer that quickly forms on its surface causes its emissivity to increase to about 0.3; either way, the IR radiation collected by a detector positioned at right angles to the foil is influenced by reflections. Consequently, in the present work, the IR camera was in fact positioned at an offset angle to the foil, rather than directly opposite the foil surface, thus preventing it from capturing a reflected image of itself. Furthermore, in order to suppress any background IR radiation and prevent imaging of the reflections of other thermal sources, the region that would have otherwise been in the line-of-sight of the camera following specular reflections from the foil was obstructed by a screen constructed from thick black cardboard (with near-unity emissivity). The screen was further backed by aluminium foil which contributed to keeping its temperature even by uniformly dispersing heat.

Having suppressed any background signals and reflections, the IR radiation originating from the screen is solely the result of its temperature T_{bg}. Thus, the total IR radiation measured by the camera I_{cam}, is the sum of the background radiation associated with the screen temperature T_{bg}, and the direct emission from the foil I_u:

$$I_{cam}=\epsilon I_u+(1-\epsilon)I_{bg}, \qquad (3)$$

such that the apparent temperature measured by the IR camera T_{app} is given by:

$$T_{\mathrm{app}}^4 = \epsilon T_{\mathrm{u}}^4 + (1-\epsilon)T_{\mathrm{bg}}^4. \qquad (4)$$

Eq. (4) can only be used to evaluate the temperature at the foil underside provided the employed IR thermograph integrates the radiation emitted over the entire IR spectrum. Typically, however, IR cameras detect radiation over a limited spectral band, in this case 3.6–5.1 μm and 8.0–9.0 μm for the Cedip and FLIR cameras respectively, and thus, the absolute temperature of the foil cannot be directly determined from the Stefan–Boltzmann law (Eq. (3)). I_{cam} can instead be linked to temperature measurements via calibration curves provided by the camera manufacturer that consider, aside from the wavelength dependence of the detector (relative spectral response), data conversion settings such as the integration time and gain, and the detector optics (lens and filters). Using the relevant calibration curve, and provided that the background temperature and emissivity of the foil are known (the first using thermocouples, the latter by comparison to a blackbody of the same temperature), $I_{bg}$and I_u can be calculated sequentially. In these experiments, Eq. (3) was used to evaluate the temperature on the underside of the foil from raw IR images following a series of calibration and validation efforts carried out using thermocouples, as well as thermal targets consisting of squares of electrical tape with known, high emissivity. In more detail, simultaneous measurements of the temperature of the tape with the IR camera and the thermocouples were used to find the correct foil temperature, and thus the correct local emissivity of the foil. As a final validation exercise, independent measurements of the film and foil temperatures were conducted by both the IR camera and a series of fine (75 μm diameter) unsheathed, butt-welded thermocouples. The root mean square (RMS) deviation between IR camera and thermocouple measurements amounted to 0.1 °C, which we consider as a representative estimate the error associated with the IR thermography measurement.

EXPERIMENTAL DEVELOPMENT

Heating Conditions

The deviation in the power supplied to the foil from changes to the foil resistance and stability of the DC power supply over a run was measured and found to be <0.1% of the nominal supplied power. In order to provide uniformly distributed heating over the foil surface-area, the use of copper electrodes that were much thicker than the foil itself, and therefore with a significantly lower electrical resistance, was necessary. The power losses in the cables, based on direct resistance measurements in which the nominal resistance of the connecting cables and the total resistance of the foil were found to be 540 μW and 18.3 mW respectively, were estimated to amount to 3% of the power delivered to the foil. The voltage distribution over the foil was examined by conducting point measurements with a multimeter. The observed maximum deviation in this test amounted to no more than 0.1% over the 300×400 mm foil area, which is expected to be the same as the non-uniformity in the imposed volumetric heating rate in the foil. Moreover, a uniform temperature distribution was observed over the foil during dry tests in which the foil was heated without establishing a film flow, and its temperature was measured with the IR camera. The standard deviation of the temperature during these dry tests did not exceed 0.1 °C.

Results from tests concerning the average temperature over the foil obtained at different heat fluxes are shown in Fig. 3. A noteworthy observation stemming from these particular tests was that heat convection to the air was small, peaking at ~1% of the of heat flux delivered to the liquid during the falling film experiments. Specifically, without a film flow, no more than $\dot{q} = 0.042$ W cm^{-2} or 50 W of thermal power were necessary to sustain a steady foil temperature of 60 °C, or about 40 °C above ambient; by comparison, nearly 5 kW were required to obtain the same foil temperature once a typical film flow was established. Given that the heat flux into the film from the foil's (wetted) top surface is three orders of magnitude higher than any heat losses to the environment from the (un-wetted) underside, the underside of the foil can therefore be considered an adiabatic boundary during the flow experiments.

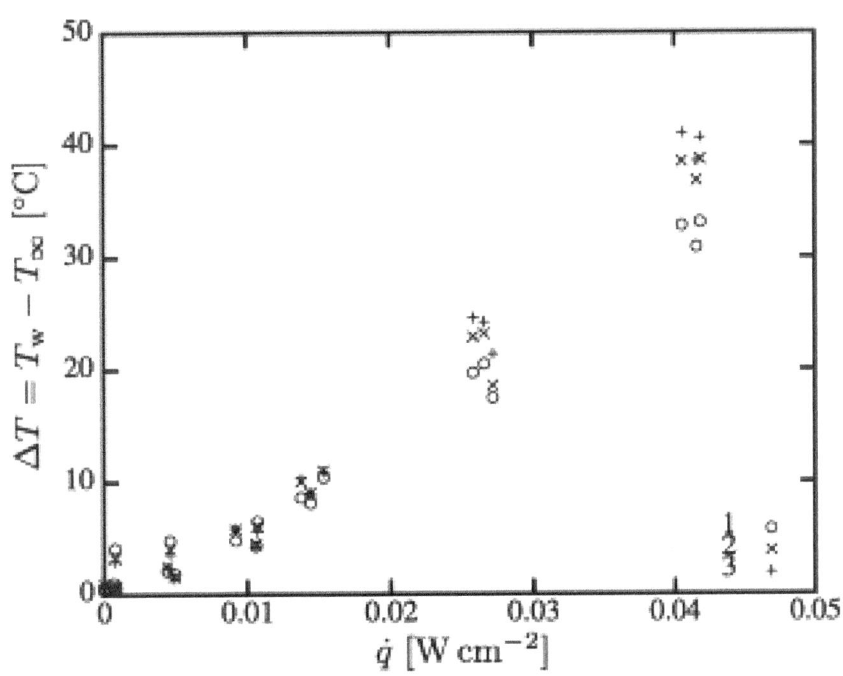

Figure 3. Foil temperature difference ΔT relative to the ambient temperature T_∞ in the absence of a film flow, plotted against the applied heat flux $\dot{q}$.

During the actual experimental runs, the liquid temperature was kept below 40 °C in order to minimize evaporation and focus the investigation on forced convective heat-transfer in liquid film flows. In a test near the upper range of achievable heat fluxes (corresponding to the highest heat losses), $\dot{q} = 3.2$ W cm^{-2} of heating power was applied to an 8 L/min water film flow, across the w=0.3 m wide test-section. Assuming complete heat transfer from the foil to the water (no heat losses to the surroundings), a streamwise temperature rise of 17.2 K m^{-1} can be estimated; however, a gradient of 16.7 K m^{-1} was recovered with the IR camera from the thermal profile of the fully developed portion of the flow. This corresponds to heat losses of ~0.1 W cm^{-2}, which amount to less than 3% of the total heat input. If we conservatively assume that all of this heat was used to evaporate liquid from the film surface, which is unlikely as other heat transfer processes take place including convective heat-transfer to the air at the film's free surface [37], the maximum evaporative flow rate would be less than 3 mL min^{-1}. This is 3–4 orders of magnitude smaller than the bulk film flow rate. Finally, the HTC between the foil and the liquid film for this flow was measured at about 2400 W K^{-1} m^{-2}, whereas the HTC between the liquid film and the surrounding air was an order of magnitude lower at ~200 W K^{-1} m^{-2} (from direct measurements). These results are in general agreement with the Nusselt theory for film flows, as well as typical HTC values obtained by other investigations in the literature (see, for example, Ref. [38]).

PLIF DATA PROCESSING

In order to extract film thickness and free-surface temperature data from raw PLIF images, a series of post-processing steps were implemented. Initially, raw images were corrected for perspective distortion by employment of a calibration graticule and a pinhole projection model available in DaVis. In more detail, a precision-machined block of acrylic with markings arranged in a regular rectangular pattern (calibration target) was carefully located at the excitation plane using micrometer stages. The accuracy of the correction procedure was approximately 50 μm, which equates to less than twice the nominal resolution of the imaging setup. A perspective distortion corrected PLIF image is presented in Fig. 4.

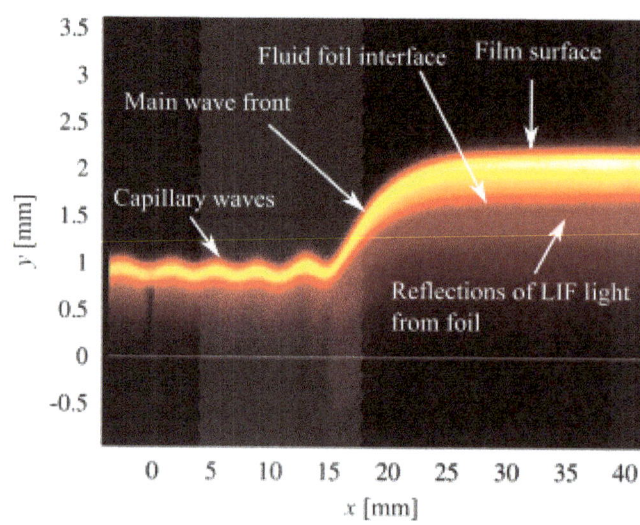

(a) Typical perspective distortion corrected PLIF image of a solitary wave. Both excitation and imaging were carried out from the gas-liquid interface side (reproduced from Ref. [15]).

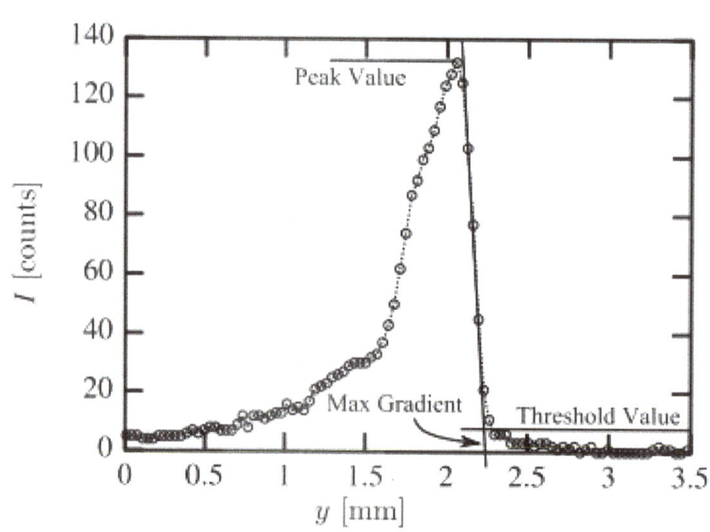

(b) PLIF intensity profile across the liquid domain showing the location of the peak intensity, the maximum gradient intercept with the $I = 0$ line, and the noise threshold level.

Figure 4. Perspective distortion corrected PLIF image and intensity profile across the gas–liquid interface and into the liquid domain.

Having corrected the raw PLIF images geometrically, a number of methods were devised to identify systematically the film free-surface. Of these, a gradient method was found to produce the most accurate film thickness estimates as well as the smoothest thickness profiles. Specifically, each image column (i.e. series of intensity values along the vertical/*y*-direction) was considered independently, so that the PLIF signal intensity was treated as a function of the *y*-position alone. Both the position and magnitude of the maximum gradient in each intensity profile (each x -pixel or image column) were then extracted, and the location of the gas–liquid interface was taken as the intercept of a tangential first order polynomial fitted through that location and a completely dark background with zero intensity I=0 (see Fig. 4(b)):

$$y_{\max,g} = \arg_y \max \frac{\partial I(x,y)}{\partial y}, \text{ and,} \tag{5}$$

$$y_{\text{int}} = y_{\max,g} + I(x, y_{\max,g}) \times \left(\max \frac{\partial I(x,y)}{\partial y} \right)^{-1}, \tag{6}$$

where $y_{max,g}$ is the location of the maximum gradient and y_{int} is the resulting interface location estimate. A similar gradient-based interface identification technique was also used in Ref. [27] with excellent results.

The fluorescence intensity near the free-surface of the film was used to measure the interface temperature in a spatially resolved manner, following a series of calibration experiments. The technique was calibrated using a number of independent PLIF acquisitions, whereby the fluid temperature was adjusted between each run and PLIF data were collected with no power input to the foil, and thus a uniform temperature throughout the liquid film. The measured PLIF signal intensities over each run were then averaged (over a number of images), normalized and plotted against the mean temperature measurement from thin (0.2 mm) thermocouples. This generated the calibration curve shown Fig. 5, which was later used to convert direct measurements of the fluorescence intensity near the free-surface to evaluations of the film's surface temperature. The relationship between the fluorescence intensity and temperature appears to be linear with a gradient of -0.015 normalized fluorescence units per °C, or -1.5% °C^{-1}.

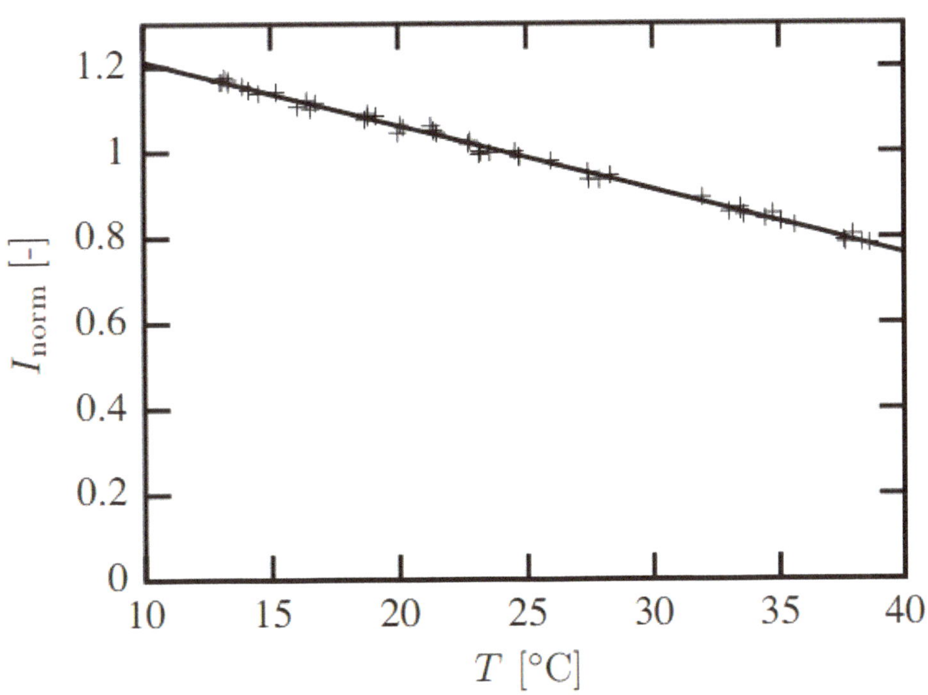

Figure 5. Calibration curve of the normalized fluorescence emission intensity I_{norm} against the liquid temperature$T$, measured using thermocouples of ±0.1 °C accuracy.

Along with the mean PLIF signal intensities used to derive the calibration curve in Fig. 5, a plot of the corresponding standard deviations of the same PLIF intensity signals (one point for each calibration run associated with a single liquid-film temperature, and hence thermocouple measurement) are presented in Fig. 6(a) as a function of temperature, while the resulting probability density function (PDF) of the difference between the PLIF and thermocouple temperature measurements is shown in Fig. 6(b). The calculated standard deviation between the thermocouple measurements and the PLIF measurements was ±0.5 °C, a value used as an estimate of the error associated with the PLIF temperature measurement. This level of accuracy is rendered satisfactory, especially considering that the manufacturer's quoted accuracy for the thermocouples employed in the calibration was ±0.1 °C.

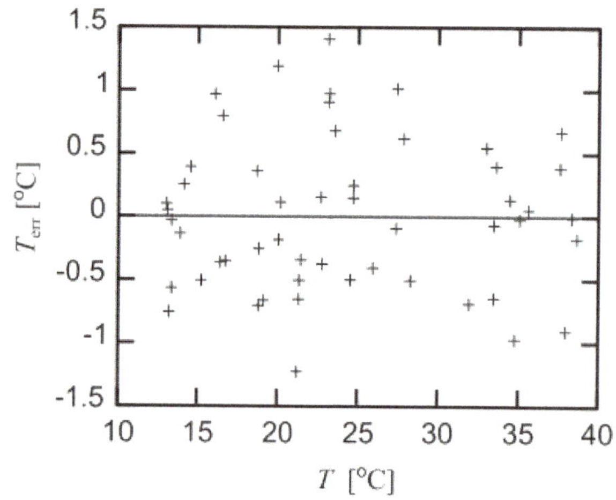

(a) Deviations corresponding to the calibration data points in Fig. 5.

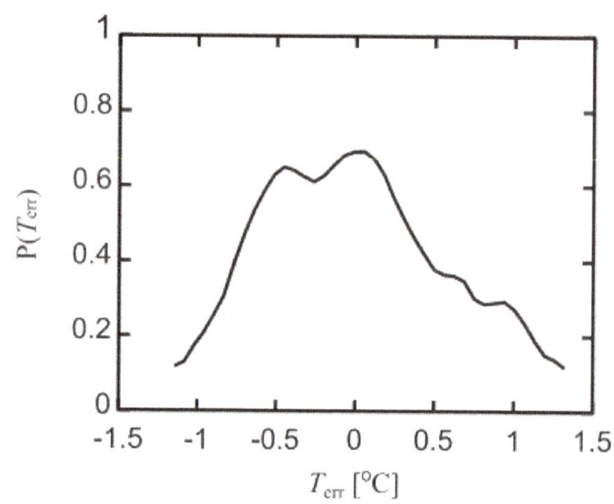

(b) PDF of the temperature measurement error, generated from the same data shown in Figs. 5 and 6(a).

Figure 6. Temperature measurement errors (discrepancy between the PLIF technique and thermocouple measurements) associated with the calibration results presented in Fig. 5.

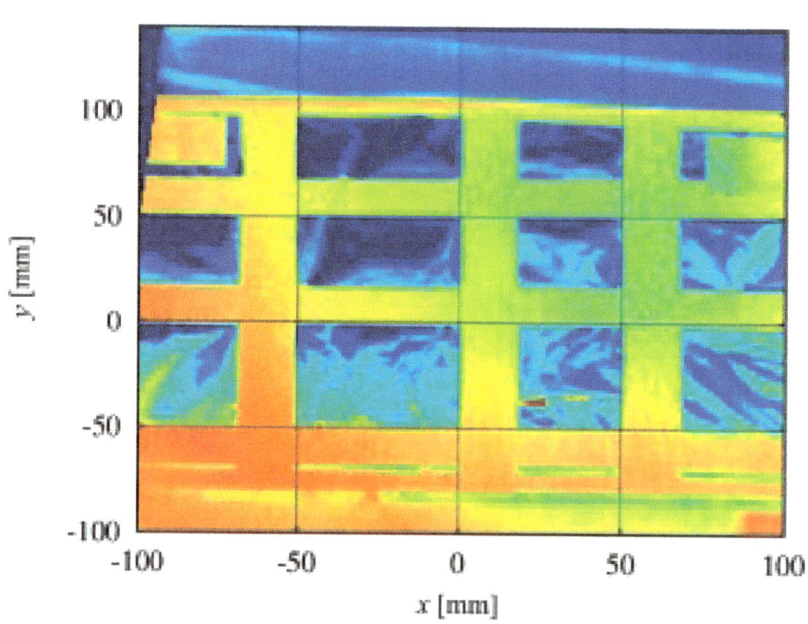

Figure 7. IR image of the calibration plate used for the perspective distortion corrections of ensuing IR images. Image shown after correction algorithms have been applied.

Additional errors associated with the PLIF-based temperature measurement technique ensue due to light distortion when the interface is strongly curved. In such cases, regions with significant film curvature typically appear brighter due to local refraction and lensing. Despite the fact that the emission intensity at shallow depths in the immediate vicinity of the interface is less sensitive to this effect, local fluorescence intensities are also relatively low in these regions, reducing the measurement signal-to-noise ratio (SNR). Therefore, the depth from the free surface to the position inside the liquid domain over which a spatially averaged fluorescence intensity was extracted when evaluating an "interfacial" fluorescence intensity value was identified manually (this varied over the experimental runs) as a compromise between measurement noise and measurement accuracy. The suitability of this selection can be gauged by comparing final, processed PLIF temperature data with direct simultaneous measurements of the free-surface temperature made by the IR camera. This was done in a series of independent quality-check runs in which the IR camera was placed over the films looking down, i.e. from the same side as the PLIF measurement.

An example simultaneous measurement is shown in Figs. 8 and 9. The mean absolute deviation between the two measurements is found to be 1.3 K, and therefore it is possible to deduce that the interfacial curvature introduces an absolute error of approximately 0.8 K to the PLIF temperature measurement. This is significantly lower than the temperature differences across the films that we are attempting to measure, which on average are in the range 10–20 K but can be locally and instantaneously 30 K, or higher, depending on the flow condition and imposed heat flux (e.g. see Fig. 9, Fig. 10 and Fig. 11).

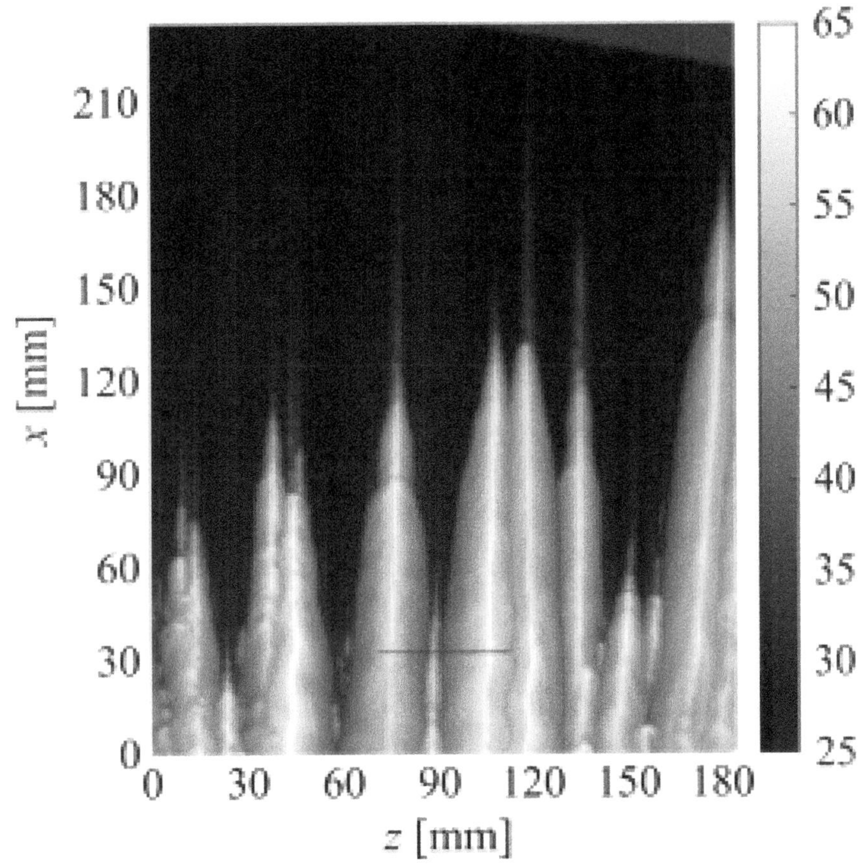

Figure 8. Corrected IR image of the liquid film free-surface temperature for *Re*=179, an imposed heat-flux of $\bar{q}$ = 3.5 W cm^{-2}, no inlet forcing, and a foil inclination angle of β=40° to the horizontal (fixed for all experiments). The vertical axis corresponds to the streamwise direction while the horizontal to the spanwise direction across the flow. From dark to bright, the colour scale represents cooler to hotter regions. The black line near the bottom of the image indicates the 'downstream' PLIF measurement position, location at x=285 mm downstream of the distribution box knife-edge at the flow inlet. The length of measurement line was approximately 40 mm at this position (x=285 mm).

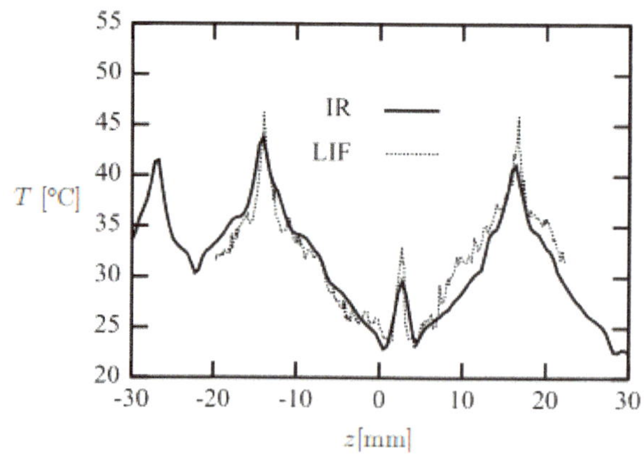

(a) Instantaneous spanwise spatial profile of film free-surface temperature obtained simultaneously by temperature-dependent PLIF and IR imaging, corresponding to the horizontal black line across the film width indicated near the bottom of Fig. 8, $x = 285$ mm downstream of the flow inlet. The z-axis indicates the spanwise direction across the flow (see Fig. 8). Same conditions as in Fig. 8.

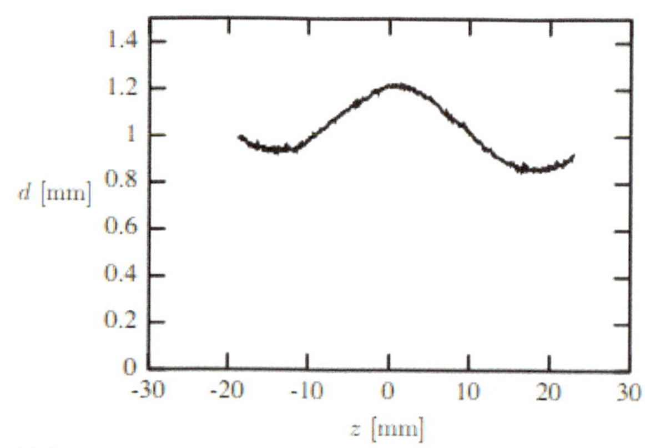

(b) Instantaneous spanwise spatial profile of film thickness obtained by PLIF, corresponding to the same horizontal line indicated near the bottom of Fig. 8 and to the free-surface temperature profile examined in Fig. 9(a), $x = 285$ mm downstream of the flow inlet. Same conditions as in Figs. 8 and 9(a)

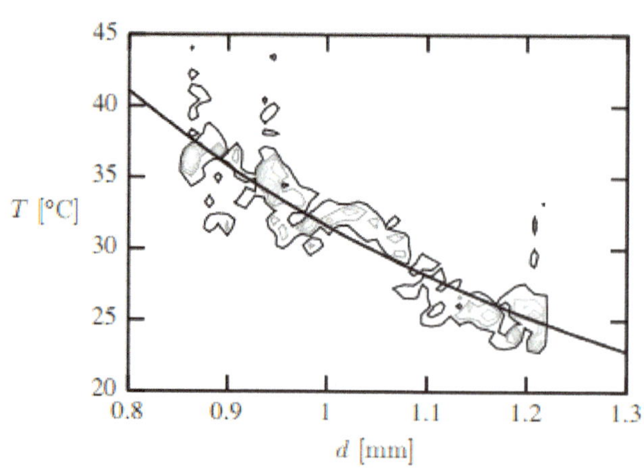

(c) Joint-PFD of instantaneous film thickness data (from Fig. 9(b)) plotted against the corresponding instantaneous film free-surface temperature data (from Fig. 9(a)) compiled over the same spatial region (line indicated in Fig. 8), $x = 285$ mm downstream of the flow inlet. A power law fit to the data was also generated using a least-squares approach and has been included to the plot.

Figure 9. Instantaneous spanwise spatial profiles of film surface temperature and thickness corresponding to the horizontal line across the film width indicated near the bottom of Fig. 8, x=285 mm from the flow inlet. The flow and heating conditions correspond to *Re*=179 and $\dot{q} = 3.5$ W cm^{-2}, respectively.

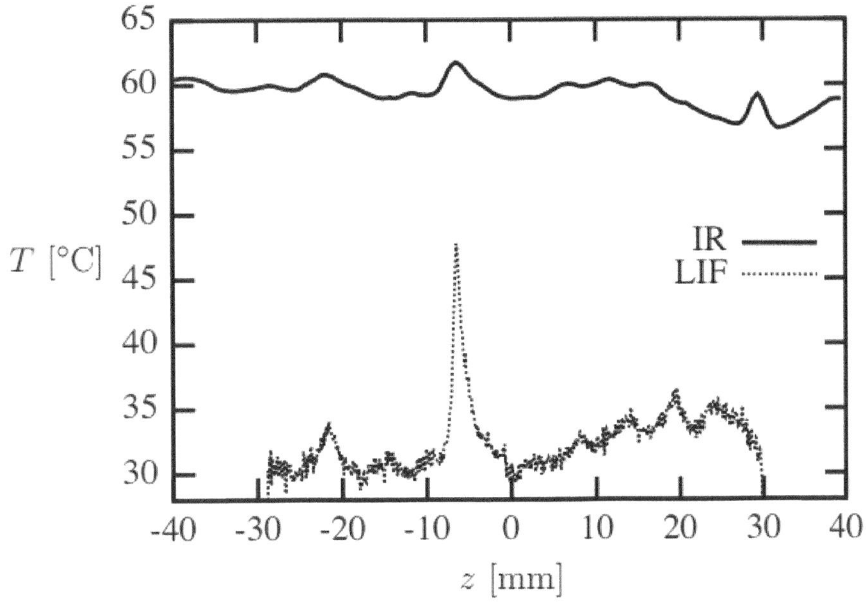

Figure 10. Instantaneous spanwise spatial profile of film free-surface temperature (from PLIF) obtained simultaneously with the corresponding instantaneous profile of heated-foil temperature (IR) over the same spatial region at the 'downstream' measurement position, x=285 mm from the flow inlet (see Fig. 8). The flow and heating conditions correspond to *Re*=179 and $\bar{q} = 3.5$ W cm^{-2}, respectively.

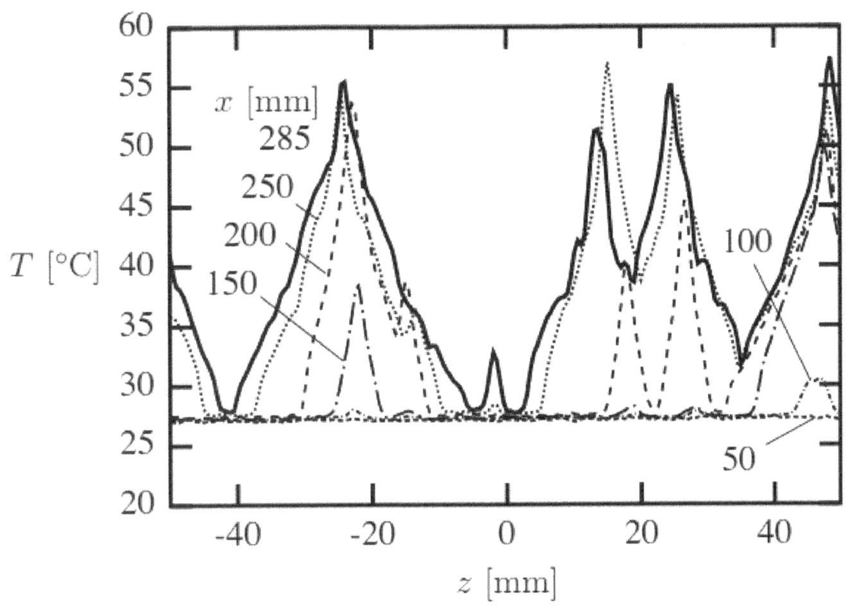

Figure11. Instantaneous spanwise spatial profiles of temperature across the film free-surface, extracted from a single IR image at different distances *x* from the flow inlet as indicated in the figure. The flow and imposed heating-conditions correspond to *Re*=179 and $\bar{q} = 3.5$ W cm^{-2}, respectively.

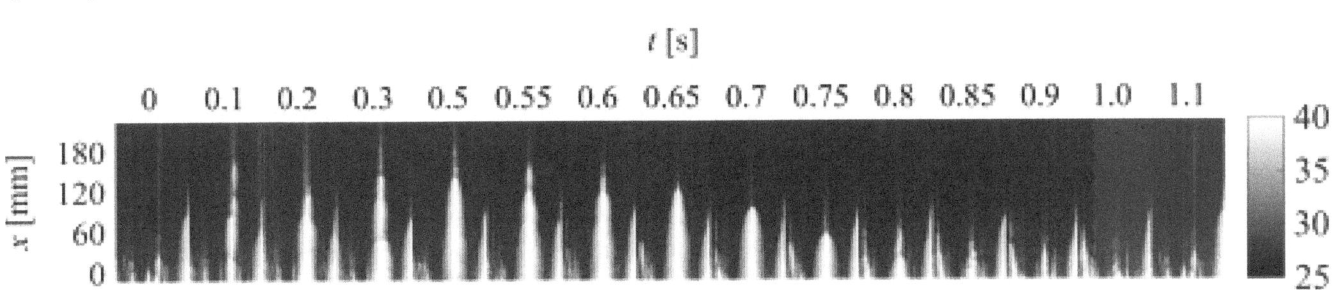

Figure 12. Temporal evolution of an oscillating thermal rivulet observed for *Re*=251, an imposed heat flux of $\bar{q} = 2.4$ W cm^{-2}, and no inlet forcing.

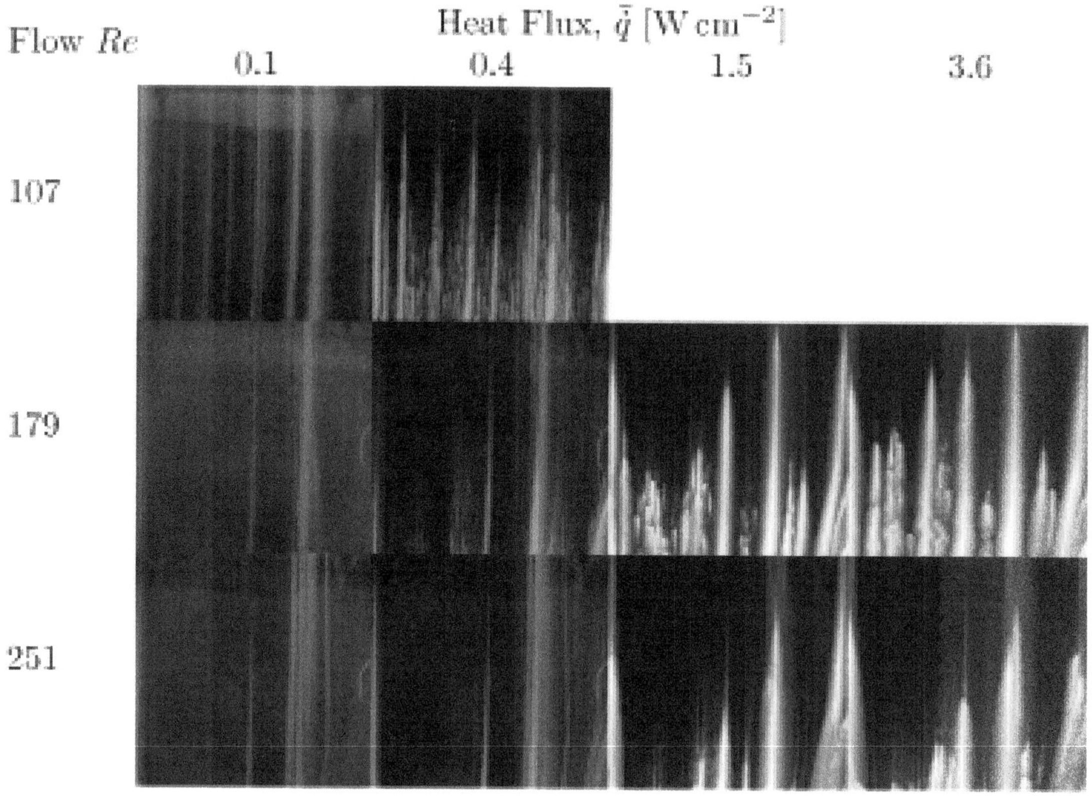

Figure 13. Instantaneous temperature distributions over the free surface of the liquid film for different flow *Re* (107, 179 and 251) and mean heat flux settings (0.1, 0.4, 1.5 and 3.6 W cm^{-2}), recovered using the IR camera. The fluid here is a mixture of 20% ethanol and 80% water, and the inclination angle of the foil is 40°. No inlet forcing was used in these experiments.

IR Thermography Data Processing

As with the PLIF measurements, perspective distortion corrections were implemented to raw IR images by employment of an affine transformation method (see, for example, Ref.[48]). Such a method is fully constrained when a minimum of four coordinates on the image plane are known. Towards that end, a graticule consisting of a thick rigid sheet covered by a reflective foil layer and marked by electrical tape was constructed and imaged, producing a strongly contrasting pattern on resulting IR images (see Fig. 7). The mapping error was reduced further by using nine rather than four control points and solving the over-constrained set of equations by means of a least squares approach. The resulting spatial alignment accuracy between simultaneous PLIF and IR images was better than 2 mm.

Following the employment of Eq. (3) in order to recover the instantaneous temperature distribution over the foil underside and the implementation of the aforementioned geometric corrections, the next processing step was to recover the temperature and heat flux distribution on the wetted foil surface (solid–liquid interface). The foil thickness was selected purposefully to be over an order of magnitude thinner than any thermal feature observed on its surface with the IR camera, and consequently, it can be assumed that heat transfer is dominated by 1-D thermal conduction through its thickness (for a detailed account of the spatial resolution of IR thermography, refer to Ref. [49]). By further assuming a constant thermal diffusivity, a model for the 1-D thermal conduction with constant heat generation by resistive heating can be used to describe the thermal response of the foil to temperature fluctuations imposed on one of its sides, in this case the top side in contact with the liquid. A detailed description of the model can be found in Refs. [14] and [15]. This response will depend upon the imposed boundary condition (B/C) on the foil underside, which as discussed in Section 3.1, can effectively be assumed to adiabatic. The thermal response that relates the solid–liquid interface temperature T_w (inner wall "w") to the measured temperature fluctuations on the underside of the foil T_u for a given frequency is given by:

$$T_{w,n} = T_{u,n}\cosh\{(1+i)a_n^*\}, \tag{7}$$

while the corresponding heat flux response is given by:

$$\dot{q}_{w,n} = \frac{(1+i)}{\sqrt{2}}\epsilon_s\sqrt{\omega_n}\sinh\{(1+i)a_n^*\}T_{u,n}, \tag{8}$$

where $a_n^* = a/\delta_{s,n}$ is a normalized foil thickness based on the thermal diffusion length in the solid $\delta_{s,n} = \sqrt{2\alpha_s/\omega_n}$ associated with the nth harmonic of the temperature fluctuations, α_s is the thermal diffusivity, $\epsilon_s = \sqrt{\kappa_s\rho_s c_s}$ the thermal effusivity, κ_s the thermal conductivity, ρ_s the density and c_s the specific heat capacity of the solid material. Typical examples of fully corrected IR measurements of the foil underside temperature are presented in Figs. 14 and 16.

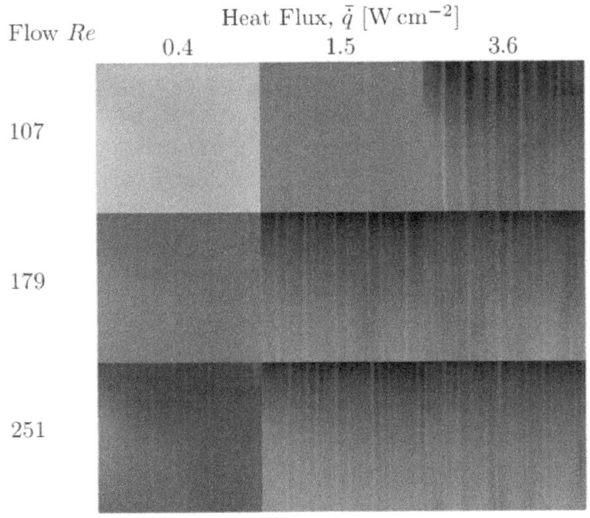

Figure 14. Instantaneous temperature distributions over the foil underside for the same fluid, inclination angle, (unforced) flow Re and heat flux settings examined in Fig. 13, recovered using the IR camera.

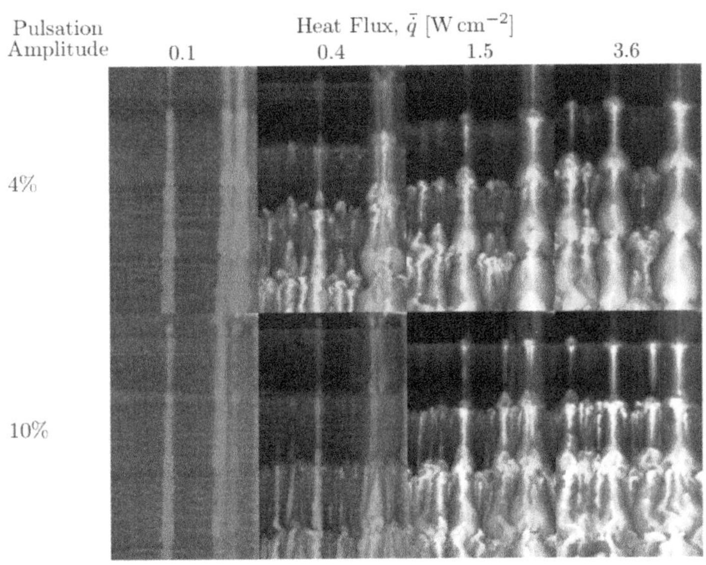

Figure 15. Instantaneous temperature distributions over the free surface of the liquid film for the same fluid, inclination angle and mean heat flux settings examined in Figs. 13 and 14. The flow here has a meanRe=179, a pulsation frequency of 1.8 Hz and two different pulsation amplitudes (4% and 10%).

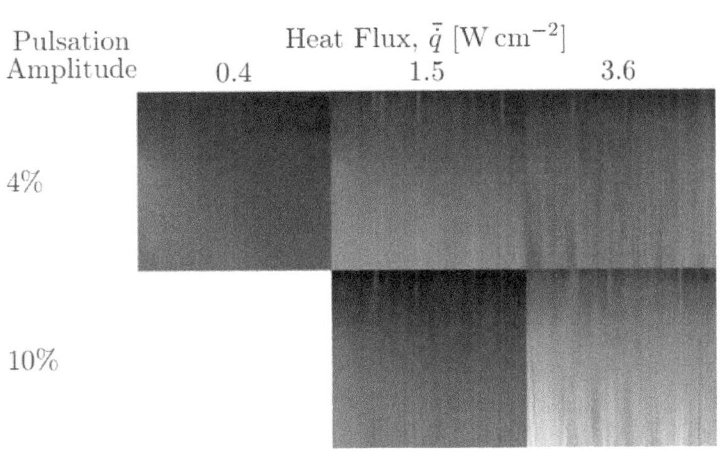

Figure 16. Instantaneous temperature distributions over the foil underside for the same fluid, inclination angle, mean flow Re, pulsation frequency and amplitudes, and heat flux settings examined in Fig. 15.

RESULTS AND DISCUSSION

Phenomenological Observations, Interfacial Features and Dynamics

Selected instantaneous results pertinent to the formation and interaction of thermal rivulet structures on the film free-surface are presented in Fig. 8, Fig. 9 and Fig. 10. Recall that for the generation of these, and all subsequent results, the liquid employed in the experimental campaign was a mixture of 20% ethanol and 80% water by volume; compared to pure water, this displays higher wettability and allows higher heating fluxes to be applied prior to film rupture. In addition, the foil inclination angle was fixed to $\beta=40°$.

Close observation of Fig. 8, generated by positioning the IR imager on the same side as the PLIF cameras over the films pointing downwards, reveals regularly spaced, finger-like structures forming close to the flow inlet and developing in the flow direction (x -direction, +ve from top to bottom). Near the inlet, these features are narrower and their temperature is comparable to the surrounding liquid, while further downstream they become progressively hotter and wider, in some cases abruptly (e.g. the large finger just to the left of the image midline). Another noteworthy observation concerns the relatively discontinuous temperature profile of these rivulets compared to the adjacent fluid which is much cooler.

The instantaneous data in Fig. 9(c) correspond to the 'downstream' PLIF measurement position identified by the horizontal line near the bottom of Fig. 8. In fact, the IR surface temperature profile in Fig. 9(a) has been extracted at this location directly from Fig. 8. This measurement position is located at a streamwise distance x=285 mm downstream of the distribution box knife-edge at the flow inlet. Measurements at a second 'upstream' PLIF position, x=185 mm from the inlet, were also performed (e.g. see Figs. 17 and 18). The length (spanwise width) of the measurement line was 40–60 mm depending on the experimental run. The PLIF data displayed here cover a 40-mm spanwise length across the flow. The flow and imposed heating conditions correspond to Re=179 and $\bar{q}=3.5$ W cm^{-2}, respectively. Fig. 9(a) presents quantitative free-surface temperature data recovered simultaneously by IR thermography and temperature-dependent PLIF, whileFig. 9(b) gives an indication of the film height obtained from the same PLIF data. In Fig. 9(c), the PLIF results (thermography and film height) are plotted against each other, demonstrating the observed relationship between the two quantities.

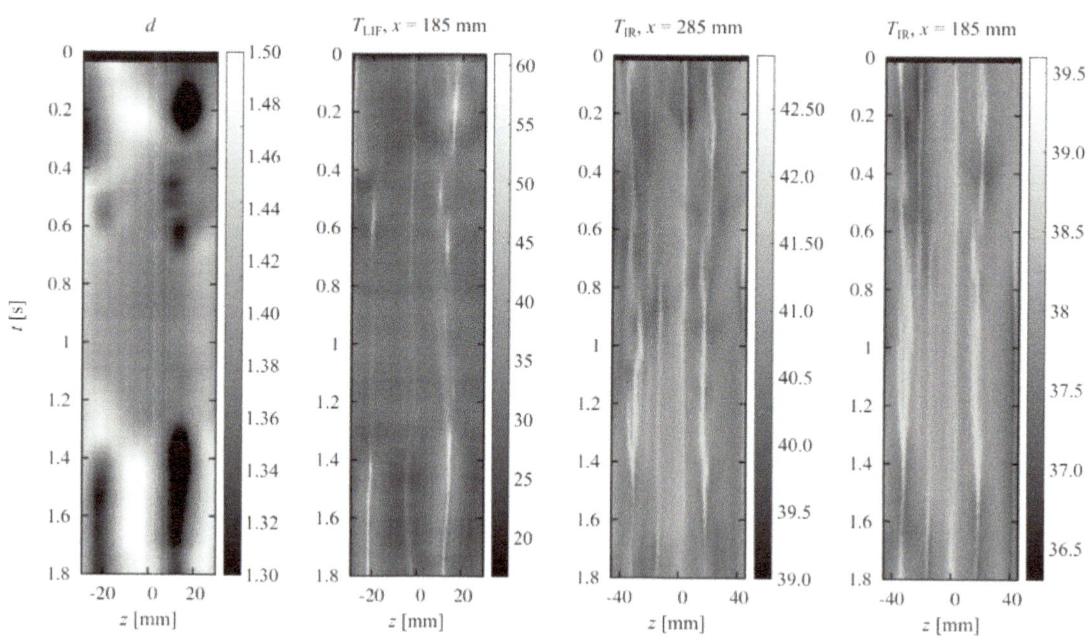

(a) Comparison between the film depth d measured at 185 mm downstream of the flow inlet by PLIF, free-surface temperature T_{LIF} measured at the same location by PLIF, and foil underside temperature T_{IR} measured by IR thermography at 285 mm and 185 mm downstream of the flow inlet, for a flow Re = 179 with no inlet forcing.

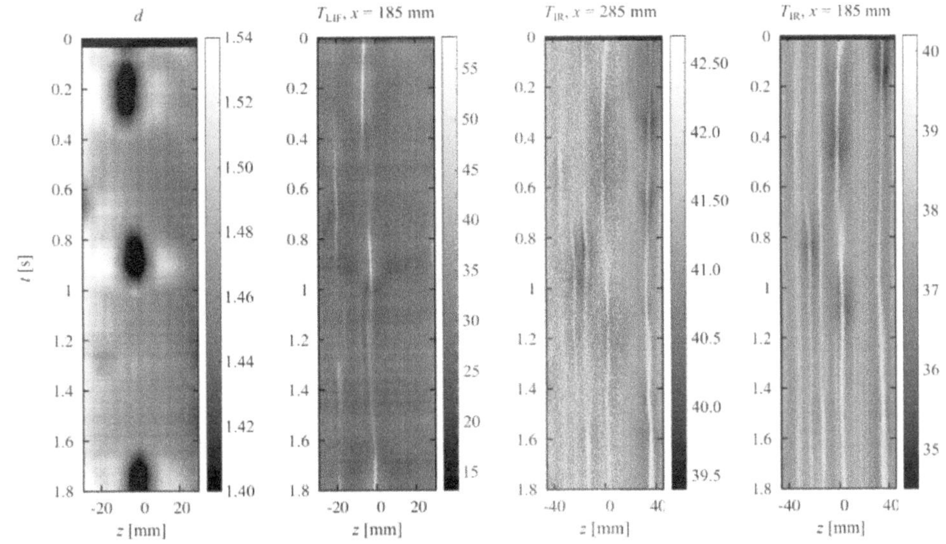

(b) Comparison between the film depth d measured at 185 mm downstream of the flow inlet by PLIF, free-surface temperature T_{LIF} measured at the same location by PLIF, and foil underside temperature T_{IR} measured by IR thermography at 285 mm and 185 mm downstream of the flow inlet, for a flow $Re = 251$ with no inlet forcing.

Figure 17. Space and time-resolved (local and instantaneous) film depth d , film free-surface temperature T_{LIF} and foil underside temperature measurements T_{IR} at two different Re and for an imposed heat-flux of $\dot{q} = 3.6$ W cm^{-2}. The images show the depth or temperature, with bright areas representing large film thicknesses or high temperatures.

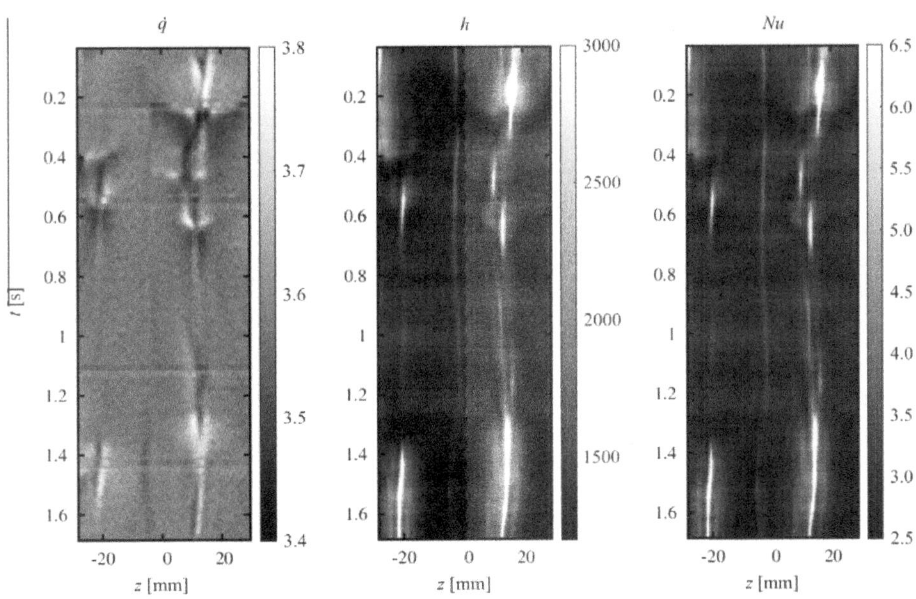

Figure 18. Local and instantaneous (space and time-resolved) heat flux $\dot{q}$, HTC h, and Nusselt number Nu for the same experimental conditions as those pertaining to Fig. 17 but a single flow $Re = 179$, $x = 185$ mm downstream of the flow inlet. The (rms) heat flux variations here amount to approximately 4% of the mean, the mean Nusselt number was about 2.8, and the HTC was in the range h=1000–3000 W m^{-2} K^{-1}.

The following remarks can be made from Fig. 9:

1. The agreement between the two optical techniques with regard to the surface temperature measurement (1.3 K, as stated earlier) is satisfactory, with the main thermal features identified and adequately resolved by both. The PLIF measurement does have a lower SNR while also offering a comparatively limited field of view.
2. The liquid film is thicker where the surface temperature is lower and vice versa (seeFig. 9(c)), with the two troughs in the film thickness measurement corresponding to the two larger peaks in the surface temperature measurement. We also note, however, that the shorter temperature peak near the centre of the measurement domain does not correspond to a film thickness trough. This observation is also apparent in Fig. 9(c), by the spread of free-surface temperature values associated with the highest film-thickness values included in the plot (leftmost region). The identification of this clearly discernible thermal feature, which was captured by both imaging techniques, was encountered consistently in these experiments.

3. From direct observations of many such image sequences, it appears that the advent of flat hot rivulet structures, such as the one discussed above, manifests itself in the development of Marangoni flows at the film surface, driven by temperature-induced surface tension gradients. In other words, local surface tension variations are believed to trigger film thickness instabilities, and hence the ensuing (Marangoni-driven) mass transfer, characterized by accumulation of fluid in the cooler regions and thinning at the hotter ones. The rivulets, whose origin can be found further upstream (i.e. the larger structures seen in Fig. 8), might have formed and evolved in this manner.
4. The sharp peak on the film free-surface temperature profile in Fig. 8 coincides with a less prominent peak on the foil temperature profile. The temperature difference between the foil and film free-surface at this location is drastically reduced, suggesting a significantly higher heat transfer coefficient compared to the surrounding liquid. This observation will be revisited extensively in the upcoming sections, and linked to the local and instantaneous film thickness.

Figure 10 proceeds to show the temperature difference across the liquid film at the same 'downstream' measurement position as above, obtained by repositioning the IR camera to observe the temperature of the underside of the foil, instead of the surface of the liquid film; the latter now measured by the LIF technique. A sharp peak (rising ~15 K from the background) can be observed in the fluid surface-temperature at $-10<z<0$, which coincides locally with a slight increase (~2–3 K) in the temperature of the foil. The temperature difference between the foil and fluid surface at this point is significantly reduced, by a factor of two, which implies that there must be a considerably higher local HTC at that point compared to the surrounding fluid. We will return to consider this in detail in Section 4.2.

The spatial evolution of these free-surface thermal features is examined in Fig. 11, where spanwise temperature profiles obtained with the IR camera are displayed at different axial locations along the streamwise direction of the flow. In more detail, instantaneous temperature profiles across the surface of the liquid film (z-direction) were extracted from the same image (and thus at the same time instant), at different streamwise distances x from the inlet. The specific x locations are indicated in the figure. The observed temperature variations across the film free-surface increase with increasing distance from the inlet, reaching a maximum near the bottom-most measurement location (x=285 mm) of nearly 20 K cm^{-1}. Furthermore, for all examined axial locations, the temperature rise between the cool fluid and the rivulet peaks along the spanwise direction of the flow appears nearly linear, with temperature differences across the liquid films reaching values up to ~30 K.

In light of the presented results and aforementioned observations, we propose the following mechanism for the formation and evolution of thermal rivulets and the associated film-height modulation across the spanwise direction of the flow. We noted earlier that the rivulet effects are most probably attributed to the emergence of Marangoni flows induced due to temperature gradients, and consequently, surface tension gradients. For such flows to develop, an initial temperature disturbance at the interface or perturbation in the flow must trigger a surface-tension inhomogeneity or gradient. This surface tension gradient then drives a spanwise flow component at the surface, which reinforces the temperature profile by the advection of hotter liquid from within the film bulk (near the foil) and towards the gas–liquid interface. A heat transfer enhancement should also ensue as a consequence of the fluid motion. We have also shown that this effect is initially (closer to the inlet) limited, manifesting itself as only a modest temperature variation at the interface. At some distance further downstream, however, this spatially limited hot-fluid region expands abruptly and hotter fluid appears to be "spilling" over to cooler, neighbouring regions. The ensuing accumulation of liquid in those regions causes the formation of rivulets (spanwise film height modulation) downstream of the initially observed thermal structures. Gravity acts as a restoring force, countering the growth of rivulets or even completely suppressing it. In such cases, significant spanwise return flow towards the lower layers of the film should be anticipated. Furthermore, gravity effects are expected to introduce a significant inclination angle dependance to the topological characteristics of these rivulets, the flow field underneath them, and by extension the heat transfer performance of these films. It should finally be noted that in order to verify the validity of the proposed mechanism, knowledge of the flow field underneath the rivulet strictured is necessary. Experiments that combine the optical techniques employed in the present study, alongside velocimetry techniques such as particle image velocimetry (PIV) and particle tracking velocimetry (PTV) could contribute greatly towards uncovering the underlying, complex flow phenomena and rivulet formation, evolution and interaction mechanisms.

The temporal evolution of the thermal features is considered in Fig. 12. The rivulets display an oscillatory behaviour with respect to the streamwise direction of the flow, an effect that would be expected to contribute to heat-transfer enhancement through stronger mixing. The high temperature fluid is observed to move towards and away from the inlet with an approximate period of 1 s, with the originally thin rivulet growing wider (0.0–0.5 s) and its distance from the inlet reaching a minimum at around 0.5 s. Finally, it gets washed back down with its edge becoming blunter until the cycle repeats (0.9–1.0 s).

Greater insight into the effects of the applied heat flux and liquid flow rate on the thermal features developing over both the film free surface and solid–liquid interface, the latter as a result of conjugation, is provided by inspection of the ensuing temperature distributions in Figs. 13 and 14. Here, the flow Re was varied between 107 and 251, while the imposed heat-flux $\bar{q}$ was adjusted in the range 0.1–3.6 W cm^{-2}. The reader will note some missing panels at higher heat fluxes and lower flow rates (Fig. 13, top right). For the conditions it was not possible to operate the experiment; following film rupture and de-wetting, there was considerable risk of damaging the foil due to the rapidly rising local temperatures. It should also be noted that the presented temperature distributions of the foil underside and liquid film free-surface were not collected simultaneously. Instead, the (single) IR camera was relocated once the desired range of experimental conditions was investigated, and the same conditions were reproduced.

Regarding the operating flow conditions (flow rate and heat flux combinations) that were successfully investigated, the presence of the previously discussed streak-like thermal features (rivulets) on the film free-surface was observed, once again, as a regular equispaced pattern along the streamwise direction of the flow. In the lower *Re* and heat flux experiments, no rivulet oscillations were observed. At the higher *Re* and heat flux flows, however, the hotter "fingers" raced up and down the x-z domain with a spatial amplitude of 5–10 cm and a frequency between 0.5 and 2 Hz. The mean distance of these features from the inlet was also affected by the flow rate and imposed heat flux, with higher flow rates pushing the oscillating structures further downstream, and higher heat fluxes moving them further upstream. The corresponding foil temperature distributions (Fig. 14) are neither uniform, nor constant, with the footprints of the free-surface rivulets clearly discernible. In this case, it is apparent that increasing either the flow rate or the imposed heat flux results to both the formation of a higher number of rivulets along the spanwise direction of the flow, as well as an increase in their amplitude relative to the background IR signal (i.e. the rivulets become hotter compared to the surrounding fluid).

Inlet flow pulsation was also instigated in some experiments by diverting a portion of the total flow through a rotating throttle valve, thus generating hydrodynamic waves which travel down the film. The purpose behind this tactic was to obtain more coherent and higher amplitude film thickness fluctuations, and therefore, better modulated HTC fluctuations. Noise-driven interfacial instabilities emerge shortly downstream of the inlet, evolve into a two-dimensional wave sequence comprising highly asymmetric solitary waves (frequency content centred around the fastest growing mode), and eventually break down into three-dimensional structures. In contrast, the imposition of monochromatic perturbations at the inlet results in wave sequences which adopt the forcing frequency, while wave growth saturates within a few wavelengths [50]. The generation of coherent wave structures (in terms of shape, amplitude and wavelength), thus, allows us to isolate and study specific aspects of the heat transfer characteristics of these waves. In addition, given the thermal boundary condition selected for these experiments (uniform substrate heating), the substrate length necessary to reach fully developed wave regimes would be prohibitive in terms of the required heating power. Finally, we would like to note that pulsed-flow arrangements have been employed by many researchers in the past, for similar reasons as in this study (Refs. [22],[51] and [52]).

Figures 15 and 16 explore the effect of heat flux (the same four values examined earlier) and pulsation amplitude (4% and 10% of the mean flow rate) on the film free-surface and foil temperature distributions respectively. In these measurements, the mean flow *Re*was fixed at 179, while the forcing (wave) frequency was set to 1.8 Hz.

Careful inspection of Fig. 16 reveals that the flow is now much more unsteady following the introduction of hydrodynamic waves. Only for the lowest pulsation amplitude and heat flux combination does the temperature distribution appear relatively regular, with little variation in the streamwise direction where typically higher temperature rivulets develop. At higher heat fluxes, more irregular higher-dimensional temperature patterns emerge, rushing down the foil due to the imposed flow rate perturbations (waves). On the film surface, the effect of pulsation on the temperature distribution is more pronounced. At the lowest heat flux, the temperature distributions appear rather uniform, with the wave spacing and smaller capillary waves just distinguishable. As the imposed heat flux is increased, regions of high surface temperature form much like before. The passing of each wave front, however, enforces strong mixing between the hotter regions and surrounding cooler liquid. The spatial extent of the hot zones along the spanwise direction of the flow seems to increase with increasing distance from the inlet, while with increasing heat flux and/or pulsation amplitude, a clear mixing enhancement is discernible by the higher number of strongly skewed, intertwined thermal rivulets.

Thermal Response Coupling to the Film Thickness

The spatiotemporal evolution of the film thickness *d* (far left) and the film free-surface temperature T_{LIF} (second from left) measured simultaneously by PLIF at the 'upstream' measurement position x = 185 mm downstream of the flow inlet over a 1.8 s time interval are plotted in Fig. 17 alongside the foil temperature T_{IR} (far right) measured by the IR camera at the same location for two flow conditions: (a) *Re*=179; and (b) *Re*=251. In addition, the foil temperature T_{IR} measured by the IR camera further downstream atx=285 mm from the flow inlet (second from right) is also presented in this figure. For the particular experimental runs, the two PLIF cameras were arranged so as to image a region along the film span, generating data equivalent to the film thickness and film free-surface sample measurement presented earlier (Fig. 9). The imposed heat-flux was set to $\bar{q} = 3.6$ W cm^{-2}. To the best knowledge of the authors, the generation of such spatiotemporally resolved (local and simultaneous) conjugate (fluid–solid) heat-transfer data is the first of its kind.

As was reported earlier in relation to Fig. 9, high film free-surface temperatures (light colour regions in Fig. 17, second from left columns) are observed in thin film-trough regions which are resolved in both space and time (dark colour regions in Fig. 17, leftmost columns) and vice versa. Moreover, the local film thinning shows a high degree of correlation with the darker (lower temperature) regions on the foil underside (Fig. 17(a) and (b), last column from the left) which are, however, observed with a slight delay compared to the corresponding trough and high temperature regions. This delay, which is more clearly observable in Fig. 17(b), appears to be around 0.1 s, and is most probably attributed to the thermal inertia of the foil. An additional observation that can be made here, is that locally higher HTCs ensue in thin film regions (troughs), as more heat is extracted from the substrate. Finally, the film thickness map of Fig. 17(a) shows troughs developing in the region z=5–20 mm between t=0.1 s and t=0.6 s and reappearing at around t=1.3 s. This oscillatory behaviour, also observed in Fig. 17(b), is believed to correspond to the one noted earlier in Fig. 12. While the troughs oscillate up and down along the streamwise direction of the flow, faint "bow-like" features appear ahead of and along their peripheries.

Figure 18 shows the evolution of the *local and instantaneous* heat flux $\dot{q}$, heat-transfer coefficient *h* and Nusselt number *Nu* in the same experiment described in Fig.

17. As expected, Nu and HTC are highest where the film is thinnest, while the aforementioned "bow-like features are once again evident. The heat flux variation amounts to approximately 4% of the imposed mean heat-flux ($\bar{q} = 3.6$ W cm^{-2}) and the mean Nuamounts to about 2.8. The HTC variations are in the range 1–3 kW K^{-1} m^{-2}.

Quantitative heat-Transfer Characterization: Mean Performance

The above observations suggest that the local and instantaneous HTC h is found here to be augmented in thinner film regions, with small amplitude fluctuations encountered even in capillary wave regions. A coupling between the film thickness and the HTC was also reported in a separate, earlier effort by the present authors (Ref. [15]) that considered measurements only at a single spanwise location, along the *streamwise* direction of the flow and over a 40 mm domain centered at x=185 mm downstream of the inlet. In Ref.[15], joint probability density functions (PDFs) of the instantaneous and local HTC against the instantaneous and local film thickness were compiled from data generated over 12 measurement runs, which had the same flow conditions (Re=300, forcing frequency of 5 Hz) but with the heat input varied from 200 W to 5 kW. A series of noteworthy observations were presented and are briefly reproduced here in order to facilitate the upcoming analysis whose aim is to consider the role of unsteadiness and the impact of the flow Re on the instantaneous and local HTC.

Firstly, one important finding from the previous effort was that the HTC, even in a time-varying instantaneous sense, was largely independent of the imposed heat flux, at least over the range of heat fluxes examined in that (and the present) experimental campaign[15]. This permits data across experimental runs with different heat fluxes (but the same otherwise flow conditions, i.e. Re , etc.) to be grouped together. Secondly, the experimental results were compared to theoretical HTC variations expected from known film thickness variations based on Nusselt flow (corresponding to Nu=2.5); such a comparison has been included also in the present results in Fig. 19 (solid line). The Nusselt solution [53] describes the steady, 1-D, fully developed, laminar film flow over an inclined foil, and therefore constitutes a useful benchmark assessment tool for the hereby investigated wavy films, allowing the effect of unsteadiness to be examined relative to the steady laminar case. The film thickness and HTC for a Nusselt flow can be obtained by considering the momentum (Navier–Stokes) and energy equations respectively, and implementing the relevant simplifications (terms are neglected based on the aforementioned assumptions). The relevant derivation can be found in Ref. [15]. It should be noted that non-negligible deviations between the analytical Nusselt predications and the experimental findings are in fact expected, since the films described in the current contribution are not steady, one-dimensional, and neither fully developed nor even laminar in certain cases. In Ref. [15] an enhancement in the heat-transfer performance of wavy films relative to the steady case was observed, as anticipated, due to unsteadiness, increased mixing and turbulence. Despite the fact that the experimental results followed the general trend described by the Nusselt relationship reasonably well (i.e. the HTC experiences an enhancement when the films become thinner), the magnitude of the HTC enhancement (mean HTC relative to the HTC at the mean film thickness) was observed to be significantly higher than the Nusselt value, by up to a factor of 2.

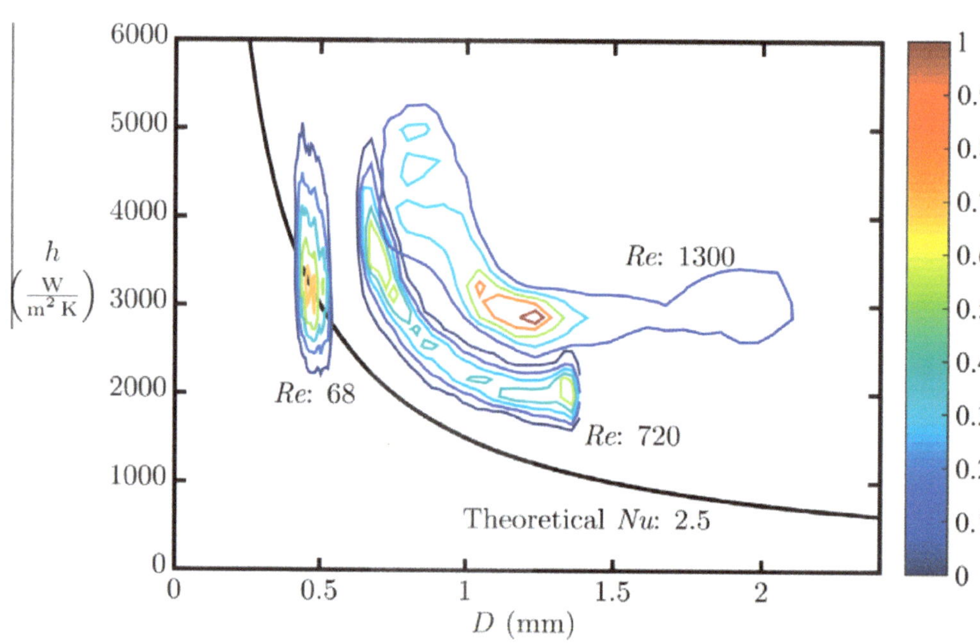

Figure 19. PDFs of the HTC, h, plotted against film thickness, d , for data acquired under three different flow conditions: $Re = 68$, $Re = 720$ and Re=1300.

New insight into the effects of unsteadiness on the observed HTC enhancement can be obtained by examining joint-PDFs generated from separate sets of data corresponding to different Re (Fig. 19). For the lowest Re flows (Re=68), the majority of the experimental data are in close proximity to the theoretical Nusselt description, but the film thickness does not show considerable variations (d≈0.5 mm) while the HTC varies by ±40% around a value of h≈3200 W m^{-2} K^{-1}.

As the *Re* increases to 720, the trend described by the Nusselt solution is reproduced, with the experimentally determined HTC values being, however, globally higher than the Nusselt trend-line. Moreover, the HTC dependence on the film thickness becomes weaker than the analytical Nusselt solution at higher film thicknesses. Further increasing the *Re* to 1300 boosts the observed HTC values even further, and accentuates the previously noted observations, with the HTC appearing almost decoupled from the film thickness, attaining a value around h≈3000 W m^{-2} K^{-1} above approximately d≈1.2 mm. This decoupling at higher film thicknesses (which presumably correspond to regions adjacent to the wave crests) and high *Re* may be potentially linked to flow recirculation in the reference frame moving with the wave speed. As the *Re* increases, the flow velocity surpasses the wave velocity, resulting in flow recirculation underneath the wave crest (where the flow velocity is typically maximized). This phenomenon has been reported in the past for both laminar and turbulent falling films [54] and [55], and has been linked to local heat transfer enhancements [5]. In downwards annular gas–liquid flows, multiple recirculation zones have been reported underneath large wave structures, referred to as disturbance waves[56].

In Refs. [34] and [35], the heat transfer enhancement experienced due to waviness was studied for a laminar water-film flowing down a heated plate inclined at 2° to the horizontal. HTCs were measured along the wave topology and normalized by the mean value at the examined axial distance away from the flow inlet (230 mm). Unfortunately, this excellent experimental effort was limited to a narrow flow-parameter range (*Re*=126), and therefore any comparison will have to be restricted to this aforementioned condition. The results indicated that an +80% enhancement compared to the mean HTC was experienced at the wave trough, while a -40% deterioration occurred below the wave crest. Behind the wave crest, and while the local film thickness exceeded the mean film thickness, HTC values lower the mean were consistently encountered. Ahead of the wave trough, and despite the fact that local film thicknesses were lower than the mean value, local HTCs once again did not exceed the mean value. Another noteworthy observation from this data is that the HTC variation followed closely (and inversely) the film thickness variation along the wave topology, a behaviour that was also observed in our experiments [15].

In reference to our present results shown in Fig. 19, we measured HTC values in the range 2100–5100 W m^{-2} K^{-1} with a mode of 3150 W m^{-2} K^{-1} for the flow with *Re*=68. These values correspond to a relative (normalized to the flow mean) HTC deterioration of (−30%) topologically emanating from below the wave crests, and a (+60%) enhancement stemming from the thinner film regions ahead of the wave crest (i.e. the wave troughs). Hence, close quantitative agreement is observed with the results presented in Refs.[34] and [35], despite the different experimental conditions and methods.

The formerly noted deviations from the Nusselt theory are re-examined in Fig. 20, where the mean flow *Nu* is plotted against the mean flow *Re* over the full range of tested pulsation amplitudes and frequencies. Along with the experimentally determined data points, a correlation produced by Wilke [57], linking the HTC, Prandtl number *Pr* and *Re*in heated falling films is also presented. Inspection of Fig. 20 reveals that whereas the lower *Re* data points (below approximately 200) fit closely the value predicted by the Nusselt theory (*Nu*=2.5), intermediate and high *Re* flows display significantly higher *Nu*numbers, matching Wilke's correlation closely when the latter is expressed in terms of the mean flow *Nu* number. A possible explanation for this substantial heat-transfer enhancement, as observed in both experimental investigations [58] and also verified by modelling efforts [59], is stronger mixing and enhanced convective effects in the main wave regions, associated with the emergence of recirculation zones.

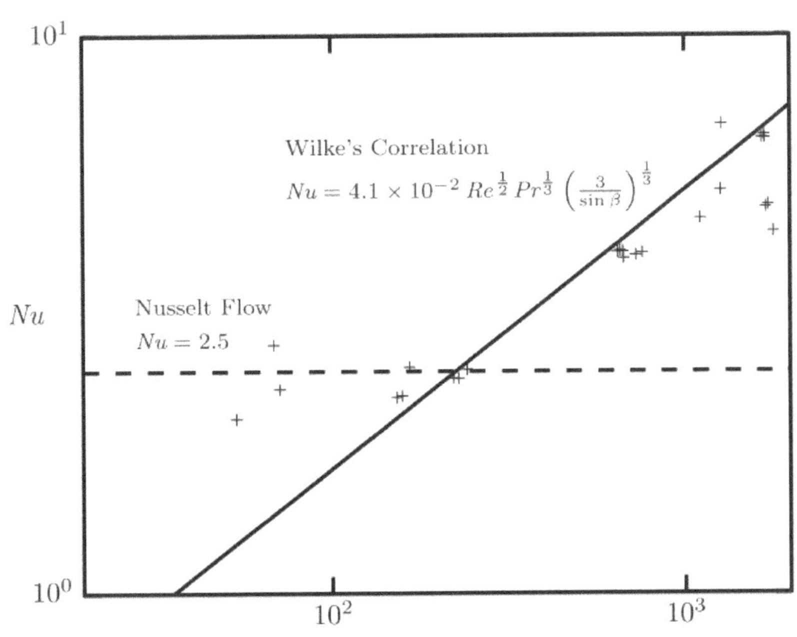

Figure 20. Mean *Nu* against the mean *Re* , compiled using data from different flow conditions spanning a range of pulsation amplitudes and frequencies for *Pr*=5.4. Along with the experimentally derived data, the correlation produced by Wilke [57] is also plotted for the examined *Re* range.

Quantitative Heat-Transfer Characterization: Fluctuating Performance

Nevertheless, the strength of the developed experimental technique is that it allows one to consider, beyond the mean heat-transfer performance of these flows, also the magnitude of the spacial and temporal variations around this mean behaviour. Fig. 21shows the standard deviation of the HTC σ_h (HTC fluctuation), normalized by the mean HTC $\bar{h}$, plotted against the standard deviation of the wave thickness σ_d (film thickness fluctuation), normalized by the mean film thickness $\bar{d}$, for two pulsation frequencies: 1.8 Hz and 3.6 Hz. These quantities (fluctuation over mean) are referred to as fluctuation intensities. Referring to either pulsation frequency data, it can be ascertained that higher HTC fluctuation intensities are associated with lower film thickness fluctuation intensities. This trend can be described for both pulsation frequency data sets by fitting power functions, suggesting that stronger waviness dampens the oscillations in the HTC relative to the mean value. By comparing the two data sets amongst each other, it can be observed that overall lower HTC fluctuation intensities ensue in the higher pulsation frequency runs. Moreover, with increasing film thickness fluctuation intensities, the disparity in the HTC behaviour between the higher wave frequency data and the lower wave frequency data diminishes.

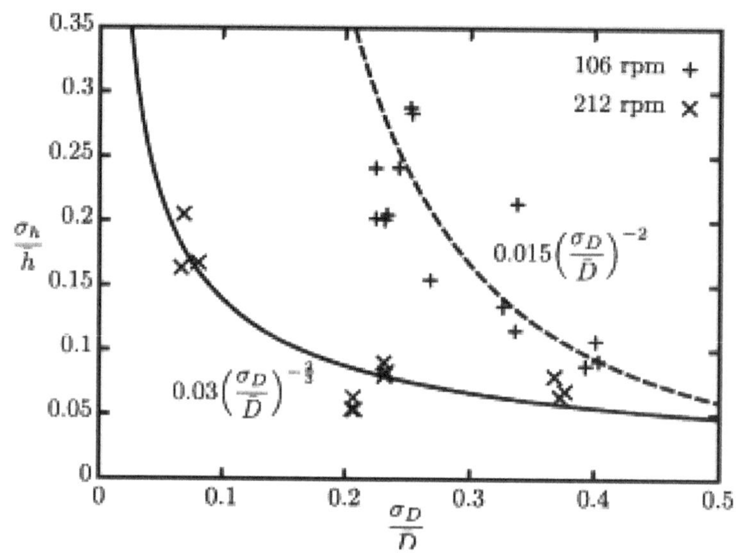

Figure 21. Fluctuation intensity of the HTC, *h*, as a function of the normalized wave amplitude, at forcing frequencies of 1.8 Hz (106 rpm) and 3.6 Hz (212 rpm).

Further, the effect of the HTC fluctuation intensity $\sigma_h/\bar{h}$ on the normalized fluctuation amplitude of the fluid interface-wall temperature difference $\sigma_{\Delta T}/\overline{\Delta T}$ is presented in Fig. 22. The temperature difference across the liquid film is observed to increase monotonically and near-linearly with the amplitude of the HTC fluctuation. The plotted data also reveals that the normalized temperature fluctuation is consistently about 25% smaller than the driving HTC fluctuation intensity. This relationship suggests that, in these particular flows, the heat flux fluctuations are small relative to those of ΔT and *h*.

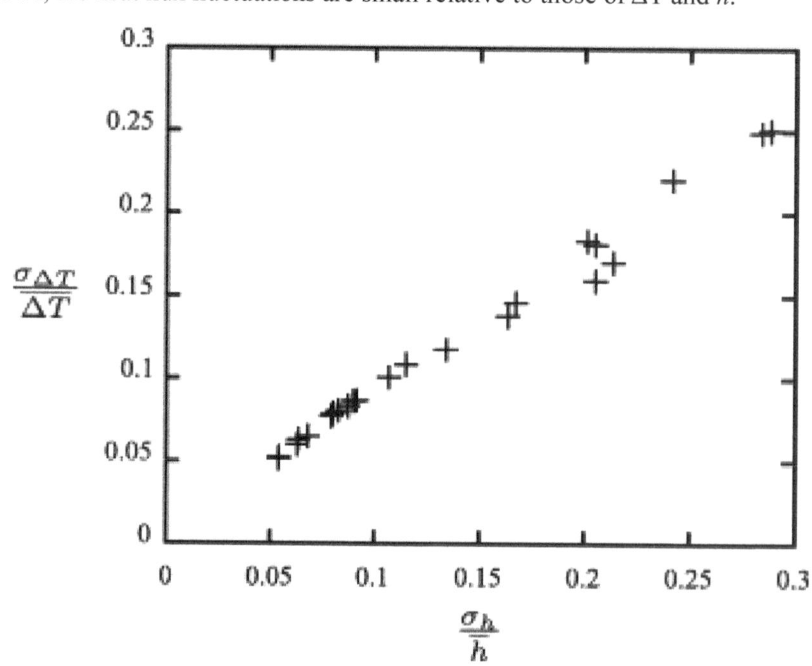

Figure 22. Fluctuation intensity of the fluid interface-wall temperature difference, ΔT, as a function of the fluctuation intensity of the HTC, *h*.

Finally, it is worth considering briefly the effect of pulsation frequency on the interfacial wave velocity, the latter calculated from cross-correlations across successive film thickness profiles. This was investigated for three different flow *Re* in the range 140–350 and two forcing frequencies, 1.8 and 3.6 Hz. The results are presented in Fig. 23. As the mean flow *Re* increases from 140 to 350, the mean interfacial wave velocity increases in the range 0.5–0.9 ms^{-1}. Moreover, a near-linear trend can be identified despite the limited number of data points; a result that is expected given that the fluid viscosity and foil inclination angle remain unaltered, and consequently, the flow *Re* scales with the bulk flow velocity. Another interesting observation is that the relationship does not go through the origin. The underlying reason for this phenomenon is believed to be the superposition of the wave velocity with the bulk flow velocity. By comparing the two data sets amongst each other, it can finally be noted that the lower forcing frequency can be associated with higher interfacial velocities. This observation can be explained on the basis that higher excitation forcing frequencies produce shallower, smoother and slower waves compared to the larger and higher-dimensional waves associated with lower frequencies.

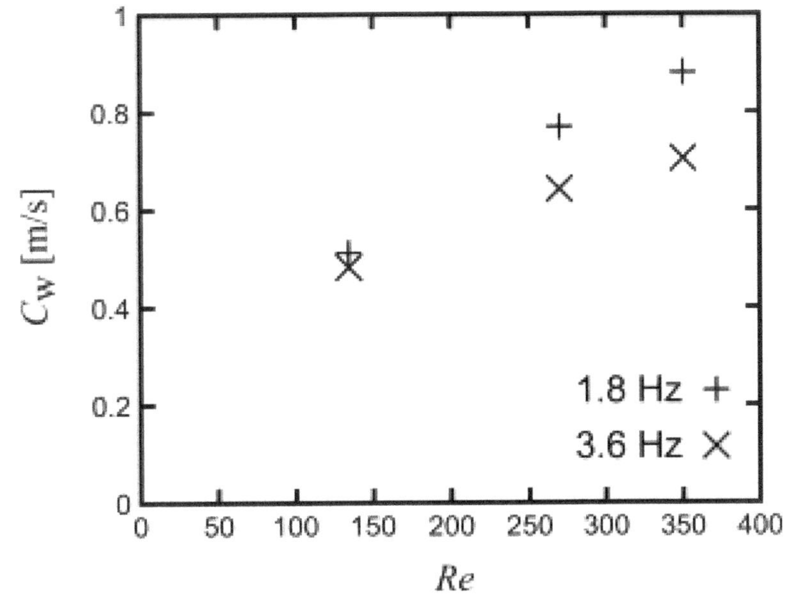

Figure 23. Average wave velocities for different mean flow *Re* over the range *Re*=140–350, and two different forcing pulsation frequencies, 1.8 and 3.6 Hz.

CONCLUSIONS

A combined PLIF and IR thermography technique was developed and employed for the detailed inspection and experimental characterization of unsteady and conjugate heat-transfer in planar, gravity-driven liquid films flowing down an inclined heated metal foil. The main challenge associated with these measurements, and in particular in the presence of finite heat-transfer conjugation effects, is that space and time-resolved information is required of both interfacial temperature variations (i.e at the solid–liquid and gas–liquid interfaces), which must be measured directly and simultaneously. Thus, simultaneous measurements were pursued and are reported herein for the local and instantaneous film thickness, free-surface temperature, foil/substrate temperature, heat flux exchanged between the heated foil and liquid film, and finally the HTC (and Nusselt number) over a range of flow conditions (i.e. Reynolds numbers, pulsation amplitudes and frequencies, heat fluxes).

The formation, interaction and spatial and temporal evolution of thermal rivulet structures on the film free-surface is examined and associated with film thickness variations. In some cases, flat thermal rivulets (i.e. not associated with film thickness fluctuations) were also identified and linked to the development of Marangoni flows on the film free surface. In more detail, local surface tension variations are believed to trigger film thickness instabilities and the ensuing (Marangoni-driven) mass transfer, characterized by accumulation of fluid in cooler regions and thinning in hotter ones. The introduction of flow pulsation had a pronounced effect on these features, instigating a clearly discernible mixing enhancement and resulting to the formation of a higher number of strongly skewed, intertwined thermal rivulets.

Following a qualitative assessment of the thermal features observed on the film free-surface and foil underside, a quantitative analysis was carried out with the ultimate goal of associating the observed interfacial wave dynamics with the heat-transfer characteristics of these flows (such as the instantaneous and local HTC variation with the mean film thickness and film thickness fluctuation intensity). These are of paramount significance as they can be used to design equipment with far greater heat and mass transfer capabilities through understanding and harnessing enhancement mechanisms. Despite describing the HTC enhancement at lower film thickness values reasonably well and also capturing the overall HTC variations due to fluctuating film thicknesses, the results presented herein suggest that the magnitude of the true HTC can be significantly higher than HTC predictions based on Nusselt theory, by up to a factor 3 in some cases. A number of unsteady flow phenomena linked to the interface waviness of are believed to contribute towards this enhancement. The average *Nu* number increased with the mean flow *Re*, while at low values (lowest flow rates), a *Nu* of around 2.5 was observed, in agreement with the steady, laminar flow theory. Evidence that the HTC becomes decoupled from the film thickness for the upper range of encountered film thicknesses, has also been presented. Finally, lower wave amplitude intensities were associated with higher HTC fluctuation intensities (relative to their respective means), while the amplitude of the wall temperature fluctuation was found to be almost linearly dependent on the amplitude of the fluctuation of the HTC.

ACKNOWLEDGMENTS

This work was supported by the UK Engineering and Physical Sciences Research Council (EPSRC) [grant numbers EP/J006041/1 and EP/K008595/1]. RM would like to acknowledge the Department of Chemical Engineering, Imperial College London and EPSRC for his DTA studentship. RM and CNM would also like to acknowledge the use made of the Cedip Titanium and FLIR SC3000 IR imagers, which were borrowed from the EPSRC Instrument Pool (EIP), and without which these measurements would not have been possible. Finally, the authors would like to extend their thanks to R. Gendarme for his helpful contribution to this work, as part of his student summer project at Imperial College. Data supporting this publication can be obtained on request from cep-lab@imperial.ac.uk.

REFERENCES

1. M. Pradas, D. Tseluiko, S. Kalliadasis, Rigorous coherent-structure theory for falling liquid films: Viscous dispersion effects on bound-state formation and self-organization, Phys. Fluids 23 (2011) 044104-1–044104-19.
2. D.A. Goussis, R.E. Kelly, Surface wave and thermocapillary instabilities in a liquid film flow, J. Fluid Mech. 223 (1991) 25–45.
3. S. Kalliadasis, E.A. Demekhin, C. Ruyer-Quil, M.G. Velarde, Thermocapillary instability and wave formation on a film falling down a uniformly heated plane, J. Fluid Mech. 492 (2003) 303–338.
4. S. Kalliadasis, A. Kiyashko, E.A. Demekhin, Marangoni instability of a thin liquid film heated from below by a local heat source, J. Fluid Mech. 475 (2003) 377– 408.
5. W. Rohlfs, B. Scheid, Phase diagram for the onset of circulating waves and flow reversal in inclined falling films, J. Fluid Mech. 763 (2015) 322–352.
6. S. Chakraborty, P.-K. Nguyen, C. Ruyer-Quil, V. Bontozoglou, Extreme solitary waves on falling liquid films, J. Fluid Mech. 745 (2014) 564–591.
7. G.F. Dietze, P.-K. Nguyen, C. Ruyer-Quil, Wavy liquid films in interaction with a confined laminar gas flow, J. Fluid Mech. 722 (2013) 348–393.
8. G.M. Skotheim, U. Thiele, B. Scheid, On the instability of a falling film due to localized heating, J. Fluid Mech. 475 (2003) 1–19.
9. H. Yu, T. Gambaryan-Roisman, P. Stephan, Numerical simulations of hydrodynamics and heat transfer in wavy falling liquid films on vertical and inclined walls, J. Heat Transfer 135 (2013) 101010-1–101010-9.
10. K. Moran, J. Inumaru, M. Kawaji, Instantaneous hydrodynamics of a laminar wavy liquid film, Int. J. Multiphase Flow 28 (2002) 731–755.
11. G.F. Dietze, F. Al-Sibai, R. Kneer, Instantaneous hydrodynamics of a laminar wavy liquid film, J. Fluid Mech. 637 (2009) 73–104.
12. E.A. Chinnov, S.S. Abdurakipov, Thermal entry length in falling liquid films at high reynolds numbers, Int. J. Heat Mass Transfer 56 (2013) 775–786.
13. S.V. Alekseenko, V.Y. Nakoryakov, B.G. Pokusaev, Wave formation on a vertical falling liquid film, Am. Inst. Chem. Eng. J. 31 (9) (1985) 1446–1460.
14. R. Mathie, C.N. Markides, Heat transfer augmentation in unsteady conjugate thermal systems – Part I: Semi-analytical 1-D framework, Int. J. Heat Mass Transfer 56 (2013) 802–818.
15. R. Mathie, H. Nakamura, C.N. Markides, Heat transfer augmentation in unsteady conjugate thermal systems – Part II: Applications, Int. J. Heat Mass Transfer 56 (2013) 819–833.
16. T.H. Lyu, I. Mudawar, Statistical investigation of the relationship between interfacial waviness and sensible heat transfer to a falling liquid film, Int. J. Heat Mass Transfer 34 (6) (1991) 1451–1464.
17. T. Ueda, H. Tanaka, Measurements of velocity, temperature and velocity fluctuation distributions in falling liquid films, Int. J. Multiphase Flow 2 (1975) 261–272.
18. T.D. Karapatsios, S.V. Paras, A.J. Karabelas, Statistical characteristics of free falling films at high Reynolds numbers, Int. J. Multiphase Flow 15 (1) (1989) 1– 21.
19. R.J. Belt, J.M.C. Van't Westende, H.M. Prasser, L.M. Portela, Time and spatially resolved measurements of interfacial waves in vertical annular flow, Int. J. Multiphase Flow 36 (2010) 570–587.
20. Y. Zhao, C.N. Markides, O.K. Matar, G.F. Hewitt, Disturbance wave development in two-phase gas–liquid upwards vertical annular flow, Int. J. Multiphase Flow 55 (2013) 111–129.
21. N. Brauner, D.M. Maron, Characteristics of inclined thin films, waviness and the associated mass transfer, Int. J. Heat Mass Transfer 25 (1) (1982) 99–110.
22. J. Liu, J.D. Paul, J.P. Gollub, Measurements of primary instabilities of film flows, J. Fluid Mech. 250 (1993) 69–101.
23. M. Vlachogiannis, V. Bontozoglou, Observations of solitary wave dynamics of film flows, J. Fluid Mech. 453 (2001) 191–215.
24. S. Alekseenko, A. Cherdantsev, M. Cherdantsev, S. Isaenkov, S. Kharlamov, D. Markovich, Application of a high-speed laser-induced fluorescence technique for studying the three-dimensional structure of annular gas–liquid flow, Exp. Fluids 453 (2012) 77–89.
25. R.G. Morgan, C.N. Markides, C.P. Hale, G.F. Hewitt, Horizontal liquid-liquid flow characteristics at low superficial velocities using laser-induced fluorescence, Int. J. Multiphase Flow 43 (2012) 101–117.
26. I. Zadrazil, O.K. Matar, C.N. Markides, An experimental characterization of downwards gas–liquid annular flow by laser-induced fluorescence: Flow regimes and film statistics, Int. J. Multiphase Flow 60 (2014) 87–102.

27. A. Charogiannis, J.S. An, C.N. Markides, A simultaneous planar laser-induced fluorescence, particle image velocimetry and particle tracking velocimetry technique for the investigation of thin liquid-film flows, Exp. Therm. Fluid Sci. 68 (2015) 516–536.
28. D. Ross, M. Gaitan, L.E. Locascio, Temperature measurement in microfluidic systems using a temperature-dependent fluorescent dye, Anal. Chem. 73 (17) (2001) 4117–4123.
29. P. Chamarthy, S.V. Garimella, S.T. Wereley, Measurement of the temperature non-uniformity in a microchannel heat sink using microscale laser-induced fluorescence, Int. J. Heat Mass Transfer 53 (15) (2010) 3275–3283.
30. L. Shi, X. Mao, A.J. Jaworski, Application of planar laser-induced fluorescence measurement techniques to study the heat transfer characteristics of parallelplate heat exchangers in thermoacoustic devices, Meas. Sci. Technol. 21 (11) (2010) 115405.
31. L. Shi, Z. Yu, A.J. Jaworski, Application of laser-based instrumentation for measurement of time-resolved temperature and velocity fields in the thermoacoustic system, Int. J. Therm. Sci. 49 (9) (2010) 1688–1701.
32. T. Karstens, K. Kobs, Rhodamine b and Rhodamine 101 as reference substances for fluorescence quantum yield measurements, J. Phys. Chem. 84 (14) (1980) 1871–1872.
33. J. Sakakibara, R.J. Adrian, Whole field measurement of temperature in water using two-color laser induced fluorescence, Exp. Fluids 26 (1–2) (1999) 7–15.
34. A. Schagen, M. Modigell, G. Dietze, R. Kneer, Simultaneous measurement of local film thickness and temperature distribution in wavy liquid films using a luminescence technique, Int. J. Heat Mass Transfer 49 (25) (2006) 5049–5061.
35. A. Schagen, M. Modigell, Local film thickness and temperature distribution measurement in wavy liquid films with a laser-induced luminescence technique, Exp. Fluids 43 (2–3) (2007) 209–221.
36. O.A. Kabov, B. Scheid, I.A. Sharina, J.-C. Legros, Heat transfer and rivulet structures formation in a falling thin liquid film locally heated, Int. J. Therm. Sci. 41 (7) (2002) 664–672.
37. E.A. Chinnov, O.A. Kabov, Marangoni effect on wave structure in liquid films, Microgravity Sci. Technol. 19 (3–4) (2007) 18–22.
38. E.A. Chinnov, E.N. Shatskii, Effect of thermocapillary perturbations on the wave motion in heated falling liquid film, Tech. Phys. Lett. 36 (1) (2010) 53–56.
39. E.A. Chinnov, E.N. Shatskii, O.A. Kabov, Evolution of the temperature field at the three-dimensional wave front in a heated liquid film, High Temp. 50 (1) (2012) 98–105.
40. K. Löffler, H. Yu, T. Gambaryan-Roisman, P. Stephan, Flow patterns and heat transfer in thin liquid films on walls with straight, meandering and zigzag minigrooves, in: Proceedings of the Sixth International ASME Conference on Nanochannels, Microchannels and Minichannels, ASME, 2008, pp. 62318/1– 62318/9.
41. K. Löffler, H. Yu, S. Hardt, T. Gambaryan-Roisman, P. Stephan, Heat transfer and evaporation of falling liquid films on surfaces with advanced threedimensional periodic structures: Experiments and numerical simulations, in: Proceedings of the ASME 3rd Joint US-European Fluids Enginering Summer Meeting and 8th International Conffrence on Nanochannels, Microchannels, Minichannels, ASME, 2010, pp. 30761/1–30761/8.
42. K. Ibrahem, M.F. Abd Rabbo, T. Gambaryan-Roisman, P. Stephan, Experimental investigation of evaporative heat transfer characteristics at the 3-phase contact line, Exp. Therm. Fluid Sci. 34 (2010) 1036–1041.
43. B. Schilder, S. Yu, N. Kasagi, S. Hardt, P. Stephan, Flow visualization and local measurement of forced convection heat transfer in a microtube, J. Heat Transfer 132 (2010) 031702/1–031702/9.
44. V.V. Lel, F. Al-Sibai, A. Leefken, U. Renz, Local thickness and wave velocity measurement of wavy films with a chromatic confocal imaging method and a fluorescence intensity technique, Exp. Fluids 39 (5) (2005) 856–864.
45. V.V. Lel, A. Kellermann, G. Dietze, R. Kneer, A.N. Pavlenko, Investigations of the Marangoni effect on the regular structures in heated wavy liquid films, Exp. Fluids 44 (2) (2008) 341–354.
46. F. Zhang, X. Zhao, J. Geng, Y.-T. Wu, Z. Zhang, A new insight into Marangoni effect in non-isothermal falling liquid films, Exp. Therm. Fluid Sci. 31 (4) (2007) 361–365.
47. F. Zhang, Y.-T. Wu, J. Geng, Z.-B. Zhang, An investigation of falling liquid films on a vertical heated/cooled plate, Int. J. Multiphase Flow 34 (1) (2008) 13–28.
48. A. Criminisi, I. Reid, A. Zisserman, A plane measuring device, Image Vision Comput. 17 (8) (1999) 625–634.
49. H. Nakamura, Frequency response and spatial resolution of a thin foil for heat transfer measurements using infrared thermography, Int. J. Heat Mass Transfer 52 (21) (2009) 5040–5045.
50. S. Kalliadasis, C. Ruyer-Quil, B. Scheid, M.G. Velarde, Falling liquid films, Springer Ser. Appl. Math. Sci. (2012).
51. T. Nosoko, P.N. Yoshimura, T. Nagata, K. Oyakawa, Characteristics of twodimensional waves on falling liquid films, Chem. Eng. Sci. 51 (5) (1996) 725–732.
52. K. Argyriadi, K. Serifi, V. Bontozoglou, Nonlinear dynamics of inclined films under low-frequency forcing, Phys. Fluids 16 (7) (2004) 2457–2468.
53. W. Nuelt, Die oberflächenkondensation des wasserdampfes. VDI 60 (1916) 541–546.
54. N. Brauner, Modelling of wavy flow in turbulent free falling films, Int. J. Multiphase flow 15 (4) (1989) 505–520.
55. I. Mudawar, R.A. Houpt, Measurement of mass and momentum transport in wavy-laminar falling liquid films, Int. J. Heat Mass Transfer 36 (17) (1993) 4151–4162.

56. I. Zadrazil, C.N. Markides, An experimental characterization of liquid films in downwards co-current gas–liquid annular flow by particle image and tracking velocimetry, Int. J. Multiphase Flow 67 (2014) 42–53.

57. W. Wilke, Wärmeübertragung an rieselfilme, VDI-Forschungsheft 490 (1962).

58. S.M. Yih, Modeling heat and mass transfer in wavy and turbulent falling liquid films, Wärme-und Stoffübertragung 21 (6) (1987) 373–381.

59. A. Miyara, Flow dynamics and heat transfer of wavy condensate film, J. Heat Transfer 123 (3) (2001) 492–500.

CHAPTER 4

PERFORMANCE ANALYSIS OF ULTRALOW GRADE WASTE HEAT UPGRADE USING ABSORPTION HEAT TRANSFORMER

Zhiwei Ma, Huashan Bao, Anthony Paul Roskilly

Sir Joseph Swan Centre for Energy Research, Newcastle University, Newcastle upon Tyne NE1 7RU, United Kingdom

ABSTRACT

The present paper aimed at exploring absorption heat transformer (AHT) to upgrade ultralow grade waste heat in the temperature range of 40–60 °C. The performance of AHTs with different configurations, including single stage, double stage and double effect, were numerically analysed and compared in terms of temperature lift, coefficient of performance (COP) and exergy coefficient of performance (COPe). The most influential and crucial factor for the studied AHTs is the recirculation flow ratio (FR), the increase of which results in an increasing temperature lift but gradually declining COP. The COPe can achieve its maximum value with a certain FR, and such a state can be considered as the optimal working condition. Within the studied waste heat temperature range, the optimal FR in single stage AHT is in the range of 10–12, at which the system can deliver 17.1~34.7 °C temperature lift with COP at 0.471~0.475. The best configuration amid the studied four different double stage AHTs has a temperature lifting capacity of 31.8~68.6 °C with a COP around 0.30. The double effect AHT compromises its temperature lifting capacity for the highest COP among all the AHTs studied, which can reach about 0.65 though necessitates relatively higher waste heat temperature and higher strong solution concentration to drive the cycle; the double effect AHT is not recommended for the upgrading of ultralow grade waste heat.

INTRODUCTION

Energy demand in industry sectors accounts for around 17% of the total energy consumption in the UK [1], and the worse thing is that the average industrial thermal waste via radiation, exhausted gas or air, cooling fluids and so on is about one sixth of this total energy demand [2]. In fact, the wasted heat still contains large amount of energy and exergy that can be reused to deliver heating, cooling or work through suitable heat recovery technologies. There has been a great number of promising and interesting technologies emerging to realise the recovery of waste heat prevailingly around 70~250 °C, like power generation via organic Rankine cycle [3] and Kalina cycle [4], refrigeration by absorption [5] and adsorption [6] technologies, and district heating [7], etc. More efforts are encouraged to address the significant though more challenging and less explored cases, the recovery and utilisation of the ultralow

Citation: Zhiwei Ma, Huashan Bao, Anthony Paul Roskilly, Performance analysis of ultralow grade waste heat upgrade using absorption heat transformer, Applied Thermal Engineering, Volume 101, 25 May 2016, Pages 350-361, ISSN 1359-4311, http://dx.doi.org/10.1016/j.applthermaleng.2016.02.002.

grade waste heat, which in the temperature range between 40 and 60 °C is way more ubiquitous and could account for about 30% of the total industrial waste heat [8]. With rational heat recovery and effective heat transformer technologies, these waste heat could play an important role on energy saving and sustainable development.

Absorption heat transformer (AHT), operating in the reverse way of absorption refrigeration, can economically upgrade low grade waste heat into useful heat at a higher temperature with only little electrical energy input for pumping work [9] and [10]. It is reported that a single stage AHT (SAHT) can recover approximately 50% of waste heat with a temperature lift of about 50 °C in principle [10], after which the useful energy can be reused in industrial processes.

The SAHT performance of using LiBr–H_2O solution and H_2O–NH_3 solution, as the two most commonly used working solutions, were compared by Kurem and Horuz [11]. The comparison showed that LiBr–H_2O solution out-performed its rival, though the temperature lift was limited by the crystallisation issue. To further enhance the performance of LiBr–H_2O based SAHT, Rivera et al. [12] used a commercial compound named Carrol, which was a mixture of LiBr and ethylene glycol in the mass ratio of 1:4.5 with water as the working solution. Carrol has higher solubility in water than LiBr does, which makes it a favourable alternative working solution for AHT. Rivera and Cerezo [13]and Rivera et al. [14] added 400 ppm 2-ethyl-1-hexanol into LiBr–H_2O solution as a heat and mass transfer enhancer so that the upgraded temperature and the coefficient of performance (COP) was improved by 5 °C and 40%, respectively. Barragan et al. [15],[16], [17] and [18] conducted a series of experimental studies using LiCl–H_2O, $CaCl_2$–H_2O, $MgCl_2$–H_2O and also ternary solutions of LiCl–$ZnCl_2$–H_2O and $CaCl_2$–$ZnCl_2$–H_2O in SAHT. The test results showed lower temperature lift by using $CaCl_2$–H_2O or $MgCl_2$–H_2O solution than using LiCl–H_2O solution due to the lower solubility of $CaCl_2$ and $MgCl_2$in water. It was also found that the addition of $ZnCl_2$ in the ternary solution could reduce the solution viscosity and eliminate the solution crystallisation. Yin et al. [19] compared the performance of using LiBr–H_2O, TFE (2, 2, 2-trifluoroethanol)–NMP (N-methyl-2-pyrrolidone), TFE–E181 (dimethylether tetraethylene glycol) and TFE–PYR (2-pyrrolidone) as working solution in SAHT. It concluded that the LiBr–H_2O based SAHT performed superiorly when the output temperature was below 150 °C, while the other three solutions were more stable at higher temperature, e.g. up to 200 °C. The authors therefore suggested a double stage system using LiBr–H_2O loop in the lower temperature cycle while using one of the other three in the higher temperature cycle to achieve higher temperature lift.

Alternatively, Horuz and Kurt [20] and Parham et al. [21] tried different configurations to improve the performance of SAHT. Three more configurations other than the conventional configuration (named as SAHT-1 in present paper) were introduced and evaluated, including SAHT-2 in which the water carrying waste heat transferred the heat to the evaporator prior to the generator; SAHT-3 in which it had an additional absorber heat exchanger on the basis of SAHT-2; and SAHT-4 in which it was based on SAHT-3 and mounted with another refrigerant heat exchanger at the inlet of evaporator. These three new configurations reduced the crystallisation risk of using LiBr–H_2O as working solution. Both studies concluded the COP value in the order of SAHT-4 > SAHT-3 > SAHT-1 > SAHT-2, and the highest COP value was about 13–14% higher than the lowest one.

Double stage AHT (DAHT) can be applied if larger temperature lift is expected, which was reported to be capable of achieving about 80 °C temperature lift with a COP of 0.35[10]. Several types of DAHT cycles have been proposed to explore the preferable configuration for optimal efficiency and effective heat transformation. The typical straightforward approach to implement DAHT (named as DAHT-1 in the present paper) is to couple two sets of SAHT units through the heat exchanging between the absorber in the low temperature loop and the evaporator in the high temperature loop [22], [23], [24],[25] and [26]. Romero et al. [24] experimentally demonstrated that the DAHT-1 could transform the heat from 80 °C to 152 °C with efficiency higher than 30%. Ji and Ishida [26]studied a DAHT-1 system where both the absorber and the generator were peculiarly designed with two or more interior compartments, and found that the temperature lift was improved by around 10.7 °C and the exergy efficiency was elevated from 48.14% to 54.95% due to the usage of these compartments.

Rivera et al. [23] proposed a different DAHT cycle (named as DAHT-2) and claimed it to be technically simpler and competent comparing to DAHT-1. DAHT-2 utilises a portion of the weak solution from the high temperature absorber as the strong solution for the low temperature absorber. Martinez and Rivera [27] conducted theoretical energy and exergy analysis on DAHT-2, and concluded that the system could generate a 74 °C temperature lift. Furthermore, they suggested that the generator was responsible for the highest thermodynamic irreversibility, which was about 40% of the total exergy destruction. Reyes et al. [28] modelled the DAHT-2 system using $CaCl_2$–H_2O and LiCl–H_2O, which could achieve typical temperature lifts around 35~40 °C.

Based on DAHT-2, Zhao et al. [29] developed a new DAHT (named as DAHT-3) that splits the strong solution from the generator into two streams – one heading for high temperature absorber and the other one for low temperature absorber. This new arrangement introduced 5~10 °C increase in temperature lift contrasted with DAHT-2, while the COP marginally dropped by 0.01. Similar analytical conclusions were also obtained by Horuz and Kurt [30].

Mostofizadeh and Kulick [31] reported another different configuration of DAHT (named as DAHT-4), which had all the weak solution from high temperature absorber received by the low temperature absorber as the second stage of absorption. Subsequently, the weaker solution exiting the low temperature absorber ended up in the generator. Zhao et al. [32] reported that the DAHT-4 had the largest COP especially when the heat output temperature, namely the absorption temperature, was required higher, as the maximum COP value could reach 0.32 with the temperature lift between 60 and 100 °C.

Double effect absorption heat transformer (DeAHT) was developed to satisfy the desire of higher system efficiency, while the heat upgrading level is inevitably compromised. Zhao et al. [33] investigated the DeAHT system using TFE–E181 and LiBr–H_2O as working solutions, and the analytical results showed a 30 °C temperature lift could be achieved with waste heat temperature at 70 °C; the corresponding COP value was as high as 0.58 and 0.64, respectively for the studied two working solutions. Gomri [34]comparatively applied the SAHT and DeAHT to the seawater desalination with heat input at the temperature of 74~96 °C and 87~96 °C, and obtained 103~158 °C and 103~123 °C upgraded heat, respectively.

Though AHTs have been widely investigated and reported, to the best of the authors' knowledge, there has been neither endeavour in recovering and upgrading ultralow grade waste heat in the temperature range between 40 and 60 °C nor the comprehensive comparative study on various AHT systems including all aforementioned cycles. Therefore, this work carried out analytical study on six different AHTs using LiBr–H_2O with primary interest in heat transformation of the ultralow grade heat from 40 to 60 °C.

WORKING PRINCIPLES OF AHTS AND ANALYSIS METHODS

The schematics of each type of AHT systems studied in this work are depicted in this section associated with the details of the predefined temperature, pressure and concentration in P–T–x diagram. All the plotted properties of LiBr–H_2O solution in vapour–liquid equilibrium states were calculated by the formulas provided in Pátek and Klomfar's work [35].

The following assumptions were predefined to simplify the calculation:

1. The analysis was carried out at steady-state conditions;
2. The solutions at the outlets of generator and absorber were at saturated state;
3. The refrigerants, water, at the outlets of condenser and evaporator were at saturated state;
4. The pumping work was neglected; therefore, the temperature and enthalpy of the fluid before and after the pump remained unchanged;
5. Throttling did not change the enthalpy of the fluid;
6. No heat loss from any well-insulated components.

SAHT

The schematic diagram of SAHT (SAHT-1) in Fig. 1(a) consists of a generator, a condenser, an evaporator, an absorber and a heat exchanger while the first two components are at low pressure level and the last three are at high pressure level. The thermodynamic cycle in P–T–x diagram is exemplified with 50 °C waste heat and 20 °C condensation temperature as shown in Fig. 1(b).

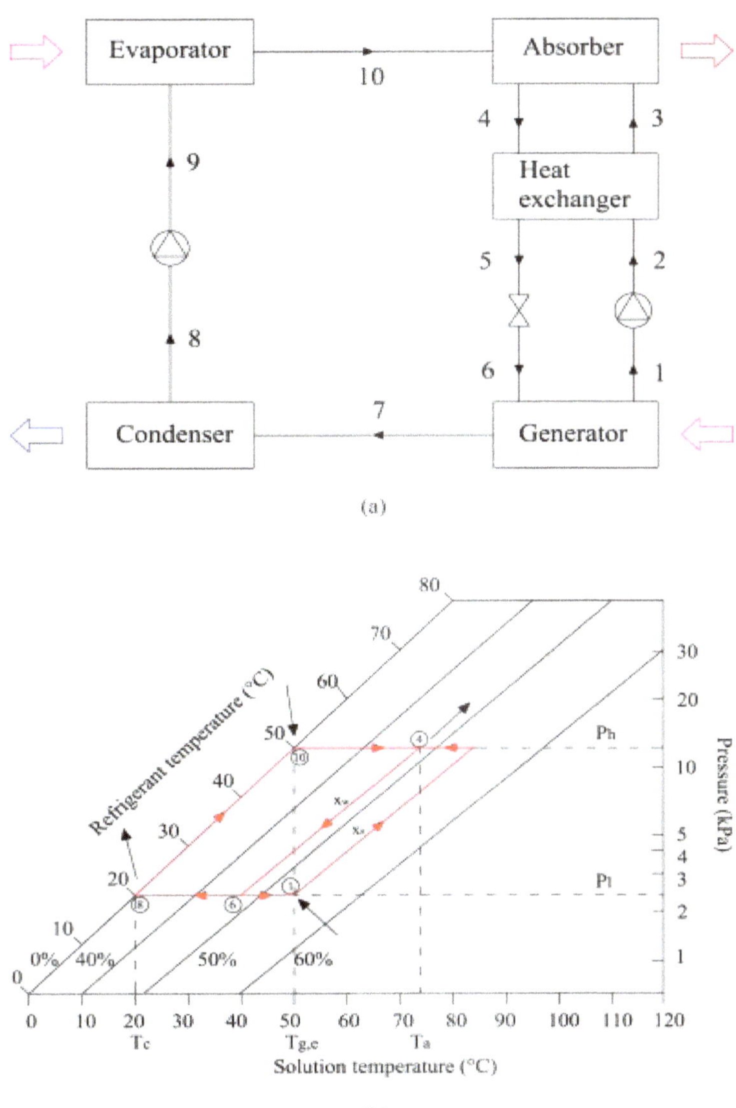

Figure 1. SAHT. (a) Schematic diagram; (b) P–T–x diagram.

The waste heat is used to separate the binary mixture in the generator by partially evaporating the refrigerant, i.e. the water, so that the solution within the generator becomes richer in LiBr and heads for the absorber. On its way to the absorber, the strong solution is firstly pumped to higher pressure (stream 1–2) and subsequently pre-heated (stream 2–3) by the return weak solution coming down from the absorber (stream 4–5) which is at higher temperature. The desorbed refrigerant vapour (point 7) from the generator is liquefied in the condenser (stream 7–8) and pumped up to higher pressure level (stream 8–9) before entering into the evaporator. The evaporator is supplied with waste heat again to generate higher pressure refrigerant vapour (point 10) which afterwards is thoroughly absorbed by the strong solution in the absorber so that the exothermic absorption process releases heat at a temperature higher than the heat source temperature due to the higher pressure. The weak solution from the absorber should be throttled (stream 5–6) after heat recovery to charge the low pressure generator.

According to the predefined assumption, the refrigerant at point 8 and point 10 are saturated liquid and saturated vapour, respectively, so that their enthalpies (h_8 and h_{10}) can be determined solely by the condensation temperature (T_c) and evaporation temperature ($T_{g,e}$). The refrigerant enthalpy remains unchanged through the pump, i.e.h_9 is equal to h_8. The enthalpy value of refrigerant at point 7 (h_7) can be calculated by considering both the refrigerant pressure and temperature equal to the generator pressure and temperature, respectively. The waste heat input to the evaporator (Q_e) can be calculated by Eq. (1).

$$Q_e = \dot{m}_r (h_{10} - h_9) \tag{1}$$

where $\dot{m}_r$ is the mass flow rate of refrigerant.

Considering the saturated states of refrigerant at point 8 and point 10, the condensing temperature T_c can be used to determine the low working pressure (P_l) across the generator and the condenser, while the waste heat temperature imposing on the evaporator determines the high working pressure (P_h) that dominates the evaporator and the absorber. Since the solution is saturated at point 1, the strong solution concentration (x_s) is then considered as the saturated solution concentration under the conditions of pressure P_l and temperature $T_{g,e}$. The recirculation flow ratio (FR) as defined in Eq. (2)should be given to determine the weak solution concentration (x_w), the flow rates of the weak solution ($\dot{m}_w$) and strong solution ($\dot{m}_s$) while combining with mass balance equations, Eqs. (3) and (4), the already attained x_s and given $\dot{m}_r$.

$$FR = \frac{\text{mass flow rate of solution leaving generator}}{\text{mass flow rate of vapour leaving generator}} = \frac{\dot{m}_s}{\dot{m}_r} \tag{2}$$

$$\dot{m}_s x_s = \dot{m}_w x_w \tag{3}$$

$$\dot{m}_r + \dot{m}_s = \dot{m}_w \tag{4}$$

Thereafter, the upgraded temperature in the absorber (T_a) can be conveniently determined as it is at intersection point of high pressure (P_h) and the weak solution concentration (x_w) as shown in Fig. 1(b).

The enthalpy of the saturated strong solution at the outlet of the generator (h_1) can be determined by its temperature ($T_1=T_{g,e}$) and the low pressure (P_l); by the same way, the outlet weak solution enthalpy (h_4) from the absorber can be determined by its temperature ($T_4=T_a$) and the high pressure (P_h). For the heat exchanger, the inlet thermal states of the counter-flow solution streams become known because the temperature and enthalpy values at point 2 equal to those of point 1 according to the assumption. With a given performance parameter UA, i.e. the multiplicative product of heat transfer coefficient and heat transfer area of the heat exchanger, the outlet temperature (T_3 and T_5) and enthalpy (h_3 and h_5) of the weak and strong solutions, those that have heat exchanging can be iteratively determined through energy equations, Eq.(5) and Eq. (6), the logarithmic temperature difference ΔT_{LMTD} in Eq. (7) and the correlation of solution enthalpy with temperature and concentration.

$$Q_{HE} = \Delta T_{LMTD} \cdot UA \tag{5}$$

$$Q_{HE} = \dot{m}_s h_3 - \dot{m}_s h_2 = \dot{m}_w h_4 - \dot{m}_w h_5 \tag{6}$$

$$\Delta T_{LMTD} = \frac{T_4 - T_3 - (T_5 - T_2)}{\ln\left(\dfrac{T_4 - T_3}{T_5 - T_2}\right)} \tag{7}$$

Then, the enthalpy of the inlet weak solution (h_6) to the generator is known as the same with that of point 5 (h_5). With all known variables on the right-hand side of energy balance equations (Eq. (8) and Eq. (9)), the heat input to the generator (Q_g) and the heat generated in the absorber (Q_a) can be obtained.

$$Q_g = \dot{m}_s h_1 + \dot{m}_r h_7 - \dot{m}_w h_6 \tag{8}$$

$$Q_a = \dot{m}_s h_3 + \dot{m}_r h_{10} - \dot{m}_w h_4 \tag{9}$$

Finally, the coefficient of performance (COP) and exergy COP (COPe) of the SAHT cycle can be calculated by Eq. (10) and Eq. (11), respectively.

$$COP = \frac{Q_a}{Q_e + Q_g} \tag{10}$$

$$COPe = \frac{Q_a\left(1 - \dfrac{T_c}{T_a}\right)}{(Q_e + Q_g)\left(1 - \dfrac{T_c}{T_{g,e}}\right)} \tag{11}$$

DAHT-1

Based on the design of SAHT, one more heat exchanger and one more absorber are added in to form the DAHT-1 as illustrated in Fig. 2(a), where three pressure levels are established to gradually elevate the temperature of the waste heat. The two absorbers are at different pressures, as tagged low pressure absorber (L-Absorber) and high pressure absorber (H-Absorber), respectively. DAHT-1 basically integrates two single effect cycles by splitting refrigerant stream from the condenser into two sub-streams, one of which is pumped to the evaporator (stream 14-15-16-17) before arriving in the L-Absorber while the other sub-stream is directly heading to the H-Absorber (stream 14-18-19-20) through a relatively more powerful pumping. Likewise, the strong solution from the generator is also split, one part of which is less pumped, heated and received by the L-Absorber (stream 7-8-9), while the rest is strongly pumped, heated and received by the H-Absorber (stream 1-2-3). The first sub-stream of the refrigerant evaporates and mixes with partial strong solution in the L-Absorber whereby the absorption heat is released at a temperature higher than the waste heat source; the second sub-stream refrigerant boils by this upgraded heat in the L-Absorber so that the generated vapour is at an even higher pressure, which subsequently thoroughly mixes with the strong solution in the H-Absorber to achieve further heat upgrading.

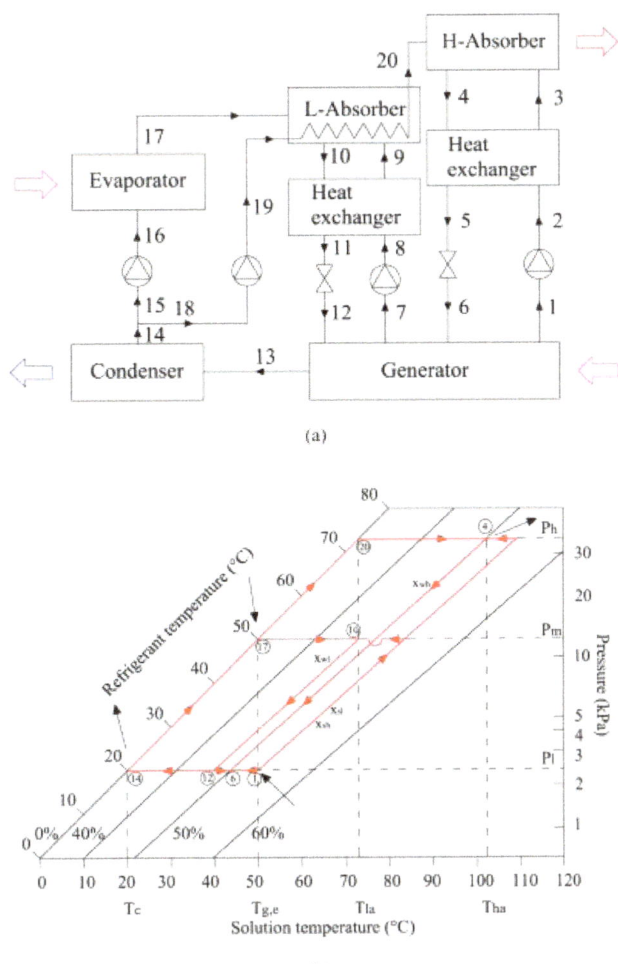

Figure 2. DAHT-1. (a) Schematic diagram; (b) P–T–x diagram.

The low pressure, P_l, and medium pressure, P_m, can be firstly determined by the condensation temperature and waste heat temperature as shown in Fig. 2(b). Four different solution concentrations are involved in this cycle, including strong solution concentrations at different pressure levels, x_{sh} and x_{sl}, and weak solution concentrations, x_{wh} and x_{wl}, at different pressure levels, in the high and low pressure loops, respectively. x_{sh} and x_{sl} could have the same value if the generator is heated uniformly, and they can be straightforward determined since they are at the intersection point of P_l and $T_{g,e}$; whereas x_{wh} and x_{wl} are unnecessarily the same but depend on the recirculation flow ratios of both high pressure loop (FR_h) and low pressure loop (FR_l), and another key parameter – the flow fraction (f_r) of the refrigerant in high pressure loop to the total refrigerant flow. All the other unknown parameters can be solved through the following steps with an iterated calculation:

1. The value of f_r is hypothesised at 0.5 (or any other value between 0 and 1), then the flow rates of the two refrigerant streams, $\dot{m}_{rl}$ and $\dot{m}_{rh}$, are obtained as $\dot{m}_r$ is already given;
2. The x_{wl}, x_{wh}, $\dot{m}_{wl}$, $\dot{m}_{wh}$, $\dot{m}_{sl}$ and $\dot{m}_{sh}$ are obtained based on the given FR_l and FR_h, the already known x_{sl}, x_{sh} and $\dot{m}_{rl}$, $\dot{m}_{rh}$ with the aid of mass balance equations in both L-Absorber and H-Absorber (Eqs. (2), (3) and (4));
3. The absorption temperature in L-Absorber ($T_{la}=T_{10}$) can be determined from Fig. 2(b) as it is at the intersection point of P_m and x_{wl};

4. The high pressure loop of the refrigeration is heated by L-Absorber (stream 19–20) and is ideally assumed to achieve the same temperature of absorption ($T_{20}=T_{la}$), then the saturation pressure of refrigerant at T_{la} is considered as the high pressureP_h in H-Absorber, as shown in Fig. 2(b);
5. Therefore, the absorption temperature in H-Absorber ($T_{ha}=T_4$) can be decided byP_h and x_{wh};
6. The Eqs. (12) and (13) are used to calculate the absorption heat released by L-Absorber ($Q_{L\text{-}Absorption}$) and the evaporation heat required by the refrigerant evaporation for H-Absorber ($Q_{H\text{-}Evaporation}$), respectively, and these two should have the same value if luckily the hypothetical f_r value is equal to the correct f_r value. If not, a new f_r should be re-hypothesised by considering the gap between $Q_{L\text{-}Absorption}$ and$Q_{H\text{-}Evaporation}$ and steps (2) to (6) should be repeated to calculate new $Q_{L\text{-}Absorption}$ and $Q_{H\text{-}Evaporation}$. The iteration will stop when reaching the convergence as $Q_{L\text{-}Absorption}$ equal to $Q_{H\text{-}Evaporation}$.

$$Q_{L\text{-Absorption}} = \dot{m}_{17}h_{17} + \dot{m}_9 h_9 - \dot{m}_{10}h_{10} \tag{12}$$

$$Q_{H\text{-Evaporation}} = \dot{m}_{20}(h_{20} - h_{19}) \tag{13}$$

DAHT-2 and DAHT-3

DAHT-2 and DAHT-3 have similar configuration, as presented in Figs. 3(a) and 4(a). Three different solutions are involved in DAHT-2 and DAHT-3, named as strong solution (x_s), medium solution (x_m), and weak solution (x_w). DAHT-2 and DAHT-3 have only one more L-Absorber compared to the SAHT. In DAHT-2, as shown in Fig. 3(a), part of the medium solution from the H-Absorber feeds into the L-Absorber (stream 4-8-9) as the absorbent that absorbs the refrigerant vapour to form the weak solution; meanwhile the rest part of the medium solution from the H-Absorber flows back to the generator (stream 4-5-6-7). In DAHT-3, as shown in Fig. 4(a), the absorbent in the L-Absorber derives from the part of the strong solution leaving the generator (stream 3-8-9). Their corresponding P–T–x diagrams are shown in Figs. 3(b) and 4(b) respectively.

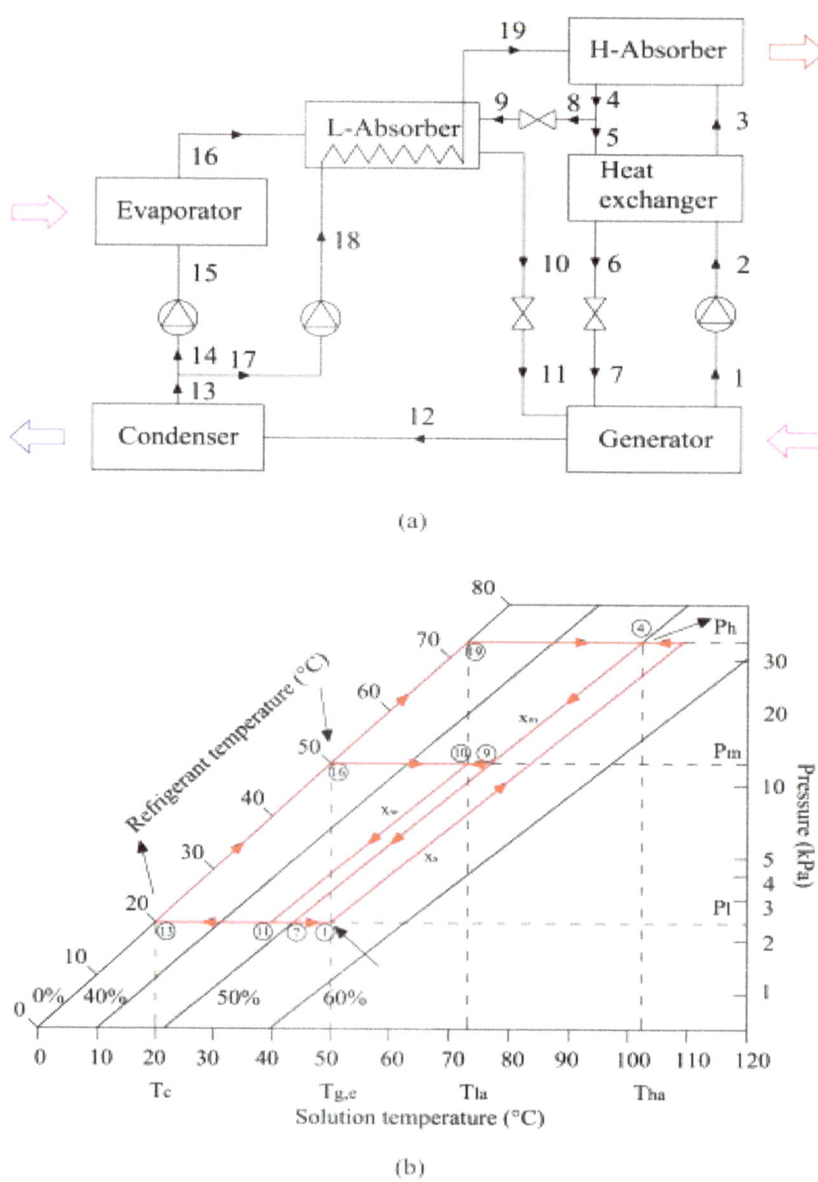

Figure 3. DAHT-2. (a) Schematic diagram; (b) P–T–x diagram.

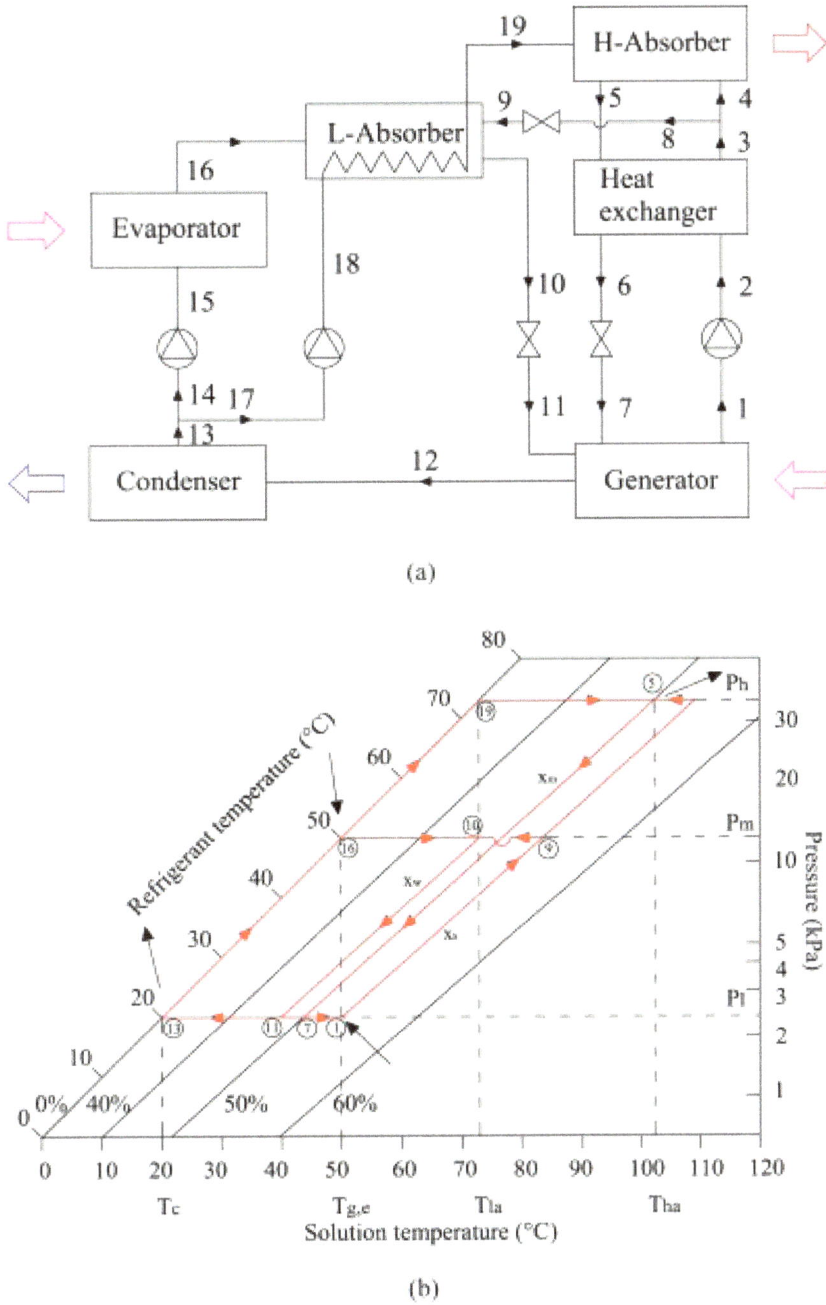

Figure 4. DAHT-3. (a) Schematic diagram; (b) P–T–x diagram.

For both DAHT-2 and DAHT-3, P_l, P_m and x_s are determined by condensation temperature T_c and waste heat temperature $T_{g,e}$. Aside from all the variables defined in the SAHT cycle, the flow fraction, f_w, of the weak solution leaving the L-Absorber to the total flow rate of solution flowing back to the generator is another indispensable parameter to solve all the puzzles of DAHT-2 and DAHT-3 cycles. The flow rates of the strong solution, the medium solution and the weak solution can be calculated with the given FR, f_w and $\dot{m}_r$. Similarly to the calculation steps for DAHT-1, the value of the flow fraction f_r is firstly set to be 0.5 to initiate the iterated calculation in this work until the final convergence is reached. The flow rates of two streams of the refrigerant are calculated firstly based on the hypothetical f_r, then the concentrations of the medium and the weak solutions can be determined by using the mass balance equations in the H-Absorber and the generator, respectively. The absorption temperature in the L-Absorber, T_{la}, is determined by the intersection point of P_m and x_w; while the high pressure P_h is the saturation pressure of refrigerant at T_{la} and the absorber temperature in the H-Absorber, T_{ha}, is at the intersection point of P_h and x_m in Figs. 3(b) and 4(b). The value of f_r must be iterated until achieving the thermal balance in L-Absorber using the similar equation as Eqs. (12) and (13).

DAHT-4

DAHT-4 shown in Fig. 5(a) has simpler configuration in terms of flow control: no splitting of solution is involved as all the medium solution coming from the H-Absorber feeds into the L-Absorber to conduct absorption (stream 5-6-7), afterwards it becomes weak solution and flows back to the generator (stream 8-9-10).

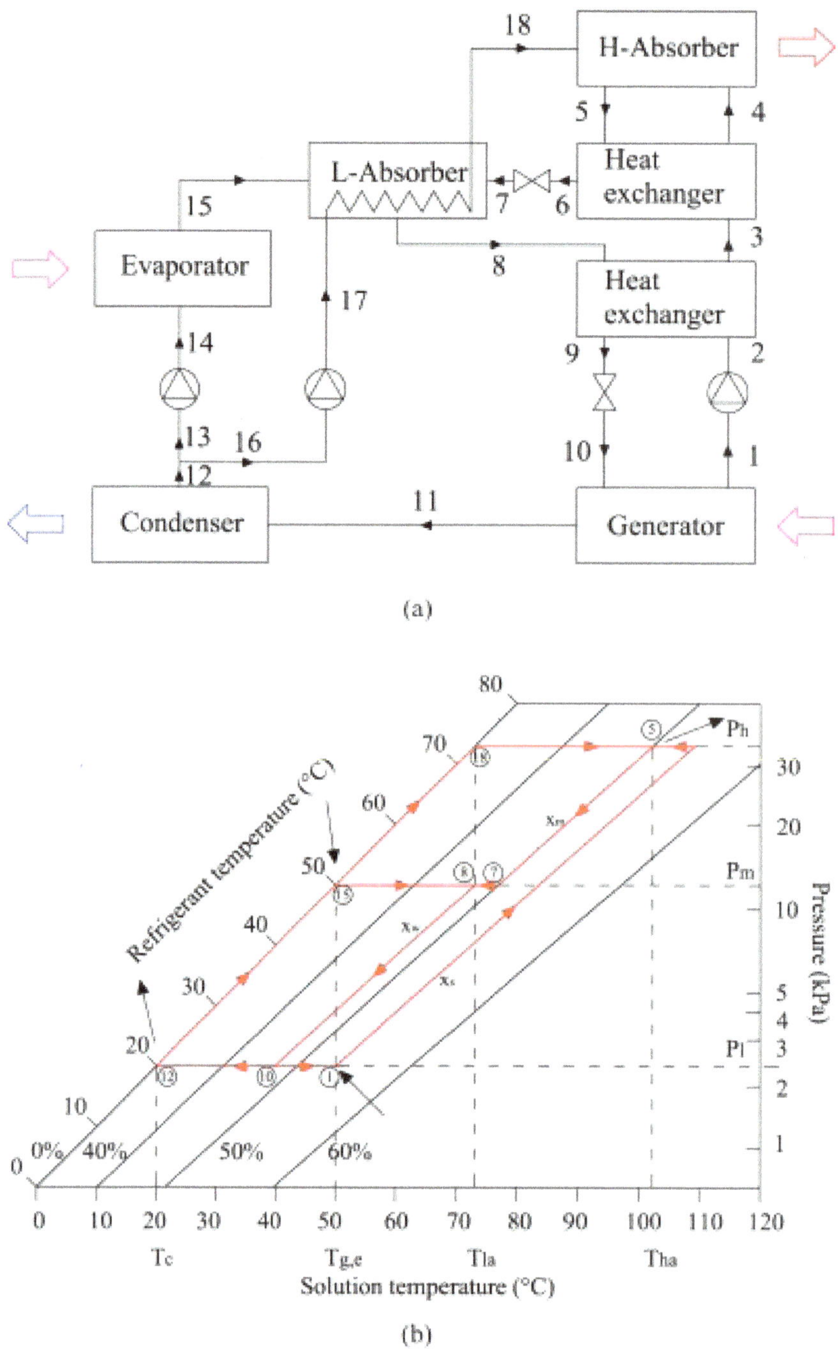

Figure 5. DAHT-4. (a) Schematic diagram; (b) P–T–x diagram.

The low pressure (P_l) and medium pressure (P_m) are determined by the equilibrium state of refrigerant at T_c and $T_{g,e}$, then x_s and x_w, $\dot{m}_s$ and $\dot{m}_w$ are calculated based on P_l, $T_{g,e}$, predefined FR, $\dot{m}_r$ and the mass balance in generator. Thereafter, the absorption temperature in L-Absorber (T_{la}) can be determined by P_m and x_w as shown in Fig. 5(b). The high absorption pressure P_h is then the saturation pressure of refrigerant at T_{la}. With an initial hypothetical value of f_r, the mass balance equation in H-Absorber is used to calculate x_m and $\dot{m}_m$, thus the high absorption temperature T_{ha} becomes known. Eventually, the entire cycle can be established by identifying the f_r through the iteration to achieve the thermal balance in L-Absorber.

DeAHT

The double effect AHT (DeAHT) shown in Fig. 6(a) has two generators at two different pressure levels instead of two absorbers in DAHTs. The waste heat is used to heat the generator at the high pressure (H-Generator) and herein generate the refrigerant vapour out from the medium solution; then the condensation heat of the vapour is used to heat up the weak solution (stream 11–12) in the low pressure L-Generator to generate low pressure refrigerant vapour. Afterwards, the cooled down refrigerant passes through a throttle device (stream 12–13) and then converges with the refrigerant vapour from L-Generator within the condenser (stream 13,14–15). The weak solution from the only absorber directly heads to the L-Generator (stream 7-8-9-10) that discharges enriched solution (the medium solution) to the H-Generator (stream 1-2-3), while the H-Generator drains the strong solution to the absorber (stream 4-5-6).

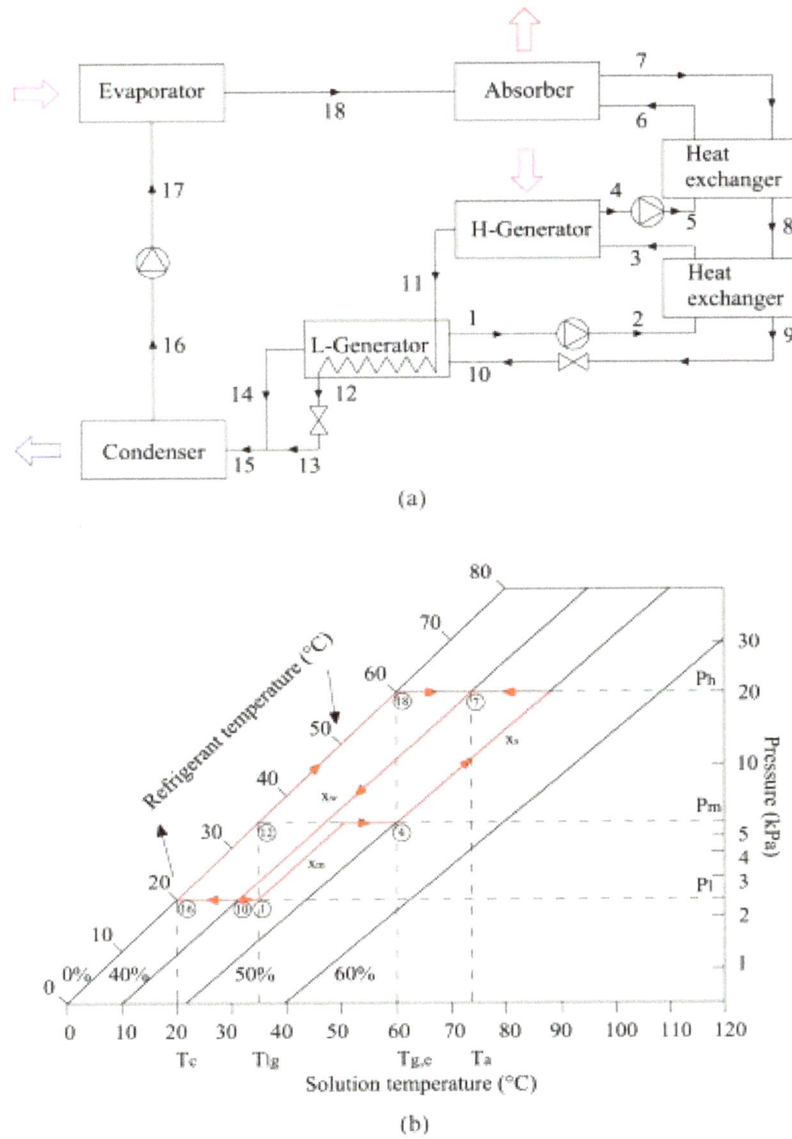

Figure 6. DeAHT. (a) Schematic diagram; (b) P–T–x diagram.

The DeAHT cycle cannot operate if the so-called strong solution is not strong enough, neither the waste heat temperature is not high enough. The present study has used a given strong solution concentration, x_s, as a start to solve all the other solution and refrigerant states. The pressures, P_l, P_m and P_h, can be determined by the saturated state of the refrigerant at T_c, the saturated state of the strong solution at $T_{g,e}$ and the saturated state of the refrigerant at $T_{g,e}$, respectively, as shown in Fig. 6(b). The temperature in the L-Generator, T_{lg}, should be the condensing temperature of the refrigerant vapour from the H-Generator, which is the refrigerant saturation temperature at pressure P_m. Thereafter, the concentration of the medium solution, x_m, is the saturation solution concentration at P_l and T_{lg}. Different from aforementioned DAHTs, the recirculation flow ratio, FR (here is the mass flow rate ratio of the strong solution to refrigerant flowing into the absorber), should be then iteratively approached to satisfy the thermal balance in the L-Generator by the following steps: with a given value of $\dot{m}_r$ and an initially assumed FR, the strong and weak solution flow rates in the absorber can be calculated; then the weak solution concentration x_w is determined by the mass balance in the absorber; a new FR can be attained by solving the thermal balance in L-Generator; with the precondition of $x_w < x_m$ (x_w and x_m are determined independently), the FR value is adjusted and iterated until acquiring a correct FR meeting all the conditions; thereafter all the other states can be worked out.

Input Parameters and Optimisation Process

The input parameters for all the calculations for different AHTs studied in this work are presented in Table 1. The mass flow rate of refrigerant, $\dot{m}_r$, is the essential variable for the full set of calculation on system performance, which was assumed to be 0.005 kg/s in this work as suggested in reference [36] for a 10 kW level absorption refrigeration system. It was found that the temperature lift is insensitive to the changes of $\dot{m}_r$ value; and the impact of the varying $\dot{m}_r$ on system COP and COPe would be offset if the heat transfer parameter UA of the heat-recovery heat exchanger increases or reduces at the same ratio as the $\dot{m}_r$ does because of the proportional changes of the involved heat in each component. Reference [36] also suggested a typical design value of recirculation flow ratio, FR, as 10.84, and used a UA value of the heat-recovery heat exchanger as 132 W/K for one of the presented examples. Hence, the UA analysed in this work ranges from 50 W/K to 450 W/K, while the FR was set in the range of 2–20 to cover the full potential region of system operation.

Table 1. Input parameters for the calculations.

	SAHT	DAHT-1	DAHT-2	DAHT-3	DAHT-4	DeAHT
$\dot{m}_r$ (kg/s)	0.005	0.005	0.005	0.005	0.005	0.005
T_c (°C)	10–30	20	20	20	20	10
FR (-)	2–20	FR_l: 1–10	2–20	2–20	2–20	-
		FR_h: 1–10				
UA (W/K)	50–450	2*75	150	150	2*75	2*75
f_w (-)	–	–	0.1–0.7	0.1–0.7	–	–
x_s (-)	–	–	–	–	–	0.48–0.6

The SAHT system performance was fully evaluated with each input parameter varying in a given range presented in Table 1 so that the most dominant factor for AHT systems could be identified. Thereafter, the influence of this factor on DAHTs and DeAHT were further discussed under the condition of 40~60 °C heat source. The relation of the energy efficiency and upgraded temperature were depicted for each heat source temperature to illustrate the influence of the varying parameters and to determine the optimal operation conditions.

RESULTS AND DISCUSSIONS

The calculation method has been validated with experimental results of SAHT reported in literature [14], as the calculated performance in this work and experimental results are compared in Fig. 7(a). The experiment was conducted under the conditions of T_g = 76.5~78.5 °C, T_e = 53.0~59.8 °C and T_c = 35.8~37.5 °C; COP_{int} was defined as the internal COP, of which the heat was calculated based on the enthalpy change of the solution and refrigerant, while the external COP, COP_{ext}, was calculated based on the heat released or absorbed by the external heat transfer fluids. The present calculation used the same operational conditions as the experiment, and the calculated COP shows good agreement with the experimental COP_{int} as can be seen in Fig. 7(a) when the FR is about 5.4–5.8.

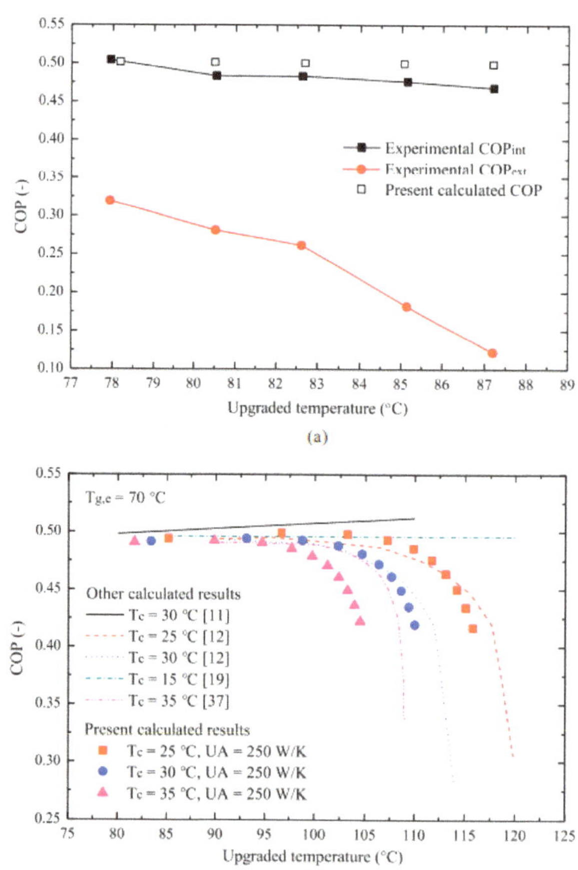

Figure 7. (a) Validation of the calculated method by experimental results of SAHT [14] with T_g = 76.5–78.5 °C, T_e = 53.0–59.8 °C and T_c = 35.8–37.5 °C; (b) comparison between present calculated COP with others [11],[12], [19] and [37], $T_{g,e}$ = 70 °C.

The COP value calculated in this work is also compared with other analysis results reported previously in references [11], [12] and [19], and 37, as shown in Fig. 7(b), while the generation temperature and the evaporation temperature at 70 °C were used in all compared cases. Kurem and Horuz [11] and Yin et al. [19] obtained almost constant COPof around 0.495~0.515 even when the temperature lift was as high as 40~50 °C. The COP results in this work show the similar variation profile against the upgraded heat temperature to those reported by Rivera et al. [12] and Zhang and Hu [37], and the COPvalues are almost the same with those in reference 12 when the UA value is assumed as 250 W/K. The COP values have no prominent change if the upgraded heat is at relatively low temperature, as shown in Fig. 7(b) from 80 °C to 95 °C, or to 100 °C, or to 105 °C, depending on different condensing temperature. Once exceeding these ranges, the COPvalues drop steeply afterwards.

SAHT

The performance of SAHT is dominated by several parameters, including waste heat temperature $T_{g,e}$, condensation temperature T_c, recirculation flow ratio FR, and the performance of heat exchanger UA.

Fig. 8 shows the calculated temperature lift and COP against the varying condensation temperature T_c from 10 °C to 30 °C at the condition of FR = 10 and UA = 150 W/K. The figure shows a temperature lift of 8.3~24.9 °C with 40 °C heat source, 16.3~34.7 °C with 50 °C heat source and 25.1~45.7 °C with 60 °C heat source. The lower condensation temperature is, the higher temperature lift can be by using a fixed waste heat temperature, which can be explained by that a lower condensation temperature allows a lower pressure in the condenser and the generator so that the waste heat can deal with stronger solution in generator, resulting in the relatively stronger weak solution in the absorber, and then the absorption temperature is higher that indicates a larger temperature lift. Unlike the temperature lift, COP value changes in a narrow range of 0.47~0.485 and is hardly affected by the condensation temperature, e.g. the fluctuation of COP value is basically less than 0.01 when the condensation temperature ranges from 10 to 30 °C regardless of waste heat temperature, which can be explained by the following reasons: (1) the flow rates of strong and weak solutions remain unchanged due to the same FR and $\dot{m}_r$; (2) the concentration difference between strong and weak solutions hardly change due to the same FR that ensures a constant value of the ratio of x_s to x_w; (3) therefore both the input heat to the generator and output heat from the absorber change little and have the same changing tendency; (4) the input energy to the evaporator changes little, hence the calculated COP by Eq. (10) is not sensitive to the varying condensation temperature.

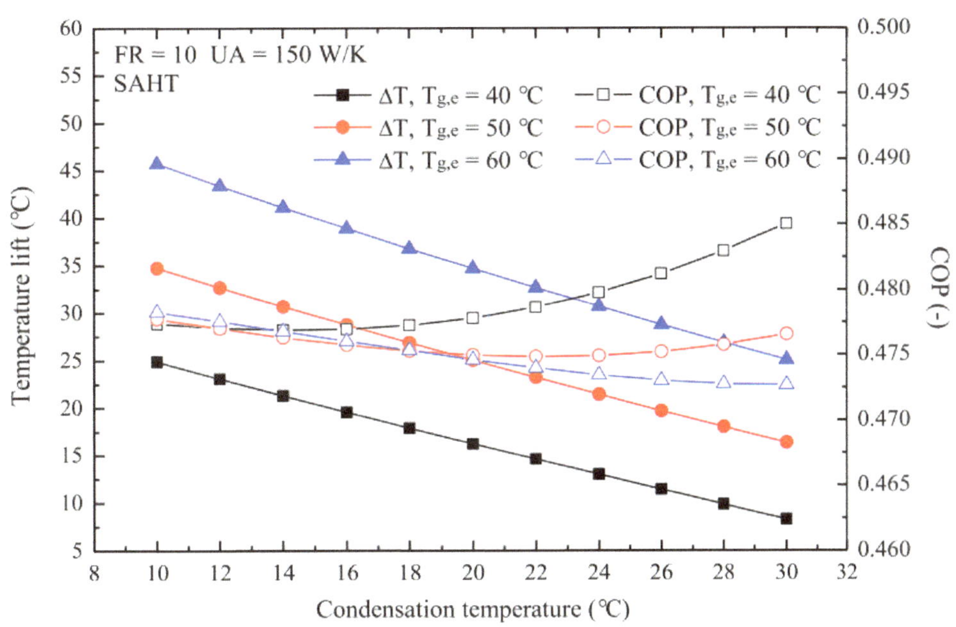

Figure 8. Temperature lift and COP with varying condensation temperature in SAHT, while FR = 10 and UA = 150 W/K.

The most direct and crucial parameter that simultaneously affects both the temperature lift and COP is the recirculation flow ratio, FR, as illustrated in Fig. 9. A higher FRsuggests a smaller concentration difference between the strong solution and the weak solution as can be derived from Eqs. (2), (3) and (4), thus the weak solution concentration is relatively higher given that the strong solution concentration is fixed. Higher concentration of weak solution indicates higher absorption temperature (T_a) in the absorber at the same pressure (P_h), i.e. larger temperature lift. Therefore, as FRincreases in the range of 2–20 while the condensation temperature and heat exchanger performance are fixed at 20 °C and 150 W/K, the temperature lifts of the waste heat at 40 °C, 50 °C and 60 °C display increasing curves within the range of 6.9~18.9 °C, 9.7~29.2 °C and 13.1~40.1 °C, respectively. Moreover, the decreasing concentration difference between strong and weak solution causes the reduction of the heat input required by the generator and heat output from the absorber, though flow rates of both solutions increase with increasing FR; nevertheless, the heat input to the evaporator remains constant with increasing FR due to the constant refrigerant flow rate and evaporation temperature. Hence the COP calculated by Eq. (10) reduces as the FR value increases, e.g. the COPis around 0.397 when FR is at 20 and the waste heat is at 60 °C, which is about 20% lower than the maximum COP value (0.494).

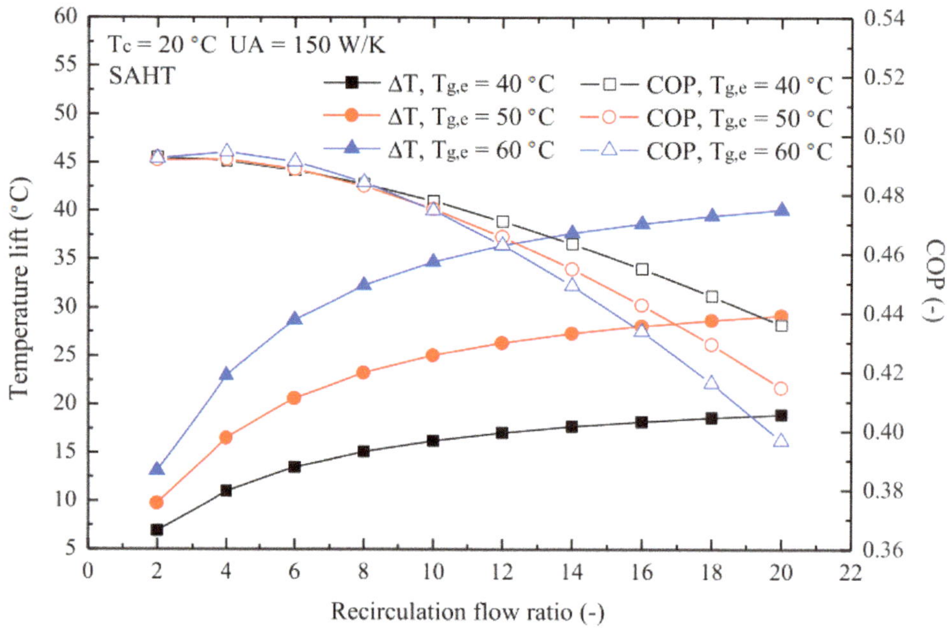

Figure 9. Temperature lift and COP with varying recirculation flow ratio in SAHT, while T_c=20 °C and UA=150 W/K.

The heat exchanger located between generator and absorber is deployed to enhance the system efficiency. The performance parameter of the heat exchanger, UA, has no impact on the temperature lift but is influential on the COP. A higher UA leads to better heat recovery, i.e. more heat input to the generator and heat output from the absorber; with constant heat input to the evaporation, the COP shows upward trend with the increasingUA as shown in Fig. 10, e.g. the COP of using 60 °C waste heat increases from about 0.452 to 0.492 as the UA increases from 50 W/K to 450 W/K when FR is at 10 and the condensation temperature is at 20 °C.

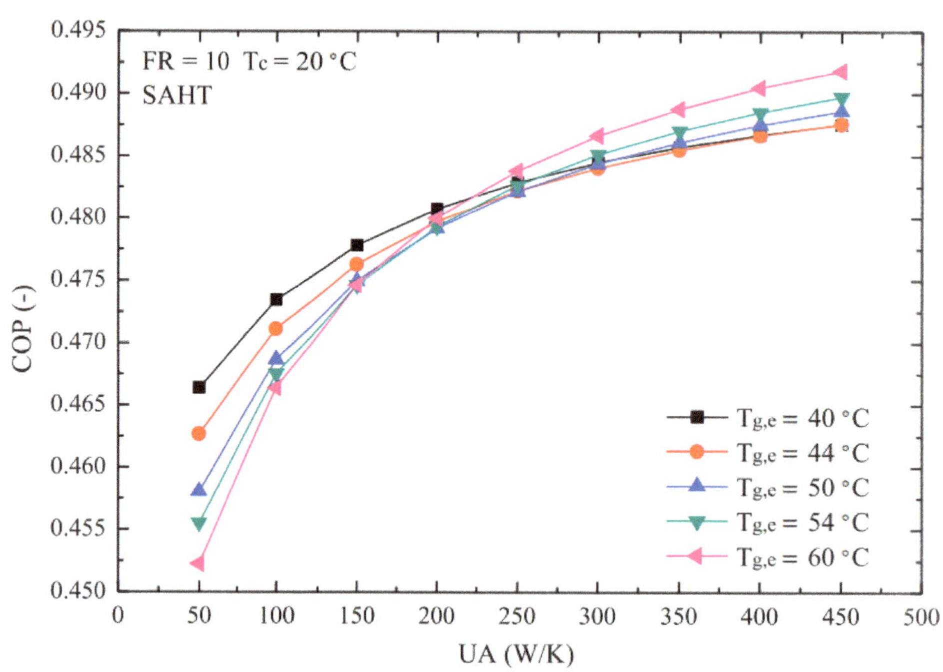

Figure 10. COP with varying heat exchanger performance in SAHT, while FR=10 and T_c=20 °C.

Fig. 11 shows the exergy coefficient, COPe, of SAHT against the upgraded temperature under the conditions of T_c=20 °C and UA=150 W/K. The values of FR are denoted as dash lines in the figure. As can be seen from the calculation equation, Eq. (11), COPereflects both COP and temperature lift. Although the COP decreases as increasing FR, the COPe surges as FR increases until 10 due to the increasing output heat temperature, followed by reaching its vertex and subsequently declining. Based on the value of COPe, the optimal FR locates between 10 and 12 for SAHT, where the COPe reaches its maximum value of 0.803~0.828 while the corresponding COP is 0.471~0.475 and the temperature lift is 17.1~34.7 °C depending on different waste heat temperature. Of course, one can pursue higher temperature lift by using larger FR if the lower COP is acceptable, vice versa.

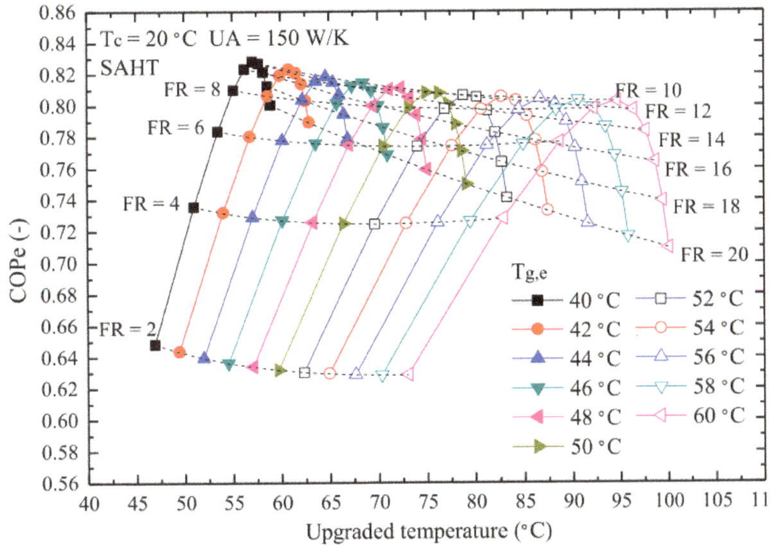

Figure 11. COPe vs. upgraded temperature with varying waste heat temperature and recirculation flow ratio in SAHT, while T_c=20°C and UA=150 W/K.

DAHTs

The performance of the four DAHTs are compared in terms of temperature lift, COP andCOPe as shown in Fig. 12(a) and (b), which present the calculation results of COP andCOPe against the upgraded temperature respectively under the condition of T_c=20°C and UA=150 W/K. The best performance in terms of the highest COPe of DAHT-2 and DAHT-3 are shown in Fig. 12 with the value of f_w varying in the range of 0.1~0.7. For all the studied cases, the optimal f_w was found at around 0.5.

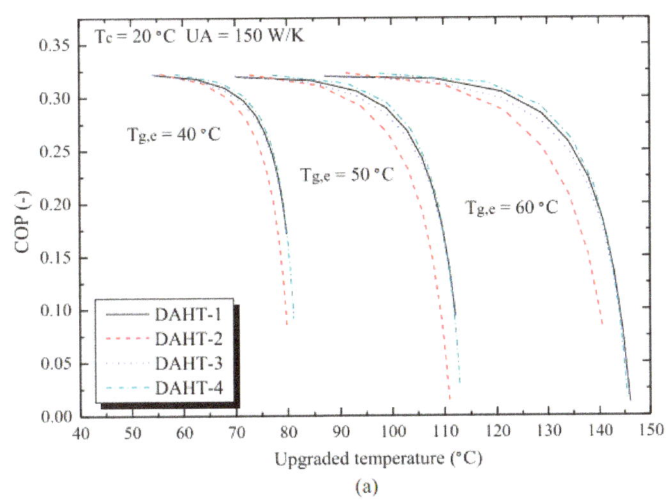

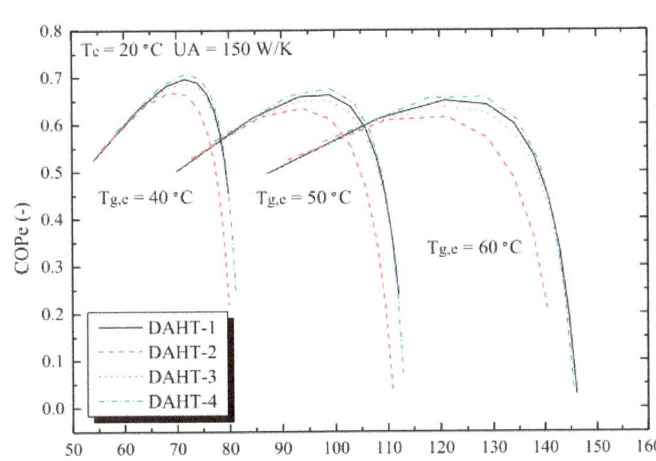

Figure 12. (a) COP and (b) COPe vs. upgraded temperature with 40°C, 50°C and 60°C waste heat in DAHTs, whileT_c=20°C and UA=150 W/K.

As shown in the figure, the DAHT cycles could produce a temperature lift of 14.1~41.2 °C by 40 °C waste heat, 20.0~62.9 °C by 50 °C waste heat and 27.1~86.1 °C by 60 °C waste heat, which are almost twice as large as that produced by SAHT. However, the larger temperature lift implies the compromised system efficiency, e.g. the maximum COPvalue is only about 0.322, whereas the minimum one could be as low as less than 0.01 while the temperature lift reaches the maximum potential. Sharing the similar varying tendency with that of SAHT, the COPe of different DAHTs in Fig. 12(b) has the maximum values from 0.614 to 0.708 depending on different waste heat temperature, which is lower than that of SAHT. Through the comparison shown in the figures, DAHT-4 exhibits the highest values of both COP and COPe among these four different DAHTs, while the DAHT-1 and DAHT-3 have slightly lower efficiencies and the DAHT-2 has the lowest. Additionally, the simplest configuration and solution flow control of the DAHT-4 further strengthens its competitive advantage over the other three.

In Fig. 13, the temperature lift and COP values of the best DAHT, DAHT-4, are plotted against recirculation flow ratio in the range of 2~20. The curves shown in the figure are similar to those shown in Fig. 9 for SAHT. The increase of FR leads to gradually increasing temperature lift but drastic COP reduction, which can be explained by the same reason as that of SAHT. The maximum COP value is around 0.322 regardless of the waste heat temperature. Comparing to SAHT, the impact of FR on the COP is amplified in DAHTs. Since the increase of FR in DAHTs implicates even smaller differences between the concentrations of strong and the medium solutions in the H-Absorber, the output heat from the absorber reduces more steeply than that in SAHT, which sometimes can even drop to null and leads to a COP near 0 as shown in the Fig. 13.

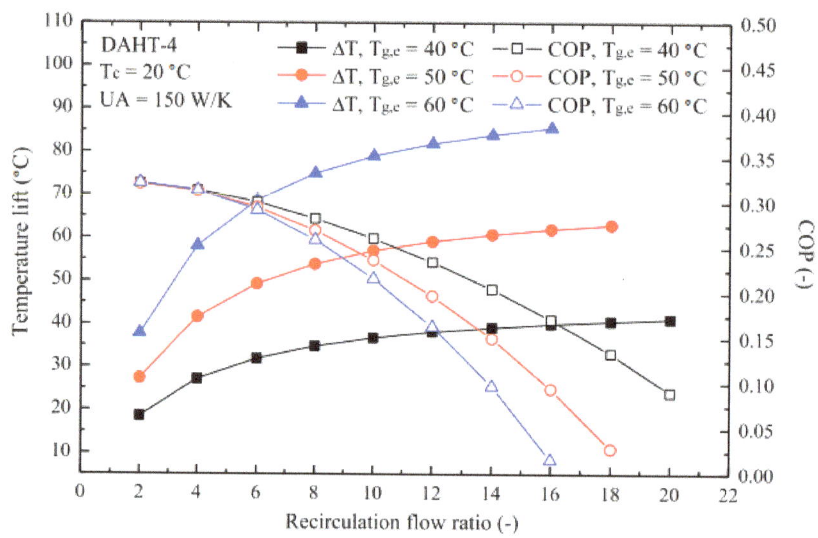

Figure 13. Temperature lift and COP with varying recirculation flow ratio in DAHT-4, while T_c = 20 °C and UA = 150 W/K.

Fig. 14 shows the COPe of DAHT-4 against the upgraded temperature. COPe has the peak value between 0.657 and 0.708 with the optimal FR at the value of 6 for all studied waste heat temperature, where the COP is around 0.292~0.301 and the temperature lift ranges from 31.8 to 68.6 °C.

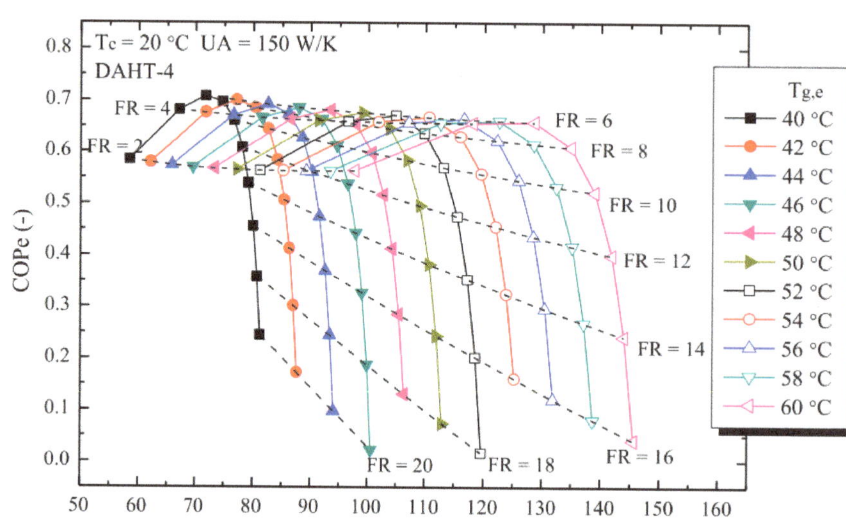

Figure 14. COPe vs. upgraded temperature with varying waste heat temperature and recirculation flow ratio in DAHT-4, while T_c = 20 °C and UA = 150 W/K.

DeAHT

As foregoing process description in Section 2.5, the operational condition in DeAHT is relatively demanding than other cycles studied in this work. It requires both the strong solution concentration and waste heat source temperature to be high enough to achieve the heat balance in the L-Generator and ensure the cycle implementation. In this work, 10 °C condensation temperature and 150 W/K heat transfer performance are applied to the calculation in order to maximise the usage of waste heat in the temperature range. In this instance, the threshold of the strong solution concentration and the waste heat temperature is 0.48 and 44 °C, respectively.

Fig. 15 shows the temperature lift and the COP of DeAHT against strong solution concentration. Due to the double effect generation, the maximum COP value of the DeAHT is about 0.65, which is about 35% and 100% higher than those of SAHT and DAHTs respectively; however, the corresponding temperature lift is much lower than those of SAHT and DAHTs. The DeAHT can only achieve a temperature lift of 0.8~13.4 °C with 44 °C waste heat, 1.8~16.2 °C with 50 °C waste heat and 5.3~24.3 °C with 60 °C waste heat. As an input parameter, higher strong solution concentration causes a lower pressure in the L-Absorber as can be seen in Fig. 6(b), thus relatively lower medium and weak solution concentrations are expected that results in a lower output temperature in absorber and then smaller temperature lift. With higher strong solution concentration, the larger concentration difference among the involved three solutions implies a lower value of FR, e.g. FR decreases from about 12.4 to 0.8 when x_s increases from 0.52 to 0.6 for 60 °C waste heat. However, the reducing FR does not lead to an increasing COP of DeAHT as SAHT and DAHT do. The reason of this contradictory result could be possibly explained as DeAHT requires quite large concentration difference when x_s is high, e.g.x_w = 0.27 and x_m = 0.38 when x_s = 0.6, then the non-linear variation of the LiBr solution enthalpy against solution concentration causes the non-linear changes of input/output heat and generally increasing system COP with x_s.

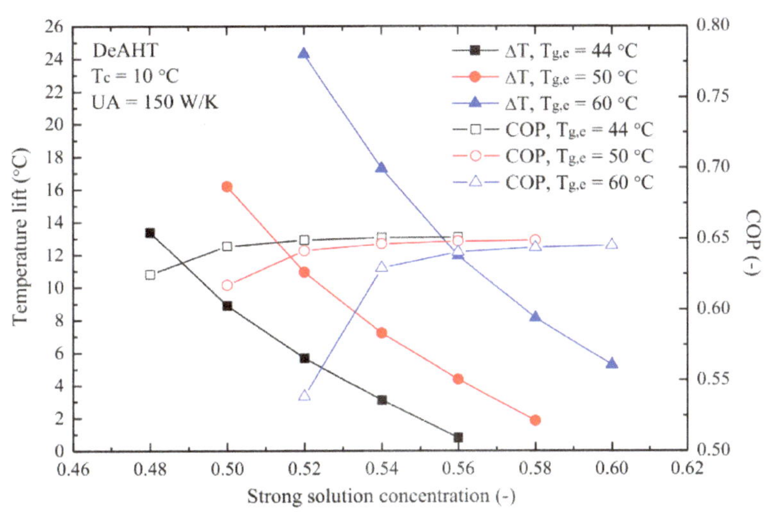

Figure 15. Temperature lift and COP with varying strong solution concentration in DeAHT, while T_c = 10 °C and UA = 150 W/K.

The COPe of DeAHT shown in Fig. 16 is in the range of 0.664~0.835 and is in function of waste heat temperature, upgraded temperature and strong solution concentration. The graphic information suggests each optimal strong solution concentration to obtain maximum COPe value when the waste heat is at 48 °C, 52 °C, 54 °C and 60 °C, respectively. The DeAHT with waste heat source at other temperatures in Fig. 16 has untapped potential under the studied condition due to the aforementioned demanding operational requirement.

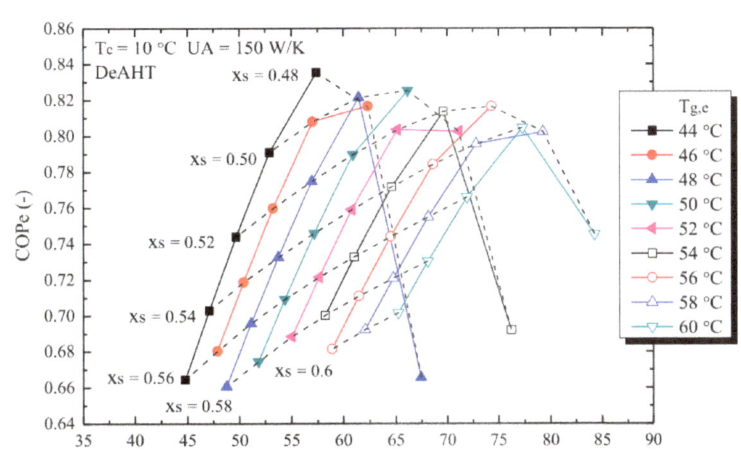

Figure 16. COPe vs. upgraded temperature with varying waste heat temperature and strong solution concentration in DeAHT, while T_c = 10 °C and UA = 150 W/K.

Comparison of AHTs

The performance comparison of SAHT, the best DAHT, i.e. DAHT-4, and DeAHT is presented in Fig. 17(a) to (c) in terms of the temperature lift, COP and COPe, respectively, while Table 2 lists the optimal working performances of these AHTs based on the maximum COPe values.

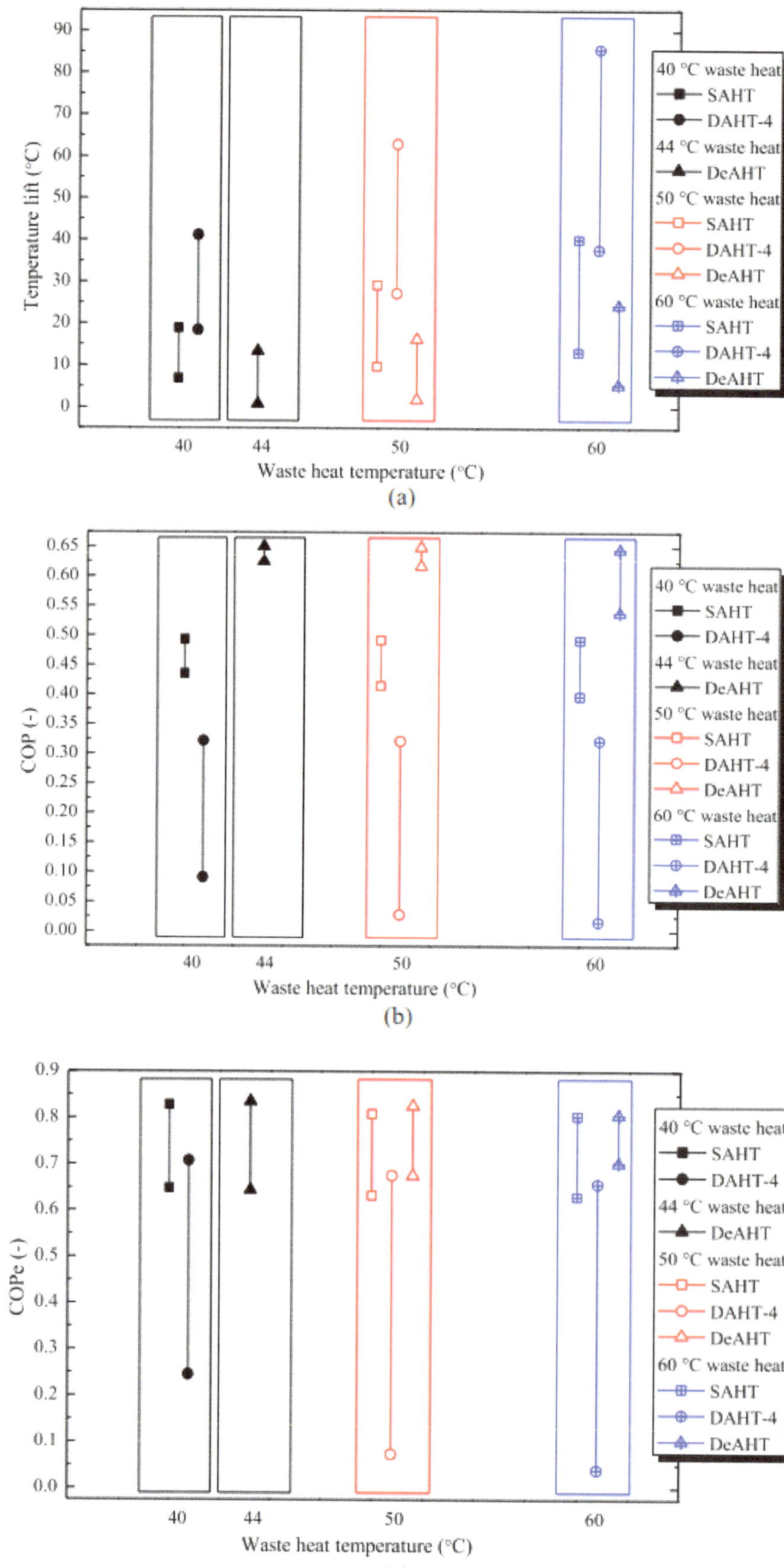

Figure 17. Performance comparison of SAHT, DAHT-4 and DeAHT. (a) Temperature lift; (b) COP; and (c) COPe.

Table 2. The optimal working performances of SAHT, DAHT-4 and DeAHT.

	SAHT			DAHT-4			DeAHT		
	40 °C	50 °C	60 °C	40 °C	50 °C	60 °C	44 °C	50 °C	60 °C
FR (-)	12	10	10	6	6	6	5.86	6.25	4.02
x_s	0.483	0.542	0.591	0.483	0.542	0.591	0.48	0.5	0.54
Temperature lift (°C)	17.1	25.0	34.7	31.8	49.1	68.6	13.4	16.2	17.3
COP (-)	0.471	0.475	0.475	0.301	0.295	0.292	0.625	0.617	0.629
COPe (-)	0.828	0.808	0.803	0.708	0.675	0.657	0.835	0.825	0.805

It is noticeable that the SAHT performance is in between the other two systems, with medium temperature lift as well as medium efficiencies. The DAHT-4 has the largest temperature lift, but with the lowest COP and COPe. The DeAHT exhibits the best heat transformation COP with the lowest temperature lift for the ultralow grade waste heat utilisation studied in this work. It should be noticed that the COPe of DeAHT is not better than that of SAHT. Because of the demanding operational conditions, relatively lower temperature lift, modest COPe and relatively complex configuration, the DeAHT is not recommended for heat transformation of 40~60 °C ultralow grade heat despite of its highest COP.

CONCLUSIONS

The present paper investigated the absorption heat transformers to upgrade the ultralow grade waste heat in the temperature range from 40 to 60 °C. The performance of a single stage, four double stage and a double effect absorption heat transformers were numerically analysed and compared while the analytical methods have been validated with experimental data from other studies and also been constracted with other numerical approaches. The conclusions are summarised as follows.

1. The condensation temperature T_c had significant effect on temperature lift but hardly influenced the system efficiency. The recirculation flow ratio, FR, was found to be the most crucial factor for the AHTs performance. The temperature lift increased while COP decreased with the increasing FR value; there was an optimal value of FR to achieve maximum COPe, which reflected the temperature lift and the COP simultaneously.
2. SAHT yielded a medium temperature lift with medium system efficiency. The optimal value of FR was in the range of 10~12, depending on different waste heat temperature, and that led to the maximum COPe of 0.803~0.828 with the COP value ranging from 0.466 to 0.475 and the temperature lift from 17.1 to 34.7 °C.
3. DAHTs sacrificed the system efficiency to achieve higher temperature lift; DAHT-4 was the best DAHT among those studied in this work, which had the optimal performance of the COPe at 0.657~0.708, the COP at 0.292~0.301 and the temperature lift at 31.8~68.6 °C when FR was at 6.
4. The DeAHT application was limited by the requirements of relatively high input temperature and high concentration of the strong solution. It achieved the highest system COP and COPe compared to the other cycles studied; however, its temperature lift was the lowest. Provided with its other limitations, the DeAHT was considered to be not suitable for the upgrading of ultralow temperature waste heat.

ACKNOWLEDGEMENTS

The authors gratefully acknowledge the support from the IAA project (Impact Acceleration Account) (EP/K503885/1) and IDRIST project (EP/M008088/1), funded by the Engineering and Physical Science Research Council of UK.

REFERENCES

1. Department of Energy and Climate Change. Energy consumption in the UK. 2015.
2. Element Energy Limited. The potential for recovering and using surplus heat from industrial, final report for DECC. 2014.
3. F. Velez, J.J. Segovia, M.C. Martin, G. Antolin, F. Chejne, A. Quijano, A technical, economical and market review of organic Rankine cycles for the conversion of low grade heat for power generation, Renew. Sustain. Energy Rev. 16 (2012) 4175–4189.
4. X. Zhang, M. He, Y. Zhang, A review of research on the Kalina cycle, Renew. Sustain. Energy Rev. 16 (2012) 5309–5318.
5. X.Q. Zhai, M. Qu, Y. Li, R.Z. Wang, A review for research and new design options of solar absorption cooling systems, Renew. Sustain. Energy Rev. 15 (2011) 4416–4423.
6. T.X. Li, R.Z. Wang, H. Li, Progress in the development of solid-gas sorption refrigeration thermodynamic cycle driven by low-grade thermal energy, Prog. Energ. Combust. Sci. 40 (2014) 1–58.
7. H. Fang, J. Xia, K. Zhu, Y. Su, Y. Jiang, Industrial waste heat utilization for low temperature district heating, Energy Policy 62 (2013) 236–246.
8. S. Brueckner, L. Miro, L.F. Cabeza, M. Pehnt, E. Laevemann, Methods to estimate the industrial waste heat potential of regions – a categorization and literature review, Renew. Sustain. Energy Rev. 38 (2014) 164–171.

9. K. Parham, M. Khamooshi, D.B.K. Tematio, M. Yari, Absorption heat transformers – a comprehensive review, Renew. Sustain. Energy Rev. 34 (2014) 430–452.
10. P. Donnellan, K. Cronin, E. Byrne, Recycling waste heat energy using vapour absorption heat transformers: a review, Renew. Sustain. Energy Rev. 42 (2015) 1290–1304.
11. E. Kurem, I. Horuz, A comparison between ammonia-water and water-lithium bromide solutions in absorption heat transformers, Int. Commun. Heat Mass Transf. 28 (2001) 427–438.
12. W. Rivera, R.J. Romero, M.J. Cardoso, J. Aguillon, R. Best, Theoretical and experimental comparison of the performance of a single-stage heat transformer operating with water/lithium bromide and water/CarrolTM, Int. J. Energy Res. 26 (2002) 747–762.
13. W. Rivera, J. Cerezo, Experimental study of the use of additives in the performance of a single-stage heat transformer operating with water–lithium bromide, Int. J. Energy Res. 29 (2005) 121–130.
14. W. Rivera, H. Martinez, J. Cerezo, R.J. Romero, M.J. Cardoso, Exergy analysis of an experimental single-stage heat transformer operating with single water/ lithium bromide and using additives (1-octanol and 2-ethyl-1-hexanol), Appl. Therm. Eng. 31 (2011) 3526–3532.
15. R.M. Barragan, C.L. Herad, V.M. Arellano, R. Best, F.A. Holland, Experimental performance of the water / lithium chloride system in a heat transformer, Int. J. Energy Res. 19 (1995) 593–602.
16. R.M. Barragan, C.L. Herad, V.M. Arellano, Experimental performance of the water/calcium chloride system in a heat transformer, Int. J. Energy Res. 20 (1996) 651–661.
17. R.M. Barragan, V.M. Arellano, C.L. Herad, R. Best, F.A. Holland, Experimental performance of the water / magnesium chloride system in a heat transformer, Int. J. Energy Res. 21 (1997) 139–151.
18. R.M. Barragan, V.M. Arellano, C.L. Herad, R. Best, Experimental performance of the ternary solutions in a heat transformer, Int. J. Energy Res. 22 (1998) 73–83.
19. J. Yin, L. Shi, M.S. Zhu, L.Z. Han, Performance analysis of an absorption heat transformer with different working fluid combinations, Appl. Energy 67 (2000) 281–292.
20. I. Horuz, B. Kurt, Absorption heat transformers and an industrial application, Renew. Energy 35 (2010) 2175–2181.
21. K. Parham, M. Yari, U. Atikol, Alternative absorption heat transformer configurations integrated with water desalination system, Desalination 328 (2013) 74–82.
22. W. Rivera, R. Best, J. Hernandez, C.L. Heard, F.A. Holland, Thermodynamic study of advanced absorption heat transformers – I. Single and two stage configurations with heat exchangers, Heat Recov. Syst. CHP 14 (1994) 173–183.
23. W. Rivera, R. Best, J. Hernandez, C.L. Heard, F.A. Holland, Thermodynamic study of advanced absorption heat transformers – II. Double absorption configurations, Heat Recov. Syst. CHP 14 (1994) 185–193.
24. R.J. Romero, S. Silva-Sotelo, J. Cerezo, First double stage heat transformer (Dsht) in Latin America, Chem. Eng. Trans. 19 (2010) 149–155.
25. S.A. Fartaj, Comparison of energy, exergy, and entropy balance methods for analysis double-stage absorption heat transformer cycles, Int. J. Energy Res. 28 (2004) 1219–1230.
26. J. Ji, M. Ishida, Behavior of a two-stage absorption heat transformer combining latent and sensible heat exchange modes, Appl. Energy 62 (1999) 267–281.
27. H. Martinez, W. Rivera, Energy and exergy analysis of a double absorption heat transformer operating with water/ lithium bromide, Int. J. Energy Res. 33 (2009) 662–674.
28. R.M.B. Reyes, V.M.A. Gomez, A. Garcia-Gutierrez, Performance modelling of single and double absorption heat transformers, Curr. Appl. Phys. 10 (2010) S244–S248.
29. Z. Zhao, F. Zhou, X. Zhang, S. Li, The thermodynamic performance of a new solution cycle in double absorption heat transformer using water/lithium bromide as the working fluids, Int. J. Refrig. 26 (2003) 315–320.
30. I. Horuz, B. Kurt, Single stage and double absorption heat transformers in an industrial application, Int. J. Energy Res. 33 (2009) 787–798.
31. C. Mostofizadeh, C. Kulick, Use of a new type of heat transformer in process industry, Appl. Therm. Eng. 18 (1998) 857–874.
32. Z. Zhao, Y. Ma, J. Chen, Thermodynamic performance of a new type of double absorption heat transformer, Appl. Therm. Eng. 23 (2003) 2407–2414.
33. Z. Zhao, X. Zhang, X. Ma, Thermodynamic performance of a double-effect absorption heat-transformer using TFE/E181 as the working fluid, Appl. Energy 82 (2005) 107–116.
34. R. Gomri, Thermal seawater desalination: possibilities of using single effect and double effect absorption heat transformer systems, Desalination 235 (2010) 112–118.
35. J. Pátek, J. Klomfar, A computationally effective formulation of the thermodynamic properties of LiBr-H2O solution from 273 to 500 K over full composition range, Int. J. Refrig. 29 (2006) 566–578.
36. K.E. Herold, R. Radermacher, S.A. Klein, Absorption Chillers and Heat Pumps. Chapter 6 Single-Effect Water/ Lithium Bromide Systems, CRC Press, 1996.
37. X. Zhang, D. Hu, Performance analysis of the single-stage absorption heat transformer using a new working pair composed of ionic liquid and water, Appl. Therm. Eng. 37 (2012) 129–135.

CHAPTER 5

DYNAMIC COUPLED THERMAL-AND-ELECTRICAL MODELLING OF SHEET-AND-TUBE HYBRID PHOTOVOLTAIC/THERMAL (PVT) COLLECTORS

Ilaria Guarracino [a], Alexander Mellor [b], Nicholas J. Ekins-Daukes [b], Christos N. Markides [a]

[a] Clean Energy Processes (CEP) Laboratory, Department of Chemical Engineering, Imperial College London, South Kensington Campus, London SW7 2AZ, UK

[b] Blackett Laboratory, Department of Physics, Imperial College London, South Kensington Campus, London SW7 2AZ, UK

ABSTRACT

In this paper we present a dynamic model of a hybrid photovoltaic/thermal (PVT) collector with a sheet-and-tube thermal absorber. The model is used in order to evaluate the annual generation of electrical energy along with the provision of domestic hot-water (DHW) from the thermal energy output, by using real climate-data at high temporal resolution. The model considers the effect of a non-uniform temperature distribution on the surface of the solar cell on its electrical power output. An unsteady 3-dimensional numerical model is developed to estimate the performance of such a collector. The model allows key design parameters of the PVT collector to vary so that the influence of each parameter on the system performance can be studied at steady state and at varying operating and atmospheric conditions. A key parameter considered in this paper is the number of glass covers used in the PVT collector. The results show that while the thermal efficiency increases with the additional glazing, the electrical efficiency deteriorates due to the higher temperature of the fluid and increased optical losses, as expected. This paper also shows that the use of a dynamic model and of real climate-data at high resolution is of fundamental importance when evaluating the yearly performance of the system. The results of the dynamic simulation with 1-min input data show that the thermal output of the system is highly dependent on the choice of the control parameters (pump operation, differential thermostat controller, choice of flow rate etc.) in response to the varying weather conditions. The effect of the control parameters on the system's annual performance can be captured and understood only if a dynamic modelling approach is used. The paper also discusses the use of solar cells with modified optical properties (specifically, reduced absorptivity/emissivity) in the infrared spectrum, which would reduce the thermal losses of the PVT collector at the cost of only a small loss in electrical output when the selective coating is applied.

Citation: Ilaria Guarracino, Alexander Mellor, Nicholas J. Ekins-Daukes, Christos N. Markides, Dynamic coupled thermal-and-electrical modelling of sheet-and-tube hybrid photovoltaic/thermal (PVT) collectors, Applied Thermal Engineering, Volume 101, 25 May 2016, Pages 778-795, ISSN 1359-4311, http://dx.doi.org/10.1016/j.applthermaleng.2016.02.056.

INTRODUCTION

Hybrid photovoltaic/thermal (PVT) collectors are devices for the conversion of solar radiation into electrical and thermal energy simultaneously. This solution is particularly interesting in residential applications, where a demand for electricity exists alongside one for low temperature heat. Liquid (water) PVT systems are interesting in cold climate regions such as the UK – the location chosen for this study – where the demand of heating and domestic hot water is almost constant during the year. It is predicted in Refs.[1] and [2] that a PVT system with a 15 m^2 collector area designed for a 3-bedroom house in the UK can cover up to 36% of the demand for hot water and up to 51% of the demand for electricity based on a four-member family.

PVT systems operate mostly under dynamic conditions, particularly where the solar irradiance fluctuates due to cloud coverage. However, most previous studies undertaken with the aims of evaluating the suitability and of assessing the potential of this technology (such as Ref. [1]) were based on quasi-steady approaches, wherein the PVT collector is assumed to operate in steady state, while other system components with a larger thermal mass – such as the hot-water tank – have a dynamic response to the time-varying inputs. A few studies (such as Ref. [3]) did take into account the dynamic response of the collector but did not go as far as estimating the temperature gradients on the PV module and the dynamic analysis is limited to a daily simulation. Similarly, in the studies reported in Refs. [4], [5] and [6], dynamic analyses of PVT collectors were also preformed but without an accompanying discussion of the interaction of the collector with other system components. Some recent experimental and numerical studies on the dynamic performance of PVT systems for DHW applications can be found in Refs. [7] and [8]where the demand of hot water was a daily or hourly average input.

Taking the dynamics of the system into account is of fundamental importance when the weather conditions change rapidly. The present paper, in fact, shows that a dynamic model, together with the use of real weather and DHW demand data, is required in order to accurately estimate the energy output of the PVT system. It will be demonstrated, based on both types of model, that quasi-steady solutions deviate significantly from dynamic solutions due to the thermal mass in the system and the inherent variability in the (real) weather data used as inputs to the simulations. Moreover, it will be shown that the use of time-averaged input data leads to an overestimation of the energy generated, as was found in Ref. [9] for a study performed on vacuum tube collectors. A more detailed discussion of these results is reported in Section 4.

In the present study a 3-D PVT collector model is used to estimate the temperature distribution over the PV panel and its influence on the panel's electrical power output. As most of the design parameters can be varied by the user, this numerical model constitutes a tool to optimize the design of sheet-and-tube water collectors focusing on a number of design parameters and operating parameters such as the number of glass covering layers, the material used for glass cover, the spacing between the tubes, the fluid flow-rate and the inlet fluid temperature for various ambient conditions. The numerical model is used to solve an energy balance by taking into account the convective and radiative losses from the collector's top surface and the optical losses due to reflection. The numerical approach is an improvement upon the aforementioned studies in that it takes into consideration the dynamic response of the collector to time-varying climate and demand inputs, and also the 3-D spatial distribution of temperature over the panel. The model can be used to generate results for hourly, daily and annual performance analyses and provide information on its transient performance. The objective is to propose better designs of this hybrid system by quantifying the influence of important design parameters, while demonstrating that a dynamic model is important for the investigation of control strategies and the interaction of the collector with other system components that require a dynamic description.

The details of the PVT collector and the wider system considered in this study are described in Section 2. This section also discusses the role of the PV module as a thermal absorber as well as an electricity generator and focuses on the emissivity of the solar cell. The numerical model is described in Section 3 and the results are reported inSection 4. The latter also contains a comparison of the annual performance of a PVT collector with standard solar cells and a PVT collector using cells with optimized optical properties.

HYBRID PVT COLLECTORS AND SYSTEMS

The greatest part of the absorbed irradiance in a PV module is converted into heat (about 60–70%). This heat is partially rejected to the environment by radiative and convective heat losses, and partially increases the temperature of the solar cell reducing its conversion efficiency [10]. A PVT system aims to improve the overall conversion efficiency of the PV panel by cooling the solar cells. Whilethe rejection of excess heat to the environment is beneficial for a standalone PV module, in a PVT module this low-grade heat is collected by the thermal absorber and recovered by a fluid stream for useful thermally driven processes such as hot water provision, space heating or absorption cooling for domestic and commercial applications.

As mentioned above, the performance of a PV module is strongly dependent on its operating temperature. This introduces additional factors that are in need of consideration. Temperature gradients on the collector surface can significantly affect its electrical efficiency because solar cells operating at higher temperatures generate less power. Therefore, a significant challenge in the design of a PVT collector is in obtaining a uniform temperature distribution over the modules. A sheet-and-tube collector is associated with a non-uniform temperature on its surface during operation. The prediction of this temperature distribution is therefore of crucial importance when selecting the best design and evaluating the thermal and electrical yield of the associated PVT system.

System description: The PVT system modelled in this work is shown in Fig. 1. It comprises a sheet-and-tube PVT collector (three configurations were considered: unglazed, single glazed and double glazed), a storage tank with an auxiliary heater, a bypass branch (as implemented in Ref. [11]), a circulation pump and thermally insulated connecting pipes. The system shown in Fig. 1 is essentially an indirect solar water-heating system in which the fluid heated in the solar collector is circulated to a heat exchanger in the water-storage tank. The bypass branch allows for the recirculation of the hot water to the collector during periods of low irradiance. A differential controller regulates the activation of the bypass branch and the circulation pump. The controller monitors the temperature in the storage tank T_t, the temperature at the collector outlet T_{f-o} and inlet T_{f-in}, the incident solar irradiance G, and the ambient temperature T_a. An external auxiliary heater ensures that the temperature of the delivered hot-water to the domestic user reaches the required value of 60 °C [12]. This configuration ensures the maximum utilization of the solar energy stored in the tank [13]. The system is designed for a terraced house in London with 15 m² roof area available for the installation. The sizing of the system (tank size and array area) was taken from Ref. [1] and the main parameters are reported in Table 1.

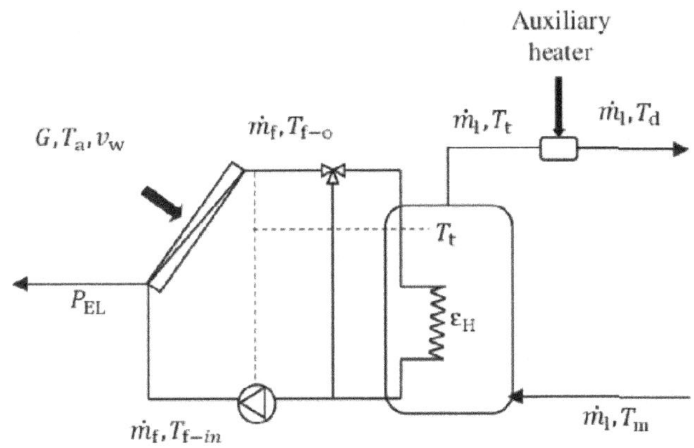

Figure 1. Schematic diagram of a PVT system for the provision of domestic hot water.

Table 1. Design parameters of the PVT system [1].

Parameter	Value
h_{lt}	1.9 W/m² K
M_t	150 kg
N_{coll}	10
T_l, T_m	20 °C, 12 °C
ε_H	0.40

Collector description: The modelled collector is a sheet-and-tube PVT/water module for the generation of electricity and domestic hot-water in the UK. The Powertherm collector from Solimpeks® has been chosen as the reference collector for this study (see Table 2for the collector specification) because of its availability on the UK market. The Powertherm is a single-glazed PVT sheet-and-tube collector with a low-iron glass cover and c-Si solar cells. The copper-sheet thermal absorber has an aperture area of 1.42 m²and is composed of 14 parallel pipes. The nominal operating flow rate is 0.02 kg/s m², which is the recommended flow rate for standard solar thermal collectors [13] and [15]. Other than the parameters reported in Table 2, information on the optical and thermal properties of the materials from which the layers of the PVT collector are manufactured (shown in Fig. 2) are required in order to fully characterize the collector. These values are listed in Table 3.

Table 2. Characteristics of the Powertherm PVT collector [14].

Geometrical parameters		
A_A	Absorber area (m²)	1.43
A_C	Aperture area (m²)	1.42
A_G	Gross area (m²)	1.40
D_o	Risers external diameter (mm)	8
δ_g	Glass cover thickness (mm)	4
δ_A	Absorber thickness (mm)	0.12
δ_i	Insulation thickness and material (mm)	50 (glass-wool), 40 (EPS)
L_1	Gross length (m)	1.66
L_2	Gross width (m)	0.86
N_p	Number of pipes	14
PV cell parameters		
N_p	Number of cells	72
W_p	Nominal power (W)	180
η_{EL} (STC)	Module efficiency (standard conditions) %	12.6
η_{ref} (STC)	Cell efficiency (standard conditions) %	17.8
Thermal characteristics		
η_{TH-0}	Zero loss collector coefficient	0.486
a_1	Heat loss coefficient	4.028
a_2	Heat loss coefficient	0.067

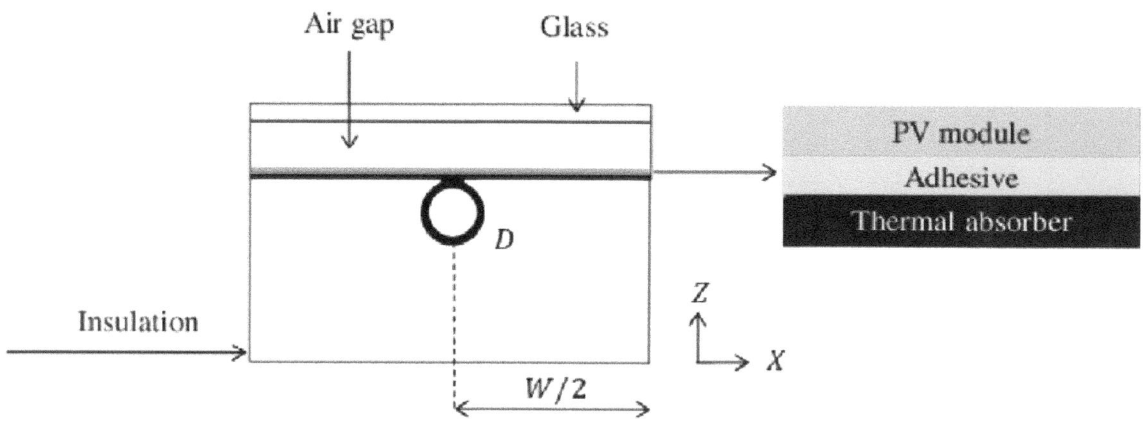

Figure 2. Cross section x-z of a single-glazed PVT module showing the pipe diameter D and the distance between two adjacent pipes W.

Table 3. Optical and thermal properties of the layers.

Layer	Parameter	Value	Unit	Refs.
Glazing	α_g	0.01		[16]
	ε_g	0.90		[16] and [17]
	τ_g	0.95		[16]
	c_g	750	J/kg K	[16]
	k_g	1.80	W/m K	[16]
Solar cell	α_{PV} Turn MathJaxon	0.93		[16]
	ε_{PV} Turn MathJaxon	0.90		[18]
	c_{PV} Turn MathJaxon	677	J/kg K	[16]
	k_{PV} Turn MathJaxon	149	W/m K	[16]
EVA	k_{EVA} Turn MathJaxon	0.35	W/m K	[16] and [17]
Adhesive	k_{gl} Turn MathJaxon	0.85	W/m K	[1] and [17]
Tedlar	k_{TED} Turn MathJaxon	0.20	W/m K	[1] and [17]
Absorber	c_A	385	J/kg K	[3]
	k_A	310	W/m K	[19]
Insulation	k_i	0.04	W/m K	[3]

A cross section on the x-z plane of the modelled PVT collector is presented in Fig. 2. The collector comprises: one or more glass covering layer(s), a PV module, a thermal absorber (aluminium or copper plate) in thermal contact with copper riser tubes, and a layer of thermal insulation. Amongst the various designs of liquid/water PVT systems, the single-cover sheet-and-tube design appears to be a particularly promising design for domestic applications [17]. The thermal efficiency of a flat-plate water PVT module based on a sheet-and-tube collector design is typically reported as being between 50% and 60% [6] and [20] at zero reduced temperature, while the annual electrical/photovoltaic efficiency is reported as being 7–15% [20], [21], [22], [23], [24] and [25].

The thermal and electrical efficiencies of a PVT collector are influenced mainly by the fluid flow-rate, the number of glass covering layers, the type of solar cells used, and the properties of the thermal absorber (material and geometry, e.g. pipe diameter D, fin-to-pipe diameter ratio W/D) that determine the temperature distribution on the absorber surface. Other than the sheet-and-tube design, a number of alternative designs have been investigated. Zondag et al. [17] analysed seven design concepts for liquid PVT collectors, considering several possible channel designs and also a free-flow design with unrestrained/unconfined fluid flowing above the absorber. All of the channel concepts investigated were found to have a slightly higher electrical efficiency than the sheet-and-tube designs due to a more uniform temperature distribution on the solar cell, while the free-flow panel was found to have a reduced efficiency due to the formation of condensate on top of the glass layer, causing additional reflection losses. On the other hand, simplicity of manufacture and cost are equally important considerations affecting the design selection.

In low temperature applications, uncovered designs allow for a higher electrical efficiency (due to reduced optical losses) while single-glazed or double-glazed designs allow for higher thermal efficiencies and higher fluid temperatures (due to reduced convection losses) at the cost of lower electrical efficiencies (due to increased optical losses and reduced PV conversion at the higher temperatures [26]). The presence of a single glass cover reduces the optical efficiency by around 5% as a consequence of reflection and transmission losses at the cover [18]. Thermal losses can also be reduced by the presence of an evacuated layer or a layer filled with a gas (e.g. Argon as discussed in Refs. [27] and [28]) together with spectrally selective glazing coatings to reduce the infrared radiation losses.

The PV module is composed of a top transparent surface (iron glass), the solar cell (c-Si), an encapsulant (ethyl vinyl acetate (EVA)), and a rear layer (Tedlar). The top glass in a PV module guarantees the required rigidity to the laminate [18]. In a PVT collector the required rigidity is already given by the thermal absorber, thus the top glass is not required, whereas some additional protection covers without mechanical rigidity but with optimal optical properties might be used, as discussed in Ref. [29].

The heat collected by the PV module is transferred to the thermal absorber by conduction, thus a good thermal contact between the PV layer and the absorber is essential. An effective solution is to laminate the whole package of top cover, PV cells and absorber together in one step. In this case an electrically insulating foil can beinterposed between the PV cell and the absorber in the lamination process, or an electrically insulating coating can be applied to the absorber top surface [30]. More often a thin layer of thermally conducting and electrically insulating adhesive material is used. Examples of adhesive materials found in the literature are:

1. Silicon adhesive (0.5 mm tick) for application in sheet-and-tube PVT with polymer absorber is used in Ref. [31].
2. Corrugated copper foil is used in Ref. [32] in a PVT with a plastic thermal absorber.
3. Aluminium-oxide-filled two-component epoxy glue was used in Ref. [33] in the construction of a sheet-and-tube PVT collector.
4. Silver-filled glue was used in the work of Ref. [34].

PV radiative losses: The provision of thermal energy for domestic applications can be enhanced if the design and operation of the PVT collector is optimized to maximize heat collection. The PV layer also acts as the thermal absorber, and thus the fraction of solar irradiance converted into thermal energy is related to the absorption factor of the solar cells. However, commercial PV modules are designed to minimize the absorption of excess heat in order to maintain a low operating temperature. In fact, mono-crystalline c-Si cells, the type considered in this study and the most commonly used for commercial applications in flat-plate PVT modules, can only absorb and convert to electricity solar irradiance at wavelengths below 1.1 μm [35].

The thermal efficiency of a PVT collector is limited by the optical properties of the solar cell, as the absorber does not have the same emissivity as a selective absorber typically used in a solar-thermal collector. The thermal efficiency of the PVT collector is expected to be lower than that of a conventional thermal collector due to: (i) the lower absorption factor of the absorber; (ii) the direct conversion of part of the incident solar irradiance into electricity, which reduces the proportion that is available in the form of heat; and (iii) the higher radiative heat losses from the absorber to the glass cover due to a higher emissivity of the solar cell compared to the emissivity of a conventional thermal absorber. The radiative losses can be suppressed by using a spectrally selective low-emissivity coating with a low emissivity in the infrared spectrum. Such a coating must be reflective at wavelengths at which thermal emission occurs (3–20 μm), but transmissive at solar wavelengths (0.3–3 μm) to allow solar radiation to be efficiently absorbed by the collector. This characteristic behaviour is exhibited by In_2O_3 : Sn (ITO) films, for example, which are typically used in energy-efficient windows for this purpose [36] and [37]. An ideal coating would cause the solar absorber to have the absorptivity/emissivity shown inFig. 3, i.e. $\varepsilon_\lambda = \alpha_\lambda = 1$ over the range of solar wavelengths and $\varepsilon_\lambda = \alpha_\lambda = 0$ over the range of emission wavelengths.

A spectrally selective absorber for solar-thermal applications is characterized by a high absorptivity over the visible spectrum and a low emissivity in the infrared and near-infrared spectrum. Such characteristics would maximize the absorption of the incident irradiance while minimizing the radiative thermal losses. The PV module in a PVT panel has a high absorptivity in the visible and infrared spectra, and a high emissivity in the infrared. Thus high radiative losses from the module enable a lower operating temperature of the solar cell and higher conversion efficiency. The absorptivity/emissivity reported in Fig. 3 has been calculated from experimental measurements of reflection on a commercial solar cell. The emissivity of the solar cell (thick red solid line) in the near infrared spectrum ($\lambda > 3$ µm) is at least 0.7, if not higher at some wavelengths, while it is zero for an ideal thermal absorber. The absorption of photons with energy less than the band-gap energy is mainly free carrier absorption, and this absorption coefficient is proportional to the carrier concentration (to the doping of the intrinsic silicon). While the absorption coefficient of the intrinsic silicon would be near zero for energies less than the band gap, this value increases after doping due to the presence of the free carrier[35] and [38]. The free carrier absorption does not lead to the generation of an electron-hole pair and it constitutes a parasitic absorption process in solar cells that is beneficial for PVT applications [39].

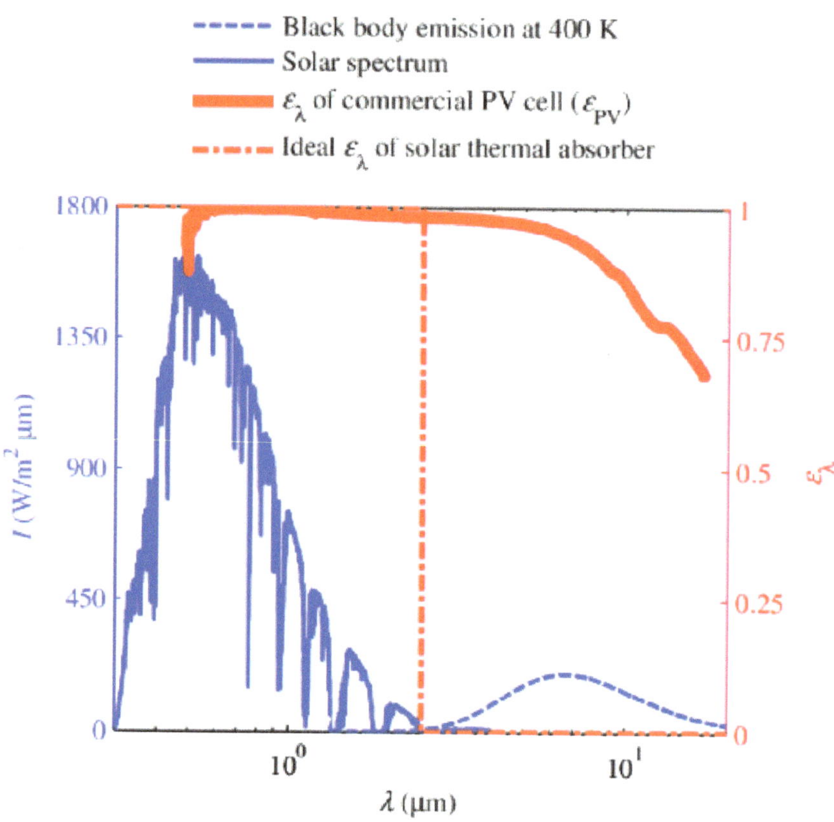

Figure 3. Emissivity ε_{PV} of a commercial silicon solar cell over the approximate range 0.3–20 µm in the visible to infrared spectrum (thick red solid line) compared with the emissivity of an ideal solar thermal absorber (red dot-dashed line). The emissivity is plotted over the incident solar spectral irradiance I (blue solid line) and the emission of a black body at 400 K (blue dashed line).

MODELLING METHODOLOGY

The numerical model developed here allows for the evaluation of the thermal and electrical energy generated by the selected PVT system to cover the demands for domestic hot water and electricity of a three-bedroom house in London, UK, and the instantaneous thermal and electrical efficiency of the system. The model uses real weather-data (solar irradiance, ambient temperature and wind speed obtained from a weather station located in London with a 1-min resolution) and a high resolution profile of domestic hot-water demand generated with the software DHWcalc [40]. The use of these high resolution inputs is found to be essential for a correct estimation of the system performance and for the analysis of the response of the system to the control algorithm.

The PVT collector is characterized by its (conventional) thermal-efficiency curve evaluated under steady-state operation and by its time-constant which indicates the response of the collector to a time-varying input. The model of the collector has been validated against experimental data obtained by other authors and against commercially available PVT and thermal collectors.

Collector Model

Modelling overview: The 3-D thermal model developed here is based on the following assumptions:

1. The thermal properties of all solid materials are constant; variations in the properties of air (as a function of temperature) were calculated by using a polynomial fit according to Ref. [41] while variations in the properties of water with temperature were calculated by using the REFPROP library [42] and [43] and it was found that these two variations introduced only minor changes to the results of interest here (e.g. < 1% in both the outlet fluid temperature and thermal output from the collector).
2. The optical properties of all relevant materials are constant.
3. The edges of the collector are well insulated, thus the edge thermal losses are negligible.
4. The temperature profile between two adjacent pipes is symmetrical, and the temperature has a maximum on the symmetry axis.

5. Water flow-rate is evenly distributed between the pipes and the thermal losses and mixing effects at the inlet and outlet manifolds are negligible [5].
6. The flow is fully developed in the tubes.
7. The effect of the friction in the pipes is neglected when calculating the temperature in the riser pipes.
8. The headers cover a small area of the collector and its effect on the temperature distribution on the absorber can be neglected.
9. The incident irradiance G, the wind speed v_w and the ambient temperature T_a are uniform boundary conditions at the surface of the PVT collector.
10. It is assumed that there is no dust or partial shading on the collector.
11. The reflection, absorption and transmission factors are calculated only for the incident solar irradiance [13], [29] and [44].
12. The electrical resistances are neglected when evaluating the electrical energy output and the electrical efficiency.

When the energy balance at the collector is solved in steady-state conditions for different ambient temperatures and incident solar irradiance values, and at a given inlet fluid temperature T_{in}, the result is the thermal efficiency curve of the collector plotted against the reduced temperature $T_r=(T_m-T_a)/G$. In order to estimate the DHW and electricity generated by the system, the model of the PVT collector is integrated within the wider dynamic model of the whole system (Fig. 1). The dynamic energy balance takes into account the masses and specific heat-capacities of each layer forming the PV module and of the mass of water stored in the tank.

PVT unit, Thermal Model Equations

The 3-D dynamic thermal model solves an energy balance equation at each layer of the PVT module. The equations are written for the element (i,j). The same equations can be used for the double-glazed and unglazed PVT collectors, and also for the conventional thermal collector. An energy-balance equation is solved numerically along the water-flow direction, y, and in the transverse direction, x, where each layer is discretized respectively into N_x and N_y number of nodes as shown in Fig. 4. The energy balance equation is solved at each finite volume having size ΔxΔyδ and the solution is a 2-D temperature distribution on the x-y plane over each layer of the PVT module and a 1-D temperature distribution on the x-z plane.

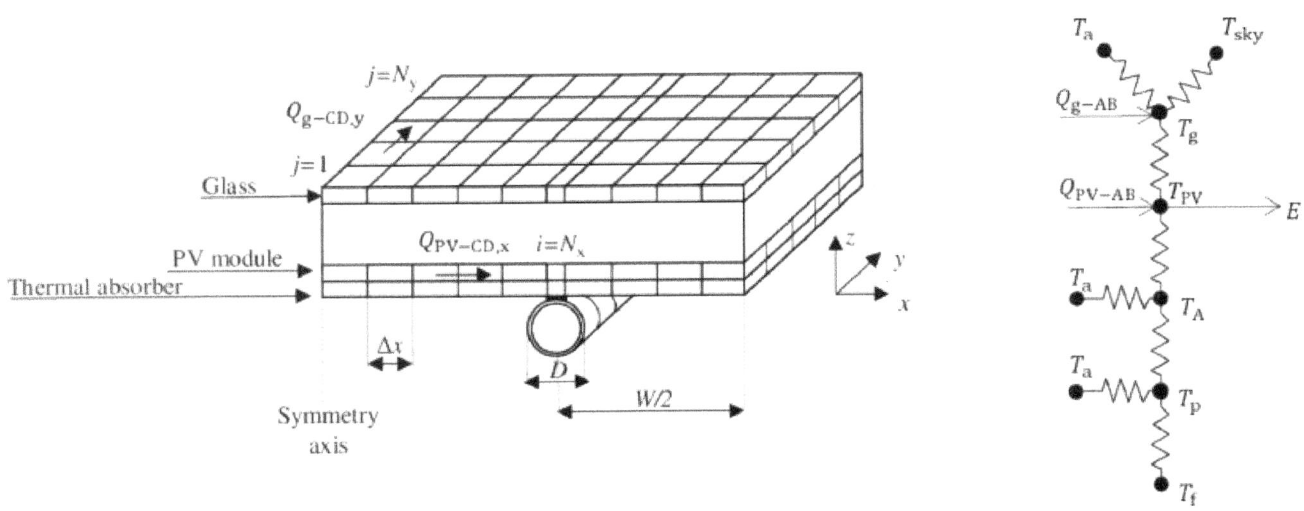

Figure 4. Sketch of the discretization used for the thermal analysis (left) and network of thermal resistances on thex-z plane between the layers of the PVT module (right).

Glass cover energy balance: The energy balance at each node (i,j) of the glass cover in Eq. (1) takes into account thermal radiation and convection losses to the ambient, thermal radiation to the PV module, and absorption of the glass cover.

$$M_g c_g \frac{dT_g(i,j)}{dt} = Q_{g\text{-}CD} - Q_{g,a\text{-}RD}(i,j) - Q_{g,a\text{-}CV}(i,j) + Q_{g,PV\text{-}RD}(i,j) + Q_{g,PV\text{-}CV}(i,j) + Q_{g\text{-}AB}(i,j). \quad (1)$$

The net conductive heat-flux $Q_{ii\text{-}CD}(i,j)$ at the node (i,j) for the layer 'ii' is the sum of the conductive flux in the x direction and the y direction, which are expressed as:

$$Q_{ii\text{-}CD,x}(i,j) = \frac{k_{ii}\delta_{ii}\Delta y}{\Delta x}[T_{ii}(i+1,j) + T_{ii}(i-1,j) - 2T_{ii}(i,j)]; \quad (2)$$

$$Q_{ii\text{-}CD,y}(i,j) = \frac{k_{ii}\delta_{ii}\Delta x}{\Delta y}[T_{ii}(i,j+1) + T_{ii}(i,j-1) - 2T_{ii}(i,j)]. \quad (3)$$

The radiative heat losses to the ambient $Q_{g,a-RD}(i,j)$ in Eq. (4) are calculated using the sky temperature T_{sky}. For clear sky conditions the sky temperature is related to the ambient temperature according to Eq. (5)[45]. Other models relate the sky temperature to the dry bulb and dew point temperatures (T_a and T_{dp}) and to the time of the day t (in hours) counted from midnight in Eq. (6) [46] and [47]. For a cloudy sky, other equations are available, which require an estimation of the cloud coverage based on the diffuse irradiance, while for an overcast day the sky temperature is calculated as in Eq. (7) [48] and [49]. The value of the collector thermal and electrical output is affected by less than 1% by the correlation chosen for calculating the sky temperature, thus Eq. (5) is chosen in this modelling work as it has been widely used in the literature (such as in Refs.[49], [50], [51], [52], [53] and [54]) and it requires no knowledge of the dew point temperature or of the cloud coverage estimation.

$$Q_{g,a-RD}(i,j)=\Delta x\Delta y\varepsilon_g\sigma\left(T_g^4(i,j)+T_{sky}^4\right); \tag{4}$$

$$T_{sky}=0.0552T_a^{1.5}; \tag{5}$$

$$T'_{sky}=T_a\left[0.711+0.0056T_{dp}+0.000073T_{dp}^2+0.013\cos\left(\frac{\pi t}{12}\right)\right]^{0.25}; \tag{7}$$

$$T_{sky}=T_a. \tag{7}$$

The convective heat losses to the ambient air $Q_{g,a-CV}(i,j)$ in Eq. (8) are calculated using the convective heat transfer coefficient h_{top}, as expressed in Eq. (9), which takes into account the forced and free convection on top of the panel [55] and [56]. The forced convection coefficient, h_w, in Eq. (10) varies linearly with wind speed (a review of empirical correlations for h_w is given in Refs. [3], [5], [7], [50], [51], [52], [53], [55], [57],[58], [59] , [60] and [61]). A number of different correlations are available in the literature for the calculation of h_w. These correlations are obtained at various testing conditions and for different geometries and the uncertainty associated to the measured value of the heat transfer coefficient ranges between 6% and 20% [62] and [63]. This large uncertainty is due to the difficulties of making the measurements involving rapid variations of the wind speed and direction and of the incident radiation. The free convection coefficient, h_{free}, in Eq. (11) is a function of Ra (Eq. (13)) calculated on the plate top-surface for the characteristic length $L=A_s/P$ and at the mid-temperature between the mean glass temperature and the ambient $\Delta T=\bar{T}_g-T_a$ [56] and [64]:

$$Q_{g,a-CV}(i,j)=\Delta x\Delta y h_{top}(T_g(i,j)+T_a); \tag{8}$$

$$h_{top}=\sqrt[3]{h_w^3+h_{free}^3}; \tag{9}$$

$$h_w=2v_w+3.8. \tag{10}$$

The correlation used for external free convection in inclined plates is:

$$\overline{h_{free}}=\frac{k_{air}}{L}\left(0.68+\frac{0.67Ra_L^{0.25}}{\left[1+\left(\frac{0.492}{Pr}\right)^{\frac{9}{16}}\right]^{\frac{4}{9}}}\right)\quad \text{for } Ra>10^9; \tag{11}$$

$$\overline{h_{free}}=\frac{k_{air}}{L}Nu^{0.25}\quad \text{for } Ra<10^9; \tag{12}$$

$$\text{with: } Ra=\left(g\cos\left(\frac{\pi}{2}-\gamma\right)\beta\Delta TL^3\rho^2c\right)/k\mu. \tag{13}$$

The radiative and convective heat fluxes between the glass and the solar cells, $Q_{g,PV-RD}(i,j)$ and $Q_{g,PV-CV}(i,j)$, according to Ref. [6], are:

$$Q_{g,PV-RD}(i,j)=\Delta x\Delta y\left(\frac{1}{\frac{1}{\varepsilon_g}+\frac{1}{\varepsilon_{PV}}-1}\right)\sigma\left(T_{PV}^4(i,j)+T_g^4\right); \tag{14}$$

$$Q_{g,PV-CV}(i,j)=\Delta x\Delta y\frac{1}{R_{gap}}\left(T_{PV}(i,j)+T_g(i)\right). \tag{15}$$

The thermal resistance R_{gap} in Eq. (16) accounts for the thin top layers on the top of the PV module (top glass and Tedlar) and for convection in the air gap.

$$R_{gap}=\frac{\delta_{EVA}}{k_{EVA}}+\frac{\delta_g}{k_g}+\frac{\delta_{PV}}{2k_{PV}}+\frac{\delta_g}{2k_g}+\frac{1}{h_{gap}}. \tag{16}$$

The convective heat transfer coefficient h_{gap} in enclosed space is expressed in Ref. [65]as:

$$h_{gap}=\frac{k_{air}}{\delta_{gap}}\left[1+1.44\left(1-\frac{1708}{Ra\cos\gamma}\right)^*\left(1-\frac{1708(\sin1.8\gamma)^{1.6}}{Ra\cos\gamma}\right)+\left(\left(\frac{Ra\cos\gamma}{5830}\right)^{0.33}-1\right)^*\right]. \tag{17}$$

In Eq. (17):

1. The brackets signified by the superscript '*' go to zero when they are negative.
2. γ is the tilt angle of the collector to the horizontal.
3. The thermal conductivity of the air k_{air} is evaluated at the temperature $T_H-\Delta T/2$, where T_H is the temperature of the hot surface and ΔT the temperature difference between the two surfaces.

The solar irradiance absorbed by the glass cover $Q_{g-AB}(i,j)$ is:

$$Q_{g-AB}(i,j)=\overline{\tau\alpha_g}G\Delta x\Delta y. \tag{18}$$

The fraction of the incident irradiance absorbed by the glass cover $\overline{\tau\alpha_g}$ and by the solar cell $\overline{\tau\alpha_{PV}}$ and the reflection losses R_g are included in Eqs. (19), (20) and (21) below[45] and [66]:

$$R_g=r_g+\frac{r_{PV}\tau_g^2}{1-r_{PV}r_g}; \tag{19}$$

$$\overline{\tau\alpha_{PV}}=\frac{(1-r_{PV})\tau_g}{1-r_{PV}\tau_g}; \tag{20}$$

$$\overline{\tau\alpha_g} = 1 - r_g - \tau_g \frac{1 - r_{PV} + r_{PV}\tau_g}{1 - r_{PV} r_g}. \quad (21)$$

PV module energy balance: The energy balance at each node (i,j) of the PV module is:

$$M_{PV} c_{PV} \frac{dT_{PV}(i,j)}{dt} = Q_{PV-CD}(i,j) - Q_{g,PV-RD}(i,j) + Q_{g,PV-CV}(i,j) + Q_{PV-AB}(i,j) - Q_{PV-A}(i,j) - E(i,j). \quad (22)$$

The conduction at node (i,j) is calculated as in Eqs. (2) and (3), while $Q_{g,PV-RD}$ and$Q_{g,PV-CV}$ are given in Eqs. (14) and (15), and Q_{PV-A} is the heat transferred by conduction from the solar cell to the thermal absorber through the layers of EVA, Tedlar and adhesive:

$$Q_{PV-A}(i,j) = \Delta x \Delta y \frac{1}{R_{CD}} (T_{PV}(i,j) - T_A(i,j)); \quad (23)$$

$$R_{CD} = \frac{\delta_{EVA}}{k_{EVA}} + \frac{\delta_{TED}}{k_{TED}} + \frac{\delta_{gl}}{k_{gl}}. \quad (24)$$

The energy absorbed by the solar cell depends on the absorption coefficient α_{PV} of the cell over the solar spectrum, and on the transmission τ_g and reflection ρ_g at the glass cover. The fraction of the absorbed irradiance that is then converted into electricity is calculated as a linear function of the cell's temperature as in Ref. [67]:

$$E(i,j) = \Delta x \Delta y G \eta_T(i,j). \quad (25)$$

In Eq. (25) the conversion efficiency η_T of the incident sunlight into electricity is assumed to decrease linearly with increasing cell operating-temperature T_{PV}. This is a typical assumption that is valid in the range of temperatures of operation of PV modules, and the most common expression is [3], [5], [7], [35], [36], [49], [53], [59], [67], [68], [69], [70], [71],[72], [73], [74], [75], [76], [77] and [78]:

$$\eta_T(i,j) = \eta_{ref}[1 - \beta_{PV}(T_{PV}(i,j) - T_{ref})]. \quad (26)$$

The conversion efficiency decreases linearly with the operating temperature as a consequence of the linear decrease of the open circuit voltage and of the fill factor with the temperature, while the short-circuit current slightly increases. The temperature coefficient β_{PV} in Eq. (26) is mainly a material property, having value of about 0.0045 K^{-1} for crystalline silicon modules [5] and [6]. η_{ref} is generally around 0.17 for c-Si cells[5] and [67].

Thermal absorber energy balance: The energy balance at each node i,j of the thermal absorber is:equation(27)

$$M_A c_A \frac{dT_A(i,j)}{dt} = Q_{A-CD}(i,j) + Q_{PV-A}(i,j) - Q_p(i,j) - Q_{loss}(i). \quad (27)$$

Here, the conduction heat flux Q_{A-CD} at the node (i,j) is calculated as in Eqs. (2) and (3),Q_{PV-A} is given in Eq. (23), Q_p in Eq. (28) is the heat transferred to the pipe and Q_{loss}accounts for the heat losses to the ambient from the rear surface through the insulation. It is assumed that: (i) the pipe wall is at uniform temperature at each node; and (ii) the fluid and the pipe temperatures vary only along the direction of the fluid flow. Q_p is given by:

$$Q_p(i,j) = \Delta x \Delta y k_b / \delta_b (T_A(i,j) - T_p(j)), \quad (28)$$

where k_b and δ_b are the bond thermal conductivity and the bond thickness.

The heat transfer due to heat losses at the rear of the panel Q_{loss} is given by:

$$Q_{loss}(i,j) = \Delta x \Delta y / R_{loss} (T_A(i,j) - T_a). \quad (29)$$

The thermal resistance R_{loss} in Eq. (30) takes into account the conduction through the insulation and the free convection at the rear of the panel calculated by:

$$R_{loss} = \frac{\delta_i}{k_i} + \frac{1}{h_{free}}. \quad (30)$$

Fluid energy balance: The bulk-fluid temperature T_f is calculated by applying the energy balance equation:

$$M_f c_f \frac{dT_f(j)}{dt} = \dot{m} c_f (T_{f-in}(j) - T_{f-o}(j)) + \pi D \Delta y h_f (T_p(i,j) - T_f(j)), \quad (31)$$

where the heat transfer coefficient depends on the flow regime (laminar, or turbulent)[64], according to Eqs. (32) and (33) for natural and for forced circulation. When the pump is not active and the fluid is not circulating in the collector, the heat transfer occurs by conduction between the pipe wall and the centre of the pipe and the heat transfer coefficient is given in Eq. (34).

$$h_f = 4.36 \frac{k_f}{D_H} \quad \text{for } Re < 2300; \quad (32)$$

$$h_f = \frac{k_f}{D_H} 0.23\, Re^{0.8} Pr^{0.4} \quad \text{for } Re > 2300; \quad (33)$$

$$h_f = \frac{2k_f}{D_H} \quad \text{for } \dot{m}_f = 0. \quad (34)$$

Pipe energy balance: The pipe temperature is calculated from:

$$M_p c_p \frac{dT_p(j)}{dt} = \frac{\Delta x \Delta y k_b}{\delta_b}(T_A(i,j) - T_p(j)) - \pi D \Delta y h_f (T_p(i,j) - T_f(j)) - \frac{\pi D}{R_{loss}}(T_p(j) - T_a). \quad (35)$$

Boundary conditions: The boundary conditions required to solve the energy balance are:

$$T_{f\text{-}in(1)} = T_{in}; \quad (36)$$

$$\left.\frac{\partial T}{\partial x}\right|_{x=0,W/2} = 0; \quad (37)$$

$$\left.\frac{\partial T}{\partial y}\right|_{x=0,L} = 0. \quad (38)$$

Thermal and electrical instantaneous efficiency: The instantaneous thermal efficiency, the electrical efficiency and the electrical power output are:

$$\eta_{TH}(t) = \frac{\dot{m} c_f (T_{f\text{-}in} - T_{f\text{-}o})}{GA}; \quad (39)$$

$$P_{EL}(t) = N_p \sum_{i=1,\, j=1}^{i=N_x,\, j=N_y} E(i,j); \quad (40)$$

$$\eta_{EL}(t) = \frac{P_{EL}}{GA}. \quad (41)$$

Storage Tank Energy Balance and Control

The storage tank has been modelled as a fully mixed tank. The tank temperature and the inlet fluid-temperature of the collector are outputs of the energy-balance equation for the storage tank. The energy balance in Eq. (42) accounts for the demand of hot water, the heat losses at the storage tank and the heat delivered from the collector array via the heat exchanger. The temperature at the outlet of the heat exchanger immersed in the storage tank is calculated using Eq. (43) and the energy for supplying domestic hot water is given in Eq. (44), while Eq. (45) is the energy loss to the environment, at room temperature T_l. The demand loop takes water from the top of the tank and replaces it with water at the utility mains temperature of $T_m = 12\,°C$ [1], thus Q_l is zero when $T_t < T_m$.

$$M_t c_t \frac{dT_t}{dt} = Q_{coll} - Q_l - Q_{loss}; \quad (42)$$

$$Q_{coll} = \dot{m}_f c_f \varepsilon_H (T_{f\text{-}o} - T_t); \quad (43)$$

$$Q_l = \dot{m}_l c_l (T_m - T_t); \quad (44)$$

$$Q_{lt} = h_{lt} A_t (T_t - T_l). \quad (45)$$

A discussion of the variations in the results when using fully mixed or stratified designs is included Section 4.4.

Input Ambient Conditions

The ambient conditions appear as inputs to the simulations performed in the present work, which uses measurements of time-varying solar irradiance, ambient temperature and wind-speed. Full information on the fluctuations in the ambient conditions (e.g. of solar irradiance during intermittent cloud cover; see Fig. 5) which appear as inputs to the model can only be captured if a small sampling interval Δt_s is used. In the present case study the smallest sampling interval (highest temporal data resolution) possible was 1-min. The time-resolved data were obtained by the authors over a one-year monitoring period (July 2014 to July 2015) from a weather station located in London, UK. The monitored parameters were measured as follows:

1. Wind speed: measured with a solid-state magnetic sensor having an accuracy of 1 m/s and a range of 0.5 m/s to 89 m/s.
2. Ambient temperature: measured with a PN junction silicon diode, in a range of −40 °C to 65 °C and with an accuracy of 0.5 °C above 20 °C or 1 °C above 20 °C.
3. Solar irradiance: measured with a precision of 5% at full scale up to 1800 W/m². The sensor is mounted on a roof-installed PV system oriented towards the south on the plane of the PV modules.

The ambient data-sampling interval of 1-min is also shorter than half of the time constantτ_c of the single-glazed collector as defined in Eq. (46), which is the shortest of the investigated collectors. In this case, if the fluctuations in the inlet conditions to the model are faster than the system dynamics as characterized by $\tau_c/2$:

$$\Delta t_s < \frac{\tau_c}{2}, \quad (46)$$

the response of (and system outputs from) (quasi-)steady state simulations will deviate significantly from results of equivalent fully dynamic simulations of the same system with the same inputs.

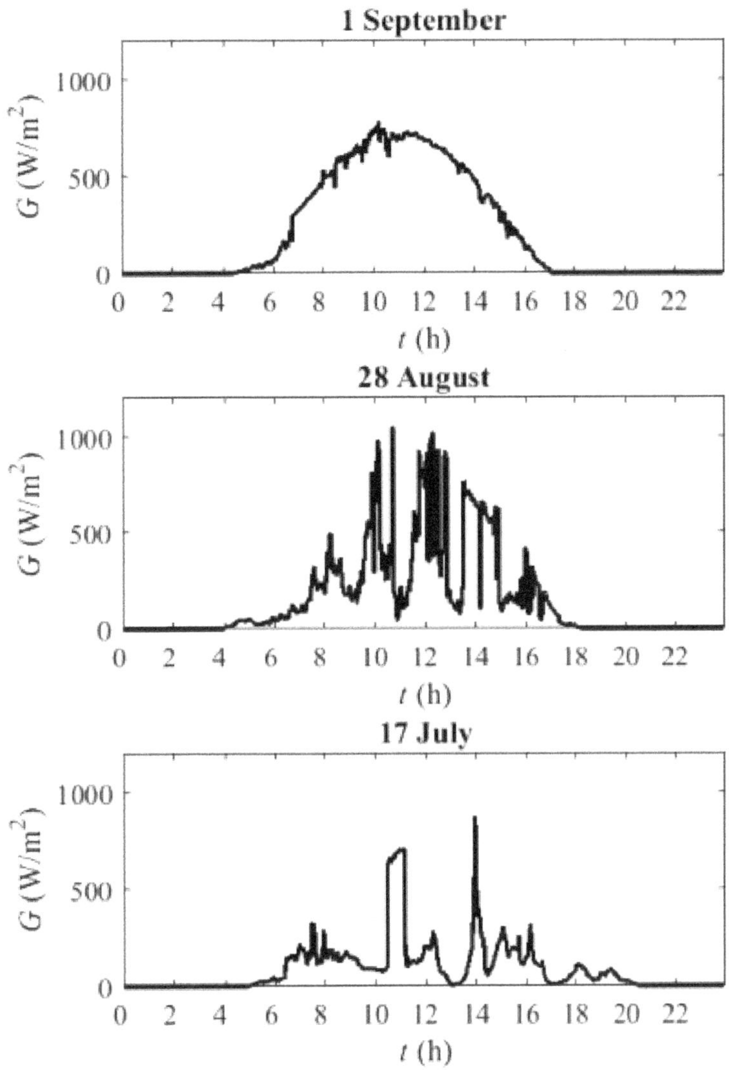

Figure 5. Solar irradiance G data at 1-min resolution collected in London in 2014. Representing three typical sky conditions: clear-sky (1st September), cloudy day/high intermittence (28th August), overcast day (17th July).

Time-averaged input data were also used for evaluating the long-term performance of the PVT system (e.g. as is done in Section 4.1; Fig. 10). This was done in order to quantify the discrepancy of the outputs from the system from the fully time-resolved result. Specifically, the results of an annual simulation obtained using 10-year averaged weather-data at a 30-min resolution have been compared with the results of an annual simulation with 1-min resolution datasets and the results are discussed in Section 4.1. The intention here is to quantify the deviations of data resolution and of using a dynamic versus a quasi-steady model. The 10-year averaged weather-data are the result of a 10-year measurement period and are available online at the Photovoltaic Geographical Information System (PVGIS) [79]. The PVGIS data are global irradiance incident on a fixed plane and ambient temperature, provided as average daily profiles for each month of the year that can be used for the estimation of solar-system performance as in the work presented in Ref. [1].

Hot-Water Demand Data

A time-varying profile of hot-water consumption is required as an input when predicting the energy demand for hot water. The performance of the solar water-heater is sensitive to the load timing and to the load day-to-day variability, as the temperature of the storage tank varies when a water draw-off event occurs. The key parameters that define the profile of hot-water demand are the average daily volume, the yearly total demand, the draw-off flow rates and the daily profile of hot water demand. The profile of hot water demand is driven by the number of occupants (which has a linear effect on the total hot water use [80]), by the appliances, and by the ambient conditions. The latter determine seasonal variations of the energy consumption due to variations in the mains temperature. It has been observed that when the mains temperature is higher, in summer, the mixed temperature required for end-uses such as showers, baths and sinks is obtained with a smaller flow rate of hot water. In the UK this translates to a seasonal variability of −16% / +8% of the demand of the energy required for hot water, having a minimum in July and a maximum in December [81].

In the present study, a statistically realistic distribution of hot-water consumption was generated with the software DHWcalc [40]. This software is based on a code used for the IEA-SHC Task 26 on solar combi-systems and can be used to evaluate and compare the performance of solar combi-systems from different European countries using realistic time-varying input data [82] and [83]. DHWcalc generates random event schedules for hot-water consumption based on:

1. Daily average consumption (for weekdays and weekend days).
2. Seasonal variation of the daily DHW consumption in percentage, described as a sine function of the hot water consumption during the year [40].
3. Flow rate, draw-off duration and portion of the total daily consumption of 4 different demand types (short load, medium load, bath and shower).
4. Household type (single house or multifamily).

DHWcalc generates profiles of DHW demand at 60, 6 or 1-min resolution, and the event flow-rate varies randomly around an average value based on a standard deviation entered by the user. The parameters used in this study are specific for the UK. A monitoring study on 124 dwellings in the UK found a mean household hot-water consumption of 122 L/day with a 95% confidence interval of 18 L/day [81], and a daily average profile as given in Fig. 6 based on the data reported in Ref. [81]. The four load types are characterized as follows:

1. Short load (toilet and kitchen sinks): the mean flow rate is 3 L/min [84] and the duration of a single drawn-off is 1 min, accounting for 30% of the daily load [81].
2. Medium load (dishwasher and washing machine): the mean flow rate is 6 L/min and the duration is 10 min [80], accounting for 10% of the daily load [81].
3. Bath: the mean flow rate is 8 L/min [84] and the duration is 10 min, accounting for 40% of the daily load [81].
4. Shower: the mean flow rate is 8 L/min [43] and the duration is 9 min, accounting for 20% of the daily load [81].

An example of a three-day profile of DHW for a single family house obtained with DHWcalc using the listed options and at a 1-min resolution is shown in Fig. 7.

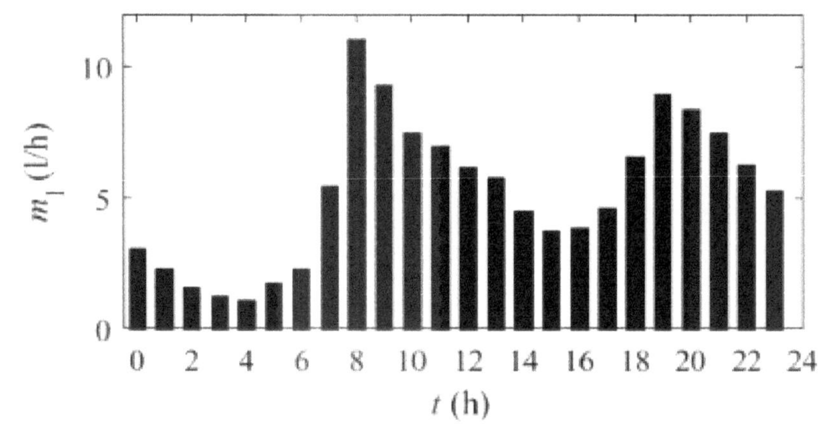

Figure 6. Daily profile of DHW demand for a single family house in the UK [75].

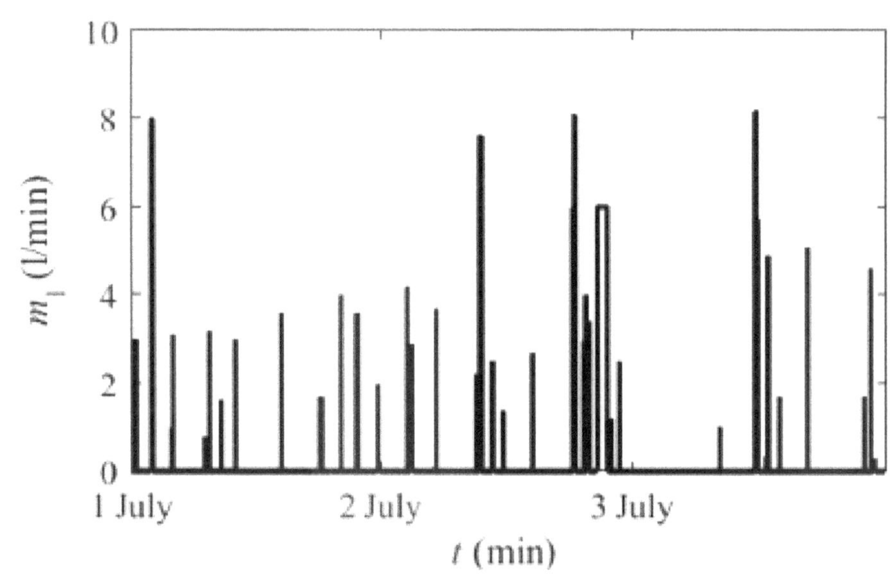

Figure 7. 3-day demand profile of DHW generated with the software DHWcalc for a single-family house in the UK.

After generating the event schedule, the results were compared with the input data to check for consistency and it was concluded that the average hourly flow rate, the daily and the annual water use match well with the user defined input data. The hot-water profile has been generated with a day-to-day standard deviation in the daily consumption of 53.3% over the year. This value can be considered acceptable if compared with the value of 48% reported in Ref. [85] as result of an investigation into daily DHW consumption in 74 dwellings. The seasonal variation of the energy demand of hot water is taken into account as a variation of the water flow rate while the mains temperature is kept constant during the simulation.

Electricity Demand

The electricity generated by the PVT system is compared with the yearly and monthly demand for electricity. The electricity demand and the electricity generation are independent of each other as a cost analysis is not an objective of this study, thus the instantaneous profiles of electricity demand are not considered here. The yearly and monthly data of electricity demand are taken from the previous study of Ref. [1]. In this paper the model developed by the Centre for Renewable Energy Systems Technology (CREST) was used for the demand profile calculation [86]. The model generated outputs at 1-min resolution for 365 days depending on the user input parameters (occupancy patterns, appliances, annual mean energy demand).

System Performance

The performance of a PVT system can be assessed in terms of its thermal output (provision of domestic hot-water) and electrical energy output. The instantaneous thermal energy provided by the system to the end-user is given by Eq. (44). Here Q_l is zero if there is no consumption of hot water or if the tank temperature is lower than the mains temperature. The second case is not encountered in this specific case because the tank is located indoors and exchanges heat with an environment at a constant temperature of 20 °C. Further, the fluid from the collector is allowed to circulate to the storage tank only if its temperature is higher than the tank temperature.

The fraction of the thermal energy demand covered by the PVT system f_Q is given in Eq.(48) and is calculated as the contribution of the storage tank to the heating/preheating of the hot water supply (as expressed in Eq. (47)) over the total energy demand of hot water at the demand temperature T_d of 60 °C:

$$Q_{TH} = 60 \cdot \sum_{t=0}^{t=T} Q_l(t); \tag{47}$$

$$f_Q = \frac{\sum_{t=0}^{t=T} Q_l(t)}{\sum_{t=0}^{t=T} m_w c_w (T_d - T_m)}. \tag{48}$$

The total electricity generated over the year is the sum of the generation of electricity at each time interval of 60 s (1 min), and is given as:

$$P_{EL}(T) = 60 \cdot \sum_{t=0}^{t=T} P_{EL}(t). \tag{49}$$

The electricity generated is compared with the energy required by the user and the fraction of the electricity demand covered by the PVT system over the year is:

$$f_E = \frac{P_{EL}(T)}{P_{ELd}}. \tag{50}$$

It is interesting to compare the profile of the electricity generated by the PVT system with the profile of the demand for electricity, but an economic analysis is beyond the scope of this study and therefore only the net values of electricity demand and generation are compared.

RESULTS AND DISCUSSION

Model Validation

The numerical model of the PVT collector was validated against the data published by the Eurofins laboratory in a technical report available online (Ref. [14]). This report presents the results of the steady-state outdoor testing of the Powertherm single glazed PVT collector. The tests were conducted at the nominal flow rate of 0.02 kg/s m² and in steady-state conditions. During the tests the incident irradiance was greater than 850 W/m², as indicated by the European standards for solar collector testing [6], [13],[15] and [87] and the inlet fluid temperature was varied between 15 °C and 54 °C. The experimental results of these steady-state tests are compared directly with the numerical predictions of the present model and the values are shown in Fig. 8.

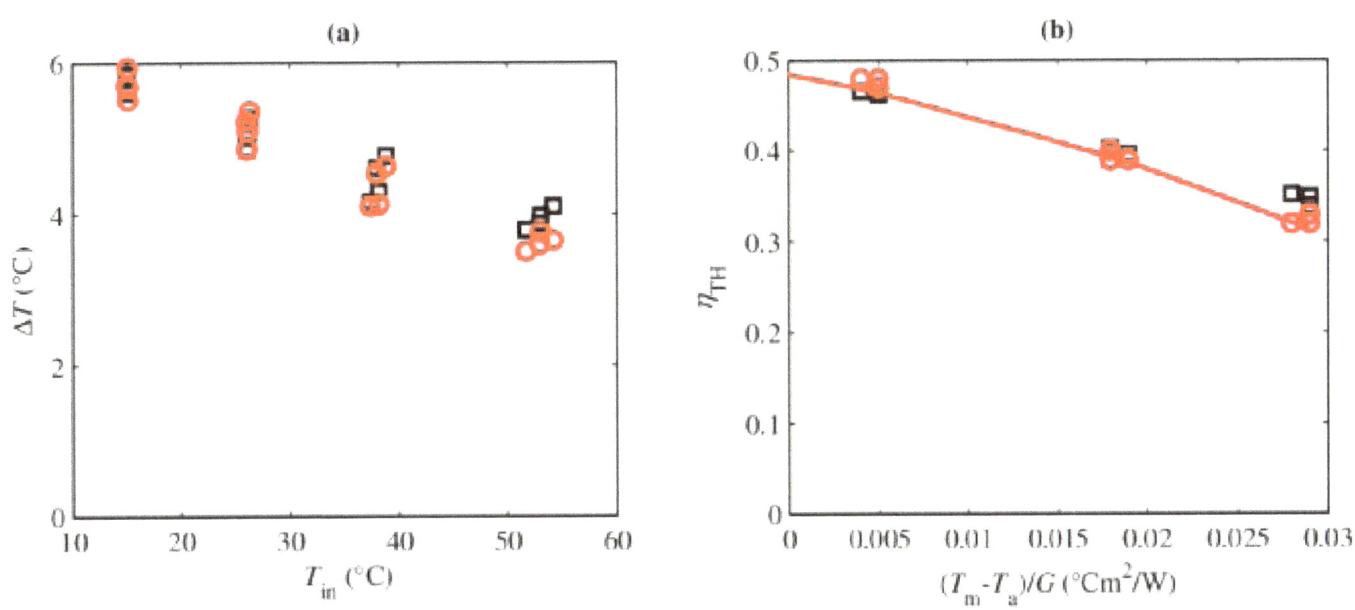

Figure 8. (a)–(b): (a) ΔT_f modelled (black squares) and experimental values (red circles) various inlet fluid temperatures T_{in}; (b) thermal efficiency modelled (black squares) and experimental (red circles). The experimental data are fitted with a 2nd-order polynomial function of the reduced temperature $T_r = (T_m - T_a)/G$, the red line is the efficiency-curve function as published in the test report in Ref. [24].

The model fits the experimental data within an average discrepancy of 8%. The calculated thermal-efficiency curve based on the aperture area for this PVT single-glazed collector is given in Eq. (51). This expression can be compared with the efficiency curve obtained by fitting the experimental data in Eq. (52).

$$\eta_{TH} = 0.483 - 4.528\frac{(T_m - T_a)}{G} - 0.014G\left(\frac{T_m - T_a}{G}\right)^2 \tag{51}$$

$$\eta_{TH} = 0.486 - 4.028\frac{(T_m - T_a)}{G} - 0.067G\left(\frac{T_m - T_a}{G}\right)^2 \tag{52}$$

The response of the system to a time-varying input was validated against the experimental results of a dynamic test of a PVT collector published by Amrizal et al. [87]. For this purpose, the parameters of the numerical models were changed in order to adapt the model to a configuration of PVT collectors that is different from the Powertherm collector considered so far. Amrizal et al. [87] tested an in-house PVT collector consisting of 2 lines of 26 c-Si cells connected in series, having a glass cover of 4 mm thickness. The dynamic behaviour of the PVT collector can be defined in terms of its time constant, which can be measured by studying the response of the collector to a step variation in solar irradiance. In practice this is achieved by covering the collector with a reflective cover until a steady state at near-zero irradiance is reached, and then uncovering the collector and measuring the response time for the collector to reach a new steady state operating condition. As defined in the EN 12975 [15], the time constant in Eq. (53) represents the elapsed time the collector needs to reach a ΔT of 63.2% (1/e) of its final value from an initial condition. The measurements of the time constant in Ref. [87] were taken according to the procedure described in Refs. [4] and [15], with a perpendicular incident irradiance at around solar noon and a constant flow rate of 0.022 kg/s m². The experimental value of the time constant obtained by Amrizal et al. was 87 ± 5 s, which is matched closely by our numerical result of 88 s.

$$\tau_c : \theta_c = \frac{T_o(t) - T_o(0)}{T_o(t) - T_o(\infty)} = 0.632 \tag{53}$$

The time constant of the Powertherm solar collector was calculated with the same numerical code and using the same approach as above by changing the geometrical parameters of the collector, resulting in a value of 138 s. The response of the PVT collector to a step change in incident solar irradiance (from 30 W/m² to 900 W/m²) is shown in Fig. 9. The outlet fluid temperature of the collector at the steady-state incident irradiance condition of 900 W/m² is 24.07 °C. The initial outlet fluid temperature of the covered collector (incident irradiance of 30 W/m²) is 18.54 °C, for a constant inlet temperature of 18.50 °C.

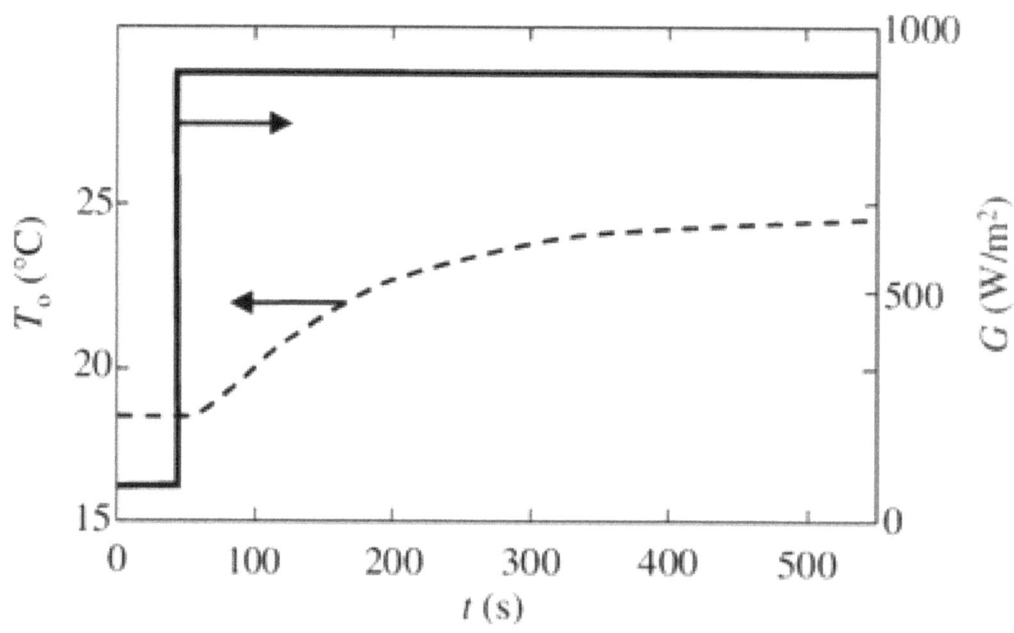

Figure 9. Response of the outlet fluid temperature $T_o(t)$ of the PVT collector (dashed line) to a step variation in the incident solar irradiance G (solid line).

The importance of using real ambient data at high resolution instead of time-averaged data can be assessed by considering the discrepancy in the results obtained when the same simulation is run with the two input datasets. For an array of single-glazed PVT collectors, the yearly electricity generation is overestimated when using time-averaged input data by 23.4% (the yearly electricity generation calculated with 1-min resolution data is 1572 kWh, or 1940 kWh when using time-average data). With regards to the thermal-energy generated, the major discrepancy is obtained in the summer period when the fraction of thermal energy demand covered by the system is overestimated by over 25% (see Fig. 10).

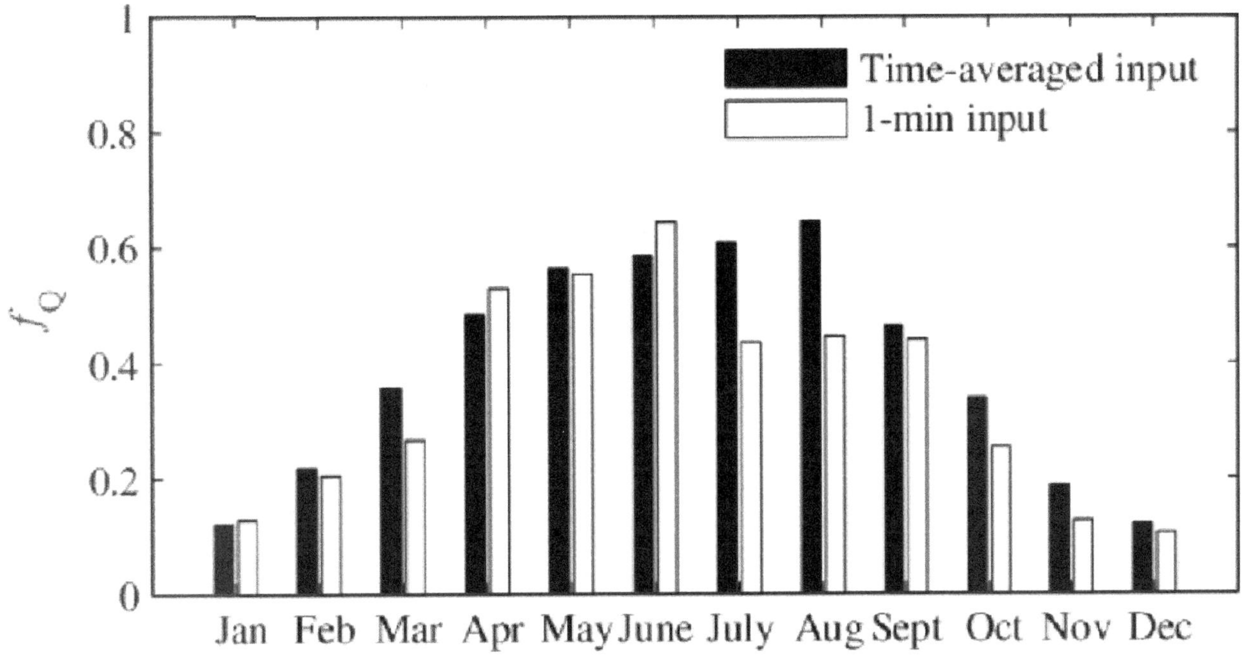

Figure 10. Comparison of the fraction of the covered thermal energy demand evaluated with 1-min resolution input data and with time-averaged input data (over 30-min intervals).

When solving the energy balance of the system using input data at 1-min resolution, a decision has to be made whether a dynamic or quasi-steady model is required. This is related to the value of the collector time-constant. In the present study, a model having a simple control strategy was tested by solving both the quasi-steady and the dynamic problem for the month of July. The results reported in Table 4 show that the quasi-steady solution overestimates the thermal energy-demand coverage fraction by up to 12.3%, while the electricity production is overestimated by 7.0%. As expected, the larger discrepancy is observed for the double-glazed collector, which has a larger thermal mass. This discrepancy can be explained by considering the activation and de-activation of the pump, which is determined by the differential controller set with a specific value of temperature-difference. When the quasi-steady solution is applied, the fluid temperature follows the profile of the solar irradiance and as soon as there is irradiance on the solar collector, the temperature reaches the required value to activate the pump. The results show that the pump is running by 57% additional hours in the case of the single-glazed collector for the quasi-steady solution.

Table 4. Comparison of the dynamic and quasi-steady solutions with the same inputs at 1-min resolution.

	Model	f_Q	P_{EL} (kWh)	Pump hours (h)
Double glazed	Dynamic	0.95	138.1	364.87
	Quasi-steady	1.07	128.4	372.87
Single glazed	Dynamic	0.68	166.7	364.53
	Quasi-steady	0.73	161.4	573.23
Unglazed	Dynamic	0.32	204.6	564.32
	Quasi-steady	0.34	203.2	371.93

Temperature Distribution on the PV Module

The thermal model allows the prediction of the temperature distribution on the solar cell, which is then used as the input to the electrical model. The top surface of the PV module is not isothermal and has a maximum between two adjacent pipes that increases in the flow direction as the coolant collects thermal energy from the absorber, as shown in Fig. 11. Knowledge of the temperature distribution on the PV module is of fundamental importance for the correct estimation of the electrical output. This numerical model constitutes a tool that can be used for the investigation of different geometries of the thermal absorber and the solar cells in order to assess the influence of the various design parameters on the temperature distribution on the PV module.

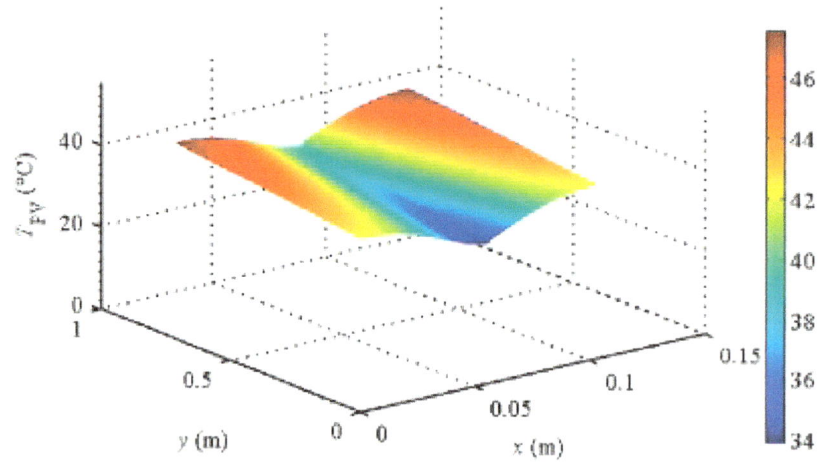

Figure 11. Temperature distribution T over the solar cell between two pipes with G = 1000 W/m², T_{f-in} = 20 °C and T_a = 20 °C (the geometrical characteristics are from the Powertherm collector). The pipe lies in the y-direction, along the middle of the surface, with the fluid inlet at x = W/2 and y = 0; here the distance between adjacent pipes is W = 0.1 m.

Comparison of Unglazed, Single-glazed and Double-glazed Collectors

The thermal/electrical performance and the dynamic behaviour of the three configurations of the PVT solar collector (unglazed, single-glazed and double-glazed) are presented in this section. A comparison of the thermal efficiencies of the three configurations is shown in Fig. 12 where the efficiency is plotted against the reduced temperature. The thermal efficiency is obtained by using the same geometrical characteristics of the Powertherm collector discussed thus far (pipe diameter over fin width ratio D/W, pipe diameter, flow rate, pipe length).

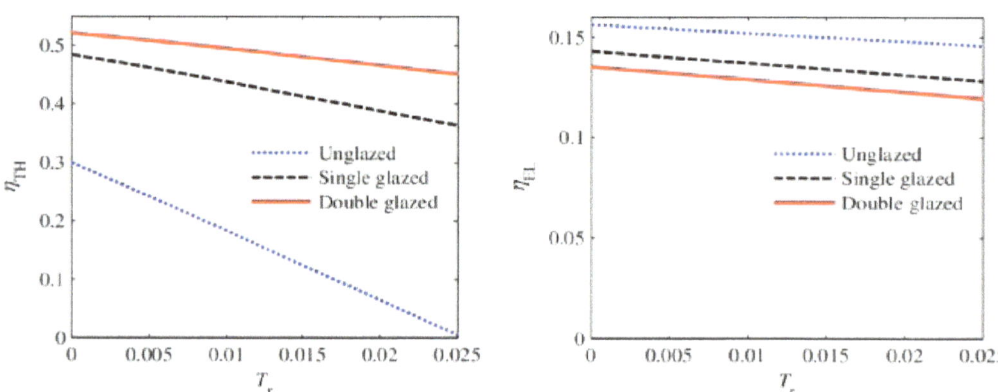

Figure 12. Comparison of the thermal and electrical efficiency for the unglazed collector, single glazed collector and double glazed collector.

As expected, the double-glazed collector shows a higher thermal efficiency than the other two configurations as the convective thermal losses to the environment are reduced by the presence of the additional glazing. The overall heat transfer coefficient to the environment calculated on the absorber area and at the absorber average temperature is 6.2 W/m² K for the double-glazed collector, 9.2 W/m² K for the single-glazed collector, and 37.8 W/m² K for the unglazed collector. This difference is evidenced by the slope of the thermal-efficiency curve. The zero-loss coefficient is reduced with additional glazing due to the reflection losses, as expected.

The fraction of incident irradiance that is reflected by the top surface amounts to 21.3% for double glazing, 16.2% for single glazing, and 10.0% for the unglazed collector. As a consequence, the fraction of the incident irradiance that is transmitted from the top glass surface to the PV cell to be absorbed reduces with the number of glazing layers. The fraction of incident irradiance that can be absorbed by the solar cell operating at the reference temperature (calculated from Eq. (20)) is 90.0% if the cell is exposed directly to the environment, 81.7% if there is a single glass-cover and 74.4% if a second cover is added. This effect is larger than the dependence of the electrical efficiency on the operating temperature (slope of the electrical-efficiency curve) and is reflected in the electrical-efficiency curve which is always higher for an unglazed collector. On the other hand, the electrical efficiency is higher for the unglazed collector when the optical losses due to reflection are reduced.

The thermal performance of the unglazed collector can be improved if a selective coating is applied to the solar cell. This reduces the emissivity of the solar cell in the infrared spectrum, thus also reducing the radiative losses as discussed in Section 1. The thermal and electrical efficiencies of the unglazed collector are shown in Fig. 13 for an emissivity ε(also equal to the absorptivity) of the solar cell in the infrared spectrum ranging from 0.9 – a typical value for a solar cell – to zero, which is the optimum value of a selective absorber. The thermal efficiency is improved by 10% with ideally no modification of electrical performance.

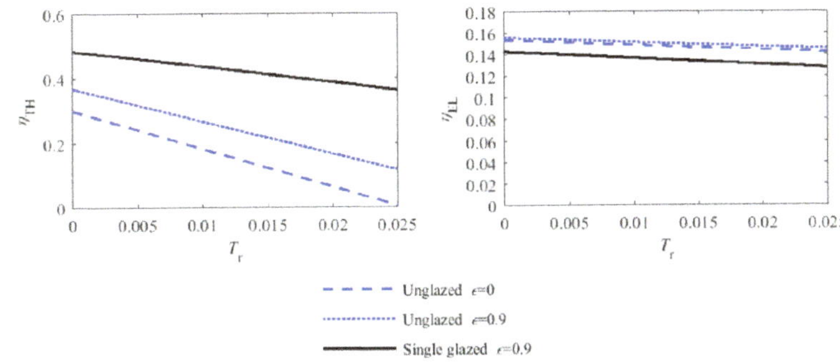

Figure 13. Comparison of the thermal and electrical efficiency of the unglazed collector when the solar cells have ε=0.9 (dotted line) and ε=0 (dashed line) with the efficiency of the single glazed collector (solid line).

PVT System Operation

This section reports the results of the dynamic model showing the predicted daily operation of a PVT system under fluctuating weather conditions and clear sky conditions. Results relating to the daily operation of the solar system in three typical weather conditions are shown in Fig. 14. The circulation of the liquid (water) in the system is determined by the control strategy as described in Section 2.3. The fluid is circulated by the pump when there is an energy gain through the collector, and the circulation is halted otherwise. Moreover, once the pump is activated, the bypass branch is de-activated (=1) and the fluid heats up the tank when the temperature at the collector is sufficiently high. Otherwise the fluid is recirculated to the collector and the bypass is active (=0).

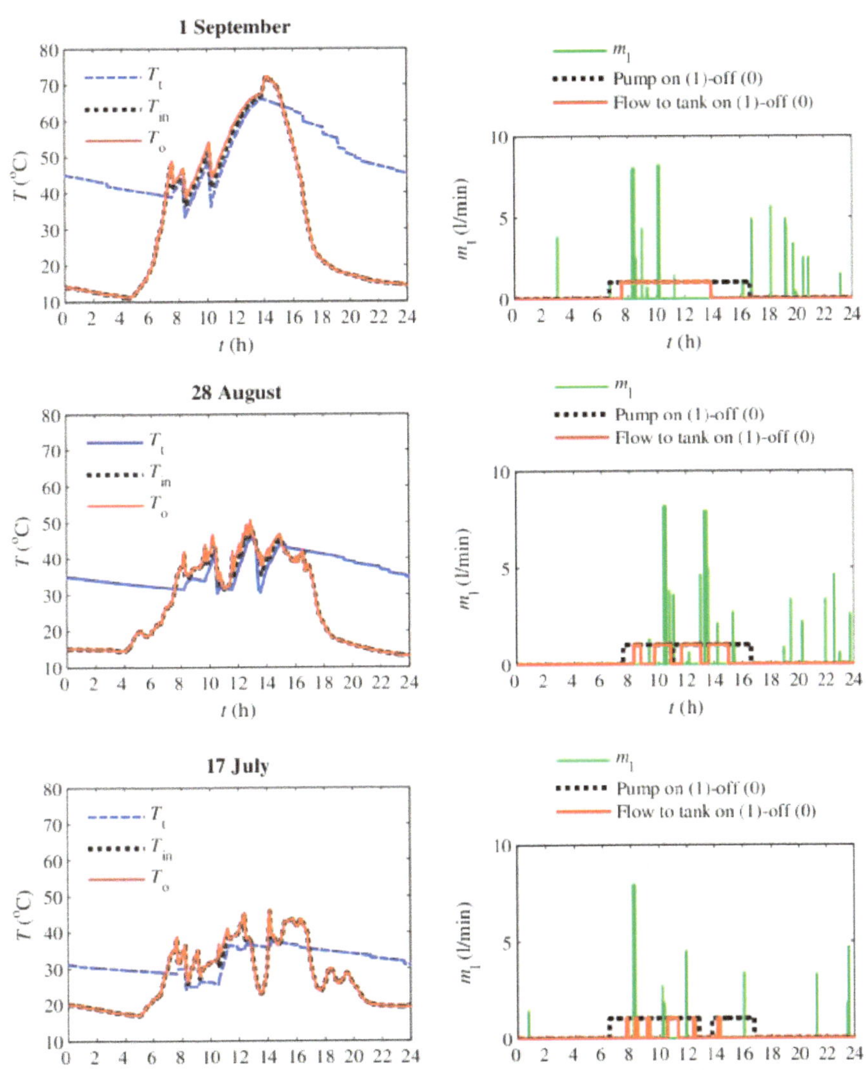

Figure 14. Temperature of the solar collector T_f at the inlet and at the outlet and temperature of the storage tank $T_t$calculated for three days with different ambient conditions and with the daily profile of hot water demandm_l. The figures also show the time of the day for which the circulation pump is on or off, the bypass branch is activated, and the collector fluid is heating the storage tank (flow to tank is on).

The operation of the solar-PVT collector is also influenced by the demand for hot water. When the demand for hot water is high, the temperature in the storage tank drops (as does the temperature of the collector) if the pump is active. On the 1st of September, during a day with a clear sky, the collector fluid circulates in the storage tank for most of the time and the operation is stable. On the 28th of August and 17th of July, which are cloudy and overcast days, the bypass branch is alternately activated and de-activated. Fluid flow to the tank is only activated for a few hours, while (mostly) the fluid is recirculated in the collector in order to increase its temperature/enthalpy. This intermittent behaviour would not be captured if average irradiance data were used. The mean daily temperature of the storage tank operating as fully mixed is also dependent on this pattern and on the ambient conditions. It is generally lower for cloudy days. Furthermore, the temperature of the storage tank is influenced by the profile of domestic hot-water demand. In the morning of the 1st of September, between 6 am and 8 am when the demand of hot water is large, the temperatures of the tank and of the collector do not follow the profile of the irradiance.

The present work employs a hot-water storage tank with a heat-exchanger (coil) design that promotes mixing, such that the tank is modelled as fully mixed (see Section 3.3). Stratified charging of the hot-water cylinder can also be employed, in which case the stratification of the fluid in the tank will be affected by the design of the coil in the tank, the design of the inlet and outlet ports, the size of the tank, etc. In order to consider the variations in the outputs of interest introduced by the design of the tank (fully mixed or stratified), a 3-node, 1-D stratified tank model with one coil heat exchanger and two ports has been tested together with the fully mixed version of the tank model. The results obtained for 1st September with the two design options are given in Table 5.

Table 5. Daily system performance when a fully mixed tank and a stratified tank are employed in the system.

Fully mixed tank	3-node, 1-D stratified tank	Relative difference [%]
Q_{TH} = 7.013 kWh	Q_{TH} = 6.76 kWh	3.8
P_{EL} = 7.87 kWh	P_{EL} = 7.96 kWh	1.1

Given the inherent uncertainty to the design of the hot-water cylinder and the small relative differences in the results, the performance (outputs) of the PVT system with a fully mixed tank discussed in this paper can be considered to extend to an equivalent PVT system featuring a stratified tank. The instantaneous electrical efficiency of the PVT system is shown in Fig. 15 along with the temperature of the fluid and of the PV module and the nominal efficiency. The nominal electrical efficiency of a PVT module is the efficiency of the panel operating in steady-state under standard conditions (G = 1000 W/m^2 and T_a = 25 °C) and empty of fluid (Ref. [7]). While the nominal electrical efficiency of the PVT module is 12.6% (red dashed line), the instantaneous value is higher than the nominal efficiency during the early morning and in the afternoon due to the low ambient temperature and incident irradiance, while it decreases to 12.2% when the incident irradiance on the solar collector is around 800 W/m^2 due to the increase of the PV cell temperature.

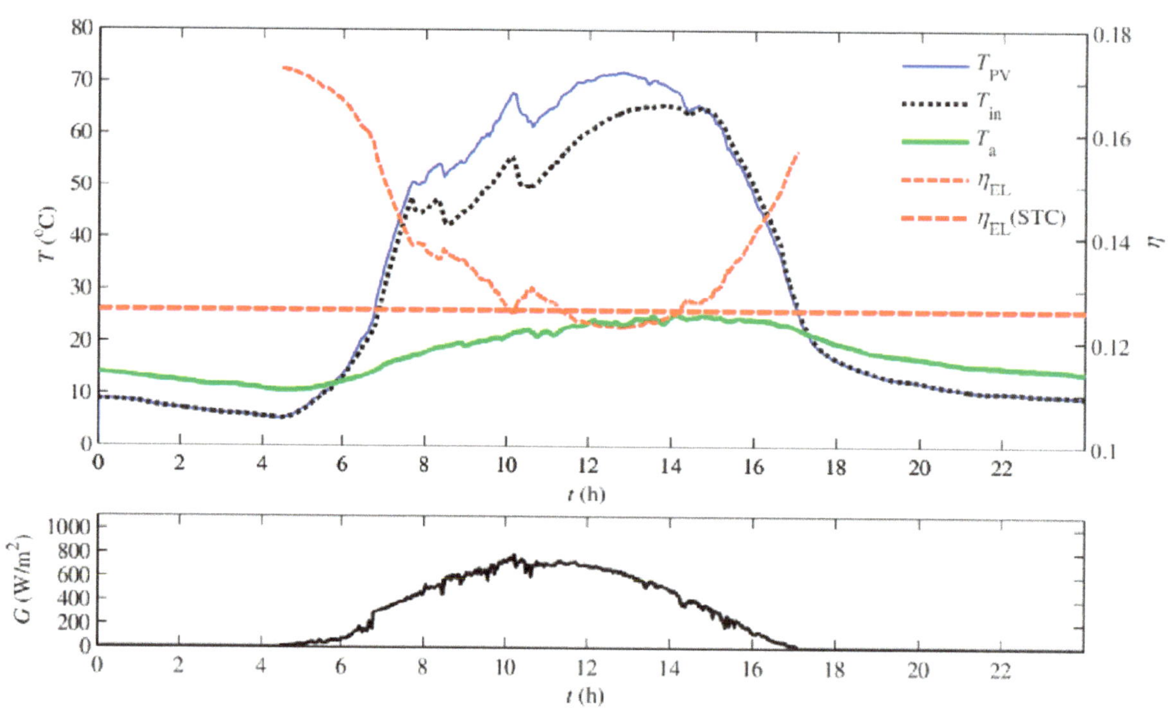

Figure 15. Instantaneous electrical efficiency of the PV module η_{EL} compared with the electrical efficiency at standard conditions η_{EL}(STC) in relation to the operating temperature of the solar cell T_{PV} and the incident irradiance G.

The instantaneous generation of electricity over two days of the year is shown in Fig. 16. The generated electricity exceeds the demand during the sunny September day, while only a small portion of demand is covered during the cloudy winter day; this significantly affects the economics of the implementation of these systems, since electricity produced during the day is more expensive than the electricity needed during the night.

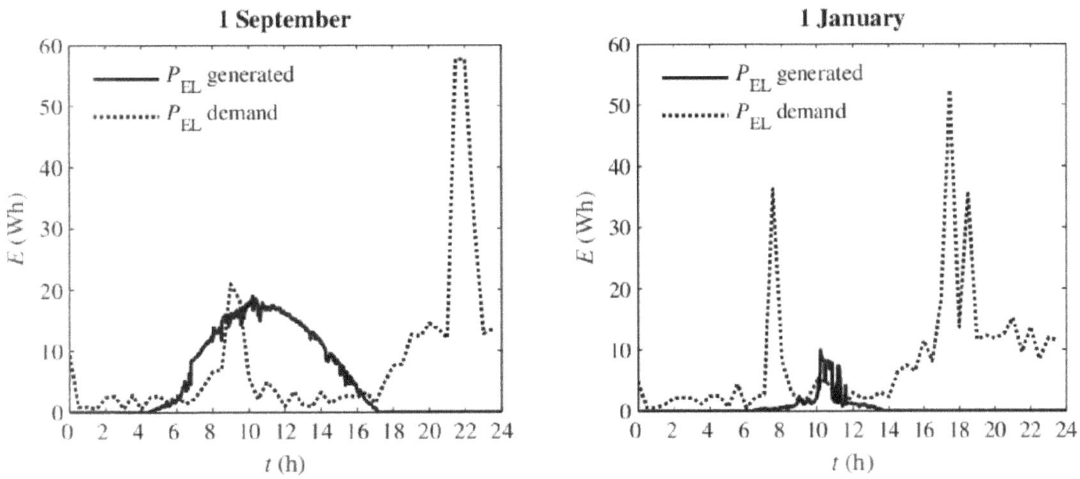

Figure 16. Instantaneous electricity generated P_{EL} compared with the instantaneous demand of electricity during two days of the year.

PVT System Monthly and Yearly Energy Yield

This section discusses the monthly performance of the PVT system in terms of the fraction of electrical energy demand f_E and of the fraction of thermal energy demand f_Q covered by the system as predicted by the dynamic model presented in this paper. The main results are shown in Fig. 17. As expected, the fraction of energy covered by the PVT system is higher during the summer months, for example, the monthly generation of electricity exceeds the demand in August if an unglazed collector is installed (as part of the system). A PVT system can cover between 25% and 50% of the total demand of domestic hot water in a UK household depending on the choice of the collector (glazed or unglazed).

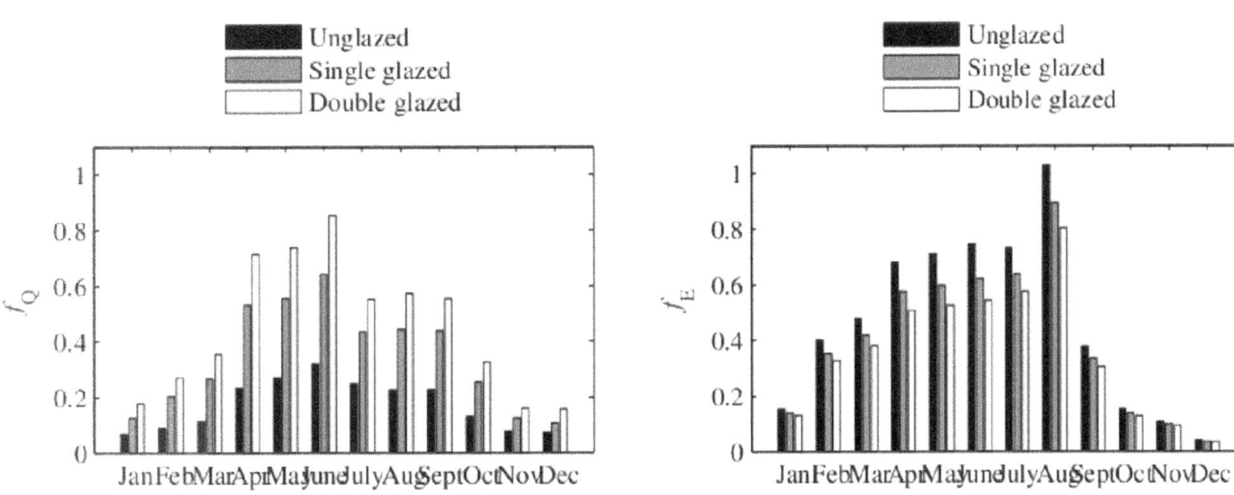

Figure 17. Fraction of the thermal energy f_Q and electricity f_E demands covered by the PVT system each month.

We can consider now the case of a PVT collector with modified solar-cell optical properties, whose aim is to reduce the radiative losses from the PVT collector. The emissivity of an ideal absorber is considered for the comparison, and ε is set to zero. In reality, it is envisaged that a value between 0.9 and the ideal value could be achieved with a selective coating. The results are shown in Fig. 18. These show that for all collectors (unglazed, single-glazed and double-glazed) the fraction of thermal energy covered would increase by more than 10%, e.g. for the unglazed collector this increases from 24.3% to 36.2% (i.e. by 12%). Slight differences in the relative improvement arise due to the different fluid temperatures in the three collectors and the operation of the differential controller that regulates the preheating of the storage tank by the collector fluid. In a solar system with unglazed collectors the tank is preheated for only few hours each day and the fluid is mainly kept recirculating through the collector array so that it can reach the set temperature. On the other hand, the electrical performance is found to deteriorate, as expected, at higher operating temperatures.

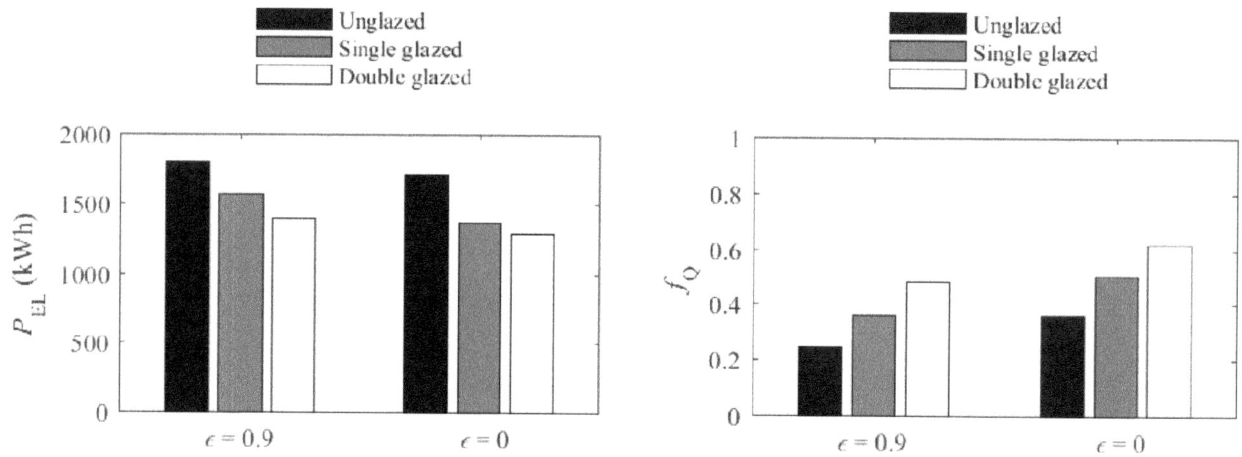

Figure 18. Fraction of thermal energy generated f_Q and electricity generated P_{EL} for one year calculated using instantaneous weather data. Here the results are reported for the case of a standard solar cell with $\varepsilon=0.9$ and with solar cells having low emissivity $\varepsilon=0$.

CONCLUSIONS

This paper describes the development of a detailed 3-D dynamic numerical model of a PVT collector with the aim of estimating the temperature distribution over the surface of the PV module and therefore of estimating the annual thermal and electrical energy outputs generated by the system in a UK domestic installation. The estimation of the temperature distribution on the surface of the PV module is required in order to accurately calculate the electricity generated. The model constitutes a useful tool for the design and optimization of the thermal absorber and for the assessment of different module and system configurations.

The model shows that there is a temperature gradient on the PVT collector which results in an efficiency variation due to a non-uniform temperature distribution on the PV cells. The module efficiency drops by only 4% below the nominal value of 12.6%, which is the electrical efficiency of the PVT module at standard conditions and empty of fluid, during the hot hours of the day when the incident irradiance reaches 800–1000 W/m^2. Due to the flow of the cooling liquid (water) the panel operates above its nominal efficiency for the rest of the day.

The model uses real weather input-data at high resolution and a high-resolution profile of domestic hot-water demand obtained with the software DHWcalc. One important conclusion concerns the importance of using real input-data at high-resolution for the correct estimation of the yearly and monthly performance of the system, as opposed to averaged data, especially if a novel control strategy that can adjust the system's outputs in response to varying demands is to be designed. In particular, it has been shown that the use of time-averaged climate data results in an overestimation of the thermal production (climate data for the year 2014–2015 are used in the present analysis). This is especially of relevance when doing an economic analysis of the system, which requires a precise knowledge of the instantaneous generation of electricity in relation to the electricity demand during the day. Moreover, when running simulations using high-resolution weather-data, some parameters of the control strategy were found to be critical, namely the cooling flow rate, the operation mode/strategy of the pump, and the temperature of activation/de-activation of the pump and bypass branch.

The effect of the emissivity of the solar cell on the thermal output of the PVT panel has also been considered. Solar cells for PVT applications can be specifically designed to increase the thermal performance of the module while maintaining a high electrical efficiency. It has been shown that if the emissivity of the solar cell is reduced, the thermal output of the PVT system can increase by 10% with almost no loss in the electrical output due to the low temperature of operation of the non-concentrated solar-thermal system.

ACKNOWLEDGEMENT

This work was supported by the UK Engineering and Physical Sciences Research Council (EPSRC) [grant number EP/M025012/1]. The authors would also like to acknowledge the EU Climate-KIC <http://www.climate-kic.org> for IG's PhD studentship. Data supporting this publication can be obtained on request from cep-lab@imperial.ac.uk.

Nomenclature

Abbreviations

c-Si	Mono-crystalline
DHW	Domestic hot water
HW	Hot water
PV	Photovoltaic
PVT	Hybrid photovoltaic and thermal collector

Symbol

A	Area (m^2)
A_A	Absorber area (m^2)
A_C	Aperture area (m^2)
A_G	Gross area (m^2)
A_S	Surface area (m^2)
A_t	Storage tank surface area (m^2)
c	Specific heat capacity (J/kg K)
c_A	Specific heat capacity of the absorber (J/kg K)
c_f	Specific heat capacity of the fluid (J/kg K)
c_g	Specific heat capacity of the glass cover (J/kg K)
c_{PV}	Specific heat capacity of the solar cell (J/kg K)
D	Diameter (m)
D_o	Riser external diameter (m)
D_H	Hydraulic diameter (m)
E	Electricity (W)
f	Fraction
f_E	Fraction of the electrical demand covered by the PVT system
f_Q	Fraction of the hot water demand covered by the PVT system
G	Irradiance (W/m^2)
g	Gravitational acceleration (m/s^2)
h	Convective heat transfer coefficient (W/m^2 K)
h_f	Convective heat transfer coefficient in the pipe (W/m^2 K)
h_{freeff}	Heat transfer coefficient due to free convection (W/m^2 K)
h_{gap}	Heat transfer coefficient due to convection in the enclosed space between the glass cover and the solar cell (W/m^2 K)
h_{lt}	Heat transfer coefficient due to convection at the storage tank (W/m^2 K)
h_w	Forced convection wind heat transfer coefficient (W/m^2 K)
h_{top}	Top convective heat transfer coefficient (glass to ambient) due to free and forced convection (W/m^2 K)
I	Solar irradiance spectrum (W/m^3)
k	Thermal conductivity (W/m K)
k_A	Thermal conductivity of the absorber (W/m K)
k_{air}	Thermal conductivity of the air (W/m K)
k_{EVA}	Thermal conductivity of the encapsulant (W/m K)
k_b	Thermal conductivity of the bond (W/m K)
k_f	Thermal conductivity of the fluid (W/m K)
k_g	Thermal conductivity of the glass cover (W/m K)
k_{gl}	Thermal conductivity of the adhesive (W/m K)
k_i	Thermal conductivity of the insulation (W/m K)
k_{PV}	Thermal conductivity of the solar cell (W/m K)
k_{TED}	Thermal conductivity of the Tedlar (W/m K)
L	Length (m)
M	Mass (kg)
M_A	Mass of the absorber (kg)
M_f	Mass of the fluid (kg)
M_g	Mass of the cover glass (kg)
M_{PV}	Mass of the solar cell (kg)
$\dot{m}_f$	Mass flow rate (kg/s)
N_p	Number of pipes
N_x, N_y	Nodes on x, y direction
P_{EL}	Electrical energy generated (W)
$P_{EL}(T)$	Electrical energy generated over the period of time T (J)
P_{ELd}	Electrical demand (J)
Pr	Prandtl number, $\mu c_p/k$
Q	Heat flow rate (W)

Q_{A-CD}	Heat transfer in the absorber between two adjacent nodes due to conduction (W)
Q_{coll}	Heat addition from the PVT collector to the storage tank through the heat exchanger (W)
Q_f	Heat addition to the fluid from the absorber due to convection (W)
Q_{g-CD}	Heat transfer in the glass cover between two adjacent nodes due to conduction (W)
$Q_{g,a-CV}$	Heat losses from the glass cover to the ambient due to convection (W)
$Q_{g,a-RD}$	Heat losses from the glass cover to the ambient due to radiation (W)
$Q_{g,PV-CV}$	Heat addition to the glass cover from the solar cell due to convection (W)
$Q_{g,PV-RD}$	Heat addition to the glass cover from the solar cell due to radiation (W)
Q_{g-AB}	Fraction of the incident irradiance absorbed by the glass cover (W)
Q_l	Enthalpy drop of the storage tank due to the flow of domestic hot water (W)
Q_{lt}	Heat losses from the storage tank due to convection (W)
Q_{loss}	Heat loss from the rear layer of the PVT collector (W)
Q_{PV-CD}	Heat transfer in the solar cell between two adjacent nodes due to conduction (W)
Q_{PV-A}	Heat addition to the absorber from the solar cell due to conduction (W)
Q_{PV-AB}	Fraction of the incident irradiance absorbed by the solar cell (W)
Q_{TH}	Thermal energy generated (J)
R	Thermal resistance (K m^2/W)
R_{CD}	Thermal resistance between the solar cell and the absorber (K m^2/W)
R_{gap}	Thermal resistance of the air gap between the glass cover and the solar cell (K m^2/W)
R_{loss}	Thermal resistance to the ambient at the rear layer of the PVT collector (K m^2/W)
r	Reflectance
r_g	Glass reflectance
r_{PV}	Solar cell reflectance
R_g	Top reflected irradiance of a single glazed system
Ra	Rayleigh number, $g\beta\rho^2 c/k\mu\Delta TL^3$
Re	Reynolds number, $\rho v D_H/\mu$
T	Temperature (K)
t	Time interval (s)
T_A	Absorber temperature (K)
T_a	Ambient temperature (K)
T_d	Required temperature of the hot water for domestic application (K)
T_f	Fluid temperature (K)
T_{f-in}	Inlet fluid temperature (K)
T_{f-o}	Outlet fluid temperature (K)
T_g	Glass cover temperature (K)
T_{in}	Inlet temperature (K)
T_m	Averaged temperature (K) / Mains water temperature (K)
T_o	Outlet temperature (K)
T_{PV}	Temperature of the solar cell (K)
T_r	Reduced temperature (K m^2/W)
T_{ref}	Reference temperature (K)
T_{sky}	Sky temperature (K)
T_t	Temperature of the storage tank (K)
v	Velocity (m/s)

v_w	Wind speed (m/s)
W	Width (m)
W_p	Nominal power (W)

Greek

α	Absorption coefficient
α_g	Glass absorption coefficient
α_{PV}	Absorption coefficient of the solar cell
β	Thermal expansion coefficient, $1/T$ (1/K)
β_{PV}	Temperature coefficient (1/K)
γ	Tilt angle of the collector to the horizontal (rad)
δ	Thickness (m)
δ_A	Absorber thickness (m)
δ_b	Bond thickness (m)
δ_{EVA}	EVA thickness (m)
δ_g	Glass cover thickness (m)
δ_{gl}	Adhesive thickness (m)
δ_i	Insulation thickness (m)
δ_{PV}	Solar cell thickness (m)
δ_{TED}	Tedlar thickness (m)
ε	Emissivity
ε_g	Glass emissivity
ε_H	Heat exchanger effectiveness
ε_{PV}	Emissivity of the solar cell
η	Efficiency
η_{ref}	Solar cell standard electrical efficiency
η_T	Electrical efficiency of the solar cell function of the operating temperature
η_{EL}	Module electrical efficiency
η_{TH}	Thermal efficiency
μ	Dynamic viscosity (N s/m^2)
ρ	Density (kg/m^3)
τ	Transmittance
τ_g	Glass transmittance
$\tau\alpha_g$	Glass combined transmission and absorption
$\tau\alpha_{PV}$	Solar cell combined transmission and absorption

REFERENCES

1. M. Herrando, C.N. Markides, K. Hellgardt, A UK-based assessment of hybrid PV and solar-thermal systems for domestic heating and power: system performance, Appl. Energy 122 (2014) 288–309.
2. M. Herrando, C.N. Markides, Hybrid PV and solar-thermal systems for domestic heat and power provision in the UK: techno-economic considerations, Appl. Energy 161 (2015) 512–532.
3. S. Bhattarai, J.-H. Oh, S.-H. Euh, G. Krishna Kafle, D. Hyun Kim, Simulation and model validation of sheet and tube type photovoltaic thermal solar system and conventional solar collecting system in transient states, Sol. Energy Mat. Sol. Cells 103 (2012) 184–193.
4. S. Fischer, W. Heidemann, H. Müller-Steinhagen, B. Perers, P. Bergquist, B. Hellström, Collector test method under quasi-dynamic conditions according to the European Standard EN 12975-2, Sol. Energy 76 (2004) 117–123.
5. T.T. Chow, Performance analysis of photovoltaic-thermal collector by explicit dynamic model, Sol. Energy 75 (2003) 143–152.
6. H.A. Zondag, D.W. de Vries, W.G.J. van Helden, R.J.C. van Zolingen, A.A. van Steenhoven, The thermal and electrical yield of a PV-thermal collector, Sol. Energy 72 (2002) 113–128.
7. H. Pierrick, M. Christophe, G. Leon, D. Patrick, Dynamic numerical model of a high efficiency PV–T collector integrated into a domestic hot water system, Sol. Energy 111 (2015) 68–81.
8. M. Rommel, D. Zenhäusern, A. Baggenstos, O. Türk, S. Brunold, Development of glazed and unglazed PVT collectors and first results of their application in different projects, Energy Procedia 70 (2015) 318–323.
9. J. Schnieders, Comparison of the energy yield predictions of stationary and dynamic solar collector models and the models' accuracy in the description of a vacuum tube collector, Sol. Energy 61 (1997) 179–190.
10. P. Dupeyrat, C. Ménézo, S. Fortuin, Study of the thermal and electrical performances of PVT solar hot water system, Energy Build. 68 (2014) 751– 755.
11. C. Cristofari, J. Canaletti, G. Notton, C. Darras, Innovative patented PV/TH Solar Collector: optimization and performance evaluation, Energy Procedia 14 (2012) 235–240.
12. C. Bales, T. Persson, External DHW units for solar combisystems, Sol. Energy 74 (2003) 193–204.

13. S.A. Kalogirou, Solar Energy Engineering Processes and Systems, Academic Press, 2013.
14. Eurofins Modulo Uno, Report of Performance Test According to EN12975-2 for Solimpeks Volther Powertherm. 10156 Torino, Italy; 2011 Nov. Technical Report No.: M1.11.NRG.0319/43724, , 2006 (accessed 25.03.06).
15. European Standard, EN 12975-2, CEN, European committee for standardisation, 2006.
16. P. Hoang, V. Bourdin, Q. Liu, G. Caruso, V. Archambault, Coupling optical and thermal models to accurately predict PV panel electricity production, Sol. Energy Mat. Sol. Cells 125 (2014) 325–338.
17. H.A. Zondag, D.W. de Vries, W.G.J. van Helden, R.J.C. van Zolingen, A.A. van Steenhoven, The yield of different combined PV-thermal collector designs, Sol. Energy 74 (2003) 253–269.
18. B. Sandnes, J. Rekstad, A photovoltaic/thermal (PV/T) collector with a polymer absorber plate. Experimental study and analytical model, Sol. Energy 72 (2002) 63–73.
19. Y. Kim, T. Seo, Thermal performances comparisons of the glass evacuated tube solar collectors with shapes of absorber tube, Renew. Energ. 32 (2007) 772–795.
20. G. Fraisse, C. Ménézo, K. Johannes, Energy performance of water hybrid PV/T collectors applied to combisystems of Direct Solar Floor type, Sol. Energy 81 (2007) 1426–1438.
21. T.T. Chow, G. Pei, K.F. Fong, Z. Lin, A.L.S. Chan, J. Ji, Energy and exergy analysis of photovoltaic–thermal collector with and without glass cover, Appl. Energy 86 (2009) 310–316.
22. J.-H. Kim, J.-T. Kim, Comparison of electrical and thermal performances of glazed and unglazed PVT collectors, Int. J. Photoenergy 2012 (2012) 7.
23. J.-H. Kim, J.-T. Kim, The experimental performance of an unglazed PV-thermal collector with a fully wetted absorber, Energy Procedia 30 (2012) 144–151.
24. H. Saitoh, Y. Hamada, H. Kubota, M. Nakamura, K. Ochifuji, S. Yokoyama, et al., Field experiments and analyses on a hybrid solar collector, Appl. Therm. Eng. 23 (2003) 2089–2105.
25. Y. Tripanagnostopoulos, T. Nousia, M. Souliotis, P. Yianoulis, Hybrid photovoltaic/ thermal solar systems, Sol. Energy 72 (2002) 217–234.
26. T. Fujisawa, T. Tani, Annual exergy evaluation on photovoltaic-thermal hybrid collector, Sol. Energy Mat. Sol. Cells 47 (1997) 135–148.
27. J. Vestlund, J.-O. Dalenbäck, M. Rönnelid, Thermal and mechanical performance of sealed, gas-filled, flat plate solar collectors, Sol. Energy 86 (2012) 13–25.
28. J. Vestlund, M. Rönnelid, J.-O. Dalenbäck, Thermal performance of gas-filled flat plate solar collectors, Sol. Energy 83 (2009) 896–904.
29. P. Dupeyrat, C. Ménézo, H. Wirth, M. Rommel, Improvement of PV module optical properties for PV-thermal hybrid collector application, Sol. Energy Mat. Sol. Cells 95 (2011) 2028–2036.
30. H.A. Zondag, Flat-plate PV-Thermal collectors and systems: a review, Renew. Sustain. Energy Rev. 12 (2008) 891–959.
31. B. Lalovic, Z. Kiss, H. Weakliem, A hybrid amorphous silicon photovoltaic and thermal solar collector, Sol. Cells 19 (1986) 131–138.
32. B.C.G. Van de Ree, Tests aan prototype onafgedekte PV/ Thermische collector, 1996.
33. D.W. De Vries, Design of a photovoltaic/thermal combi-panel (Ph.D.), Eindhoven University, 1998.
34. A. Suzuki, S. Kitamura, Combined photovoltaic and thermal hybrid collector, Jpn. J. Phys. 2 (1979) 79–83.
35. R. Santbergen, R.J.C. van Zolingen, The absorption factor of crystalline silicon PV cells: a numerical and experimental study, Sol. Energy Mat. Sol. Cells 92 (2008) 432–444.
36. I. Hamberg, C.G. Granqvist, Evaporated Sn-doped In2O3 films: basic optical properties and applications to energy-efficient windows, J. Appl. Phys. 60 (1986) R123–R160.
37. A. Solieman, M.A. Aegerter, Modeling of optical and electrical properties of In2O3 : sn coatings made by various techniques, Thin Solid Films 502 (2006) 205–211.
38. M.A. Green, Silicon solar cells: advanced principles & practice, Centre for photovoltaic devices and systems, University of New South Wales, 1995.
39. D.A. Clugston, P.A. Basore, Modelling free-carrier absorption in solar cells, Prog. Photovoltaics 5 (1997) 229–236.
40. U. Jordan, K. Vajen, DHWcalc: program to generate domestic hot water profiles with statistical means for user defined conditions, in: Proc. ISES Solar World Congress, Orlando (US), 8.–12.8, CD, 2005.
41. S.C. Mullick, S.K. Samdarshi, An improved technique for computing the top heat loss factor of a flat-plate collector with a single glazing, J. Sol. Energy Eng. 110 (1988) 262–267.
42. W. Wagner, A. Pruss, The IAPWS formulation 1995 for the thermodynamic properties of ordinary water substance for general and scientific use, J. Phys. Chem. Ref. Data 31 (2002) 387–535.
43. E.W. Lemmon, M.L. Huber, M.O. McLinden, S.R.D.P. National Institute of Standards and Technology (Ed.), NIST Standard Reference Database 23: Reference Fluid Thermodynamic and Transport Properties-REFPROP, Version 9.1, Gaithersburg, 2013.
44. S. Krauter, R. Hanitsch, Actual optical and thermal performance of PV-modules, Sol. Energy Mat. Sol. Cells 41–42 (1996) 557–574.
45. J.A. Duffie, W.A. Beckman, Solar Engineering of Thermal Processes, Wiley, New York, 1980.
46. M. Martin, P. Berdahl, Characteristics of infrared sky radiation in the United States, Sol. Energy 33 (1984) 321–336.
47. P. Berdahl, M. Martin, Emissivity of clear skies, Sol. Energy 32 (1984) 663–664.

48. A.D. Jones, C.P. Underwood, A thermal model for photovoltaic systems, Sol. Energy 70 (2001) 349–359.

49. G. Notton, C. Cristofari, M. Mattei, P. Poggi, Modelling of a double-glass photovoltaic module using finite differences, Appl. Therm. Eng. 25 (2005) 2854–2877.

50. K. Touafek, A. Khelifa, M. Adouane, Theoretical and experimental study of sheet and tubes hybrid PVT collector, Energy Convers. Manag. 80 (2014) 71–77.

51. O. Rejeb, H. Dhaou, A. Jemni, Parameters effect analysis of a photovoltaic thermal collector: case study for climatic conditions of Monastir, Tunisia, Energy Convers. Manag. 89 (2015) 409–419.

52. K.E. Amori, M.A. Abd-AlRaheem, Field study of various air based photovoltaic/ thermal hybrid solar collectors, Renew. Energ. 63 (2014) 402–414.

53. Y.B. Assoa, C. Ménézo, Dynamic study of a new concept of photovoltaic–thermal hybrid collector, Sol. Energy 107 (2014) 637–652.

54. O. Rejeb, H. Dhaou, A. Jemni, A numerical investigation of a photovoltaic thermal (PV/T) collector, Renew. Energ. 77 (2015) 43–50.

55. S. Armstrong, W.G. Hurley, A thermal model for photovoltaic panels under varying atmospheric conditions, Appl. Therm. Eng. 30 (2010) 1488–1495.

56. D. Torres-Lobera, S. Valkealahti, Inclusive dynamic thermal and electric simulation model of solar PV systems under varying atmospheric conditions, Sol. Energy 105 (2014) 632–647.

57. C.D. Ho, H.M. Yeh, R.C. Wang, Heat-transfer enhancement in double-pass flat-plate solar air heaters with recycle, Energy 30 (2005) 2796–2817.

58. R. Kumar, M.A. Rosen, Performance evaluation of a double pass PV/T solar air heater with and without fins, Appl. Therm. Eng. 31 (2011) 1402–1410.

59. G. Evola, L. Marletta, Exergy and thermoeconomic optimization of a watercooled glazed hybrid photovoltaic/ thermal (PVT) collector, Sol. Energy 107 (2014) 12–25.

60. G.N. Tiwari, R.K. Mishra, S.C. Solanki, Photovoltaic modules and their applications: a review on thermal modelling, Appl. Energy 88 (2011) 2287–2304.

61. S.A. Kalogirou, Solar thermal collectors and applications, Prog. Energ. Combust. Sci. 30 (2004) 231–295.

62. S. Sharples, P.S. Charlesworth, Full-scale measurements of wind-induced convective heat transfer from a roof-mounted flat plate solar collector, Sol. Energy 62 (1998) 69–77.

63. S.E.G. Jayamaha, N.E. Wijeysundera, S.K. Chou, Measurement of the heat transfer coefficient for walls, Build. Environ. 31 (1996) 399–407.

64. F.P. Incropera, Fundamentals of Heat and Mass Transfer, John Wiley & Sons, 2011.

65. H. Buchberg, I. Catton, D.K. Edwards, Natural convection in enclosed spaces – a review of application to solar energy collection, J. Heat Transfer 98 (1976) 182–188.

66. R. Viskanta, D.L. Siebers, R.P. Taylor, Radiation characteristics of multiple-plate glass systems, Int. J. Heat Mass Transf. 21 (1978) 815–818.

67. E. Skoplaki, J.A. Palyvos, On the temperature dependence of photovoltaic module electrical performance: a review of efficiency/power correlations, Sol. Energy 83 (2009) 614–624.

68. S.C. Solanki, S. Dubey, A. Tiwari, Indoor simulation and testing of photovoltaic thermal (PV/T) air collectors, Appl. Energy 86 (2009) 2421–2428.

69. J.K. Tonui, Y. Tripanagnostopoulos, Improved PV/T solar collectors with heat extraction by forced or natural air circulation, Renew. Energ. 32 (2007) 623–637.

70. P. Dupeyrat, C. Ménézo, M. Rommel, H.-M. Henning, Efficient single glazed flat plate photovoltaic–thermal hybrid collector for domestic hot water system, Sol. Energy 85 (2011) 1457–1468.

71. A. Tiwari, M.S. Sodha, Performance evaluation of solar PV/T system: an experimental validation, Sol. Energy 80 (2006) 751–759.

72. A. Ibrahim, M.H. Ruslan, S. Mat, M.Y. Othman, A. Zaharim, K. Sopian, Predicting the characteristics of a special designed Photovoltaic Thermal collector absorber (PVT), in: Proceedings of the 3rd WSEAS International Conference on Energy Planning, Energy Saving, Environmental Education, EPESE '09, Renewable Energy Sources, RES '09, Waste Management, WWAI '09 pp. 336–341, 2009.

73. M.N. Abu Bakar, M. Othman, M. Hj Din, N.A. Manaf, H. Jarimi, Design concept and mathematical model of a bi-fluid photovoltaic/thermal (PV/T) solar collector, Renew. Energ. 67 (2014) 153–164.

74. F. Shan, F. Tang, L. Cao, G. Fang, Dynamic characteristics modeling of a hybrid photovoltaic–thermal solar collector with active cooling in buildings, Energy Build. 78 (2014) 215–221.

75. R. Dubey, P. Batra, S. Chattopadhyay, A. Kottantharayil, B.M. Arora, K.L. Narasimhan, et al., Measurement of temperature coefficient of photovoltaic modules in field and comparison with laboratory measurements, in: Photovoltaic Specialist Conference (PVSC), 2015 IEEE 42nd pp. 1–5, 2015.

76. D.L. King, J.A. Kratochvil, W.E. Boyson, Temperature coefficients for PV modules and arrays: measurement methods, difficulties, and results, in: Photovoltaic Specialists Conference, 1997, Conference Record of the Twenty-Sixth IEEE pp. 1183–1186, 1997.

77. G. Makrides, B. Zinsser, G.E. Georghiou, M. Schubert, J.H. Werner, Temperature behaviour of different photovoltaic systems installed in Cyprus and Germany, Sol. Energy Mat. Sol. Cells 93 (2009) 1095–1099.

78. J.L. Gray, The physics of the solar cell, in: Handbook of Photovoltaic Science and Engineering, John Wiley & Sons, Ltd, 2011, pp. 82–129.

79. M. Šúri, T.A. Huld, E.D. Dunlop, H.A. Ossenbrink, Potential of solar electricity generation in the European Union member states and candidate countries, Sol. Energy 81 (2007) 1295–1305.

80. R. Hendron, J. Burch, Development of standardized domestic hot water event schedules for residential buildings, in: ASME 2007 Energy Sustainability Conference pp. 531–539, 2007.

81. E.S.T. England, Measurement of domestic hot water consumption in dwellings/ monitoring/housing professionals/publications/home (England), 2008.

82. U. Jordan, K. Vajen, Influence of the DHW load profile on the fractional energy savings: a case study of a solar combi-system with TRNSYS simulations, Sol. Energy 69 (2001) 197–208.

83. U. Jordan, Statistically generated DHW-load profiles on a one-miunte time scale, in: IEA SHACP Task 26, 2000.

84. AECB Water Standards, Delivering buildings with excellent water and energy performance, Volume 1: The water standard. , 2009 (accessed 25.03.16).

85. J. Widén, M. Lundh, I. Vassileva, E. Dahlquist, K. Ellegård, E. Wäckelgård, Constructing load profiles for household electricity and hot water from time-use data – Modelling approach and validation, Energy Build. 41 (2009) 753–768.

86. I. Richardson, M. Thomson, D. Infield, C. Clifford, Domestic electricity use: a high-resolution energy demand model, Energy Build. 42 (2010) 1878–1887.

87. N. Amrizal, D. Chemisana, J.I. Rosell, Hybrid photovoltaic–thermal solar collectors dynamic modeling, Appl. Energy 101 (2013) 797–807.

CHAPTER 6

THERMAL EFFECTIVENESS AND MASS USAGE OF HORIZONTAL MICRO-FINS UNDER NATURAL CONVECTION

Leonardo Micheli [a], K.S. Reddy [b], Tapas K. Mallick [a]

[a] Environment and Sustainability Institute, University of Exeter, Penryn Campus, Penryn, Cornwall TR10 9FE, UK

[b] Heat Transfer and Thermal Power Laboratory, Department of Mechanical Engineering, Indian Institute of Technology Madras, Chennai 600 036, India

ABSTRACT

In recent times, the micro-technologies have gained prominence in various engineering applications. The micro-technologies are already in use for cooling purposes in several systems, but the information on the thermal performance of micro-fins under natural convective heat transfer conditions is yet limited. The correlations between heat transfer coefficients and fin geometry have already been investigated, but are not sufficient to optimize the design of the micro-finned arrays. For this reason, the present investigation gives an overview of micro-fins behavior taking into account, for the first time, different heat sink metrics: the fin effectiveness and the mass specific heat transfer coefficient. The results of an original experimental investigation are merged with the data available in literature. Natural convective micro-fins are able to achieve overall fin effectivenesses higher than 1.1. Even if not always beneficial in terms of heat transfer, micro-fins are found always positive in terms of the material usage. In this light, micro-fins can be considered advantageous in those applications that require a minimized weight of the heat sinks. Moreover, a limited effect due to the orientation is observed.

INTRODUCTION

Extended surfaces/fins are commonly used to enhance the heat transfer from a system to the surrounding fluid. The application of fins has been investigated in numerous studies and is currently employed for many different purposes, such as electronics, industrial processes and energy generation. Fins in natural convection are considered as passive coolers, since they do not require input of mechanical or electrical power. Fins act by exploiting both the natural convective motion of a fluid due to a thermal gradient and the radiative heat transfer [1]. Passive coolers are generally considered more reliable and less exposed to cooling failures [2] than active coolers, which need instead external energy in input to perform.

Industries and consumers are always for the production of more efficient, more compact and low cost products. In this light, micro-technologies have gained much interest in the last decades, because of the better performance achieved and the limited space and material required compared to macro-

Citation: Leonardo Micheli, K.S. Reddy, Tapas K. Mallick, Thermal effectiveness and mass usage of horizontal micro-fins under natural convection, Applied Thermal Engineering, Volume 97, 25 March 2016, Pages 39-47, ISSN 1359-4311, http://dx.doi.org/10.1016/j.applthermaleng.2015.09.042.

scaled solutions. For this reason, micro-cooling technologies, including micro-finned arrays in forced convection conditions, have been extensively investigated [3], [4] and [5]. Despite that, only a limited number of researches on naturally convective micro-fins can be found in literature. Kim et al. [6] investigated vertically orientated micro-fins and demonstrated that macro-fin heat transfer correlations could not be used to describe the behavior of micro-scaled systems. Mahmoud et al. [7] experimentally sorted out, for the first time, the correlations between micro-fin geometries and heat transfer coefficients. Even if lower than those of a flat surface, the heat transfer coefficients for horizontal upward facing fins were found to increase when the fin spacing was increased and the fin height was decreased. Recently, Micheli et al. [8] reported that the heat transfer coefficient is enhanced by the thickness of the fins. Shokouhmand and Ahmadpour [9] numerically demonstrated the non-negligible contribution of radiation, which accounted for more than 20% of the global micro-fins heat transfer.

Earlier researches mainly focused on the heat transfer coefficients. In real world applications, instead, engineers and system designers look for thermal resistance, compactness, weight and cost of the heat sinks. The present work aims to study, for the first time, the effective thermal enhancement due to the application of micro-fins on a horizontal flat surface by analyzing additional heat sink metrics, such as the individual and overall fin effectiveness and the mass specific heat transfer coefficient. The correlations between geometry and heat sink metrics are explored and commented in the present paper to contribute to optimize the design of micro-fins. The arrays are studied in upward and downward orientations: despite the upward conditions are known to be the best for natural cooling [10], different orientations might be required in some applications[11] and, for this reason, need to be investigated.

HEAT SINK METRICS

The previous works [6] and [7] studied the heat transfer coefficient of micro-fins arrays (h_{fins}), which is described as follows:

$$h_{fins} = \frac{Q_{fins}}{S_{fins}\cdot(T_{fins}-T_a)} \qquad (1)$$

where Q_{fins} is the heat dissipated through the fins by convection, S_{fins} and T_{fins} are respectively the surface and the temperature of the finned surface, and T_a is the ambient temperature. In accordance with previous investigations on micro-fins, T_{fins} is considered constant along both the fin height and the length of the array [7] and [8]. Because of this assumption, no analysis of the fins efficiency has been carried out. Therefore, since no temperature gradient is considered across the finned surface, h_{fins} represents the average of the heat transfer coefficients of the various surfaces of the fins.

The aim of a fin is to increase the heat transfer from a surface to a fluid by increasing the thermal exchanging surface. In practical applications, it is required to understand the effective heat transfer enhancement introduced by the fins if compared to the original flat surface. The heat transfer coefficient measures the thermal property per unit of surface and it is not an indicator of the thermal performance of the heat sink, because it does not take into account the surface extension obtained when the fins are introduced. For this reason, a second parameter, the individual fin effectiveness (ε_f), is generally used [12] to compare the behavior of a single fin and that of a flat surface:

$$\varepsilon_f = \frac{Q_f}{h_{flat}\cdot A_f\cdot(T_{flat}-T_a)} \qquad (2)$$

where Q_f is the heat transferred by a single fin, A_f is its cross-sectional area, and h_{flat} and T_{flat} are respectively the heat transfer coefficient and the temperature of the flat plate. Q_f is obtained as:

$$Q_f=h_{fins}\cdot(2\cdot L\cdot H+2\cdot t\cdot H+A_f)\cdot(T_{fins}-T_a) \qquad (3)$$

where L is the fin length, H is fin height and t is the fin thickness (Fig. 1). Incropera et al. [13] defines the individual fin effectiveness as the ratio of the fin heat transfer rate to the heat transfer rate that would exist without the fin. ε_f is expected to be as large as possible. A fin array is made of a number of fins: for this reason the overall fin effectiveness (ε_{fins}) has to be considered as well, in order to have a most accurate measure of the performance of the whole fin array [14], [15] and [16]. The overall fin effectiveness (ε_{fins}) is obtained as the ratio of heat transfer with fins to that without fins over the same area of a flat surface of length L and width W:

$$\varepsilon_{fins} = \frac{Q_{fins}}{Q_{flat}} \qquad (4)$$

where Q_{flat} is the heat transferred by the flat plane. The overall fin effectiveness directly compares the heat transferred by the fin array and by the unfinned surface: if $\varepsilon_{fins}>1$, the fins have enhanced the thermal behavior of the surface.

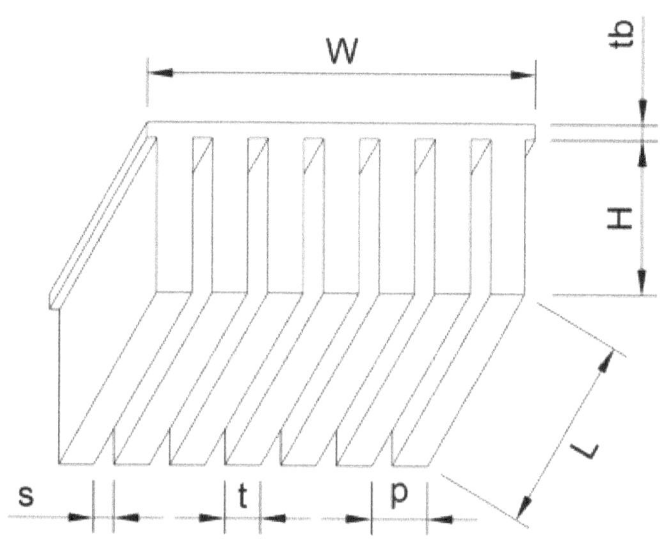

Figure 1. Horizontal plate micro-fins geometry with typical dimensions.

Micro-fins are usually obtained through material subtractions: along with the effects on the heat transfer, they reduce the mass of the heat sink. This feature becomes particularly important in portable or tracked systems, such as concentrating photovoltaics, where a reduced weight means a reduced load for the tracker [17]. The mass specific heat transfer coefficient measures the effectiveness with which fin material is utilized in the promotion of heat transfer [18] and is expressed as:

$$h_m = \frac{Q_{fins}}{\rho \cdot V_{fins} \cdot (T_{fins} - T_a)} \quad (5)$$

where ρ is the density of the fin material and V_{fins} is the volume of the whole micro-finned heat sink.

Table 1. Dimensions of the fin arrays, in mm. L=W=50 mm.

Fin geometry no.	Height (H)	Pitch (p)	Thickness (t)	Spacing (s)	Base thickness (tb)
#1	0.6	0.4	0.2	0.2	0.8
#2	0.6	1.0	0.2	0.8	0.8
#3	0.6	0.6	0.4	0.2	0.8
#4	0.6	1.2	0.4	0.8	0.8
#5	0.6	0.8	0.4	0.4	0.8
#6	0.6	1.6	0.8	0.8	0.8
#7	0.8	0.4	0.2	0.2	0.6
#8	0.8	0.8	0.4	0.4	0.6
#9	1.0	0.6	0.2	0.4	0.4
#10	1.0	0.6	0.4	0.2	0.4

EXPERIMENTAL INVESTIGATION

Micro-fin Geometries

A micro-fin is any extended surface with at least one micro-scaled dimension. In the present work, ten plate fin geometries have been considered: the fin dimensions are resumed in Table 1, according to the nomenclature shown in Fig. 1, and numbered from #1 to #10. The fin arrays are diced on 1.4 mm-thick, 50 mm × 50 mm-sized square silicon wafers and their thermal behaviors are compared with that of a flat unfinned silicon wafer.

Experimental Setup

The experimental setup described in Ref. 8 is used, as shown in Fig. 2. The heat is generated by 5 cm × 5 cm-sized electrical heaters (Omega KHLV-202/2.5), regulated using a DC power supply (Weir 413D). The heaters are bonded to the back surface of the silicon wafers through a thin conductive adhesive (3M tape 966, 0.18 W/mK). The power in input (Q_{in}) is calculated by multiplying the input voltage and the input current, measured using two digital multimeters (Fluke 115 and Fluke 8050).

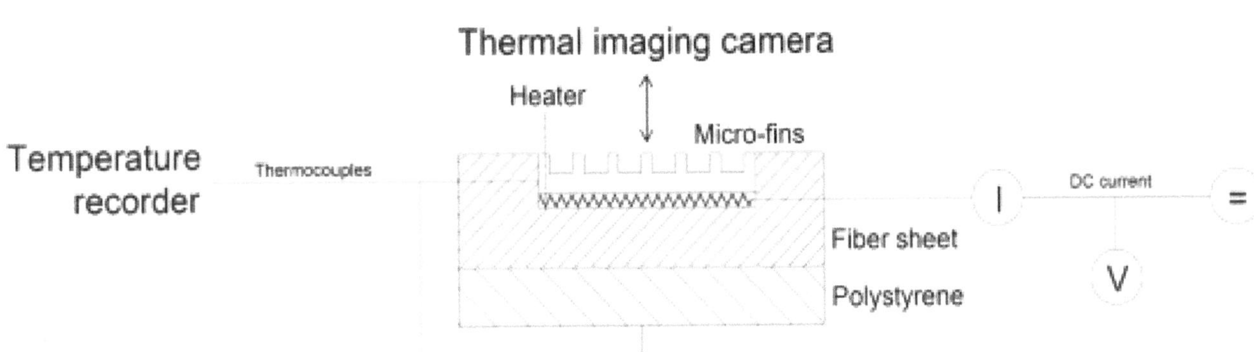

Figure 2. Schematic of experimental setup.

In order to minimize the thermal losses from any surface other than the fins, each wafer is placed in a 1-cm thick case made of fiber thermal material (0.05 W/mK), covered on the back by a 1 cm-thick polystyrene block (0.03 W/mK). Measures are taken at steady state conditions, 20 minutes after the test has started, using some K type thermocouples placed on the back and the sides of the insulating case and a thermal imaging camera to determine the fin array temperature. The temperature of the fins is considered constant across the whole volume of each fin array, as well as the heat transfer coefficients [6],[7] and [8]: indeed, due to the limited size of the fins, no gradient of temperature could be defined across the array. Therefore, the maximum temperature recorded by the thermal imaging camera at steady state has been used for the calculations, as shown in Fig. 3. The temperature of the surrounding air is measured averaging the outputs of three K type thermocouples located in proximity of the experimental setup.

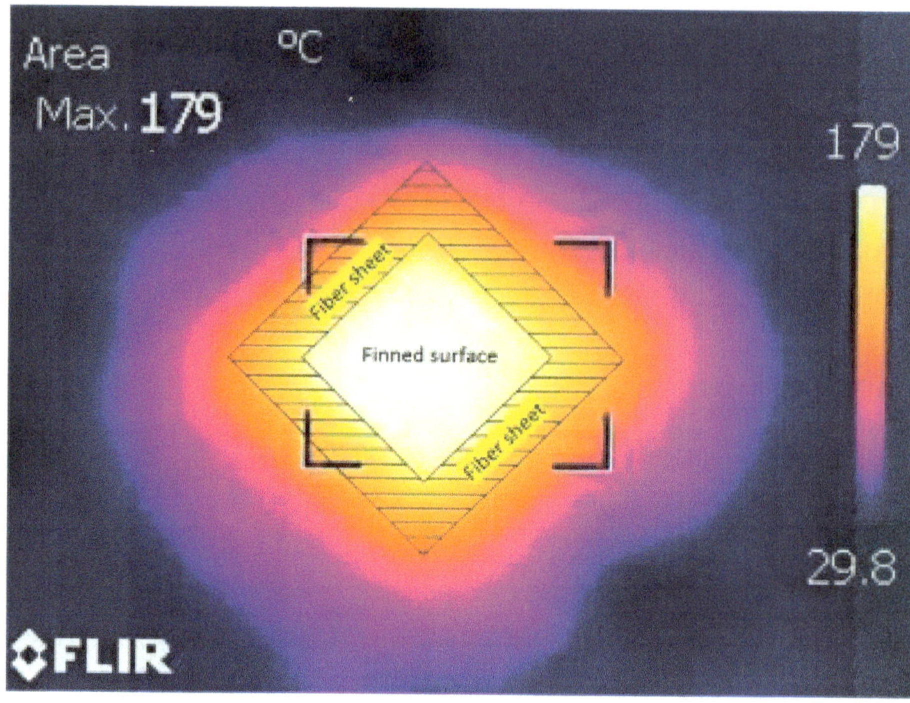

Figure 3. Front shot of the fin array in upward position taken from the thermal imaging camera. The finned surface and the fiber sheet have been labeled.

Power inputs of 1.0 kW/m², 2.0 kW/m², 2.5 kW/m², 3.0 kW/m², 3.5 kW/m² and 4.0 kW/m² are supplied to the heater. Each test is repeated three times and the average value of the outputs is considered. The fin behavior is investigated both in upward facing and in downward facing conditions.

The power dissipated by convection from the fins array is calculated after the radiative heat transfer from the fins (Q_r) and the losses that happen on the back and the sides of the samples (Q_{losses}) [8]:

$$Q_{fins}=Q_{in}-Q_r-Q_{losses} \tag{6}$$

The radiative component is estimated by using the following equation [7]:

$$Q_r = \sum_i \varepsilon \cdot \sigma \cdot S_i \cdot F_{i,a} \cdot (T_s^4 - T_a^4) \tag{7}$$

where ε is the emissivity of fin material, σ is the Stefan–Boltzmann constant ($5.67 \cdot 10^{-8}$ W/m²K⁴), S_i is the area of the correspondent i-surface of the fins, $F_{i,a}$ is the view factor between the i-surface of the fins and the ambient, and T_s and T_a are respectively the temperatures of the i-surface and the ambient. The view factors of each surface are calculated according to the methodology suggested by Ref. 10.

The Q_{losses} are due to the heat dissipation happening on the unfinned surfaces of the experimental setup: these are the radiative and the convective thermal exchanges taking place on the sides and on the back of the insulating case. The losses have been calculated according to the procedures reported in [7] and [8]: an average of 26% of the power in input is dissipated through the insulated case made of fiber thermal material and polystyrene.

Uncertainty and Experimental Validation

Taking into account the propagation of error for independent variables [7], the uncertainties on the calculation of the heat transfer coefficient and the heat transfer coefficients are found to be lower than ±8.3% [8]. The uncertainties on the measurements of the fin surface, the fin and the ambient temperatures have been considered in the calculation. In particular, the thermocouple measurements depend on the emissivity of the material, which has been calculated from the reflectivity of the wafer, measured using a spectrometer (Perkimeter Lamda 1050) [8]. On the other hand, calculating the overall fin effectiveness raises the uncertainty up to ±10.3%, since the uncertainty on the measurements made for the flat surface has to be considered as well, and it accounts for 6.1%.

Some of the overall fin effectivenesses presented in Section 4.2 are obtained by processing data available in graphical format in Refs. 7. In Ref. 7, the dimensions of the fins are clearly stated in the text, whereas the heat transfer coefficients and the temperatures are presented in figures only. These data, necessary to calculate the overall fin effectiveness, have then been extracted using *Engauge Digitizer 4.1* (M. Mitchell, Engauge Digitizer, http://digitizer.sourceforge.net). This is open source software package able to convert graphs into numbers. The authors of Ref. 7 reported an experimental uncertainty of ±9.4% on their data. Repeating the digitalization twice, a repeatability uncertainty of ±1.0% is found for the present work. Moreover, the size of the markers in the graphs introduces an uncertainty that corresponds to a maximum ±3.1%. Overall, the uncertainty on the digitalized heat transfer coefficients rises then to 9.9%.

The present experimental setup has been previously validated in Ref. 8. As a further proof of validation, the Nusselt numbers experimentally obtained in this investigation are compared with those predicted by referenced correlations for both upward and downward orientations. The experimental setup is firstly validated according to the correlation reported in Ref. 19 for upward-facing horizontal square plates in air. The mean heat transfer correlation is expressed as:

$$Nu = 1.05 \cdot Ra^{0.215} \quad (8)$$

where the Rayleigh number is calculated as [19]:

$$Ra = \frac{g \cdot \beta \cdot (T_s - T_a) \cdot W^3 \cdot Pr}{\nu^2} \quad (9)$$

where g is the gravitational acceleration, β is the volumetric thermal expansion of air, Pr is the Prandtl number, and ν is the kinematic viscosity of air. All the properties of air are extracted from the tables presented by Ref. 16 and are evaluated for an air temperature of $(T_s+T_a)/2$, with the exception of the volumetric thermal expansion coefficient, evaluated instead at ambient temperature [7] and [10]. The numerical equation is found to consistently underestimate the experimental heat transfer coefficients, with an average discrepancy of 19%. This value falls within the ±20% range reported by the author [19]. It has to be mentioned that considerable uncertainties can take place when natural convection correlations are adopted: a ±20% uncertainty span is considered as acceptable [10], though in some previous investigations, the differences between experimental and numerical data have been found to amount up to 50% [19].

Secondly, the fin arrays in downward facing orientation are similarly validated using the equation proposed by Ref. 11 for downward-facing horizontal square plates in air:

$$Nu = \left[1 + 0.24 \cdot e^{(-0.0025 L^*)}\right] \cdot 0.46 \cdot Ra_L^{0.20} \quad (10)$$

where Ra_L is the Rayleigh number expressed as function of the fin array half length, and L* is the characteristic length, defined as:

$$L^* = \frac{L}{2 \cdot \sqrt[3]{\alpha \cdot \nu / g}} \quad (11)$$

where α is the thermal diffusivity of air.

RESULTS AND DISCUSSION

Individual fin Effectiveness

Micro-fins achieve lower heat transfer coefficients than flat plates [7]. This is due to the air gaps between the micro-fins which act similarly to an insulating layer [6], reducing the rate of heat transferred per unit of surface on the side walls and the base of the fins. The reduction in heat transfer coefficient can be balanced by the increase in thermal exchanging surface: the variation in heat transfer due to the fins can be measured through the individual and the overall fin effectivenesses.

In order to grant a real benefit, fins are generally required to achieve individual effectivenesses greater than 2 [13]. In this experimental investigation, the individual fin effectiveness ranges between 1.25 and 3.32. In particular, ε_f is found to be enhanced by low fin thicknesses and high fin spacing. As shown in Fig.4, the maximum individual fin effectiveness is achieved when the ratio t/s is minimized, whereas the individual fin effectiveness lowers if t/s increases. Increased fin spacings favor the thermal exchange because larger air volumes are present between the fins and, this way, air is more likely to exchange heat by convection than by conduction. Thus, in larger fin spacings, more heat is transferred through the walls of the fins, improving the amount of heat globally dissipated by the fins. The same behavior and a similar range of ε_f are found when the micro-fins are downward orientated: the individual fin effectiveness ranges between 1.19 and 3.30. It means that the effect of the orientation on the individual fin effectiveness can be considered as negligible.

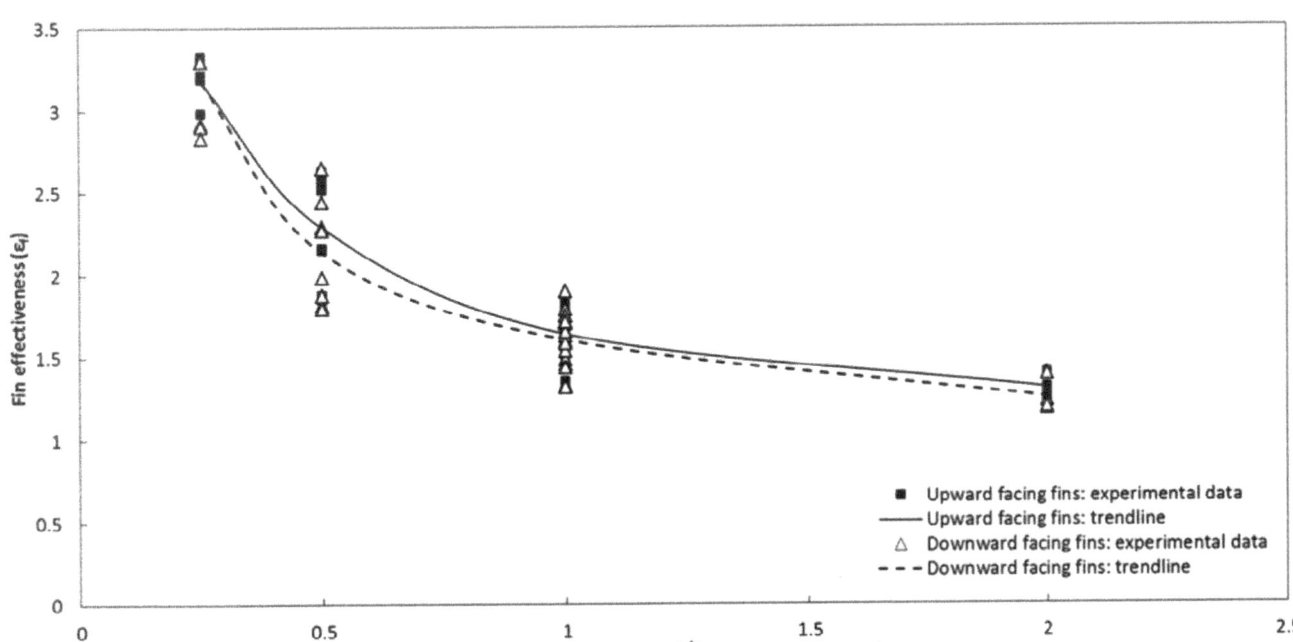

Figure 4. The behavior of individual effectiveness of the fins depending on the t/s ratio for upward facing fins (■) and downward facing fins (Δ).

Overall Fin Effectiveness

The thermal performance of the fin arrays depends on the individual fin effectiveness and on the number of fins. For this reason, the overall fin effectiveness is generally considered a better index to measure the performance of a finned surface than the individual fin effectiveness [16]. Kim et al. [6] measured a heat transfer enhancement up to 10% after the installation of micro-fins, in agreement with the findings of previous researches on micro-fins in forced flow conditions. The overall fin effectiveness was found to increase when the fin spacing was decreased. This was explained because a limited spacing decreased the volume of air compared to that of the higher-conductive fin material, increasing the overall heat transfer. Within the range considered for that work (s spanned between 30 and 360 μm), enlarging the fin spacing did not benefit to the convective heat transfer, since air was still mostly acting as a solid conductive layer. For this reason, in those conditions, increasing the volume of the more conductive silicon benefitted to the heat transfer.

Despite the wider spacing considered, the limited enhancement is confirmed in the present experimental investigation, where the overall fin effectiveness is found to range between 0.98 and 1.02. This means that the introduction of micro-fins for natural convection is not necessarily positive for the overall heat transfer. The same results are obtained by analyzing the data reported by Ref. 7, where the overall fin effectiveness ranges between a maximum of 1.14 and a minimum of 0.86. In Fig. 5, a summary of the overall effectivenesses presented in this work and in Ref. 7 is reported. The average overall effectiveness is found as 0.985.

The results shown in Fig. 4 and Fig. 5 are not in contradiction. A good individual fin effectiveness does not necessarily lead to a good overall fin effectiveness. This is due to the spacing between the fins, which, at micro-scale, becomes a particularly critical parameter: a reduced spacing forces the air to thermally behave similarly to a solid, strongly affecting the heat transfer from the base and from the sides of the fins. Therefore, the big discrepancy between the individual fin effectiveness and the overall fin effectiveness remarks the complexity of heat transfer at micro-scale. The dimensions of fins and the spacings between them strongly affect the performance of an array of the fins, whereas it has only a limited effect on the heat transfer of a single fin. For this reason, the overall fin effectiveness is found to be lower than the individual fin effectiveness.

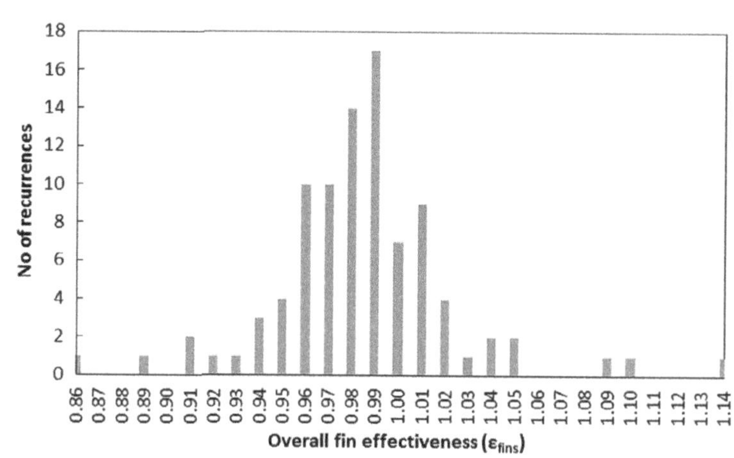

Figure 5. Number of recurrence per overall fin effectiveness.

Due to the high viscous forces and to its low thermal conductivity, the air confined in the narrow volumes between two fins behaves as an insulating layer. For this reason, the heat transfer coefficient is found to decrease when micro-fins are added to a flat plane surface [7] and [8]. The drop in thermal convection can be balanced by the increase in surface: the overall fin effectiveness in natural convection can be as high as 1.14. Although an enhancement in heat transfer coefficient has been proved when the base thickness is decreased and the spacing is increased, conflicting results have been obtained when the overall fin effectiveness is analyzed. The correlations between the overall fin effectiveness and the fin geometries have not been sorted out yet and are essential to design optimized micro-finned heat sinks. It has been clearly shown, instead, that the individual effectiveness of a single fin is enhanced when the fin thickness is lowered and the fin spacing is increased. The micro-fins tested in the present work shows individual effectiveness higher than 3, both in upward and downward facing conditions.

In real applications, a heat sink exchanges heat by both convection and radiation. The heat sink metrics used to investigate thermal performance of fin heat sink generally do not take into account the contribution of radiation [20]. It has already been shown that radiation should not be neglected when a naturally convective micro-finned array is designed [8] and [9]. Compared to a flat surface, the contribution of radiation in a finned array is influenced by an increased thermal exchanging surface and by some limiting view factors. For this reason, the combined contribution of radiation and convection in micro-fin arrays should be taken into account in future works.

Mass Specific Heat Transfer

The mass specific heat transfer coefficient measures the thermal performances per unit of mass of the heat sink. Even if the traditional heat transfer coefficient is negatively affected, the mass specific heat transfer coefficient is found to be enhanced by the introduction of fins. It means that the benefit in weight reduction due to the material subtraction is more effective than the change in the thermal performance.

The correlations between the mass specific heat transfer and the fin geometries are not coherent with those valid for the heat transfer coefficients. It is found that the specific mass heat transfer increases when increasing the spacing of the fins (Fig. 6), and/or increasing the height (Fig. 7). The increase in fin height means a drop in the heat sink weight, because the base thickness is decreased.

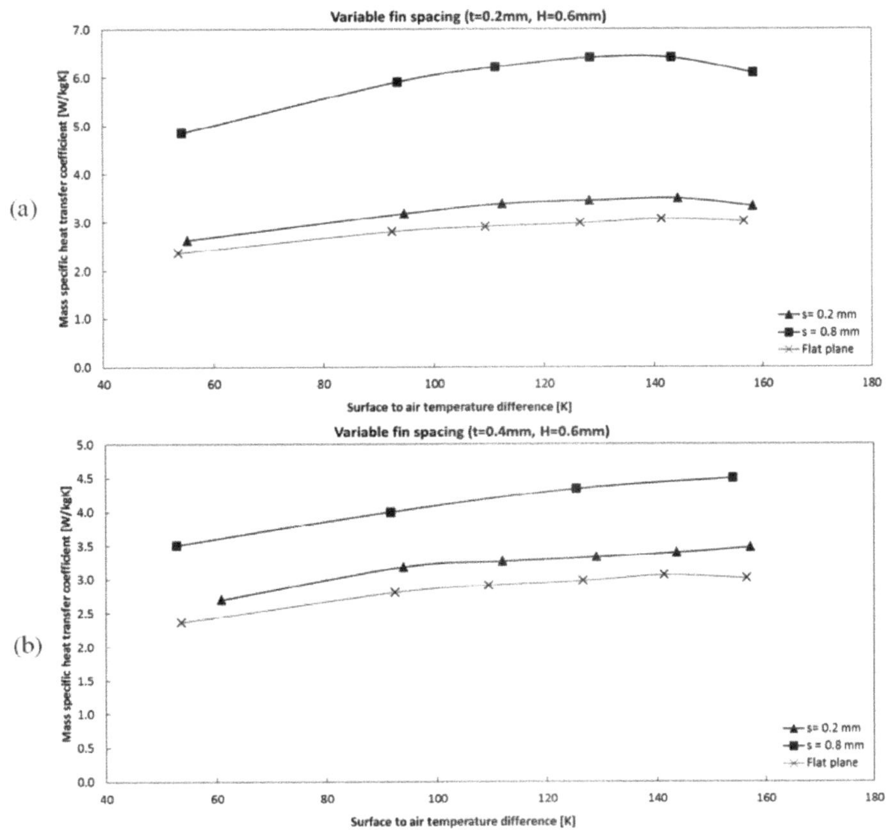

Figure 6. The effects of fin spacing for upward-facing fin arrays on the mass specific heat transfer coefficient: (a) t=0.2 mm and H=0.6 mm; (b) t=0.4 mm and H=0.6 mm.

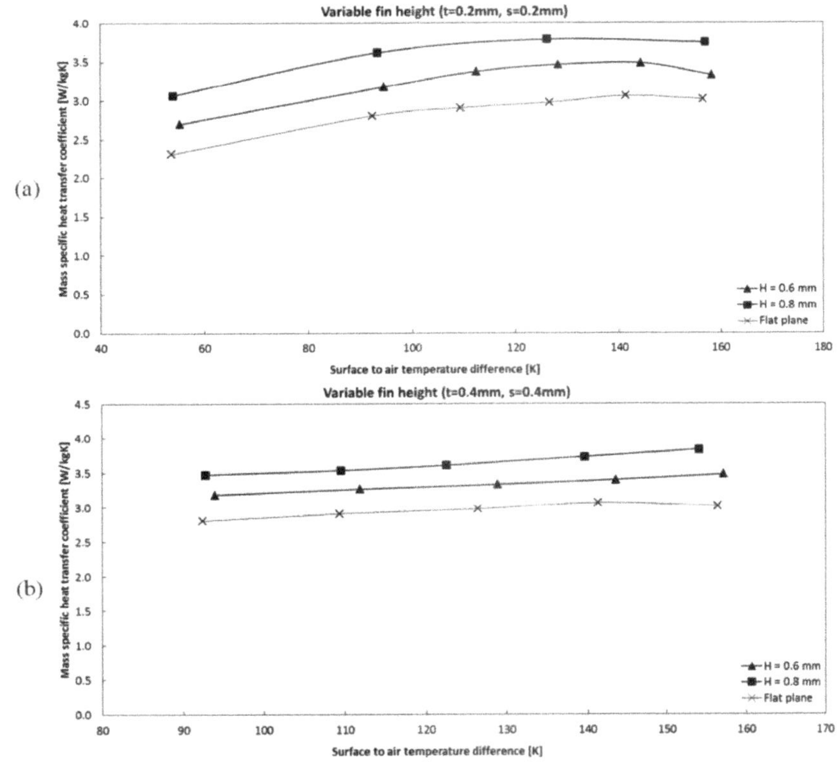

Figure 7. The effects of fin height for upward-facing fin arrays on the mass specific heat transfer coefficient: (a) t=0.2 mm and s=0.2 mm; (b) t=0.4 mm and s=0.4 mm.

These results lead to the conclusion that the drop in weight, instead of an enhancement in heat transfer, is the most important benefit obtained by dicing micro-fins on a flat cooling surface. For the same reason, the mass specific heat transfer is found to increase when the fin thickness increases (Fig. 8).

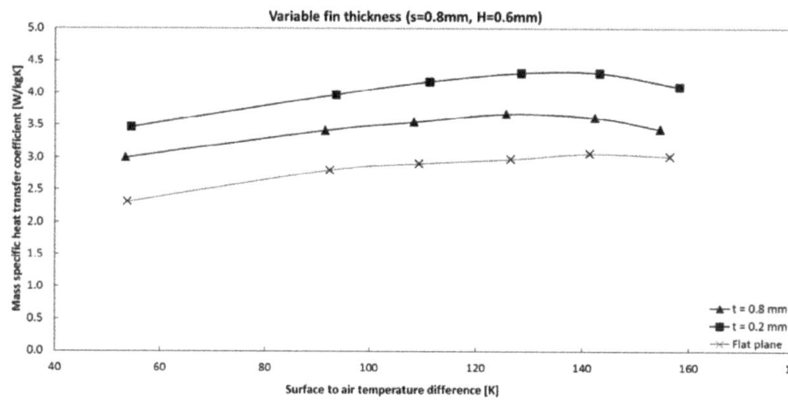

Figure 8. The effects of fin thickness for upward-facing fin arrays on the mass specific heat transfer coefficient: s=0.8 mm and H=0.6 mm.

The introduction of micro-fins leads to an improvement up to 50% of mass specific heat transfer coefficient compared to the flat surface. As expected, the maximum enhancements are found in conditions of maximum spacing and minimum thickness. This makes micro-finned heat sinks particularly preferable for moved systems, where the weight is a parameter to be limited, in those applications where the use of an active cooling system is not considered advantageous in terms of costs, volume and/or energy consumption.

The presented results are found to be consistent when the fins are downward oriented: the mass specific heat transfer coefficient increases when the fin spacing increases (Fig. 9), the fin height increases (Fig. 10), and the fin thickness decreases (Fig. 11). Compared to the upward conditions, the mass specific heat transfer coefficients are found to decrease by 13–27% for downward-facing micro-fin arrays.

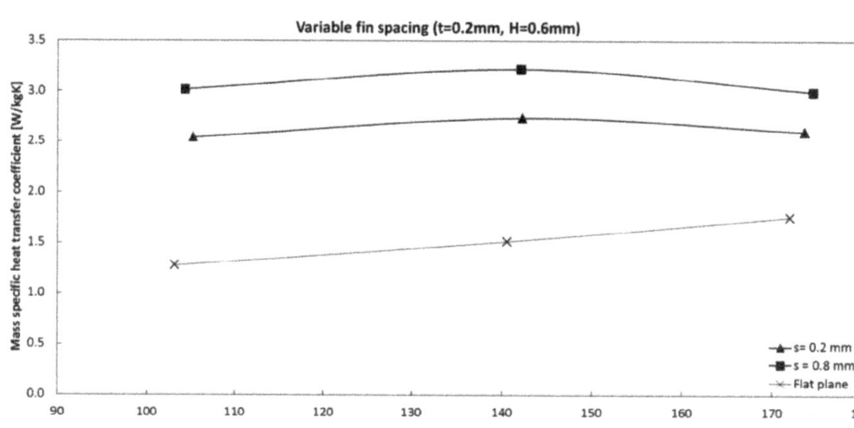

Figure 9. The effects of fin spacing for downward-facing fin arrays on the mass specific heat transfer coefficient: t=0.2 mm and H=0.6 mm.

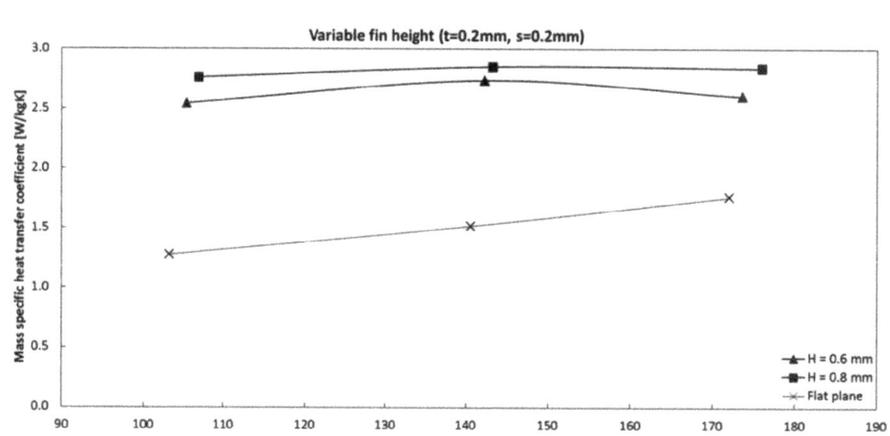

Figure 10. The effects of fin height for downward-facing fin arrays on the mass specific heat transfer coefficient: t=0.2 mm and s=0.2 mm.

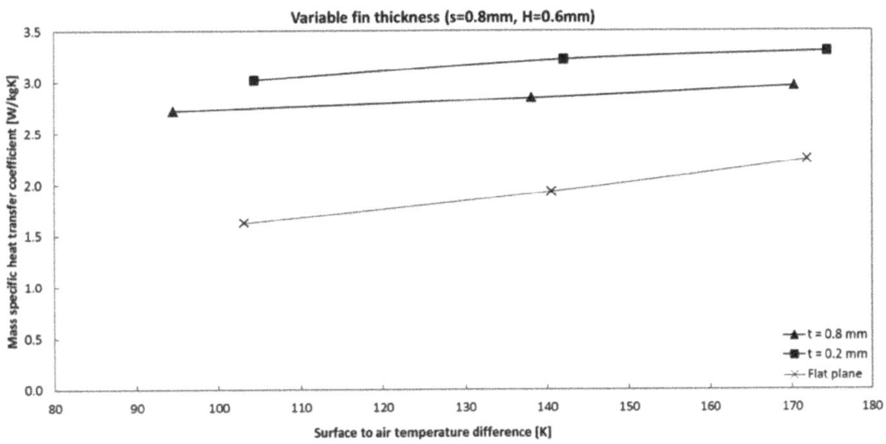

Figure 11. The effects of fin thickness for downward-facing fin arrays on the mass specific heat transfer coefficient: s = 0.8 mm and H = 0.6 mm.

The results presented in the work can be good guidelines to identify the optimal design of micro-finned heat sinks. Natural convective micro-fins can find application in several fields: they are found to be a suitable passive cooling solution for those systems that require a limited heat sink's weight, such as concentrating photovoltaics. In this light, more investigations on different geometries, materials and conditions are necessary to fully understand the benefit that can be obtained by naturally convective micro-fins. In particular, in order to make it possible to predict the micro-fin performance in a wider range of applications, a study on natural convective heat transfer of micro-finned arrays for different tilt angles is recommended. The contribution of radiation should not be neglected in future works. Potential mass reductions and heat transfer enhancements are not the only benefits that a heat sink should achieve. Therefore, investigations on the costs, the applicability and the reliability of micro-finned array have to be carried out to make of this technology an effective solution for micro-scaled coolers.

CONCLUSIONS

Natural convective micro-fins are a high-potential, widely unexplored technology. Only a limited number of studies are available and these are mainly focused on the heat transfer coefficients, a metric that cannot be used for an exhaustive characterization of a heat sink. In this light, the present paper investigated the thermal performance of horizontal micro-fins under natural convective conditions using different heat sink metrics and showed that micro-fins can enhance the heat transfer of a flat surface. The experimental setup used in this work has been described and the results of the investigation have been presented. Micro-fins achieved individual effectiveness higher than 3, beneficing of high spacings and low thicknesses. The overall heat exchange of a surface has been found to be enhanced up to 14% by using an opportunely-designed micro-finned array, whereas in some cases, the micro-fins have been found to lower the thermal performance. Due to the limited variation of overall fin effectiveness and to the non-negligible experimental uncertainty, no clear correlations between geometry and overall fin effectiveness have been found. Further studies need to be carried out to understand how to optimize the design of a micro-fins array for natural convective applications. For the first time, the mass specific heat transfer coefficient has been investigated to understand the effects of the micro-fins on the weight of the heat sink. The introduction of micro-fins is found to be particularly beneficial to the material usage: a variation up to 50% in mass specific heat transfer has been observed when micro-fins are used. The orientation of the arrays did not show any qualitative effect on the correlations between the fin geometries and the thermal performance. Downward facing fins did have a lower heat transfer management, but showed thermal behaviors similar to those of upward facing fins.

ACKNOWLEDGEMENTS

This work has been carried out as a part of BioCPV project, jointly funded by EPSRC, UK(Ref No: EP/J000345/1) and DST, India (Ref No: DST/SEED/INDO-UK/002/2011). The financial support provided by both the agencies is duly acknowledged.

Nomenclature

A_f Cross-sectional area of a fin [m^2]

$F_{i\text{-}a}$ View factors between the surfaces i and the ambient

G Standard gravitational acceleration [m/s^2]

H Height of a fin [m]

h_{fins} Heat transfer coefficient of the micro-finned surface [W/m^{2}K]

h_{flat} Heat transfer coefficient of the unfinned surface [W/m^{2}K]

h_m Mass specific heat transfer coefficient [W/kg K]

L Length of the fins [m]

L* Characteristic length

P Pitch of the fin array [m]

Q_f Heat dissipated by a single fin through convection [W]

Q_{fins} Heat dissipated by fins through convection [W]

Q_{flat} Heat dissipated by the unfinned surface through convection [W]

Q_{in} Heat supplied by the DC supply [W]

Q_{losses} Heat dissipated by the surfaces other than the fins [W]

Q_r	Heat dissipated by fins through radiation [W]
S	Spacing between adjacent fins [m]
S_{fins}	Area of the finned surface [m^2]
S_i	Area of the i-surface of the fins [m^2]
T	Thickness of a fin [m]
T_a	Ambient temperature [K]
tb	Thickness of the fin array's base [m]
T_{fins}	Temperature of the finned surface [K]
T_{flat}	Temperature of the unfinned surface [K]
T_s	Surface's temperature [K]
V_{fins}	Volume of the finned surface [m^3]
W	Width of the fin array [m]

Greek symbols

α	Thermal diffusivity [m^2/s]
β	Coefficient of volumetric expansion [1/K]
ε	Emissivity
ε_f	Individual fin effectiveness
ε_{fins}	Overall fin effectiveness
ν	Kinematic viscosity of air [m^2/s]
ρ	Density [kg/m^3]
σ	Stefan–Boltzmann constant [W/m^2 K^4]

Dimensionless Numbers

Nu	Nusselt number
Pr	Prandtl number
Ra	Rayleigh number

REFERENCES

1. L. Micheli, N. Sarmah, X. Luo, K.S. Reddy, T.K. Mallick, Opportunities and challenges in micro- and nano-technologies for concentrating photovoltaic cooling: a review, Renew. Sustain. Energy Rev. 20 (2013) 595–610, doi:10.1016/ j.rser.2012.11.051.
2. Y.-S. Tseng, H.-H. Fu, T.-C. Hung, B.-S. Pei, An optimal parametric design to improve chip cooling, Appl. Therm. Eng. 27 (2007) 1823–1831, doi:10.1016/ j.applthermaleng.2007.01.012.
3. D.P. Kulkarni, D.K. Das, Analytical and numerical studies on microscale heat sinks for electronic applications, Appl. Therm. Eng. 25 (2005) 2432–2449, doi:10.1016/j. applthermaleng.2004.12.010.
4. H. Shafeie, O. Abouali, K. Jafarpur, G. Ahmadi, Numerical study of heat transfer performance of single-phase heat sinks with micro pin-fin structures, Appl. Therm. Eng. 58 (2013) 68–76, doi:10.1016/j. applthermaleng.2013.04.008.
5. J. Yeom, M.A. Shannon, 3.16 Micro-coolers, in: Y. Gianchandani, O. Tabata, H. Zappe (Eds.), Compr. Microsystems, Elsevier, New York, 2007, pp. 499–550.
6. J.S. Kim, B.K. Park, J.S. Lee, Natural convection heat transfer around microfin arrays, Exp. Heat Transf. 21 (2008) 55–72, doi:10.1080/08916150701647835.
7. S. Mahmoud, R. Al-Dadah, D.K. Aspinwall, S.L. Soo, H. Hemida, Effect of micro fin geometry on natural convection heat transfer of horizontal microstructures, Appl. Therm. Eng. 31 (2011) 627–633, doi:10.1016/j. applthermaleng.2010 .09.017.
8. L. Micheli, K.S. Reddy, T.K. Mallick, General correlations among geometry, orientation and thermal performance of natural convective micro-finned heat sinks, Int. J. Heat Mass Transf. 91 (2015) 711–724, doi:10.1016/ j.ijheatmasstransfer.2015.08.015.
9. H. Shokouhmand, A. Ahmadpour, Heat transfer from a micro fin array heat sink by natural convection and radiation under slip flow regime. Proc. World Congr. Eng., vol. 2, 2010.
10. N.V. Suryanarayana, Engineering Heat Transfer, West Publishing Company, 1995.
11. A. Dayan, R. Kushnir, G. Mittelman, A. Ullmann, Laminar free convection underneath a downward facing hot fin array, Int. J. Heat Mass Transf. 47 (2004) 2849–2860, doi:10.1016/j.ijheatmasstransfer.2004.01.003.
12. P.A. Razelos, Critical review of extended surface heat transfer, Heat Transf. Eng. 24 (2010) 11–28, doi:10.1080/714044411.
13. F.P. Incropera, D.P. DeWitt, T.L. Bergman, A.S. Lavine, Fundamentals of Heat and Mass Transfer, Wiley, 2007.
14. M.M. Rathore, R.R. Kapuno, Engineering Heat Transfer, Jones & Bartlett Learning, 2011.
15. V.D. Rao, S.V. Naidu, B.G. Rao, K.V. Sharma, Heat transfer from a horizontal fin array by natural convection and radiation – A conjugate analysis, Int. J. Heat Mass Transf. 49 (2006) 3379–3391, doi:10.1016/j. ijheatmasstransfer.2006 .03.010.
16. Y.A. Çengel, Introduction to Thermodynamics and Heat Transfer, second ed., McGraw-Hill, 2008.
17. L. Micheli, S. Senthilarasu, K.S. Reddy, T.K. Mallick, Applicability of silicon micro-finned heat sinks for 500× concentrating photovoltaics systems, J. Mater. Sci. 50 (2015) 5378–5388, doi:10.1007/s10853-015-9065-2.
18. A. Bar-Cohen, M. Iyengar, A.D. Kraus, Design of optimum plate-fin natural convective heat sinks, J. Electron. Packag. 125 (2003) 208, doi:10.1115/ 1.1568361.
19. M. Corcione, Heat transfer correlations for free convection from upward-facing horizontal rectangular surfaces, Trans. Heat Mass. Transf. 2 (2007) 48–60.
20. Y.K. Khor, Y.M. Hung, B.K. Lim, On the role of radiation view factor in thermal performance of straight-fin heat sinks, Int. Commun. Heat Mass. Transf. 37 (2010) 1087–1095, doi:10.1016/j.icheatmasstransfer. 2010.06.012.

CHAPTER 7

DYNAMIC TEMPERATURE RISE MECHANISM AND SOME CONTROLLING FACTORS OF WET CLUTCH ENGAGEMENT

Zhang Zhigang[1,2], Shi Xiaohui,[1] and Guo Dong[1]

[1]School of Automobile Engineering, Chongqing University of Technology, Chongqing 400054, China

[2]Key Laboratory of Safety Design and Reliability Technology for Engineering Vehicle, Changsha University of Science & Technology, Hunan 410114, China

ABSTRACT

The friction transmission model of wet clutch is established to analyze the friction transmission mechanism of its engagement. The model is developed by applying both the average flow model and the elastic contact model between the friction disk and separator plate. The key components during wet clutch engagement are the separator plate, friction disk, and lubricant. The one-dimension transient models of heat transfer in radial direction for the three components are built on the basis of the heat transfer theory and the conservation law of energy. The friction transmission model and transient heat transfer models are coupled and solved by using the Runge-Kutta numerical method, and the radial temperature distribution and their detailed parametric study for the three components are conducted separately. The simulation results show that the radial temperature for the three components rises with the increase of radius in engagement. The changes in engagement pressure, lubricant viscosity, friction lining permeability, combined surface roughness RMS, equivalent elasticity modulus, difference between dynamic and static friction coefficients, and lubricant flow have important influence on the temperature rise characteristics. The proposed models can get better understanding of the dynamic temperature rise characteristics of wet clutch engagement.

INTRODUCTION

Wet clutch is one of the most essential elements of automobile transmission system and plays a crucial role in the vehicle starting and shifting process [1]. Wet clutch is used for transmitting engine power to the wheel by mechanical friction between separator plate and friction disk. Consequently, the sliding friction heating is produced on the friction pairs of wet clutch and results in the severe temperature rise and thermal stress in separator plate and friction disk. As a result, high temperature will change the physical features of lubricant and the local thermal ablation on surfaces of separator plate and friction disk and even leads to the crack, warp, and sintering of friction pair and the deterioration of lubricant [2–4].

So far, much research has been carried out in this field, but most of publications only pay attention to analyzing the temperature field and stress field by using the finite element software. For instance, Zagrodzki and Truncone [5] investigated

Citation: Zhang Zhigang, Shi Xiaohui, and Guo Dong, "Dynamic Temperature Rise Mechanism and Some Controlling Factors of Wet Clutch Engagement," Mathematical Problems in Engineering, vol. 2016, Article ID 6530213, 12 pages, 2016. doi:10.1155/2016/6530213.

the generation of hot spots during short-term engagement and presented the theoretical analysis based on finite element model of thermoelastic contact problem and experimental stand tests. Their study shows that severe hot spots can be produced during short-term clutch operation with high initial sliding speed and some small geometric imperfections of the disk, and the hot spots can be effectively mitigated even if instantaneous sliding at high speed cannot be avoided. Jen and Nemecek [6] developed an analytical model to simulate the temperature rise of plates within wet clutch for one engagement. And the temperature rises predicted by the model were validated with the experimental data. Tatara and Payvar [7] developed a general thermal numerical model in two dimensions to simulate the heat transfer in a grooved wet clutch and characterized one cycle by three stages including an engagement period, a locked stage, and a cooling period. The finite difference technique was employed to solve the heat conduction equation. Jang et al. [8] developed a model to investigate the effect of radial grooves and waffle-shape grooves on the performance of a wet clutch and investigated the effects of groove geometry on the temperature and the engagement characteristics. Yang et al. [9] proposed the research methods of thermomechanical coupling problem of wet multidisc clutch and investigated the thermo-mechanical coupling phenomenon and the failure mechanism of separator plate of wet multidisc clutch. Their results show that the circumferential tensile thermal stress appeared in the region close to the inner edge, and the radial tensile thermal stress appeared in the region close to the outer edge. Zhang et al. [2, 10] developed a three-dimensional model of the friction pair by using the nonlinear FEM software ABAQUS and investigated the factors to influence the distribution of temperature field and stress field for one engagement. Results indicated that high relative rotary speed resulted in higher rise of temperature and stress on interior areas and increased the radial stress gradient on the contact surfaces, and the temperature field and stress field on the friction disk were more uniform than those on the steel disk. These previous works achieved certain understanding of the temperature field of wet clutch. Natsumeda and Miyoshi [11] carried out theoretical analysis on the engagement of wet clutch on the basis of two-dimensional average flow model for the analysis of oil film squeeze motion and mainly analyzed the effect of temperature on the frictional torque of wet clutch engagement. However, few studies focus on the dynamic friction thermal mechanism of wet clutch to predict the temperature rise characteristics and controlling factors, which are important for the thermal characteristics of wet clutch.

Therefore, the purpose of this paper is to numerically investigate the friction thermal mechanism in different engagement stages of wet clutch based on the friction transmission mechanism of wet clutch. The friction transmission model and transient heat transfer models are coupled and solved by using the numerical method. Key parametric factors are conducted by the proposed models.

MODEL DEVELOPMENT

According to the differences of generation mechanisms of the transmitted torque and axial load [9], the engagement of wet clutch is divided into three stages: squeezing fluid stage, boundary lubrication stage, and asperity contact stage. In squeezing fluid stage, engagement pressure is alone borne by oil film, and transmitted torque only consisted of the viscous shear torque of oil film. In boundary lubrication stage, engagement pressure is together borne by the oil film pressure and the asperity contact of friction pair, and transmitted torque involves the viscous shear torque of oil film and the asperity friction torque of friction pair, and with wet clutch engaging, contact and friction of asperities gradually play a dominant role in the formation of pressure and torque. Finally, in asperity contact stage, the asperity contact between friction pairs bears all of the engagement pressure, and transmitted torque is completely composed of asperity friction torque. Friction transmission mechanism of wet clutch is modeled in the following, respectively.

Oil Film Model

The engagement process of wet clutch can be simplified as the physical model shown in Figure 1. Before engagement, the oil film will separate the friction disk and separator plate and perform cooling lubrication on the both. At this time, the angular velocity is ω_A for the separator plate and ω_B for the friction disk, and the engagement pressure P_s acts on the friction pair. For the reason of simplicity, many reasonable assumptions were proposed in the following analysis. For example, surface of wet clutch friction was assumed to have isotropic roughness and the asperity height was assumed to have a Gaussian distribution. Although these reasonable assumptions would reduce the accuracy of the model, the model could still be used to analyze the controlling factors of temperature in engagement of wet clutch. The one-dimension average flow model of wet clutch lubricant oil film is established under the cylindrical-coordinate system by use of the average flow model [12, 13] and in consideration of the axial symmetric characteristics of the friction disk and separator plate:

$$\frac{d}{dr}\left[r\phi_r\left(h^3 + 12\Phi d\right)\frac{dP_h}{dr}\right] = 12\eta r\frac{d\bar{h}_T}{dt}. \tag{1}$$

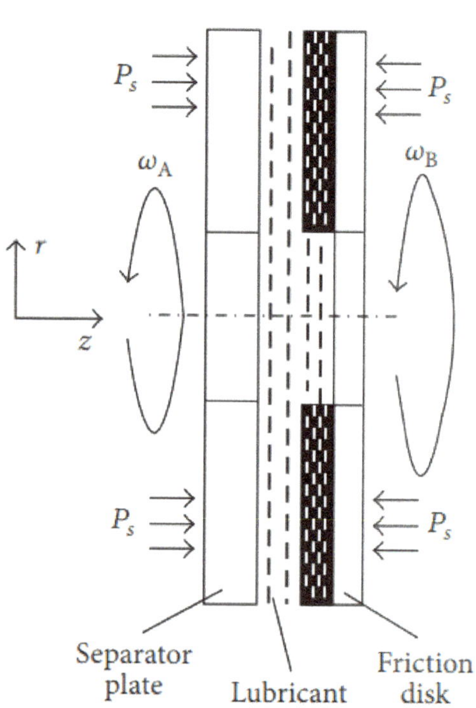

Figure 1: Physical model of wet clutch engagement.

The relation between the average gap $\bar{h}_T$ of friction pairs and the oil film thickness h can be obtained by assuming that the surface roughness peaks of both friction disk and separator plate follow the Gaussian probability density distribution with mean value of zero:

$$\frac{d\bar{h}_T}{dt} = \left\{\frac{1}{2}\left[1 + \operatorname{erf}\left(\frac{h}{\sqrt{2}\sigma}\right)\right]\right\}\frac{dh}{dt} = g(h)\frac{dh}{dt}. \quad (2)$$

The boundary condition of oil film pressure is as follows:

$$P_h(r = a) = 0,$$
$$P_h(r = b) = 0,$$
$$h(t = 0) = h_0. \quad (3)$$

By utilizing the boundary conditions of oil film pressure at the inner and outer diameters of friction pair, it can be obtained that the oil film pressure distributes along radial direction through integration against r in (1):

$$P_h(r) = \frac{3\eta}{\phi_r(h^3 + 12\Phi d)}\left[r^2 + \frac{b^2 - a^2}{2\ln(a/b)}\ln\frac{r}{b} - b^2\right] \cdot g(h)\frac{dh}{dt}. \quad (4)$$

Asperity Contact Model

The actual contact area A_c at the asperity contact part of the wet clutch friction pair can be derived by the use of the G-W elastic contact model under the assumed condition that the surface roughness peak of the wet clutch friction pair follows the Gaussian density function:

$$A_c = \pi\lambda\gamma\sigma\left[\frac{1}{\sqrt{2\pi}}e^{-(h/\sqrt{2}\sigma)^2} + \frac{h}{2\sigma}\left(\operatorname{erf}\left(\frac{h}{\sqrt{2}\sigma}\right) - 1\right)\right]. \quad (5)$$

The asperity contact pressure P_c is borne on the unit actual contact area of the wet clutch friction pair:

$$P_c = E \cdot \varepsilon = E \cdot \frac{A_c}{\pi(b^2 - a^2)}. \quad (6)$$

Engagement Pressure and Frictional Torque

According to the characteristics of the wet clutch engagement process, the engagement pressure is successively borne by the oil film pressure and asperity contact pressure, and the bearing capacity and transmitted torque composition of the wet clutch are different under different engagement stages; thus, the following can be obtained as per the balance conditions of bearing capacity during engagement process of wet clutch:

$$\pi(b^2 - a^2)P_0 = \iint P_h dA + \iint P_c dA. \quad (7)$$

The following can be obtained in accordance with the transmitted torque composition during the engagement process of wet clutch and in combination with the torque balance conditions during the same process:

$$I\frac{d\omega_{rel}}{dt} = \iint\left[\left(\phi_f - \phi_{fs}\right)\frac{r^2\eta\omega_{rel}}{h} + f_c r P_c\right]dA. \quad (8)$$

THERMAL MODEL OF WET CLUTCH

It is known through analysis on friction transmission mechanism of wet clutch that the transmitted torque during the engagement process of wet clutch consists of viscous shear torque of oil film and asperity friction torque of asperity contact and that, with wet clutch engaging, the transmitted torque of wet clutch gradually transitions from complete viscous torque of oil film to complete asperity friction torque of asperity contact. The changes in the transmitted torque composition cause the changes in the friction heat mechanism between friction pairs during the engagement process of wet clutch, that is, gradual transition from sliding friction power of complete viscous shear to sliding friction power of complete asperity friction. Therefore, during the engagement process of wet clutch, the heat flux density q between friction pairs is correspondingly composed of the heat flux density of viscous shear and asperity contact friction, which can be indicated as

$$q = \left[\left(\phi_f - \phi_{fs}\right)\frac{r^2\eta\omega_{rel}}{h} + f_c r P_c\right] \cdot \omega_{rel}. \quad (9)$$

During the engagement process of wet clutch, the friction heat generated between friction pairs dissipates on the separator plate, friction disk, and lubricant, which causes temperature rise of the separator plate, friction disk surface, and lubricant. Therefore, the study on the dynamic temperature rise characteristics of wet clutch is mainly to analyze the temperature variation role of the separator plate, friction disk, and lubricant.

To facilitate the analysis and calculation, the following assumptions are made: (1) the thermal radiation effect of friction pair is not taken into account and the heat exchange between separator plate A and inner and outer cylindrical surfaces of friction lining B is ignored because the engagement time of wet clutch is short; (2) the axial conduction of temperature on the separator plate A and friction lining B is ignored because the both are thin-wall parts; (3) the central plane of core plate C can be deemed as a thermal isolation plane and the temperature gradient between the friction linings and core plate is ignored because the heat conductivity coefficient of friction linings is relatively small while the heat conductivity coefficient of core plate is relatively large and the core plate is of symmetric structure; (4) only the radial heat conduction of temperature on separator plate and friction disk, heat transfer between friction pairs, and forced-convection heat transfer on friction surface are considered. Meanwhile, to study the radial temperature distribution role of friction pair, the friction pair is divided equally into 3 zone nodes along radial direction (r_3 outer zone, r_2 middle zone, and r_1 inner zone), and the average temperature of each zone is taken as the zone node temperature. The simplified model of thermal analysis on wet clutch friction pair is shown in Figure 2. The heat transfer control equations for the separator plate, friction disk, and lubricant in the engagement process of wet clutch are established, respectively, by the use of the heat transfer theory.

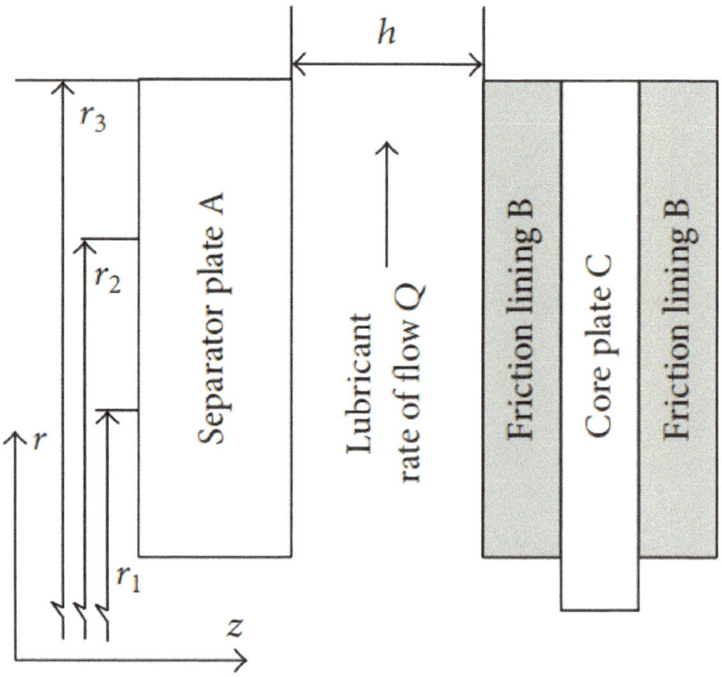

Figure 2: Model of thermal analysis for the friction pairs of wet clutch.

Under the conditions of cylindrical-coordinate system, the one-dimensional transient equation of heat transfer of separator plate A in radial direction can be established:

$$\rho_A c_A \frac{dT_A}{dt} = k_A \left(\frac{1}{r} \frac{dT_A}{dr} + \frac{d^2 T_A}{dr^2} \right). \quad (10)$$

The one-dimensional transient equation of heat transfer of friction disk B in radial direction is established:

$$\rho_B c_B \frac{dT_B}{dt} = k_B \left(\frac{1}{r} \frac{dT_B}{dr} + \frac{d^2 T_B}{dr^2} \right). \quad (11)$$

During the engagement process of wet clutch, there is forced-convection heat transfer between lubricant and separator plate and friction disk, whose thermal boundary condition can be indicated as

$$\rho_o c_o Q \frac{dT_o}{dr} = 2\pi r_p h_f \left[(T_A - T_o) + (T_B - T_o) \right]. \quad (12)$$

Because the physical property parameters such as specific heat capacity c, density ρ, and heat conductivity coefficient k of the separator plate and friction disk are different, the heat generated by friction during the engagement process of wet clutch is not equally distributed between the separator plate and friction disk; the distribution depends on the heat flux distribution coefficient ρ [2] determined by material density c, specific heat capacity β, heat conductivity coefficient k, and other parameters:

The convective heat transfer coefficient h_f is affected by many factors such as temperature, pressure, and relative angular velocity during the engagement process. To facilitate the study, only the effect of relative angular velocity ω_{rel} is considered herein; thus, the convective heat transfer coefficient [14] can be indicated as

RESULTS AND DISCUSSION

The Runge-Kutta numerical integration method is used in this paper to calculate numerical integration through combining (4) and (8), by which the oil film thickness h and relative angular velocity ω_{rel} of friction pair at each moment can be obtained; the step size for numerical calculation is 0.001 s, and the iteration termination condition is relative angular velocity ω_{rel}=0.001. The viscous torque T_h, asperity friction torque T_c, and total transmitted torque T can be obtained with the oil film thickness and change rate of relative angular velocity; thus, the heat flux density of friction pair at each moment can be gotten by the use of (9). Finally, by using the heat transfer control equations (10), (11), and (12) of wet clutch, the average temperature of 3 zones of separator plate, friction disk, and lubricant during the engagement process of wet clutch can be obtained. If it is assumed that the angular velocity ω_A of separator plate is zero constantly, the initial angular velocity of friction disk is ω_0, the friction disk gradually compresses the separator plate, and the initial temperature of wet clutch is 45°C during simulation, the initial condition of simulation is as shown in Table 1.

Table 1: Initial conditions for calculation.

Parameters	Value
Inner radius of friction pairs a/m	0.064
Outer radius of friction pairs b/m	0.085
Thickness of friction lining d/m	0.001
Surface roughness σ/m	8.41×10^{-6}
Permeability of friction lining Φ/m^2	4×10^{-12}
Equivalent elastic modulus E/Pa	2.7×10^8
Density of roughness peak λ/m^2	7×10^7
Asperity tip radius R/m	8×10^{-4}
Initial film thickness h_0/m	8.8×10^{-4}
Viscosity of lubricant η/Pa·s	0.0681
Maximum engagement pressure P_0/Pa	6×10^4
Initial angular speed ω_0/rad/s	1200
Mass moment of inertia I/kg·m^2	0.56

Influence of Engagement Pressure

As shown in Figure 3, the engagement pressure directly affects the temperature of friction disk, separator plate, and lubricant. With the increase of the engagement pressure, the engagement time is shortened, and the rise rate and maximum of temperature in each zone of friction pair and lubricant increase. This is because the transmitted torque and heat flux density will be increased along with the increase of engagement pressure. Meanwhile, the engagement time contraction reduces the lubricant volume inflowing to the friction pair and thus increases the temperature of the friction pair and lubricant.

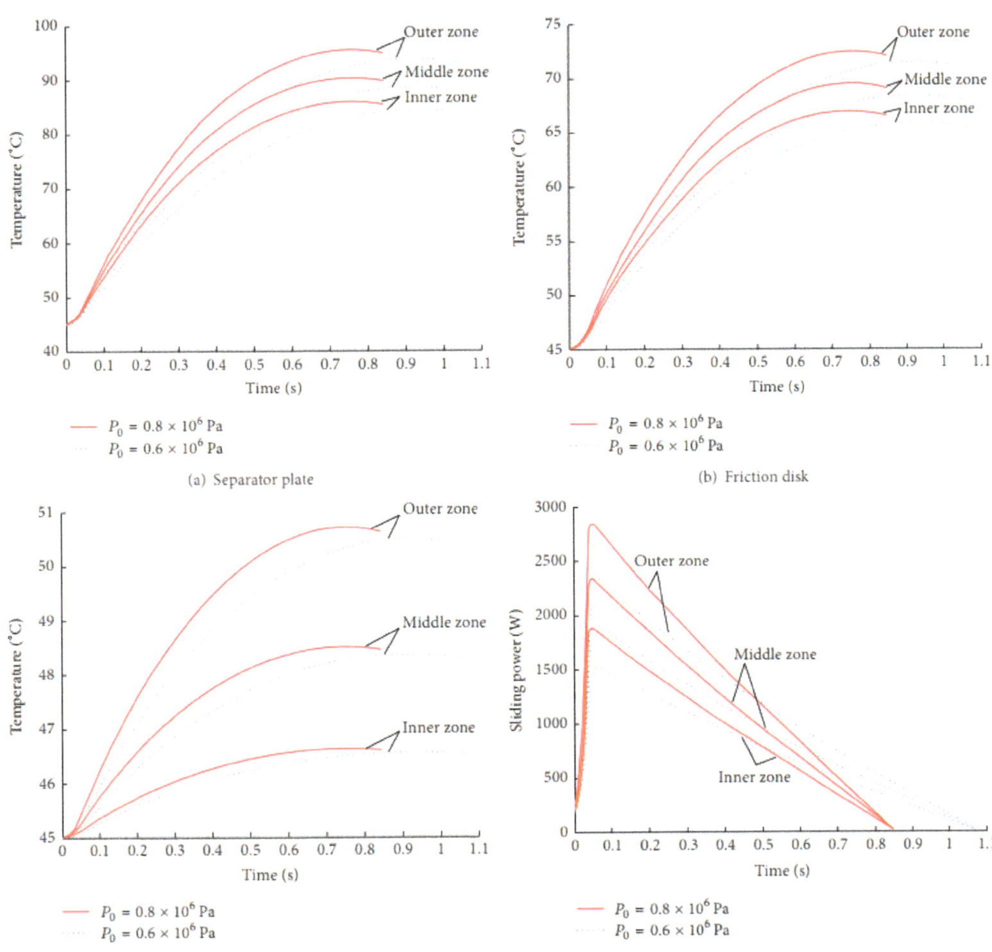

Figure 3: Effect of engagement pressure on temperature.

Influence of Lubricant Viscosity

Figure 4 shows the role of the influence of changes in lubricant viscosity in the temperature rise characteristics of wet clutch. A larger lubricant viscosity will result in a shorter engagement time as well as a higher temperature rise rate and maximum temperature for each zone of friction pair and lubricant. The influence of lubricant viscosity on the outer zone temperature of friction pair and lubricant is more apparent. This is because the viscous shear torque increases at squeezing stage and boundary lubrication stage due to the increase of lubricant viscosity and thus the corresponding viscous shear heat flux density increases. Meanwhile, the viscosity of lubricant is relatively large with bad mobility, which further weakens the cooling effect of lubricant.

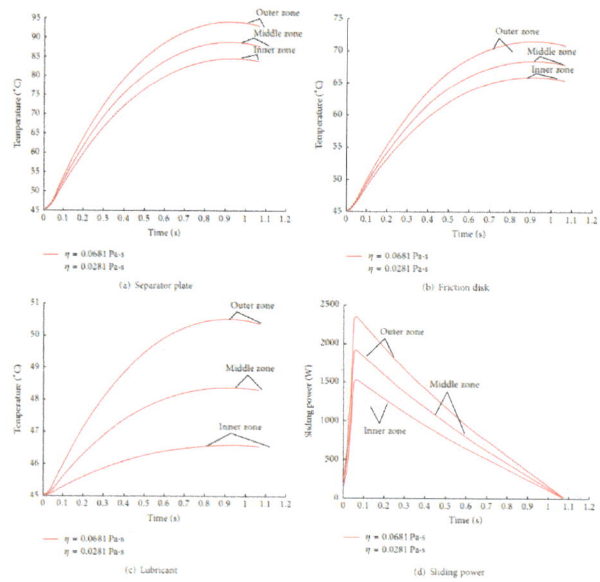

Figure 4: Effect of lubricant viscosity on temperature.

Influence of Friction Lining Permeability

As shown in Figure 5, the permeability of friction lining has important influence on the temperature rise characteristics of friction pair and lubricant at the squeezing stage and boundary lubrication stage during engagement of wet clutch. With the increase of friction lining permeability, the temperature rise rate for each zone of friction pair and lubricant increases obviously; however, the influence on the maximum temperature of friction pair and lubricant is relatively small. This is because the lubricant is easier to permeate into cellular structure of friction lining material with the increase of friction lining permeability, which largely shortens the squeezing period, and the friction pair can enter the boundary lubrication stage of heat dissipation mutually by the viscous shear heat flux and asperity friction heat flux more early, which apparently increases the heat generation rate of wet clutch.

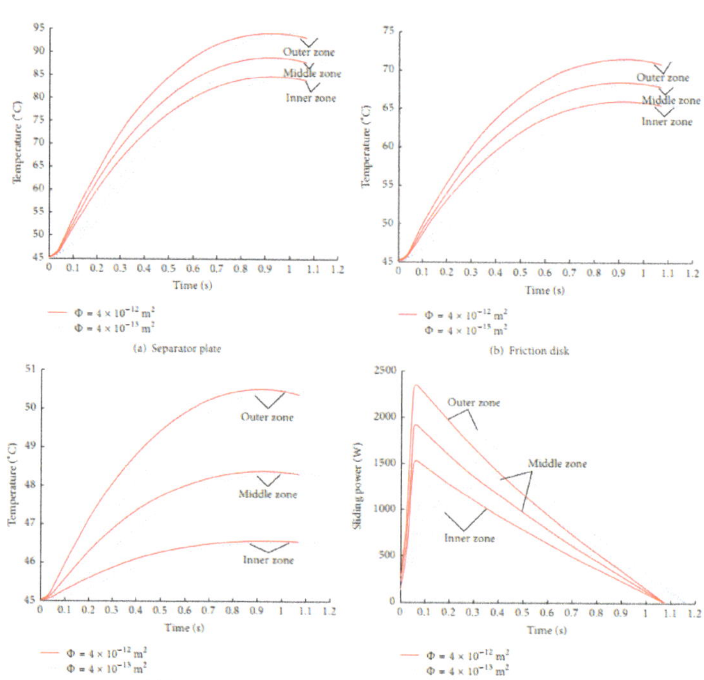

Figure 5: Effect of friction lining permeability on temperature.

Influence of Combined Surface Roughness RMS

As shown in Figure 6, the larger the combined surface roughness RMS (Root Mean Square) of friction pair is, the longer the engagement time and the smaller the temperature rise rate and maximum temperature of friction pair and lubricant will be. This is because larger combined surface roughness RMS of friction pair will result in rougher surface of friction pair and early contact of individual asperity, which slows down the squeezing velocity of lubricant between friction pairs and decreases the viscous shear torque of lubricant. Meanwhile, with the increase of the combined surface roughness RMS of friction pair, the density of asperity contact points on friction pair surface and the asperity friction torque will be smaller, and the increase rate of sliding friction power will be slower.

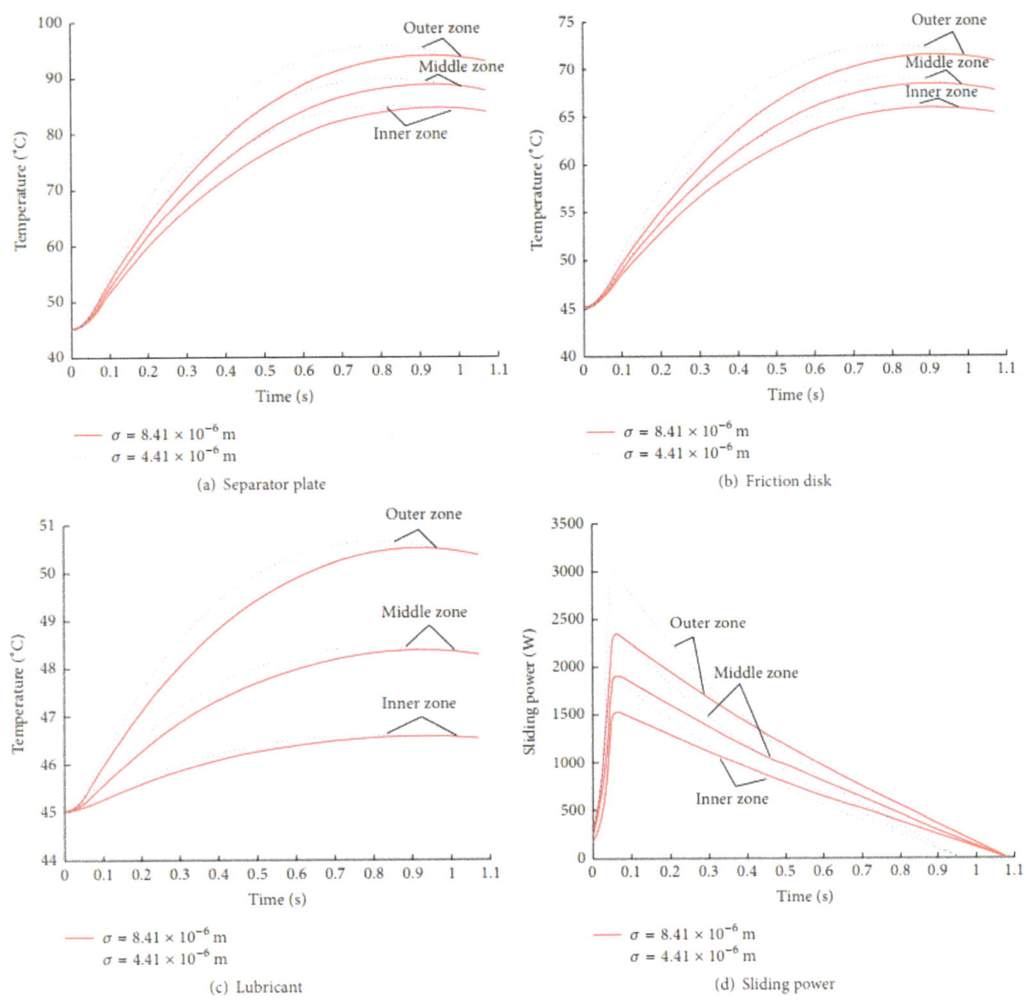

Figure 6: Effect of combined surface roughness RMS on temperature.

Influence of Equivalent Elasticity Modulus

As shown in Figure 7, the larger the equivalent elasticity modulus of friction pair is, the shorter the engagement time and the larger the temperature rise rate and maximum temperature of each zone of friction pair and lubricant will be. This is because of the notion that, with the increase of the equivalent elasticity modulus of friction pair, the asperity friction torque will increase correspondingly and the asperity friction heat flux density will also increase under the same contact conditions of asperity, causing increase in the sliding friction power during engagement of wet clutch. Meanwhile, because the engagement time is shortened, the time of cooling effect of lubricant is also shortened, causing increase of maximum temperature of friction pair and lubricant.

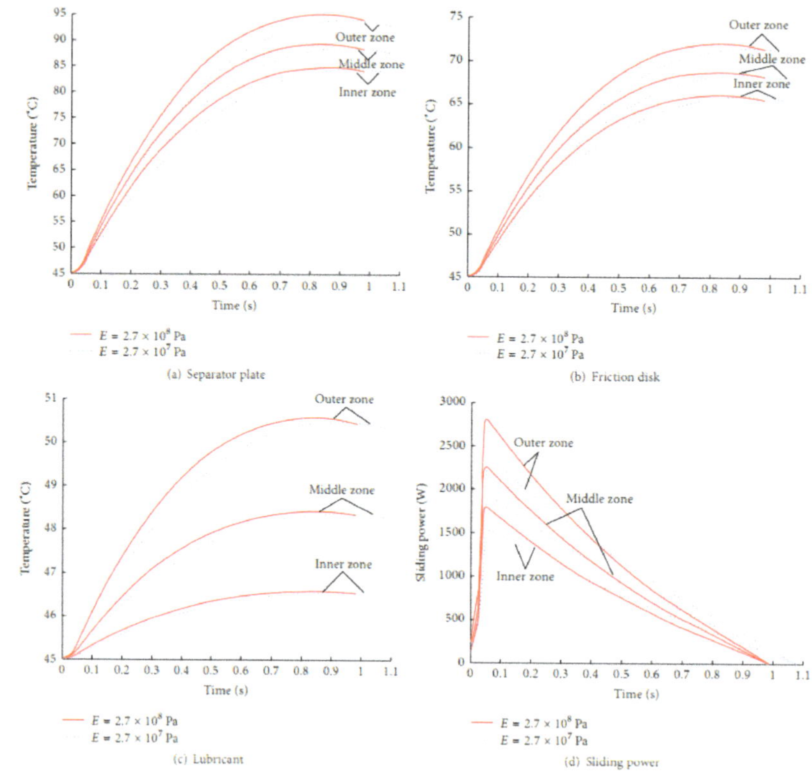

Figure 7: Effect of equivalent elasticity modulus on temperature.

Influence of Difference between Dynamic and Static Friction Coefficients

Figure 8 shows the influence of the difference between dynamic and static friction coefficients on the temperature rise characteristics of friction pairs and lubricant of wet clutch. As shown in Figure 8, with the decrease of the difference between dynamic and static friction coefficients, the engagement time is shortened, and the temperature rise rate and maximum temperature of each zone of friction pair and lubricant increase.

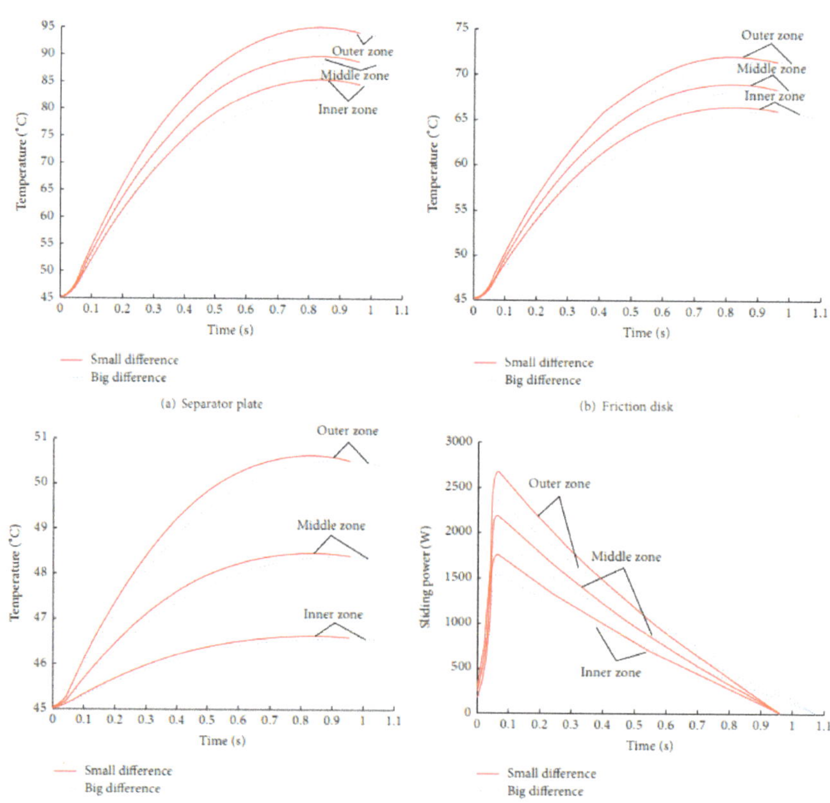

Figure 8: Effect of difference between static and dynamic friction coefficient on temperature.

Influence of Lubricant Flow

As shown in Figure 9, the change of lubricant flow only has small influence on the sliding friction power during the engagement of wet clutch; however, it certainly affects the temperature rise rate and maximum temperature of friction pair in asperity contact stage and has relatively large effect on the temperature rise characteristics of lubricant. With the increase of lubricant flow, the temperature rise rate and maximum temperature of outer and middle zones of friction pair will decrease, while the effect on the inner zone temperature is relatively small; the temperature drop of lubricant is most obvious. This is because larger lubricant flow will result in more lubricant flowing through the gap between friction pairs in unit time, larger thermal capacity of lubricant, larger convective heat exchange amount between friction pair and lubricant, and thus lower temperature rise rate and maximum temperature of friction pair and lubricant.

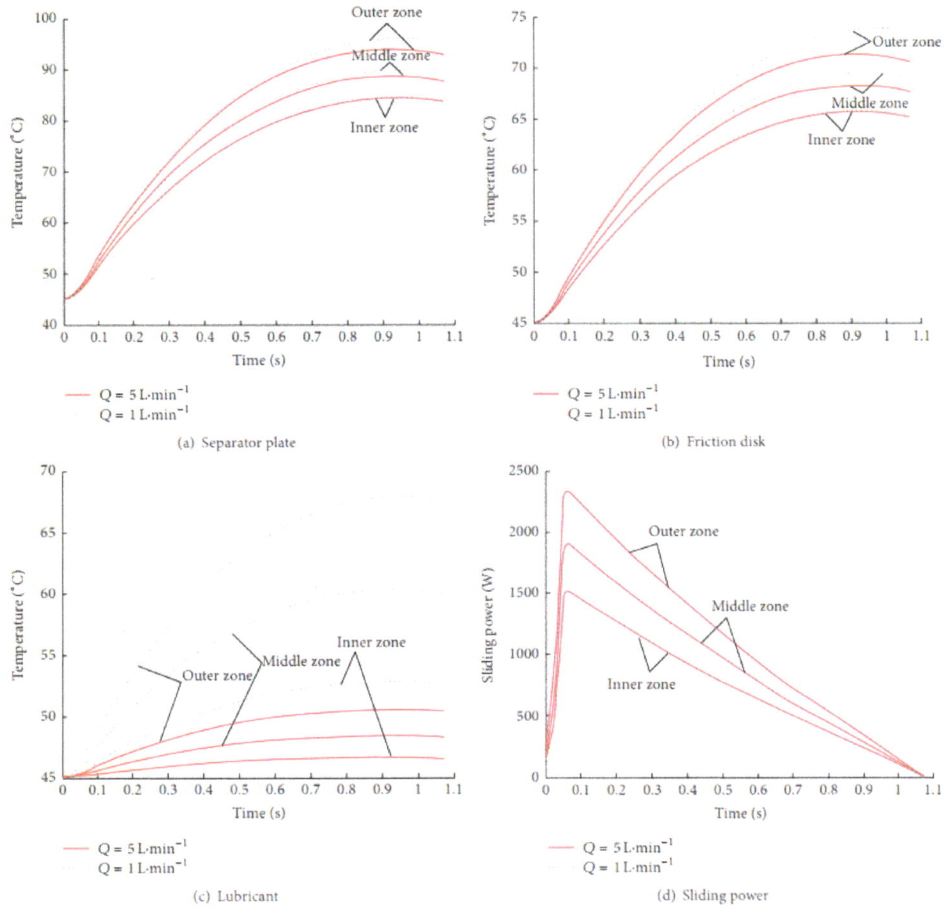

Figure 9: Effect of lubricant flow on temperature.

CONCLUSIONS

Numerical simulation of dynamic temperature rise characteristics of wet clutch has been conducted. Effects of parameters, such as engagement pressure, lubricant viscosity, friction lining permeability, surfaces roughness RMS, equivalent modulus, friction coefficient, and lubricant flow, on temperature rise in wet clutch engagement were separately studied. Numerical results have led to the following conclusions:(1)During the engagement process of wet clutch, the radial temperature distribution of separator plate, friction disk, and lubricant will rise along with the increase of radius. (2)The changes in engagement pressure, lubricant viscosity, friction lining permeability, combined surfaces roughness RMS, equivalent elasticity modulus, difference between dynamic and static friction coefficients, and lubricant flow have important effects on the temperature rise characteristics of the engagement of wet clutch.(3)With the increase of engagement pressure, the rise rate and maximum of temperature on different zone in friction disk, separator plate, and lubricant obviously increase.(4)With the increase of lubricant viscosity, the rise rate and maximum of temperature on different zone in friction disk, separator plate, and lubricant slightly increase. However, the change of viscosity has significant effect on temperature of outer zone in friction pairs and lubricant film. (5)With the increase of friction lining permeability, the rise rate of temperature in the squeezing and boundary lubrication stages during engagement has a significant increase, but the maximum temperature of friction pairs and lubricant almost remains the same.(6)The combined surface roughness RMS and equivalent elasticity modulus have the same effect on the temperature rise characteristics. With the increase of them, the rise rate and maximum of temperature both rise obviously.(7) With the decrease of the difference between dynamic and static friction coefficients, the rise rate and maximum of temperature significantly increase.(8)The change of lubricant flow has an important influence on the rise rate of temperature in asperity contact stage and on the maximum temperature. And with the increase of lubricant flow, the lubricant temperature dramatically declines.

Nomenclature

r: Radial variable (m)
t: Time (s)
ε: Compressive strain
E: Elastic coefficient for compressive strain
A_n: Nominal disk area (m^2)
h: Oil film thickness (m)
h_0: Initial film thickness (m)
Φ: Permeability of friction lining (m^2)
d: Thickness of friction lining (m)
P_h: Hydrodynamic pressure (Pa)
η: Viscosity of lubricant (Pa·s)
$\overline{h}_T$: Average gap between friction pairs (m)
σ: Combined surfaces roughness RMS between friction pairs (m)
erf(): Gauss error function
λ: Asperity tip density
γ: Asperity tip radius (m)
a: Inner radius of friction pairs (m)
b: Outer radius of friction pairs (m)
ϕ_f, ϕ_{fs}, and ϕ_r: Patir-Cheng factors [15, 16]
ω_{rel}: Angular speed difference between friction pairs (rad/s)
f_c: Dynamic friction coefficient, $f_c = 0.13 \pm 0.008 \log \omega_{rel}$
N_u: Nusselt number
k_f: Thermal conductivity of lubricant (W/(m·K))
ω: Angular speed (rad/s)
P_s: Pressure (Pa)
h_f: Convective coefficient ($W/m^2 \cdot K$)
q: Heat flux ($J/m^2 \cdot s$)
c: Specific heat (J/(kg·K))
ρ: Density (kg/m^3)
k: Thermal conductivity (W/(m·K)).

Subscript

A: Separator plate
B: Friction lining
o: Lubricant.

Competing Interests

The authors declare that they have no competing interests.

ACKNOWLEDGMENTS

This project is supported by the National Natural Science Foundation of China (51205432) and Hunan Province Key Laboratory of Safety Design and Reliability Technology for Engineering Vehicle (KF1502).

REFERENCES

1. B. Depraetere, G. Pinte, W. Symens, and J. Swevers, "A two-level Iterative Learning Control scheme for the engagement of wet clutches," Mechatronics, vol. 21, no. 3, pp. 501–508, 2011.
2. J.-L. Zhang, B. Ma, Y.-F. Zhang, and H.-Y. Li, "Simulation of thermal characteristic of wet shift clutch,"Journal of Jilin University (Engineering and Technology Edition), vol. 41, no. 2, pp. 321–326, 2011.
3. T. Deng, F. B. Hu, and D. Y. Sun, "An analysis on the thermoelastic instability of wet multi-disc clutch,"Automotive Engineering, vol. 34, no. 10, pp. 918–922, 2012.
4. T. M. Cameron, T. McCombs, and S. Tersigni, "Flash temperature in clutches," SAE Technical Paper 2005-01-3890, 2005.
5. P. Zagrodzki and S. A. Truncone, "Generation of hot spots in a wet multidisk clutch during short-term engagement," Wear, vol. 254, no. 5-6, pp. 474–491, 2003.
6. T.-C. Jen and D. J. Nemecek, "Thermal analysis of a wet-disk clutch subjected to a constant energy engagement," International Journal of Heat and Mass Transfer, vol. 51, no. 7-8, pp. 1757–1769, 2008.
7. R. A. Tatara and P. Payvar, "Multiple engagement wet clutch heat transfer model," Numerical Heat Transfer, Part A: Applications, vol. 42, no. 3, pp. 215–231, 2002.
8. J. Y. Jang, M. M. Khonsari, and R. Maki, "Three-dimensional thermohydrodynamic analysis of a wet clutch with consideration of grooved friction surfaces," Journal of Tribology, vol. 133, no. 1, Article ID 011703, 2011.
9. W. Yang, G.-D. Lu, H.-S. Lv, Y. Liu, and L.-H. Lin, "Thermo-mechanical coupling analysis of dual-steel-disc of wet multi-disk clutch," Journal of Chongqing University, vol. 34, no. 9, pp. 26–32, 2011.
10. J.-L. Zhang, B. Ma, Y.-F. Zhang, and H.-Y. Li, "Study on the factors affecting temperature field and stress field of the wet shift clutch," Transaction of Beijing Institute of Technology, vol. 30, no. 6, pp. 660–664, 2010.
11. S. Natsumeda and T. Miyoshi, "Numerical simulation of engagement of paper based wet clutch facing,"Journal of Tribology, vol. 116, no. 2, pp. 232–237, 1994.
12. E. J. Berger, F. Sadeghi, and C. M. Krousgrill, "Analytical and numerical modeling of engagement of rough, permeable, grooved wet clutches," Journal of Tribology, vol. 119, no. 1, pp. 143–148, 1997.
13. Y. Yang, R. Lam, and T. Fujii, "Prediction of torque response during the engagement of wet friction clutch," SAE Technical Paper 981097, 1998.
14. G. Chen, B. Kevin, and C. Edward, "Real time virtual temperature sensor for transmission clutches," SAE Technical Paper No. 2011-01-0230, SAE International, 2011.
15. N. Patir and H. S. Cheng, "Application of average flow model to lubrication between rough sliding surfaces," Journal of Lubrication Technology, vol. 101, no. 2, pp. 220–230, 1979.
16. N. Patir and H. S. Cheng, "Prediction of the bulk temperature in spur gears based on finite element temperature analysis," A S L E Transactions, vol. 22, no. 1, pp. 25–36, 1979.

CHAPTER 8

MAXIMUM TEMPERATURE AND RELAXATION TIME IN WET SURFACE GRINDING FOR A GENERAL HEAT FLUX PROFILE

J. L. González-Santander

Universidad Católica de Valencia "San Vicente mártir", C/ Guillem de Castro 94, 46001 Valencia, Spain

ABSTRACT

We solve the boundary-value problem of the heat transfer modeling in wet surface grinding, considering a constant heat transfer coefficient over the workpiece surface and a general heat flux profile within the friction zone between wheel and workpiece. We particularize this general solution to the most common heat flux profiles reported in the literature, that is, constant, linear, parabolic, and triangular. For these cases, we propose a fast method for the numerical computation of maximum temperature, in order to avoid the thermal damage of the workpiece. Also, we provide a very efficient method for the numerical evaluation of the transient regime duration (relaxation time).

INTRODUCTION

Grinding is a machining industrial process of metallic plates that consist in an abrasive wheel rotating at a high speed over the surface of a workpiece. The abrasion that produces the grinding wheel removes the surface material of the metallic plate being ground. During the grinding process, most of the energy is converted into heat, which accumulates within the contact zone between the wheel and the workpiece [1]. The high temperatures reached during this machining process may produce an unacceptable decrease in the quality of workpieces, such as burning or residual stresses [2]. In order to avoid this thermal damage of the workpiece, liquid coolant is usually injected into it, so power generation by friction is reduced and cooling by convection occurs. Therefore, a theoretical model that can predict the maximum temperature in wet grinding for the online monitoring process is demanded.

In surface grinding, the heat transfer inside the workpiece usually is modeled by a strip heat source infinitely long and of 2ℓ width (m in SI units), which moves at a speed $\vec{v}_f = v_f\vec{i}$ (m s^{-1}) over a semi-infinite solid surface (see Figure 1). Setting the Cartesian coordinate system fixed to wheel, as it is shown in Figure 1, the temperature field of the workpiece T(t,x,z) with respect to the room temperature T_0 (K) must satisfy the convective heat equation [3, §1.7(2)]:

Citation: J. L. González-Santander, "Maximum Temperature and Relaxation Time in Wet Surface Grinding for a General Heat Flux Profile," Mathematical Problems in Engineering, vol. 2016, Article ID 5387612, 14 pages, 2016. doi:10.1155/2016/5387612.

$$\frac{\partial T}{\partial t} = k\left(\frac{\partial^2 T}{\partial x^2} + \frac{\partial^2 T}{\partial z^2}\right) - v_f \frac{\partial T}{\partial x}, \tag{1}$$

where k is the thermal diffusivity ($m^2 s^{-1}$). Since initially the workpiece is at room temperature, (1) is subjected to an homogeneous initial condition:

$$T(0, x, z) = 0. \tag{2}$$

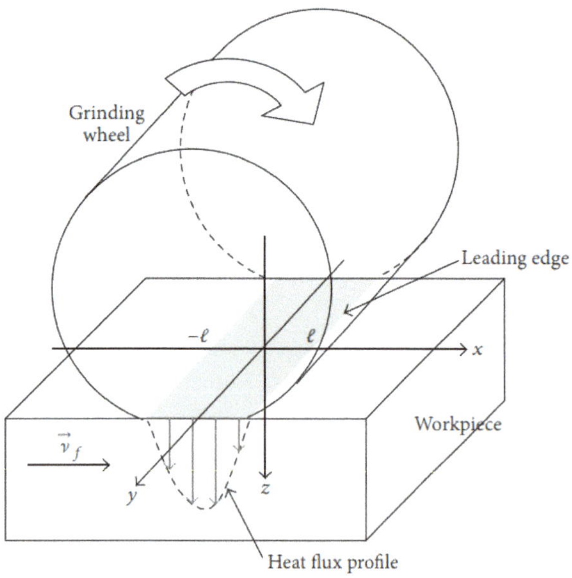

Figure 1: Setup in surface down-grinding.

Wet grinding can be modeled assuming a constant heat transfer coefficient h ($W m^{-2} K^{-1}$) on the workpiece surface [4]. Assuming as well a dimensionless heat flux profile f(x) within the contact area between wheel and workpiece, we have the following boundary condition:

$$k_0 \frac{\partial T}{\partial z}(t, x, 0) = hT(t, x, 0) - qf(x)\,\theta(\ell - x)\,\theta(\ell + x), \tag{3}$$

where θ(x) denotes the Heaviside function, k_0 ($W m^{-1} K^{-1}$) is the thermal conductivity, and q ($W m^{-2}$) is the average heat flux entering into the workpiece along the contact width 2ℓ; thus,

$$\frac{1}{2\ell}\int_{-\ell}^{\ell} f(x)\,dx = 1. \tag{4}$$

In the case of dry grinding, that is, h=0, boundary-value problem (1)–(3) can be solved in integral form for the case of a constant heat flux profile [5, Eqn.22]. Also, considering a general heat flux profile, a closed form expression for the surface temperature in the stationary regime, that is, $\lim_{t\to\infty} T$ (t, x, 0), can be found [6].

For the wet case, that is, h>0, (1)–(3) has been solved for constant [4] and linear [7] heat flux profiles, both in the stationary regime. The scope of this paper is to generalize these results and calculate, in terms of an integral expression, the time-dependent temperature field T(t, x, z) satisfying (1)–(3) for any constant h≥0, considering any heat flux profile. From this general expression, we will recover the expressions for the particular cases found in the literature (constant [8, 9] and linear [7] heat flux profiles), calculating also solutions for other heat flux profiles found in the literature, such as triangular [10–15] and parabolic [16]. These latter solutions do not seem to be reported in the literature. Further, we will provide a simple and rapid method for the calculation of the maximum temperature reached in the workpiece and some approximated expressions for the duration of the transient regime, for all the heat flux profiles considered (i.e., constant, linear, triangular, and parabolic). These approximate expressions seem to be novel as well. It is worth noting that the computation of the relaxation time is essential for analyzing the transient regime in which the wheel is engaged and disengaged from the workpiece (cut in and cut out) [17].

This paper is organized as follows. Section 2 derives the solution of the value boundary problem stated in (1)–(3) for any analytic heat flux profile f(x). Section 3 particularizes the general solution of the previous section to the most common heat flux profiles reported in the literature, that is, constant, linear, and parabolic. We also give an expression for a triangular heat flux profile, although in this case f(x) is not an analytic function. Section 4 provides a very efficient method for searching the maximum temperature in the workpiece for the different heat flux profiles considered. In Section 5, for each heat flux profile considered, we propose the root searching of an equation in order to evaluate numerically the duration of the transient regime. For this purpose, we provide analytical approximations to these roots, to be used as starting iteration point in the root finding. Section 6 presents some numerical simulations of the workpiece surface temperature in the stationary regime as well as the maximum temperature computation by using the expressions derived in previous sections. We also plot the surface temperature evolution during the transient regime. Our conclusions are summarized in Section7.

GENERAL HEAT FLUX PROFILE SOLUTION

In order to solve the problem given in (1)–(3), we can build the solution by using the method of Green's function. For this purpose, let us denote the Cartesian coordinates fixed to the workpiece as $X'Y'Z'$ (see Figure2), where, at t=0, both coordinates systems are overlapped.

Figure 2: Relationship between the coordinates systems, one fixed to the wheel and the other one to the workpiece.

Green's function $G(t, \vec{r}\,'; t_0, \vec{r}\,'_0)$ is interpreted in this case as the temperature field rise evaluated at $(t, \vec{r}\,') = (t, x', y', z')$ in a semi-infinite body (z>0) in which, at position $\vec{r}\,'_0 = (x'_0, y'_0, z'_0)$, an instantaneous point heat source of energy E (J) appears at the instant t_0<t and where the surface (z=0) is subject to a constant heat transfer coefficient h [3, §14.9 (4)]:

$$G\left(t,\vec{r}';t_0,\vec{r}'_0\right) = \frac{E}{4\pi\rho ck\left(t-t_0\right)}\exp\left(-\frac{\left(x'-x'_0\right)^2+\left(y'-y'_0\right)^2}{4k\left(t-t_0\right)}\right)$$

$$\cdot\left\{\frac{\exp\left(-\left(z'-z'_0\right)^2/4k\left(t-t_0\right)\right)+\exp\left(-\left(z'+z'_0\right)^2/4k\left(t-t_0\right)\right)}{2\sqrt{\pi k\left(t-t_0\right)}}\right.$$

$$\left.-\tilde{h}\exp\left(\tilde{h}\left(z'+z'_0\right)+\tilde{h}^2k\left(t-t_0\right)\right)\operatorname{erfc}\left(\tilde{h}\sqrt{k\left(t-t_0\right)}+\frac{z'+z'_0}{2\sqrt{k\left(t-t_0\right)}}\right)\right\},$$

(5)

Where ρ is the density of the body ($\mathrm{kg\,m^{-3}}$) and its specific heat ($\mathrm{J\,kg^{-1}K^{-1}}$) and where we have defined theradiation coefficient ($\mathrm{m^{-1}}$) as

$$\tilde{h} = \frac{h}{k_0}.$$

(6)

A formal derivation of (5) is given in [18, Appendix].

According to Figure 2, a point P of the heat source has (x_0, y_0, 0) coordinates in XYZ and $(x'_0, y'_0, 0)$ in $X'Y'Z'$. The relationship between both coordinates systems is

$$x' = x + v_f\left(t-t_0\right),$$

$$y' = y,$$

$$z' = z.$$

(7)

Therefore, the temperature rise in the workpiece at coordinates (t, x, y, z) due to a point P of the heat source with respect to the coordinate system fixed to the wheel XYZ is given by

$$G\left(t,x,y,z\right) = \frac{E}{4\pi k_0\left(t-t_0\right)}$$

$$\cdot\exp\left(-\frac{\left(x+v_f\left(t-t_0\right)-x_0\right)^2+\left(y-y_0\right)^2}{4k\left(t-t_0\right)}\right)$$

$$\cdot\left\{\frac{\exp\left(-z^2/4k\left(t-t_0\right)\right)}{2\sqrt{\pi k\left(t-t_0\right)}}-\tilde{h}e^{\tilde{h}z+\tilde{h}^2k\left(t-t_0\right)}\right.$$

$$\left.\cdot\operatorname{erfc}\left(\tilde{h}\sqrt{k\left(t-t_0\right)}+\frac{z}{2\sqrt{k\left(t-t_0\right)}}\right)\right\}.$$

(8)

The derivation of the solution of (1)–(3) by using (8) follows three steps:

(i) Superposition in space to give temperature due to instantaneous line source, which acts on the surface z=0 parallel to the -axis.

(ii) Superposition of line sources to give temperature due to an instantaneously acting strip source on the surface.

(iii) Superposition in time to give temperature due to continuously acting strip source.

Instantaneous Line Source

Let us consider now an instantaneous line heat source parallel to the -axis. Let us assume in (8) that the energy per unit length of this heat source q_L (J m^{-1}) is constant:

$$dE = q_L dy_0; \tag{9}$$

thus, integrating in $y_0 \in (-\infty, \infty)$, we obtain the field temperature due to an instantaneous line heat source as

$$T_{\text{line}}(t,x,z) = \frac{q_L}{2\pi\rho ck(t-t_0)} \cdot \exp\left(-\frac{\left(x+v_f(t-t_0)-x_0\right)^2}{4k(t-t_0)}\right) \cdot \left\{\exp\left(\frac{-z^2}{4k(t-t_0)}\right) - \tilde{h}\sqrt{\pi k(t-t_0)}e^{\tilde{h}z+\tilde{h}^2 k(t-t_0)} \cdot \operatorname{erfc}\left(\tilde{h}\sqrt{k(t-t_0)} + \frac{z}{2\sqrt{k(t-t_0)}}\right)\right\}. \tag{10}$$

Instantaneous Strip Source

Let us consider now a strip source of variable heat strength along the x-axis, according to the following profile:

$$q_L = q_B f(x_0)\, dx_0, \tag{11}$$

where q_B(J m^{-2}) is the energy that this heat source leaves at the instant t_0 per surface unit of the strip source. Notice that the function in (11) is dimensionless, as in (3). Also, f takes into account the heat flux profile entering into the workpiece within the contact zone with the wheel. Substituting now (11) in (10) and integrating over the strip width, $x_0 \in (-\ell, \ell)$, we have the following temperature field in the workpiece:

$$T_{\text{strip}}(t,x,z) = \frac{q_B}{2\pi\rho ck(t-t_0)} \left\{\exp\left(\frac{-z^2}{4k(t-t_0)}\right) - \tilde{h}\sqrt{\pi k(t-t_0)}e^{\tilde{h}z+\tilde{h}^2 k(t-t_0)} \cdot \operatorname{erfc}\left(\tilde{h}\sqrt{k(t-t_0)} + \frac{z}{2\sqrt{k(t-t_0)}}\right)\right\} \cdot \int_{-\ell}^{\ell} f(x_0) \cdot \exp\left(-\left[\frac{x-x_0+v_f(t-t_0)}{2\sqrt{k(t-t_0)}}\right]^2\right) dx_0. \tag{12}$$

Assuming that the f function is analytic, we can expand it in its Taylor series, so that, taking into account also that f is dimensionless, we have

$$f(x_0) = g\left(\frac{x_0}{\ell}\right) = \sum_{m=0}^{\infty} \frac{g^{(m)}(0)}{m!}\left(\frac{x_0}{\ell}\right)^m; \tag{13}$$

thus, the integral given in (12) can be written as

$$\int_{-\ell}^{\ell} f(x_0)\exp\left(-\left[\frac{x-x_0+v_f(t-t_0)}{2\sqrt{k(t-t_0)}}\right]^2\right)dx_0$$
$$=\sum_{m=0}^{\infty}\frac{g^{(m)}(0)}{\ell^m m!}$$
$$\cdot\int_{-\ell}^{\ell} x_0^m\exp\left(-\left[\frac{x-x_0+v_f(t-t_0)}{2\sqrt{k(t-t_0)}}\right]^2\right)dx_0. \quad (14)$$

According to the result given in Appendix (A.9), we have

$$\int_{-\ell}^{\ell} x_0^m\exp\left(-\left[\frac{x-x_0+v_f(t-t_0)}{2\sqrt{k(t-t_0)}}\right]^2\right)dx_0$$
$$=\frac{\left(-2\sqrt{k(t-t_0)}\right)^{m+1}}{2}\sum_{n=0}^{m}\binom{m}{n}$$
$$\cdot\left(-\frac{x+v_f(t-t_0)}{2\sqrt{k(t-t_0)}}\right)^{m-n} F_n(u,t-t_0)\Big|_{u=x-\ell}^{x+\ell}, \quad (15)$$

Where

$$F_n(u,t)=\operatorname{sign}^{n+1}\left(\frac{u+v_f t}{2\sqrt{kt}}\right)\gamma\left(\frac{n+1}{2},\left[\frac{u+v_f t}{2\sqrt{kt}}\right]^2\right) \quad (16)$$

and where $\gamma(a,z)$ denotes the lower incomplete gamma function [19, Eqn. 45:1:2]. Substituting back these results, we arrive at

$$T_{\text{strip}}(t,x,z)=\frac{q_B}{2\pi\rho c\sqrt{k(t-t_0)}}\left\{\exp\left(\frac{-z^2}{4k(t-t_0)}\right)\right.$$
$$-\tilde{h}\sqrt{\pi k(t-t_0)}e^{\tilde{h}z+\tilde{h}^2k(t-t_0)}$$
$$\left.\cdot\operatorname{erfc}\left(\tilde{h}\sqrt{k(t-t_0)}+\frac{z}{2\sqrt{k(t-t_0)}}\right)\right\}$$
$$\cdot\sum_{m=0}^{\infty}g^{(m)}(0)\left[\frac{x+v_f(t-t_0)}{\ell}\right]^m\sum_{n=0}^{m}\frac{1}{n!(m-n)!}$$
$$\cdot\left(\frac{-2\sqrt{k(t-t_0)}}{x+v_f(t-t_0)}\right)^n F_n(u,t-t_0)\Big|_{u=x-\ell}^{x+\ell}. \quad (17)$$

Continuous Strip Source

Let q ($W m^{-2}$) be the energy per time unit and per surface unit that the strip heat source leaves over the workpiece surface. Therefore, during an infinitesimal interval of time dt_0, we have

$$q_B = q dt_0. \tag{18}$$

Substitution of (18) in (17) and integration over time $t_0 \in (0, t)$(remember that $t_0<t$) result in

$$T(t,x,z) = \frac{q}{2\pi\rho c}\int_0^t \left[\frac{\exp\left(-z^2/4kt'\right)}{\sqrt{kt'}} - \bar{h}\sqrt{\pi kt'}e^{\bar{h}z+\bar{h}^2kt'}\operatorname{erfc}\left(\bar{h}\sqrt{kt'} + \frac{z}{2\sqrt{kt'}}\right)\right] \cdot \left[\sum_{m=0}^{\infty} g^{(m)}(0)\left(\frac{x+v_f t'}{\ell}\right)^m \sum_{n=0}^{m}\frac{1}{n!\,(m-n)!} \cdot \left(\frac{-2\sqrt{kt'}}{x+v_f t'}\right)^n F_n\left(u,t'\right)\Big|_{u=x-\ell}^{x+\ell}\right] dt', \tag{19}$$

where we have performed the change of variables $t' = t - t_0$. Considering now the dimensionless variables

$$X = \frac{x}{s},$$

$$Z = \frac{z}{s},$$

$$L = \frac{\ell}{s},$$

$$H = \bar{h}s,$$

$$\tau = \frac{v_f\sqrt{t}}{2\sqrt{k}},$$

$$\mathcal{T} = \frac{\pi\rho c v_f}{2q}T, \tag{20}$$

where s is a characteristic length

$$s = \frac{2k}{v_f}, \tag{21}$$

we finally arrive at

$$\mathcal{T}(\tau,X,Z) = \int_0^\tau \left[\exp\left(\frac{-Z^2}{4w^2}\right) - \sqrt{\pi}Hwe^{HZ+H^2w^2} \cdot \operatorname{erfc}\left(Hw + \frac{Z}{2w}\right)\right]\left[\sum_{m=0}^{\infty}\frac{g^{(m)}(0)}{L^m} \cdot \sum_{n=0}^{m}\frac{(-2w)^n\left(X+2w^2\right)^{m-n}}{n!\,(m-n)!}\mathcal{F}_n(u,w)\Big|_{u=X-L}^{X+L}\right]dw, \tag{22}$$

where

$$\mathcal{F}_n(u,w) = \operatorname{sign}^{n+1}\left(\frac{u}{2w}+w\right)\gamma\left(\frac{n+1}{2},\left[\frac{u}{2w}+w\right]^2\right). \tag{23}$$

PARTICULAR CASES

In this section, we derive, from general expression (22), particular expressions for the most common heat flux profiles considered in the literature (i.e., constant, linear, triangular, and parabolic). For the constant and linear cases, we obtain expressions already given in the literature, as mentioned before.

Constant Heat Flux Profile

For a constant heat flux profile satisfying (4), we have

$$f_0(x) = g_0\left(\frac{x}{\ell}\right) = 1. \tag{24}$$

Thus, in (22), we have only the summand which corresponds to m=0, so, from (23) and using the property given in Appendix (A.12), we have

$$\mathcal{F}_0(u,w) = \operatorname{sign}\left(\frac{u}{2w}+w\right)\gamma\left(\frac{1}{2},\left[\frac{u}{2w}+w\right]^2\right) = \sqrt{\pi}\operatorname{erf}\left(\frac{u}{2w}+w\right). \tag{25}$$

Therefore, (22) reduces to

$$\mathcal{T}^{(0)}(\tau,X,Z) = \sqrt{\pi}\int_0^\tau \left[\exp\left(\frac{-Z^2}{4w^2}\right) - \sqrt{\pi}Hwe^{HZ+H^2w^2}\operatorname{erfc}\left(Hw+\frac{Z}{2w}\right)\right] \cdot \left[\operatorname{erf}\left(\frac{X+L}{2w}+w\right) - \operatorname{erf}\left(\frac{X-L}{2w}+w\right)\right]dw, \tag{26}$$

where the superscript (0) denotes that we are considering a constant heat flux profile. Particularizing (26) to H=0 and performing the change of variables s=2w², we obtain the following expression for the case of dry grinding:

$$\mathcal{T}^{(0)}_{\mathrm{dry}}(\tau, X, Z) = \frac{1}{2}\sqrt{\frac{\pi}{2}}\int_0^{2\tau^2} \exp\left(\frac{-Z^2}{4s}\right) \cdot \left[\operatorname{erf}\left(\frac{X+L+s}{\sqrt{2s}}\right) - \operatorname{erf}\left(\frac{X-L+s}{\sqrt{2s}}\right)\right]\frac{ds}{\sqrt{s}}, \quad (27)$$

which is the result given in [5, Eqn.22]. Undoing the change of variables and performing the limit $\tau \rightarrow \infty$, we get the stationary regime for the dry case [5, Eqn.7]:

$$\lim_{\tau\to\infty}\mathcal{T}^{(0)}_{\mathrm{dry}}(\tau, X, Z) = \sqrt{\pi}\int_0^{\infty} \exp\left(\frac{-Z^2}{4w^2}\right) \cdot \left[\operatorname{erf}\left(\frac{X+L}{2w}+w\right) - \operatorname{erf}\left(\frac{X-L}{2w}+w\right)\right] dw \quad (28)$$

$$= \int_{X-L}^{X+L} e^{-w} K_0\left(\sqrt{w^2+Z^2}\right) dw, \quad (29)$$

where $K_0(x)$ is the zeroth-order modified Bessel function of the second kind [20, Sect. 9.6]. Therefore, taking into account (29) in (26), we obtain the stationary regime for the wet case as

$$\lim_{\tau\to\infty}\mathcal{T}^{(0)}(\tau, X, Z) = \int_{X-L}^{X+L} e^{-w} K_0\left(\sqrt{w^2+Z^2}\right) dw - \pi H e^{HZ}\int_0^{\infty} w e^{H^2w^2} \operatorname{erfc}\left(Hw + \frac{Z}{2w}\right) \cdot \left[\operatorname{erf}\left(\frac{X+L}{2w}+w\right) - \operatorname{erf}\left(\frac{X-L}{2w}+w\right)\right] dw, \quad (30)$$

which is the expression given by [4]. Recently, in [21], the expression given in (30) particularized on the surface (Z=0) can be expressed in closed form as

$$\lim_{\tau\to\infty}\mathcal{T}^{(0)}(\tau, X, 0) = \sqrt{\pi}\sum_{n=0}^{\infty}\frac{\left(-H/\sqrt{2}\right)^n}{\Gamma\left((n+1)/2\right)} \mathrm{Yu}_{n/2}(u)\Big|_{X-L}^{X+L}, \quad (31)$$

where

$$\mathrm{Yu}_{\nu}(x) = \begin{cases} \frac{x|x|^{\nu} e^{x}}{2\nu+1}\left[K_{\nu}(|x|) + \operatorname{sign}(x) K_{\nu+1}(|x|)\right], & x \neq 0, \quad (32) \\ 0, & x = 0. \end{cases} \quad (32)$$

It is worth noting that (31) does not converge for high Biot numbers, that is, H>1.

Linear Heat Flux Profile

For a linear heat flux profile satisfying (4), we have

$$f_1(x) = g\left(\frac{x}{\ell}\right) = 1 + \frac{x}{\ell}; \quad (33)$$

thus, (22) is reduced to

$$\mathcal{T}^{(1)}(\tau, X, Z) = \frac{1}{L}\int_0^{\tau} dw \left[\exp\left(\frac{-Z^2}{4w^2}\right)\right.$$

$$\left. - \sqrt{\pi} H w e^{HZ+H^2w^2} \operatorname{erfc}\left(Hw + \frac{Z}{2w}\right)\right]$$

$$\cdot \left[\sqrt{\pi}\left(X + L + 2w^2\right)\operatorname{erf}\left(\frac{u}{2w} + w\right)\right.$$

$$\left. + 2w \exp\left(-\left[\frac{u}{2w} + w\right]^2\right)\right]_{u=X-L}^{X+L}, \tag{34}$$

where the superscript (1) indicates a linear heat flux profile and where we have applied (25). Also, according to (A.14), the following result has been applied:

$$\mathcal{F}_1(u, w)\Big|_{u=X-L}^{X+L} = \gamma\left(1, \left[\frac{u}{2w} + w\right]^2\right)\Bigg|_{u=X-L}^{X+L}$$

$$= -\exp\left(-\left[\frac{u}{2w} + w\right]^2\right)\Bigg|_{u=X-L}^{X+L}. \tag{35}$$

Performing in (34) the change of variables r=2w$_2$ and the limit $\tau \to \infty$, we obtain the following expression for the stationary regime, which is given in [7, Eqns.3.46-47]:

$$\lim_{\tau\to\infty}\mathcal{T}^{(1)}(\tau, X, Z) = \frac{1}{2L}\int_0^{\infty}\exp\left(\frac{-Z^2}{2r}\right)\left\{\sqrt{\frac{\pi}{2u}}(X\right.$$

$$+ L + r)\left[\operatorname{erfc}\left(\frac{X+L+r}{\sqrt{2r}}\right) - \operatorname{erfc}\left(\frac{X-L+r}{\sqrt{2r}}\right)\right]$$

$$+ \exp\left(-\frac{(X+L+r)^2}{2u}\right)$$

$$\left. - \exp\left(-\frac{(X-L+r)^2}{2r}\right)\right\} dr$$

$$- \frac{\pi H}{4L}\int_0^{\infty}\exp\left(\frac{-Z^2}{2r}\right)\operatorname{experf}\left(\frac{Hr}{\sqrt{2r}}\right)\left\{(X + L + r)\right.$$

$$\cdot \left[\operatorname{erfc}\left(\frac{X-L+r}{\sqrt{2r}}\right) - \operatorname{erfc}\left(\frac{X+L+r}{\sqrt{2r}}\right)\right]$$

$$+ \sqrt{\frac{2r}{\pi}}\left[\exp\left(-\frac{(X+L+r)^2}{2r}\right)\right.$$

$$\left.\left. - \exp\left(-\frac{(X-L+r)^2}{2r}\right)\right]\right\} dr, \tag{36}$$

where the following function has been defined:

$$\operatorname{experf}(x) = \exp\left(x^2\right)\operatorname{erfc}(x).$$

Triangular Heat Flux Profile

For a triangular heat flux profile satisfying (4) (see Figure 3), we have

$$f_\Delta(x) = g_\Delta\left(\frac{x}{\ell}\right) = \begin{cases} \dfrac{2}{\lambda+1}\left(\dfrac{x}{\ell}+1\right), & x \in (-\ell, \ell_{\max}) \\ \dfrac{2}{\lambda-1}\left(\dfrac{x}{\ell}-1\right), & x \in (\ell_{\max}, \ell), \end{cases} \tag{38}$$

where we have defined the following dimensionless parameter:

$$\lambda = \frac{\ell_{\max}}{\ell} \in [-1, 1]. \tag{39}$$

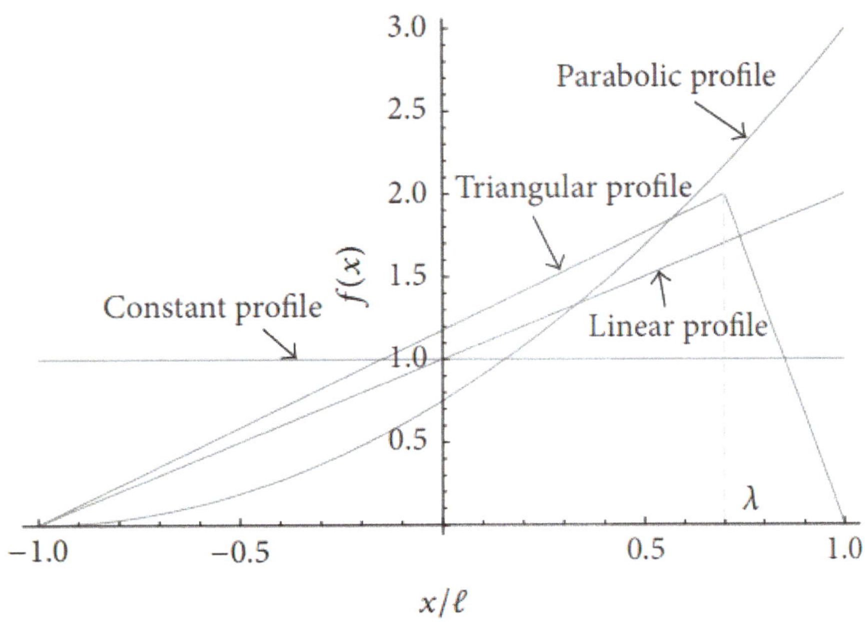

Figure 3: Heat flux profiles considered.

Although we cannot expand (38) in its Taylor series, we can redo the calculation of Section 2.2, performing the integral given in (12) as follows:

$$\int_{-\ell}^{\ell} f(x_0)\exp\left(-[A+Bx_0]^2\right)dx_0 = \frac{2}{\lambda+1}\int_{-\ell}^{\ell_{\max}}\left(\frac{x_0}{\ell}+1\right)\exp\left(-[A+Bx_0]^2\right)dx_0 + \frac{2}{\lambda-1}\int_{\ell_{\max}}^{\ell}\left(\frac{x_0}{\ell}-1\right)\exp\left(-[A+Bx_0]^2\right)dx_0, \tag{40}$$

where we have set

$$A = \frac{x+v_f(t-t_0)}{2\sqrt{k(t-t_0)}},$$

$$B = \frac{-1}{2\sqrt{k(t-t_0)}}. \tag{41}$$

By using formulas (A.13) and (A.15) given in Appendix, some algebra leads to the following result:

$$\int_{-\ell}^{\ell} f(x_0) \exp\left(-[A + Bx_0]^2\right) dx_0$$

$$= \frac{1}{\ell B^2} \left[\frac{\varphi(A - B\ell)}{\lambda + 1} + 2\frac{\varphi(A + B\ell_{\max})}{\lambda^2 - 1} - \frac{\varphi(A + B\ell)}{\lambda - 1} \right], \tag{42}$$

where we have defined

$$\varphi(x) = \sqrt{\pi} x \operatorname{erf}(x) + \exp\left(-x^2\right). \tag{43}$$

Substituting (42) into (12), integrating over time $t_0 \in (0, t)$, and performing the change of variables $t' = t - t_0$, we get the time-dependent temperature field due to a triangular heat flux profile as

$$T^{\Delta}(t, x, z) = \frac{2q}{\pi \rho c} \int_0^t \left\{ \frac{\exp\left(-z^2/4kt'\right)}{\sqrt{kt'}} - \sqrt{\pi}\tilde{h} e^{\tilde{h}z + \tilde{h}^2 kt'} \operatorname{erfc}\left(\tilde{h}\sqrt{kt'} + \frac{z}{2\sqrt{kt'}}\right) \right\}$$
$$\cdot \frac{\sqrt{kt'}}{\ell} \left\{ \frac{\varphi\left(\left(x + \ell + v_f t'\right)/2\sqrt{kt'}\right)}{\lambda + 1} + \frac{2\varphi\left(\left(x - \lambda\ell + v_f t'\right)/2\sqrt{kt'}\right)}{\lambda^2 - 1} - \frac{\varphi\left(\left(x - \ell + v_f t'\right)/2\sqrt{kt'}\right)}{\lambda - 1} \right\} dt',$$

which, in dimensionless variables (20), is rewritten as

$$\mathcal{T}^{\Delta}(\tau, X, Z) = \frac{4}{L} \int_0^{\tau} \left[\exp\left(\frac{-Z^2}{4w^2}\right) - \sqrt{\pi} H w e^{HZ + H^2 w^2} \operatorname{erfc}\left(Hw + \frac{Z}{2w}\right) \right]$$
$$\cdot w \left[\frac{\varphi((X + L)/2w + w)}{\lambda + 1} + \frac{2\varphi((X - \lambda L)/2w + w)}{\lambda^2 - 1} - \frac{\varphi((X - L)/2w + w)}{\lambda - 1} \right] dw. \tag{44}$$

Parabolic Heat Flux Profile

For a parabolic heat flux profile satisfying (4) (see Figure 3), we have

$$f_2(x) = g_2\left(\frac{x}{\ell}\right) = \frac{3}{4}\left(1 + \frac{x}{\ell}\right)^2; \tag{46}$$

thus, (22) is reduced to

$$\mathcal{T}^{(2)}(\tau, X, Z) = \frac{3}{2} \int_0^{\tau} \left[\exp\left(\frac{-Z^2}{4w^2}\right) - \sqrt{\pi} H w e^{HZ + H^2 w^2} \operatorname{erfc}\left(Hw + \frac{Z}{2w}\right) \right]$$
$$\cdot \left[\left(\frac{1}{2} + \frac{X + 2w^2}{L} + \frac{\left(X + 2w^2\right)^2}{2L^2} \right) \mathcal{F}_0(u, w) - \frac{2w}{L}\left(1 + \frac{X + 2w^2}{L}\right) \mathcal{F}_1(u, w) + \frac{2w^2}{L^2} \cdot \mathcal{F}_2(u, w) \right]_{u = X - L}^{X + L} dw, \tag{47}$$

where the superscript (2) denotes that we are considering a parabolic heat flux profile. By using the properties given in Appendices (A.12), (A.14), and (A.17), we arrive at

$$\mathcal{T}^{(2)}(\tau, X, Z) = \frac{3}{4L^2} \int_0^{\tau} \left[\exp\left(\frac{-Z^2}{4w^2}\right) - \sqrt{\pi} H w e^{HZ + H^2 w^2} \operatorname{erfc}\left(Hw + \frac{Z}{2w}\right) \right]$$
$$\cdot \left[\sqrt{\pi}\left(2w^2 + \left(L + 2w^2 + X\right)^2\right) \operatorname{erf}\left(\frac{u}{2w} + w\right) + 2w\left(2L + 2X + 2w^2 - u\right) \cdot \exp\left(-\left[\frac{u}{2w} + w\right]^2\right) \right]_{u = X - L}^{X + L} dw. \tag{48}$$

MAXIMUM TEMPERATURE

From a physical point of view, the maximum temperature $\mathcal{T}_{\max}$ must be reached at the stationary regime $\tau \to \infty$, because the longer the heat source is acting, the greater the temperature in the workpiece is. Moreover, the location of maximum temperature $X_{\max}$ must be on the surface Z=0, within the contact zone between wheel and workpiece, $X_{\max} \in [-L, L]$. It is worth noting that recently the latter has been mathematically proved in [22] for constant and linear heat flux profiles. Therefore, since $\mathcal{T}(\tau, X, Z)$ is a differentiable function in X, a convenient way to evaluate maximum temperature is to solve numerically all points X^* that satisfies

$$\lim_{\tau \to \infty} \frac{\partial \mathcal{T}(\tau, X^*, 0)}{\partial X} = 0, \quad X^* \in [-L, L], \tag{49}$$

and then the maximum is given by

$$\mathcal{T}_{\max} = \max_{X^*} \left[\lim_{\tau \to \infty} \mathcal{T}(\tau, X^*, 0) \right]. \tag{50}$$

In Section 6, we will check numerically that for the most common heat flux profiles proposed in the literature (constant, linear, triangular, and parabolic) the root X^* is unique, so $X^* = X_{\max}$.

For a constant heat flux profile, (49) reads

$$0 = \int_0^{\infty} \left[\frac{1}{w} - \sqrt{\pi} H e^{H^2 w^2} \operatorname{erfc}(Hw) \right] \cdot \exp\left(-\left[\frac{u}{2w} + w\right]^2\right) \Big|_{u = X^* - L}^{X^* + L} dw. \tag{51}$$

Similarly, for a linear heat flux profile, we have

$$0 = \int_0^{\infty} \left[1 - \sqrt{\pi} H w e^{H^2 w^2} \operatorname{erfc}(Hw)\right] \cdot \left[\sqrt{\pi}\left(X^* + L + 2w^2\right) \operatorname{erf}\left(\frac{u}{2w} + w\right) - \frac{X^* + L - u}{w} \exp\left(-\left[\frac{u}{2w} + w\right]^2\right)\right]_{u=X^*-L}^{X^*+L} dw. \quad (52)$$

For a triangular profile, we obtain

$$0 = \int_0^{\infty} \left[1 - \sqrt{\pi} H w e^{H^2 w^2} \operatorname{erfc}(Hw)\right] \cdot \left[\frac{\operatorname{erf}\left(\left(X^* + L\right)/2w + w\right)}{\lambda + 1} + \frac{2 \operatorname{erf}\left(\left(X^* - \lambda L\right)/2w + w\right)}{\lambda^2 - 1} - \frac{\operatorname{erf}\left(\left(X^* - L\right)/2w + w\right)}{\lambda - 1} \right] dw, \quad (53)$$

and for a parabolic one

$$0 = \int_0^{\infty} \left[1 - \sqrt{\pi} H w e^{H^2 w^2} \operatorname{erfc}(Hw)\right] \cdot \left[\sqrt{\pi}\left(X^* + L + 2w^2\right) \operatorname{erf}\left(\frac{u}{2w} + w\right) + \left(\frac{\left(X^* + L - u\right)^2}{2w} + 2w\right) \cdot \exp\left(-\left[\frac{u}{2w} + w\right]^2\right)\right]_{u=X^*-L}^{X^*+L} dw. \quad (54)$$

RELAXATION TIME

The stationary regime is asymptotically reached at $\tau \to \infty$; that is,

$$\lim_{\tau \to \infty} \frac{\partial \mathcal{T}(\tau, X, Z)}{\partial \tau} = 0. \quad (55)$$

In order to avoid thermal damage, the most important point is the location of the maximum temperature X_{max}. Therefore, for estimating how rapidly the stationary regime is reached (i.e., the relaxation time of the transient regime), we can solve the following equation for $\overline{\tau} \in (0, \infty)$:

$$\frac{\partial \mathcal{T}(\overline{\tau}, X_{max}, 0)}{\partial \tau} = \eta \approx 0. \quad (56)$$

In this section, we are going to provide approximated explicit solutions of (56) for the different heat flux profiles considered. Each of these approximate solutions will be used as starting iteration point for the numerical root searching of (56).

Constant Heat Flux Profile

For a constant heat flux profile, according to (26), (56) is written as

$$\eta = \sqrt{\pi}\left[1 - \sqrt{\pi} H \overline{\tau} e^{H^2 \overline{\tau}^2} \operatorname{erfc}(H\overline{\tau})\right] \cdot \left[\operatorname{erf}\left(\frac{X_{max} + L}{2\overline{\tau}} + \overline{\tau}\right) - \operatorname{erf}\left(\frac{X_{max} - L}{2\overline{\tau}} + \overline{\tau}\right)\right]. \quad (57)$$

Let us solve (57) approximately. On the one hand, we find in the literature [20, Eqn.7.1.23] the following asymptotic expansion:

$$\sqrt{\pi} z e^{z^2} \operatorname{erfc}(z) \approx 1 + \sum_{m=1}^{\infty} (-1)^m \frac{(2m-1)!!}{\left(2z^2\right)^m}, \quad z \longrightarrow \infty; \quad (58)$$

thus, expanding up to m=1 and taking $z = H\overline{\tau}$, we have

$$1 - \sqrt{\pi} H \overline{\tau} e^{H^2 \overline{\tau}^2} \operatorname{erfc}(H\overline{\tau}) \approx \frac{1}{2H^2 \overline{\tau}^2}, \quad \overline{\tau} \longrightarrow \infty. \quad (59)$$

On the other hand, expanding in Taylor series the following function up to the first order, we have

$$\operatorname{erf}\left(\overline{\tau} + \frac{X_{max} \pm L}{2\overline{\tau}}\right) \approx \operatorname{erf}(\overline{\tau}) + \frac{X_{max} \pm L}{\sqrt{\pi}\,\overline{\tau}} e^{-\overline{\tau}^2}, \quad \overline{\tau} \longrightarrow \infty; \quad (60)$$

thus,

$$\sqrt{\pi}\left[\operatorname{erf}\left(\frac{X_{max} + L}{2\overline{\tau}} + \overline{\tau}\right) - \operatorname{erf}\left(\frac{X_{max} - L}{2\overline{\tau}} + \overline{\tau}\right)\right] \approx \frac{2L}{\overline{\tau}} e^{-\overline{\tau}^2}, \quad \overline{\tau} \longrightarrow \infty. \quad (61)$$

Substituting (59) and (61) in (57), we get the following approximated equation:

$$\overline{\tau}^3 e^{\overline{\tau}^2} \approx \frac{L}{H^2 \eta}. \quad (62)$$

By using the Lambert W function [23], (62) can be solved explicitly, obtaining the following approximated expression for the relaxation time:

$$\overline{\tau} \approx \sqrt{\frac{3}{2} \mathrm{W}\left(\frac{2}{3}\left[\frac{L}{H^2 \eta}\right]^{2/3}\right)}. \quad (63)$$

Notice that (63) does not depend on X_{max}, so we do not have to compute (57) if we want an estimation of the relaxation time for a constant heat flux profile.

Linear Heat Flux Profile

Considering now a linear heat flux profile, according to (34), (56) is written as

$$\eta L = \left\{ \sqrt{\pi} \left(X_{\max} + L + 2\overline{\tau}^2 \right) \left[\operatorname{erf} \left(\frac{X_{\max} + L}{2\overline{\tau}} + \overline{\tau} \right) \right. \right.$$
$$\left. - \operatorname{erf} \left(\frac{X_{\max} - L}{2\overline{\tau}} + \overline{\tau} \right) \right]$$
$$- 2\overline{\tau} \left[\exp \left(- \left[\frac{X_{\max} + L}{2\overline{\tau}} + \overline{\tau} \right]^2 \right) \right.$$
$$\left. \left. - \exp \left(- \left[\frac{X_{\max} - L}{2\overline{\tau}} + \overline{\tau} \right]^2 \right) \right] \right\} \left[1 - \sqrt{\pi} H \overline{\tau} e^{H^2 \overline{\tau}^2} \right.$$
$$\left. \cdot \operatorname{erfc} \left(H \overline{\tau} \right) \right]. \tag{64}$$

In order to solve (64) approximately, let us expand in Taylor series the following function up to first order:

$$\exp \left(- \left[\overline{\tau} + \frac{X_{\max} \pm L}{2\overline{\tau}} \right]^2 \right) \approx e^{-\overline{\tau}^2} - \left(X_{\max} \pm L \right) e^{-\overline{\tau}^2},$$
$$\overline{\tau} \longrightarrow \infty; \tag{65}$$

thus,

$$\exp \left(- \left[\frac{X_{\max} + L}{2\overline{\tau}} + \overline{\tau} \right]^2 \right)$$
$$- \exp \left(- \left[\frac{X_{\max} - L}{2\overline{\tau}} + \overline{\tau} \right]^2 \right) \approx 2Le^{-\overline{\tau}^2},$$
$$\overline{\tau} \longrightarrow \infty. \tag{66}$$

Now, substituting (59), (61), and (66) in (64), we arrive at

$$\overline{\tau}^3 e^{\overline{\tau}^2} \approx \frac{X_{\max} + L}{H^2 \eta} \tag{67}$$

that can be solved explicitly as

$$\overline{\tau} \approx \sqrt{\frac{3}{2} \mathrm{W} \left(\frac{2}{3} \left[\frac{X_{\max} + L}{H^2 \eta} \right]^{2/3} \right)}. \tag{68}$$

Triangular Heat Flux Profile

Let us consider now a triangular heat flux profile, so, according to (45), (56) reads as

$$\frac{\eta L}{4} = \left[\frac{\varphi \left(\left(X_{\max} + L \right) / 2\overline{\tau} + \overline{\tau} \right)}{\lambda + 1} \right.$$
$$+ \frac{2\varphi \left(\left(X_{\max} - \lambda L \right) / 2\overline{\tau} + \overline{\tau} \right)}{\lambda^2 - 1}$$
$$\left. - \frac{\varphi \left(\left(X_{\max} - L \right) / 2\overline{\tau} + \overline{\tau} \right)}{\lambda - 1} \right] \overline{\tau} \left[1 \right.$$
$$\left. - \sqrt{\pi} H \overline{\tau} e^{H^2 \overline{\tau}^2} \operatorname{erfc} \left(H \overline{\tau} \right) \right]. \tag{69}$$

Equation (69) can be solved approximately expanding in the Taylor series the following function up to first order, taking into account the φ function definition (43) and asymptotic formulas (60) and (65):

$$\varphi\left(\frac{X}{2\overline{\tau}}+\overline{\tau}\right)=\sqrt{\pi}\left(\frac{X}{2\overline{\tau}}+\overline{\tau}\right)\operatorname{erf}\left(\frac{X}{2\overline{\tau}}+\overline{\tau}\right)+\exp\left(-\left[\frac{X}{2\overline{\tau}}+\overline{\tau}\right]^{2}\right)\approx\sqrt{\pi}\left(\frac{X}{2\overline{\tau}}+\overline{\tau}\right)\operatorname{erf}(\overline{\tau})+\left(1+\frac{X^{2}}{2\overline{\tau}^{2}}\right)e^{-\overline{\tau}^{2}},\quad \overline{\tau}\longrightarrow\infty. \tag{70}$$

Taking into account (59) and (70) in (69), after some algebra, we arrive at the following approximated equation:

$$\overline{\tau}^{3}e^{\overline{\tau}^{2}}\approx\frac{2L}{H^{2}\eta} \tag{71}$$

that can be solved explicitly as

$$\overline{\tau}\approx\sqrt{\frac{3}{2}W\left(\frac{2}{3}\left[\frac{2L}{H^{2}\eta}\right]^{2/3}\right)}. \tag{72}$$

As in the constant case, (72) does not depend on X_{max}; moreover, it does not depend on λ either.

Parabolic Heat Flux Profile

Finally, for a parabolic heat flux profile, according to (48), (56) is expressed as

$$\frac{4L^{2}}{3}\eta=\left\{2\overline{\tau}\left[\left(X_{max}+L+2\overline{\tau}^{2}\right)\cdot\exp\left(-\left[\frac{X_{max}+L}{2\overline{\tau}}+\overline{\tau}\right]^{2}\right)-\left(X_{max}+3L+2\overline{\tau}^{2}\right)\cdot\exp\left(-\left[\frac{X_{max}-L}{2\overline{\tau}}+\overline{\tau}\right]^{2}\right)\right]+\sqrt{\pi}\left[2\overline{\tau}^{2}+\left(X_{max}+L+2\overline{\tau}^{2}\right)^{2}\right]\left[\operatorname{erf}\left(\frac{X_{max}+L}{2\overline{\tau}}+\overline{\tau}\right)-\operatorname{erf}\left(\frac{X_{max}-L}{2\overline{\tau}}+\overline{\tau}\right)\right]\right\}\left[1-\sqrt{\pi}H\overline{\tau}e^{H^{2}\overline{\tau}^{2}}\cdot\operatorname{erfc}(H\overline{\tau})\right]. \tag{73}$$

Taking into account asymptotic formulas (59), (61), and (65), some algebra leads (73) to the following approximate equation:

$$\frac{4L}{3}\eta\approx\frac{(X_{max}+L)^{2}+4X_{max}\overline{\tau}^{2}}{H^{2}\overline{\tau}^{3}}e^{-\overline{\tau}^{2}}\approx\frac{4X_{max}}{H^{2}\overline{\tau}}e^{-\overline{\tau}^{2}},\quad \overline{\tau}\longrightarrow\infty, \tag{74}$$

which can be solved as

$$\overline{\tau}\approx\sqrt{\frac{1}{2}W\left(\frac{18X_{max}}{H^{4}L^{2}\eta^{2}}\right)}. \tag{75}$$

NUMERICAL RESULTS

Table 1 shows three sets of parameters (SI units) for the numerical simulations. Data set 1 considers a titanium alloy VT20 workpiece, whose thermal properties are given in [24]. The grinding regime for this simulation can be found in [25]. Data sets 2 and 3 consider plain carbon steel and aluminum oxide, Al_2O_3 (sapphire), as workpiece material, respectively. The thermal properties and the grinding regimes for these data sets can be found in [26].

Table 1: Simulation parameters in SI units.

		Data 1	Data 2	Data 3
	Material	VT20	Steel	Sapphire
Workpiece	k_0	13	60.5	46
	k	4.23×10^{-6}	1.77×10^{-5}	1.51×10^{-5}
	h	27.29×10^4	4.1×10^5	1.3×10^5
Grinding regime	2ℓ	2.66×10^{-3}	1.4×10^{-3}	2.5×10^{-3}
	q	5.89×10^7	1.4×10^7	1.8×10^7
	v_f	0.53	3.3×10^{-2}	3.3×10^{-2}
	T_0	300	300	300
	H	0.335	7.29	2.59
Dimensionless parameters	L	83.41	0.65	1.36
	$\mathcal{T}$	23.02	79.23	114.3
	η	10^{-3}	10^{-3}	10^{-3}

Figure 3 presents the different heat flux profiles considered. All of them have been normalized according to (4); thus, they have got the same heat flux on average. Therefore, we can compare how the distribution of the heat flux within the contact zone affects the graph of the workpiece surface temperature. Also, for all the numerical simulations, we have taken a value of $\lambda=0.7$ for the triangular profile (45).

Table 2 shows the value of maximum temperature T_{max} and its location x_{max}/ℓ within the friction zone. For this purpose, we have computed the roots of (51)–(54) by using Brent's method [27], taking as starting interval [-L,L]. Notice that, for every data set, the maximum temperature is ordered depending on the heat flux profile. This order, from lowest to highest, corresponds to a constant, linear, triangular, and parabolic heat flux profile, respectively. This qualitative behavior of maximum temperature agrees with the heat flux distribution of each profile (see Figure 3), in which maximum temperature is higher where input of heat is also higher. Therefore, the location of maximum temperature agrees with the location of maximum heat flux. Thus, considering the same input data, the ordering of these locations from left to right with respect to the type heat flux of profile is as follows: constant, triangular, linear, and parabolic; that is, the more the heat flux profile is skewed to the leading edge, the nearer the maximum temperature is located in that edge. It is worth noting that the numerical evaluation of maximum temperature by the root searching of (51)–(54) is much faster than applying directly Mathematica's NMaximize command to the surface temperature in the stationary regime, that is, $\lim_{\tau\to\infty}\mathcal{T}(\tau, X, 0)$. For instance, taking data set 1 as input parameters and a constant heat flux profile,NMaximize command is $\approx$ 125 times slower than the numerical root finding of (51). Moreover, NMaximizecommand fails if we take a triangular heat flux profile for all the data sets of Table 1.

Table 2: Location and value of maximum temperature.

		Data 1	Data 2	Data 3
Constant profile	x_{max}/ℓ	−0.973	−0.301	−0.501
	T_{max} [K]	477.6	329.1	407.6
Linear profile	x_{max}/ℓ	0.609	0.687	0.605
	T_{max} [K]	537.3	341.3	442.4
Triangular profile	x_{max}/ℓ	0.459	0.530	0.459
	T_{max} [K]	551.6	343.6	450.4
Parabolic profile	x_{max}/ℓ	0.748	0.799	0.751
	T_{max} [K]	612.8	353.4	481.4

Figures 4–6 show the surface temperature (Z=0) in the stationary regime ($\tau \to \infty$) for the three data sets given in Table 1, respectively. For the numerical evaluation of these plots, we have used the formulas given in Section 3: (26) for constant heat flux profile, (34) for linear profile, (45) for triangular profile, and (48) for a parabolic one. It is worth noting that the numerical integration of the linear and the parabolic cases is quite efficient if we use the double exponential strategy [28]. The shapes of the graphs are quiet similar to the different data sets of input parameters, except at the leading and trailing edges ($x/\ell = \pm 1$). This is because the heat source velocity v_f in Figure 4 is much faster than in Figures 5 and 6. Also, Figures 4–6 show that the skewness of the temperature curves follows the corresponding heat flux profile.

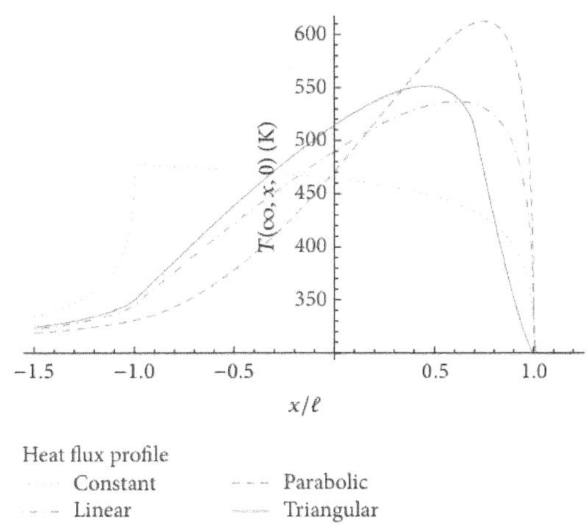

Figure 4: Surface temperature for a VT20 titanium alloy workpiece (data 1).

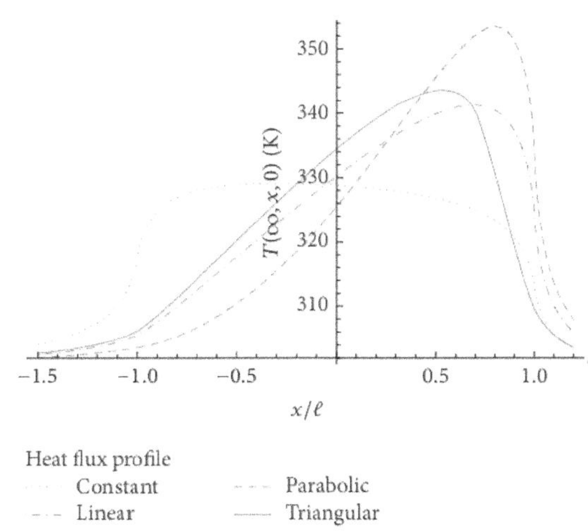

Figure 5: Surface temperature for a carbon steel workpiece (data 2).

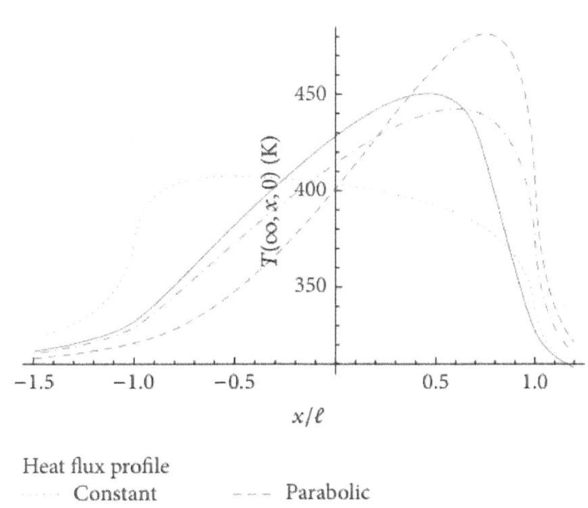

Figure 6: Surface temperature for a sapphire workpiece (data 3).

Table 3 shows the relaxation times for the different heat flux profiles considered. For this purpose, we have used the relaxation times approximations given in (63), (68), (72), and (75) for the different heat flux profiles, as starting iteration points for the root searching of (57), (64), (69), and (73), respectively. In general, the approximation formulas provide the order of magnitude of the exact root. Also, the convergence to the exact value is extremely rapid (≈0.01 s).

Table 3: Relaxation times for the different heat flux profiles considered.

	$\bar{t}$ [s]	Data 1	Data 2	Data 3
Constant profile	Exact	5.90×10^{-3}	9.89×10^{-2}	2.00×10^{-1}
	Approx.	6.06×10^{-4}	1.11×10^{-1}	1.92×10^{-1}
Linear profile	Exact	1.66×10^{-3}	1.04×10^{-1}	1.82×10^{-1}
	Approx.	1.08×10^{-4}	5.60×10^{-3}	1.77×10^{-2}
Triangular profile	Exact	1.90×10^{-3}	1.05×10^{-1}	1.83×10^{-1}
	Approx.	6.42×10^{-4}	1.36×10^{-1}	2.19×10^{-1}
Parabolic profile	Exact	1.25×10^{-3}	1.05×10^{-1}	1.83×10^{-1}
	Approx.	5.31×10^{-4}	2.10×10^{-1}	2.79×10^{-1}

Figures 7–9 show the temperature evolution on the workpiece surface, taking a triangular heat flux profile for the three data sets of Table 1. The black solid line in these figures corresponds to the stationary regime and the gray solid line to the relaxation time (see Table 3). In the case of Figures 8 and 9, the plot corresponding to the relaxation time and the stationary regime are overlapped. For these cases, this is due to the fact that the relaxation time is greater than the time that the heat source needs to cover the friction zone width; that is, $\bar{t} > 2\ell/v_f$. The plots are exponentially discretized over time because the transient regime occurs very rapidly. Notice that the first stages of the temperature evolution resemble the triangular heat flux profile, this being more clear for data set 1 in Figure 7.

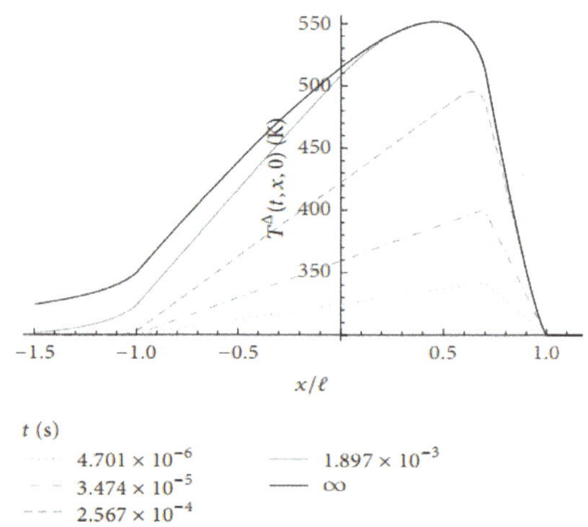

Figure 7: Evolution of the surface temperature in a titanium alloy VT20 workpiece (data set 1).

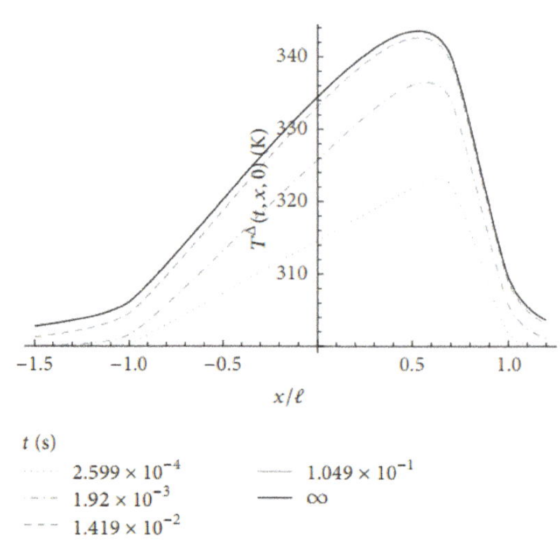

Figure 8: Evolution of the surface temperature in a carbon steel workpiece (data set 2).

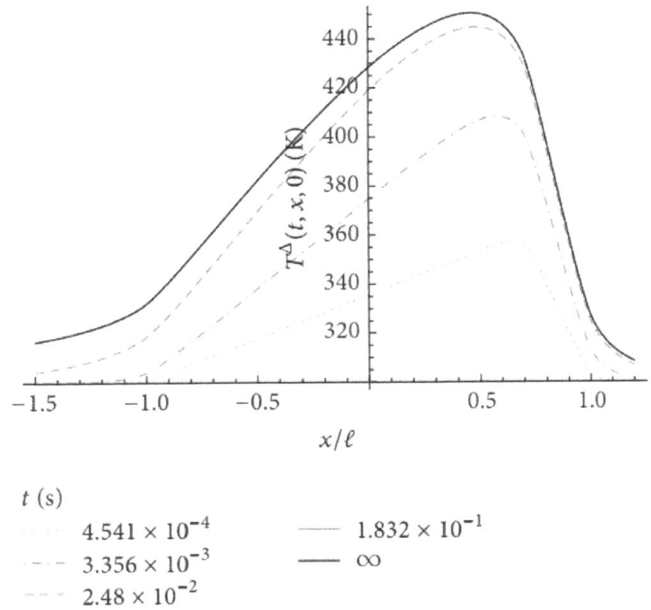

Figure 9: Evolution of the surface temperature in a sapphire workpiece (data set 3).

CONCLUSIONS

By using the Green function, we have solved the boundary-value problem that models the heat transfer in wet surface grinding, considering an arbitrary heat flux profile and a constant heat transfer coefficient on the workpiece surface. We have particularized this result to the most common heat flux profiles reported in the literature, that is, constant, linear, parabolic, and triangular. In constant case (26) and linear case (34), we recover expressions given in the literature. Nonetheless, the triangular case (45) and parabolic case (48) do not seem to be found in the literature.

From the analytical expressions found for the surface temperature in the stationary regime for the different heat flux profiles considered, we propose the root searching of some equations in integral form (see (51)–(53)), for the maximum temperature computation. It turns out that this method is very fast (≈1 s) and also much more rapid than the direct numerical maximization of the surface temperature in the stationary regime. Therefore, it provides a very useful aid to prevent thermal damage in the online monitoring grinding process.

Finally, we have provided a very rapid method (≈0.01 s) for the computation of the relaxation time for the different heat flux profiles considered. For the initial guess of this root searching, we have given analytical expressions that in general provide the order of magnitude of the relaxation time.

APPENDIX

Let us calculate the integral

$$V_m(a,b) = \int_{\ell_1}^{\ell_2} x^m \exp\left(-[a+bx]^2\right) dx, \quad m = 0, 1, \ldots; \tag{A.1}$$

performing the change of variables u=a+bx, we get

$$V_m(a,b) = \frac{1}{b^{m+1}} \int_{a+b\ell_1}^{a+b\ell_2} (u-a)^m e^{-u^2} du. \tag{A.2}$$

Taking into account Newton's formula [19, Eqn. 6:14:1], we can rewrite (A.2) as

$$V_m(a,b) = \frac{1}{b^{m+1}} \sum_{k=0}^{m} \binom{m}{k} (-a)^k \int_{a+b\ell_1}^{a+b\ell_2} u^k e^{-u^2} du. \tag{A.3}$$

Now, consider the integral

$$I(\alpha,k) = \int_0^{\alpha} u^k e^{-u^2} du. \tag{A.4}$$

When $\alpha > 0$, perform the change of variables s=u^2, so

$$I(\alpha,k) = \frac{1}{2}\int_0^{\alpha^2} s^{(k-1)/2} e^{-s} ds = \frac{1}{2}\gamma\left(\frac{k+1}{2}, \alpha^2\right), \tag{A.5}$$

Where

$$\gamma(\nu,z) = \int_0^z t^{\nu-1} e^{-t} dt \tag{A.6}$$

denotes the lower incomplete gamma function [19, Eqn. 45:3:1]. Notice that, performing the change of variables s-u, we have

$$I(-\alpha,k) = \int_0^{-\alpha} u^k e^{-u^2} du = (-1)^{k+1} \int_0^{\alpha} s^k e^{-s^2} du = (-1)^{k+1} I(\alpha,k). \tag{A.7}$$

Therefore, for $\alpha \in \mathbb{R}$, we have

$$I(\alpha,k) = \frac{\text{sign}^{k+1}(\alpha)}{2} \gamma\left(\frac{k+1}{2}, \alpha^2\right). \tag{A.8}$$

Applying (A.8) to (A.3) results in

$$V_m(a,b) = \frac{1}{2b^{m+1}} \sum_{k=0}^{m} \binom{m}{k} (-a)^k F_k(x)\Big|_{a+b\ell_1}^{a+b\ell_2}, \tag{A.9}$$

where we have defined

$$F_k(x) = \operatorname{sign}^{k+1}(x)\,\gamma\left(\frac{k+1}{2}, x^2\right). \tag{A.10}$$

According to the formula [19, Sect. 45:4]

$$\gamma\left(\frac{1}{2}, x\right) = \sqrt{\pi}\,\operatorname{erf}\left(\sqrt{x}\right), \quad x \geq 0, \tag{A.11}$$

and knowing that $\operatorname{erf}(-x) = -\operatorname{erf}(x)$ [19, Eqn. 40:5:1] and $x = \operatorname{sign}(x)|x|$ [19, Eqn. 8:0:1], for $k = 0$, (A.10) is reduced to

$$F_0(x) = \operatorname{sign}(x)\,\gamma\left(\frac{1}{2}, x^2\right) = \sqrt{\pi}\,\operatorname{erf}(x), \tag{A.12}$$

So

$$\begin{aligned} V_0(a,b) &= \int_{\ell_1}^{\ell_2} \exp\left(-[a+bx]^2\right) dx \\ &= \frac{\sqrt{\pi}}{2b} \operatorname{erf}(x)\Big|_{a+b\ell_1}^{a+b\ell_2}. \end{aligned} \tag{A.13}$$

For k=1, applying directly [19, Sect. 45:4], we have

$$F_1(x) = \gamma(1,x) = 1 - e^{-x}, \tag{A.14}$$

So

$$\begin{aligned} V_1(a,b) &= \int_{\ell_1}^{\ell_2} x \exp\left(-[a+bx]^2\right) dx \\ &= \frac{-1}{2b^2}\left[\sqrt{\pi}a\,\operatorname{erf}(x) + e^{-x^2}\right]_{a+b\ell_1}^{a+b\ell_2}. \end{aligned} \tag{A.15}$$

Similarly, for K=2, taking into account (A.12) and (A.14) and applying the recursion formula [19, Eqn. 45:5:1]

$$\gamma(\nu+1, x) = \nu\gamma(\nu, x) - x^{\nu} e^{-x}, \tag{A.16}$$

we arrive at

$$\begin{aligned} F_2(x) &= \operatorname{sign}(x)\,\gamma\left(\frac{3}{2}, x^2\right) \\ &= \operatorname{sign}(x)\left[\frac{1}{2}\gamma\left(\frac{1}{2}, x^2\right) - |x|\exp\left(-x^2\right)\right] \\ &= \frac{\sqrt{\pi}}{2}\operatorname{erf}(x) - x\exp\left(-x^2\right), \end{aligned} \tag{A.17}$$

So

$$\begin{aligned} V_2(a,b) &= \int_{\ell_1}^{\ell_2} x^2 \exp\left(-[a+bx]^2\right) dx \\ &= \frac{1}{2b^3}\left[\sqrt{\pi}\left(a^2 + \frac{1}{2}\right)\operatorname{erf}(t) + (2a - t)\,e^{-t^2}\right]_{a+b\ell_1}^{a+b\ell_2}. \end{aligned} \tag{A.18}$$

Competing Interests

The author declares that there are no competing interests regarding the publication of this paper.

ACKNOWLEDGMENTS

The author wishes to thank the financial support received from Generalitat Valenciana under Grant GVA/2015/007.

REFERENCES

1. S. Malkin and C. Guo, Grinding Technology: Theory and Application of Machining with Abrasives, Industrial Press, New York, NY, USA, 2008.
2. J. B. J. W. Hegeman, J. T. M. De Hosson, and G. De With, "Grinding of WC—Co hardmetals," Wear, vol. 248, no. 1-2, pp. 187–196, 2001.
3. H. S. Carslaw and J. C. Jaeger, Conduction of Heat in Solids, Oxford Science Publications, Oxford, UK, 1986.
4. N. R. DesRuisseaux and R. D. Zerkle, "Temperature in semi-infinite and cylindrical bodies subjected to moving heat surfaces and surface cooling," Journal of Heat Transfer, vol. 92, pp. 456–464, 1970.
5. J. C. Jaeger, "Moving sources of heat and the temperature at sliding contracts," Proceedings of the Royal Society of New South Wales, vol. 76, pp. 204–224, 1942.
6. J. L. González-Santander, "Calculation of an integral arising in dry flat grinding for a general heat flux profile. Application to maximum temperature evaluation," Journal of Engineering Mathematics, vol. 88, pp. 137–160, 2014.
7. W. J. Sauer, Thermal aspects of grinding [PhD dissertation], Carnegie Mellon University, Pittsburgh, Pa, USA, 1971.
8. M. C. Shaw, "Temperatures in cutting and grinding," American Society of Mechanical Engineering, vol. 146, pp. 17–24, 1990.
9. A. S. Lavine, S. Malkin, and T. C. Jen, "Thermal aspects of grinding with CBN wheels," CIRP Annals—Manufacturing Technology, vol. 38, no. 1, pp. 557–560, 1989.
10. C. Guo and S. Malkin, "Analysis of energy partition in grinding," Journal of Engineering for Industry, vol. 117, no. 1, pp. 55–61, 1995.
11. L. C. Zhang, T. Suto, T. H. Noguchi, and T. Waida, "An overview of applied mechanics in grinding,"Manufacturing Review, vol. 4, pp. 261–273, 1992.
12. M. Mahdi and L. Zhang, "Applied mechanics in grinding—VI. Residual stresses and surface hardening by coupled thermo-plasticity and phase transformation," International Journal of Machine Tools and Manufacture, vol. 38, no. 10-11, pp. 1289–1304, 1998.
13. L. C. Zhang, T. Suto, H. Noguchi, and T. Waida, "Applied mechanics in grinding. Part II: modelling of elastic modulus of wheels and interface forces," International Journal of Machine Tools and Manufacture, vol. 33, no. 2, pp. 245–255, 1993.
14. C. Guo, Y. Wu, V. Varghese, and S. Malkin, "Temperatures and energy partition for grinding with vitrified CBN wheels," CIRP Annals, vol. 48, no. 1, pp. 247–250, 1999.

15. I. Zarudi and L. C. Zhang, "A revisit to some wheel—workpiece interaction problems in surface grinding," International Journal of Machine Tools and Manufacture, vol. 42, no. 8, pp. 905–913, 2002.

16. W. B. Rowe, S. C. E. Black, B. Mills, H. S. Qi, and M. N. Morgan, "Experimental investigation of heat transfer in grinding," CIRP Annals—Manufacturing Technology, vol. 44, no. 1, pp. 329–332, 1995.

17. J. González-Santander, "Analytic solution for maximum temperature during cut in and cut out in surface dry grinding," Applied Mathematical Modelling, vol. 40, no. 3, pp. 2356–2367, 2016.

18. J. L. González-Santander, J. M. Isidro, and G. Martín, "An analysis of the transient regime temperature field in wet grinding," Journal of Engineering Mathematics, vol. 90, pp. 141–171, 2015.

19. K. Oldham, J. Myland, and J. Spanier, An Atlas of Functions, Springer, New York, NY, USA, 2nd edition, 2008.

20. M. Abramowitz and I. Stegun, Handbook of Mathematical Functions, National Bureau of Standards, Washington, DC, USA, 1972.

21. J. L. González-Santander and G. Martín González, "Closed form expression for the surface temperature in wet grinding. Application to maximum temperature evaluation," Journal of Engineering Mathematics, vol. 90, pp. 173–193, 2015.

22. J. L. González-Santander and G. Martín, "A theorem for finding maximum temperature in wet grinding,"Mathematical Problems in Engineering, vol. 2015, Article ID 150493, 13 pages, 2015.

23. R. M. Corless, G. H. Gonnet, D. E. G. Hare, D. J. Jeffrey, and D. E. Knuth, "On the Lambert W function,"Advances in Computational Mathematics, vol. 5, no. 4, pp. 329–359, 1996.

24. V. I. Murav'ev, A. V. Yakimov, and A. V. Chernyshev, "Effect of deformation, welding, and electrocontact heating on the properties of titanium alloy VT20 in pressed and welded structures," Metal Science and Heat Treatment, vol. 45, no. 11-12, pp. 419–422, 2003.

25. D. L. Skuratov, Y. L. Ratis, I. A. Selezneva, J. Pérez, P. Fernández de Córdoba, and J. F. Urchueguía, "Mathematical modelling and analytical solution for workpiece temperature in grinding," Applied Mathematical Modelling, vol. 31, no. 6, pp. 1039–1047, 2007.

26. A. S. Lavine, "A simple model for convective cooling during the grinding process," Journal of Engineering for Industry, vol. 110, no. 1, pp. 1–6, 1988.

27. R. P. Brent, "An algorithm with guaranteed convergence for finding a zero of a function," in Algorithms for Minimization without Derivatives, chapter 4, Prentice-Hall, Englewood Cliffs, NJ, USA, 1973.

28. H. Takahasi and M. Mori, "Double exponential formulas for numerical integration," Publications of the Research Institute for Mathematical Sciences, vol. 9, pp. 721–741, 1973.

CHAPTER 9

FINITE DIFFERENCE ANALYSIS OF TRANSIENT HEAT TRANSFER IN SURROUNDING ROCK MASS OF HIGH GEOTHERMAL ROADWAY

Yuan Zhang[1,2], Zhijun Wan[1,2], Bin Gu[1], Hongwei Zhang[1], and Peng Zhou[1]

[1]School of Mines, China University of Mining & Technology, No. 1 University Road, Xuzhou 221116, China

[2]Key Laboratory of Deep Coal Resource Mining of Ministry of Education, China University of Mining & Technology

ABSTRACT

Based on finite difference method, a mathematical model and a numerical model written by Fortran language were established in the paper. Then a series of experiments were conducted to figure out the evolution law of temperature field in high geothermal roadway. Research results indicate that temperature disturbance range increases gradually as the unsteady heat conduction goes on and it presents power function relationship with dimensionless time. Based on the case analysis, there is no distinct expansion of temperature disturbance range after four years of ventilation, when the temperature disturbance range R=13.6.

INTRODUCTION

At present, solid mineral resource exploitation can reach 4 km underground [1]. With the rapid development of China's economy, the production and consumption of solid mineral resources increase year by year, resulting in the gradual exhaustion of shallow resources and escalation of mining depth. Taking coal as an example, in the past decade, the annual increasing rate of coal production was 8.6%, and the yield reached 366.9 billion t at 2013 [2]. The shallow minerals were exhausted at the majority of mining districts because of long-time, high-intensity, and large-scale mining activities, and the mining deepens downward 8 to 12 m every year [3]. Up till 2013, there were about 200 state-owned coal mines exploiting 800 m beneath the ground, among which 47 coal mines conducted their mining activities under −1000 m. Most of these coal mines are located in mideastern China such as Shandong, Jiangsu, and Anhui province. The average mining depth is −1086 m and maximum value is −1501 m in Suncun coal mine of Xinwen Mining Group, Shandong province [4]. It is predicted that in east, middle, north, and northeast China, most coal mines will conduct their deep exploiting activities from −1000 m to −1500 m by 2020.

With increasing mining depth, heat hazard usually takes place and becomes more and more prominent. High geothermal

Citation: Yuan Zhang, Zhijun Wan, Bin Gu, Hongwei Zhang, and Peng Zhou, "Finite Difference Analysis of Transient Heat Transfer in Surrounding Rock Mass of High Geothermal Roadway," Mathematical Problems in Engineering, vol. 2016, Article ID 8951524, 7 pages, 2016. doi:10.1155/2016/8951524.

temperature is a decisive cause of heat hazard in deep mining. Generally, the geothermal temperature gradient of China's coal seams is 2.5~3°C/hm, but at abnormal geothermal areas it can reach 4~5°C/hm. It is thus estimated that, with the depth and temperature of constant temperature zone of 30 m and 15°C, respectively, the original rock temperature can reach 40~45°C at −1000 m [5]. Actually, the rock temperature at −1000 m of many heat hazard mines is close to or exceeds the upper limit. For example, in Sanhejian mine of Xuzhou Mining Group the original rock temperature can reach 46.8°C, in Suncun mine of Xinwen Mining Group it is 45°C, and in Zhaolou mine of Yanzhou Mining Group it is 43°C [6]. Heat hazard is prone to occur under the combined influence of high geothermal temperature, heat of air compression, and heat dissipation of machine. According to incomplete statistics, mining activities in about 150 coal mines are affected by heat hazard in China, and their workface temperature is usually in the range of 30~40°C. Heat has already become the sixth mine hazard after roof falling, gas, water inrush, fire, and mine dust. Beside deteriorating the air environment of mines, high geothermal temperature can change the physical and mechanical properties of rocks and further affect the stability of surrounding rock and heat conduction process [6–9].

Temperature that is not disturbed by human engineering activities is named original rock temperature [10, 11]. Once the roadway excavation of high geothermal mine was conducted, convective heat transfers between surrounding rock and low-temperature airflow and instantly breaks the balanced temperature distribution in strata; the geothermal energy constantly transferring to airflow may result in underground air temperature increase. Meanwhile, the surrounding rock temperature decreases from surface to deeper area by and by, and the cooling scale of surrounding rock continuously expands, which is considered as temperature disturbance range (TDR). The rock inside TDR is called heat regulating circle (HRC). Because of the heat resistance in the HRC, the temperature decreasing amplitude declines with the distance going deeper until reaching a temperature boundary which is relatively unchanging, and its value is similarly equal to the original rock temperature. Instead of losing huge amount of geothermal heat quickly in a short time, the HRC can release the heat into the airflow little by little so as to buffer the heat dissipation. In this perspective, research on temperature distribution of HRC has important theoretical and practical significance.

Temperature distribution of the surrounding rock can be measured by thermocouples in holes drilling, and this shortcut method has been successfully applied in Huaibei and Pingdingshan Mining Groups in China. The field measure process indicates that, with the ventilation moves on, the surrounding rock temperature drops constantly with a smaller and smaller decreasing amplitude [12, 13]. Other scholars measured the original rock temperature by drilling holes deep or shallow. Besides, with the development of computer technology, numerical method has also been applied in temperature field studies of surrounding rocks; for instance, by numerical computation, Zhang concluded that, under the coupling effect of airflow and seepage, the symmetrical distribution characteristics of temperature field are influenced by seepage field rather than the airflow [14]. Meanwhile, Zhang et al. suggested that the temperature distribution fits exponential function in radical direction; the temperature gradient and heat flux at roadway surface are maximum; the relation between surrounding rock temperature and ventilation time similarly meets Hill function, and the fitting formula of the HRC was acquired [15]. According to Fan and Zhang, the scale of HRC was calculated by mathematical model and its asynchronous semiexplicit difference form; however, because of the specific research purpose, general rules of the TDR were not summarized [16]. The research of [17] points out that thermal diffusivity greatly affects both the TDR and temperature distribution of surrounding rock, while roadway geometry size and convective heat transfer coefficient of surrounding rock and airflow have little impact on the TDR. Though the characteristics and evolution laws of roadway temperature distribution have been fully reported in existing literatures, topics concerned about TDR, especially the general rules of the HRC expansion, still demand further studies.

In this paper, the authors established a mathematical model to reveal the expansion laws of the HRC based on the FDM and summarized the general rules of the TDR.

MATHEMATICAL MODEL OF ROADWAY TEMPERATURE FIELD

To facilitate the evolution laws study of roadway temperature distribution, the following assumptions were proposed at first [18]:

(1) The roadway cross section is round ($r=r_0$) and its surrounding rock is a hollow cylinder with infinite radius ($r \to \infty$).

(2) The surrounding rock is continuous, homogeneous, and isotropous, and its thermophysical parameters are constants.

(3) Initially when t=0, the surrounding rock temperature at each point is uniform and equal to its original rock temperature (t_0).

(4) The airflow temperature only changes along the axial direction and remains uniform on the same radical cross section. The heat transfer conditions are identical at the circumferential direction.

Meanwhile, it is assumed that there is an ideal heat insulation layer with infinitesimal thickness, infinitely great external diameter, and an inner diameter equal to roadway diameter; it also has the properties of homogeneity, continuity, and isotropy and heat insulation. In this case, if a roadway with length of L was divided into infinite segments by the above layer and assured that inner bore of ideal heat insulation layer was overlapped to roadway cross section, then each segment of the roadway surrounding rock would be infinitely thin heat conduction slice without heat transfer between adjacent slices. Consequently, the heat would only transfer along radical direction and the conduction would become one-dimensional. These assumptions neglect heat transfer in axial direction in the surrounding rock mass. In fact, the heat flux in axial direction is negligibly small, especially when the rock mass temperature has not been disturbed on a large scale.

Based on the above assumptions, the law of conservation of energy, and Fourier's law, taking a microunit from the assumed surrounding rock slice, differential equations (mathematical model) of heat conduction under one-dimensional cylindrical coordinate system were established, as shown in Figure 1 and (1). One has

$$\frac{\partial t}{\partial \tau} = a\left(\frac{1}{r}\frac{\partial t}{\partial r} + \frac{\partial^2 t}{\partial r^2}\right), \quad (1)$$

where $a = \lambda/\rho c$ stands for the thermal diffusivity of roadway surrounding rock; t stands for the surrounding rock temperature; τ stands for time; r stands for the radical coordinate of surrounding rock.

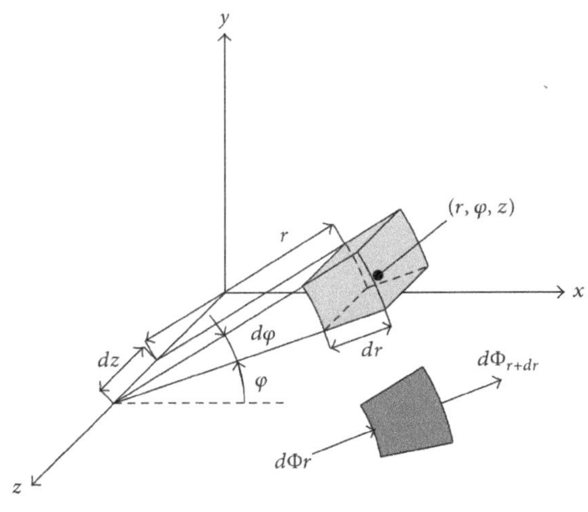

Figure 1: Differential volume element of heat conduction in cylindrical coordinate system.

Derived from law of conservation of energy and Fourier's law, (1) is a general formula of material heat conduction. To describe the heat conduction phenomenon of surrounding rock, the following monodromy conditions should be limited:

(1) Geometrical condition: it is semi-infinite plate with inner radius of r_0.

(2) Physical condition: rock density ρ_s, specific heat capacity c_s, and heat conductivity λ_s are constant and free from the influence of temperature variation; no internal heat source exists in the isotropous surrounding rock.

(3) Initial condition: $t(r, 0)=t_0$.

(4) Boundary condition: the boundary condition of surrounding rock heat conduction reflects the interaction between surrounding rock and its ambience during the process of heat conduction; it reflects the characteristics of heat conduction on surrounding rock boundary. As a semi-infinite hollow round slice of infinitely great diameter, the temperature at the area infinitely far away from the circle center tends to approach the original rock temperature unlimitedly; that is

$$\text{when } r \longrightarrow \infty, \quad t = t_0. \quad (2)$$

The other boundary of roadway surrounding rock is the surface, where $r=r_0$. It is difficult to acquire the temperature function (the first boundary condition) due to complex boundary conditions and complicated process of convective heat transfer between surrounding rock and airflow at roadway surface. Many scholars assumed that the rock temperature is constant and unchanging at roadway surface, which was obviously appropriate. If the heat flux of roadway surface was given (the secondary boundary condition), all problems would be settled easily, but either rock temperature field or surface heat flux coefficient function should be known as a premise. The third boundary condition is most extensively utilized because of the continuous convective heat conduction between surrounding rock surface and airflow. For convective heat transfer under steady airflow condition and temperature, the surface convective heat transfer coefficient h can be seen as constant and the function of airflow temperature f_s with time and space is considered as known; then

$$-\lambda_s\left(\frac{\partial t}{\partial n}\right)_w = h\left(t_w - t_f\right), \quad (3)$$

where n is the exterior normal of roadway surface. The surface rock temperature t_w and temperature gradient $(\partial t/\partial n)_w$ are unknown.

The following mathematical model for surrounding rock temperature field is established as a consequence:

$$\frac{\partial t}{\partial \tau} = a\left(\frac{1}{r}\frac{\partial t}{\partial r} + \frac{\partial^2 t}{\partial r^2}\right),$$
$$t(r,\tau)|_{\tau=0} = t_0,$$
$$\lambda\frac{\partial t}{\partial r}\bigg|_{r=r_0} = h\left(t - t_f\right),$$
$$t(\infty,\tau) = t_0. \quad (4)$$

FDM OF SURROUNDING ROCK TEMPERATURE FIELD

Regional Discrete and Difference Scheme

The established finite difference grid figure was illustrated in Figure 2, where meshing space is set as Δr , time step is set as $\Delta\tau$, and then time and space coordinate can be expressed as

$$r = i\Delta r, \quad i = 0,1,2,3,\ldots,N, \quad L = N\Delta r,$$
$$\tau = k\Delta\tau, \quad k = 0,1,2,3,\ldots, \quad (5)$$

where temperature of regional nodes $(i\Delta r, k\Delta\tau)$ is denoted as t_i^k.

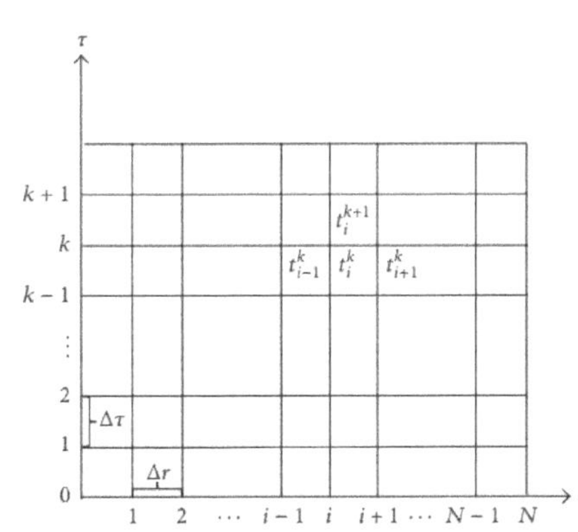

Figure 2: Finite difference mesh.

Unsteady unit $\partial t/\partial \tau$ in (4) applied backward difference scheme while $\partial t/\partial \tau$ and $\partial^2 t/\partial r^2$ adopted central difference scheme; then

$$\frac{\partial t_i^{k+1}}{\partial \tau} = \frac{t_i^{k+1} - t_i^k}{\Delta \tau} + O(\Delta \tau),$$
$$\frac{\partial t_i^{k+1}}{\partial r} = \frac{t_{i+1}^{k+1} - t_{i-1}^{k+1}}{2\Delta r},$$
$$\frac{\partial^2 t_i^{k+1}}{\partial r^2} = \frac{t_{i-1}^{k+1} - 2t_i^{k+1} + t_{i+1}^{k+1}}{\Delta r^2}, \tag{6}$$

where $O(\Delta\tau)$ which represents the bottom step of the remainder is one order and can be omitted.

Introducing (6) into (5) gives

$$\frac{t_i^{k+1} - t_i^k}{\Delta \tau} = a\left(\frac{1}{r}\frac{t_{i+1}^{k+1} - t_{i-1}^{k+1}}{2\Delta r} + \frac{t_{i-1}^{k+1} - 2t_i^{k+1} + t_{i+1}^{k+1}}{\Delta r^2}\right) \tag{7}$$

which can also be written as

$$t_i^k = \left(\frac{a\Delta\tau}{2r\Delta r} - \frac{a\Delta\tau}{\Delta r^2}\right)t_{i-1}^{k+1} + \left(1 + 2\frac{a\Delta\tau}{\Delta r^2}\right)t_i^{k+1} - \left(\frac{a\Delta\tau}{2r\Delta r} + \frac{a\Delta\tau}{\Delta r^2}\right)t_{i+1}^{k+1}. \tag{8}$$

Introducing α, β, and γ, then

$$\alpha = \frac{a\Delta\tau}{2r\Delta r} - \frac{a\Delta\tau}{\Delta r^2},$$
$$\beta = 1 + 2\frac{a\Delta\tau}{\Delta r^2},$$
$$\gamma = -\left(\frac{a\Delta\tau}{2r\Delta r} + \frac{a\Delta\tau}{\Delta r^2}\right). \tag{9}$$

Introducing (9) into (8) gives

$$t_i^k = \alpha t_{i-1}^{k+1} + \beta t_i^{k+1} + \gamma t_{i+1}^{k+1}. \tag{10}$$

The matrix form is

$$\begin{bmatrix} \beta_0 & \gamma_0 & 0 & 0 & 0 \\ \alpha_1 & \beta_1 & \gamma_1 & 0 & 0 \\ 0 & \cdots & \ddots & \cdots & 0 \\ 0 & 0 & \alpha_{N-2} & \beta_{N-2} & \gamma_{N-2} \\ 0 & 0 & \cdots & \alpha_{N-1} & \beta_{N-1} \end{bmatrix} \begin{bmatrix} t_1^{k+1} \\ t_2^{k+1} \\ \vdots \\ t_{N-2}^{k+1} \\ t_{N-1}^{k+1} \end{bmatrix} = \begin{bmatrix} t_1^k - \alpha_0 t_0^{k+1} \\ t_2^{k+1} \\ \vdots \\ t_{N-2}^k \\ t_{N-1}^k - \gamma_{N-1} t_{N-1}^{k+1} \end{bmatrix}. \tag{11}$$

Setting A as coefficient matrix gives

$$A = \begin{bmatrix} \beta_0 & \gamma_0 & 0 & 0 & 0 \\ \alpha_1 & \beta_1 & \gamma_1 & 0 & 0 \\ 0 & \cdots & \ddots & \cdots & 0 \\ 0 & 0 & \alpha_{N-2} & \beta_{N-2} & \gamma_{N-2} \\ 0 & 0 & \cdots & \alpha_{N-1} & \beta_{N-1} \end{bmatrix}. \tag{12}$$

Equation (12) is a tridiagonal matrix, which can be solved by Thomas Algorithm method. Discrete initial conditions give

$$t_i^0 = t_0. \tag{13}$$

In terms of boundary conditions, the left boundary meets the third boundary condition while the right boundary meets the first boundary condition. When $r=r_0$, in order to ensure the discrete precision uniformity with extensive function, one order central difference is applied to the third boundary condition $\lambda(\partial t/\partial r) = h(t - t_f)$ and it gives

$$\lambda\frac{t_1^{k+1} - t_{-1}^{k+1}}{2\Delta r} = h\left(t_0^{k+1} - t_f\right). \tag{14}$$

Combining (9) and (14) gives

$$t_0^{k+1} = \frac{((\alpha + \gamma)/\alpha)\, t_1^{k+1} - (1/\alpha)\, t_0^k + (2\Delta r h/\lambda)\, t_f}{(2\Delta r h/\lambda) - (\beta/\alpha)}. \tag{15}$$

By using the same analysis method, the first boundary condition of the left boundary $t=t_0$ will be discretized as

$$t_N^k = t_0. \tag{16}$$

Stability of Difference Scheme

Because of marching method is involved to solve the FDM of unsteady heat conduction, if deviation was introduced from step n, then deviation in step n+1 would be inevitably produced and even increased. If this deviation could be controlled, disappears gradually, or has a boundary, then the established FDM is steady. In this paper, Fourier series is adopted to assure the stability of the difference scheme.

Setting $t_i^{n+1} = v^{n+1}\exp(ikjh)$ introducing it to (9) give

$$v^{n+1} = v^n \frac{1}{\alpha\exp(-kjh) + \beta + \gamma\exp(kjh)}. \tag{17}$$

Excessive divisor is acquired as

$$G(r,k) = \frac{1}{\alpha\exp(-kjh) + \beta + \gamma\exp(kjh)}. \tag{18}$$

Setting $\lambda_1 = a\Delta\tau/2r\Delta r$ and $\lambda_2 = a\Delta\tau/\Delta r^2$ and combining (18) give

$$G(r,k) = \frac{1}{1 + 2\lambda_2 - 2\lambda_2\cos(kh) - 2i\lambda_1\sin(kh)}. \tag{19}$$

To all λ_1 and λ_2, $|G(r,k)| \leq 1$ which meets Von Neumann condition, the difference scheme equation (8) is steady absolutely.

Based on the above difference scheme, a numerical program for unsteady temperature field of roadway surrounding rock is written by Fortran language.

NUMERICAL TEST PROGRAM

According to the correspondent mathematical model, simulated roadway has a round cross section, with radius r_0 of 2 m which is determined by the conversion of hydraulic diameter. Generally, the thickness of HRC would not exceed 40 m, so the external boundary is set as 60 m after proper expansion. Besides, as the thermal environment of simulated roadway is incapable of resulting in large variation of the thermophysical properties of surrounding rock, rock density ρ_s, specific heat capacity c_s, and heat conductivity λ_s can be regarded as constants, where λ_s =1.2 W/m·K and a=0.8×10^{-6} m^2/s ($a = \lambda_s/\rho c$).

From the assumptions of surrounding rock heat conduction models, the inner boundary matches the third boundary condition, while the external boundary belongs to the first boundary condition. The original rock temperature is set as 37°C. In general, the convective heat transfer coefficient h of turbulent airflow condition is 20–100 W/m^2·K, and it is rarely affected by the airflow temperature because of its slight temperature variation in roadways. The value of h can be calculated by following equations:

$$\mathrm{Bi} = 0.0115\frac{\lambda_f}{\lambda_s}(av)^{-0.4}(ud)^{0.8} \approx 21.5\frac{u^{0.8}}{\lambda_s},$$

$$\mathrm{Bi} = \frac{hr_0}{\lambda_s}. \quad (20)$$

By conversion,

$$h \approx 21.5\frac{u^{0.8}}{r_0} = 10.75u^{0.8}, \quad (21)$$

where λ_f is the airflow heat conductivity coefficient, 0.026 W/m·K; λ_s is the surrounding rock heat conductivity coefficient, 1.2 W/m·K; a is the coefficient of thermal diffusivity, 0.8 × 10^{-6} m^2/s; v is the airflow kinematic viscosity coefficient, 14.4 × 10^{-6} m^2/s; and d is the hydraulic diameter, 4 m.

In the simulation process, the airflow temperature is constantly 20°C and the external boundary temperature of roadway surrounding rock is equal to the original rock temperature; that is, t=t_0. Based on the above assumptions, a one-dimensional hollow cylinder slice model with inner radius of 2 m and external radius of 62 m was built up. To shorten the simulation test time, the model is divided into 600 grids by equal step length of 0.1 m. The simulation time is 3 years long.

The test program is illustrated in Table 1.

Table 1: Test program.

Parameter	Value
Original rock temperature	$t_0 = 37°C$
Roadway radius	$r_0 = 2$ m
Airflow temperature	$t_f = 20°C$
Heat conductivity coefficient	$\lambda_s = 1.2$ W/m·K
Thermal diffusivity coefficient	$a = 0.8 \times 10^{-6}$ m^2/s
Airflow velocity	$u = 2.87$ m/s
Convective heat transfer coefficient on the surface	$h = 25$ W/m^2·K
Biot	Bi = 41.63

RESULTS AND DISCUSSION

Numerical Test Results

Figure 3 presents the cloud picture of roadway temperature field in 30000 h. It is obvious that the TDR extends continuously with the progress of ventilation and reaches its maximum at about 25 m, but detailed characteristics cannot be observed from the figure. To detect the evolution laws of rock temperature field in detail, the experiment data was rehandled. To obtain the universal law between TDR, space, and time, dimensionless method was applied to the above factors, where the dimensionless temperature $\Theta = (t - t_f)/(t_0 - t_f)$, the dimensionless radius R=r/r_0, and the dimensionless time $Fo = a\tau/r_0^2$.

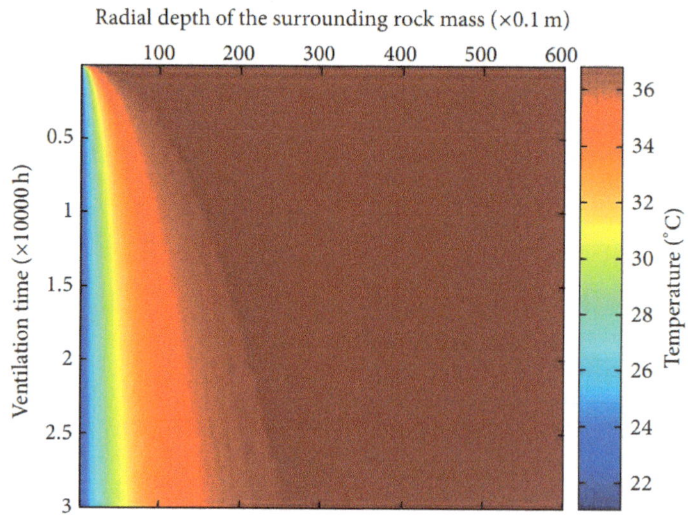

Figure 3: General characteristics of temperature field of roadway surrounding rock.

The relationship between TDR and its corresponding dimensionless time is illustrated in Figure 4.

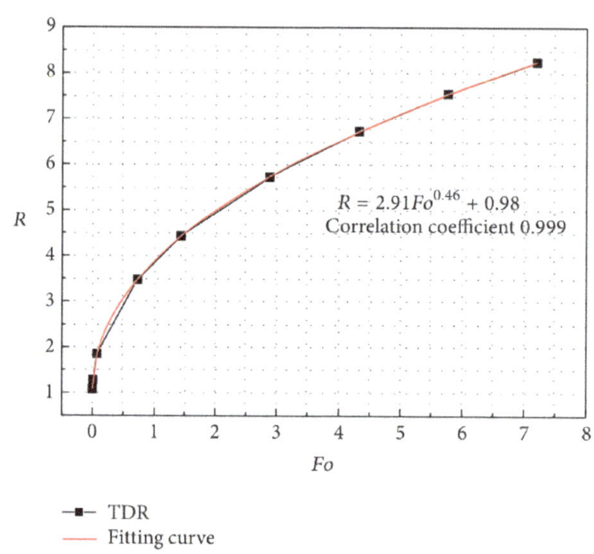

Figure 4: Relationship between temperature disturbance range and dimensionless time.

Figure 4 elaborates that, with the progress of unsteady heat conductivity, TDR extends gradually with a declining rate (the range boundary is lower than 1% of original rock temperature). It is clear that the TDR presents exponential function with dimensionless time Fo

$$R = 2.91Fo^{0.46} + 0.98. \quad (22)$$

Certain deviation exists in (22) because theoretically when Fo is equal to 0,R should be 1, and the temperature disturbance depth is 0. The deviation may be produced by calculation precision error. For example, if the step length is set as 0.1 m when dividing the spatial grids, experimental data only exists at the nodes integral times of 0.1 m, which would cause a deviation of 0.1 m in TDR statistics.

In order to acquire a more practical formula for TDR calculation, the fit function was defined as follows:

$$R = A\sqrt{Fo} + 1. \quad (23)$$

Introducing (23) to fit the data in Figure 4 which gives A=2.75 can get a goodness of 0.9988 and a standard error of 0.02, which can meet the needs of engineering application. Introducing $Fo = a\tau/r_0^2$ to (23) gives

$$(R - 1)\,r_0 = 2.75\sqrt{a\tau}, \quad (24)$$

where $(R-1)r_0$ is the temperature disturbance depth, m; a is the thermal diffusivity coefficient of surrounding rock, m^2/s; τ is the heat conduction time, s.

To calculate the maximum range of temperature disturbance, the paper set 6 months or so as an interval (Fo=3) and plotted the relation curve between dimensionless temperature and depth of surrounding rock in Figure 5.

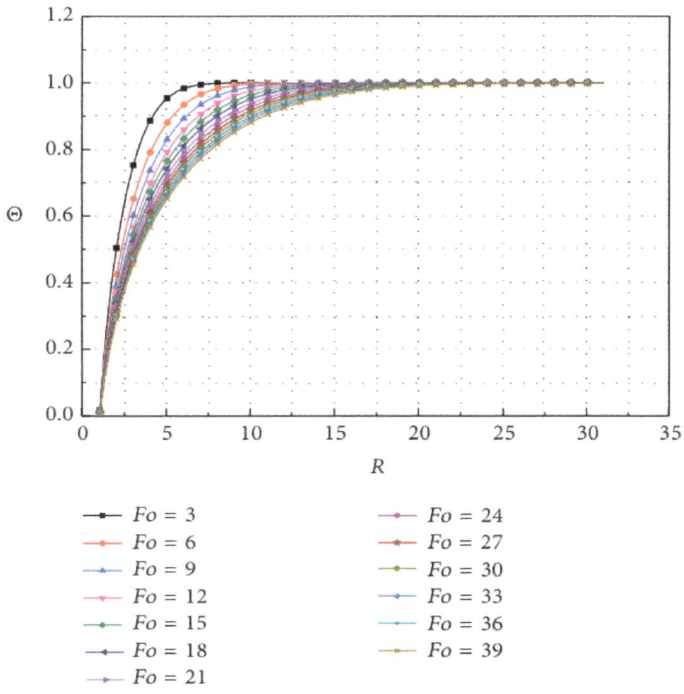

Figure 5: Variation trend of temperature disturbance with depth.

It is clear that the rock temperature changes distinctly during the first 3 years, then this variation gradually diminishes. Four years later when the TDR is about 13.6 (25.2 m), there is no apparent expansion. The value is similar to the theoretical value calculated by (24) (14.47), which provides evidence for the steady temperature field calculation of roadway surrounding rock and drilling detection of original rock temperature.

Discussions on HRC

HRC is a key concept in studying the roadway temperature field. Initially, HRC was proposed in view of the effect of periodic variation of surface air temperature on surrounding rock temperature distribution when the surface air was supplied into the underground. In opposite, low-temperature airflow would be heated by roadway surrounding rock temperature, and high-temperature airflow would be cooled. In a narrow sense, HRC is the surrounding rock affected by periodical changing air temperature. Afterwards, this concept was also introduced to the research of cooling range of high geothermal roadway.

At present, HRC has no uniform definition. Some scholars hold the opinion that the HRC is a static concept, which is the maximum scale of surrounding rock whose temperature has an apparent variation after the heat conduction of roadway surrounding rock is basically steady, while others think that the HRC is a dynamic concept, and it would extend constantly with the continuous progress of heat transfer. In this paper, the authors adopt the latter opinion. Strictly, the basis of the HRC is a temperature disturbance scale affected by ventilation. Therefore, this paper defines the TDR as a scale with temperature difference more than 1% of original rock temperature in the progress of unsteady heat transfer.

Studies of the roadway temperature disturbance range have realistic significance. For example, the depth of drilling hole for measuring rock temperature should be larger than the TDR when conducting in situ measurement of the temperature field. In this case, the external boundary is the maximum value of the TDR because of small variation of rock temperature after long-time ventilation. So the rock temperature field at this time can be simplified as steady. The calculation results in this paper present exponential function between TDR and ventilation time. Seriously, there is no fixed boundary of the temperature disturbance district because the TDR will increase with the progress of ventilation. However, as the order of the power function between the TDR and ventilation is less than 1, with longer ventilation time, the increasing rate of the TDR declines continuously. In that way, there is a relatively steady area of the temperature disturbance district and it can be called "maximum temperature disturbance range" after long-time ventilation. In fact, in mining practice, the judgment standard for the TDR is to seek a point that its temperature keeps on being less than 1% of the original rock temperature in an allowed long interval, rather than to find the external boundary when the surrounding rock temperature is steady.

CONCLUSIONS

In this paper, a Fortran language source procedure for surrounding rock temperature field calculation and the corresponding evolution laws was written on the basis of one-dimensional heat conductivity mathematical model and the FDM. The results show that the TDR of roadway surrounding rock extends constantly with the progress of ventilation. Meanwhile, it presents power function with dimensionless time of unsteady heat conduction. The findings will provide evidence for original rock temperature measurement in underground mines.

Competing Interests

There is no conflict of interests for all authors of this paper.

ACKNOWLEDGMENTS

This work was financially supported by Fundamental Research Funds for Central Universities (2014-6121), National Natural Science Foundation of China (51504236), and the Priority Academic Program Development of Jiangsu Higher Education Institutions (PAPD).

REFERENCES

1. L. M. Surhone, M. T. Tennoe, and S. F. Henssonow, Tautona Mine, Betascript Publishing, Beau Bassin, Mauritius, 2011.
2. National Bureau of Statistics of China, National economy and society developed statistical bulletin 2014 [EB/OL], http://www.stats.gov.cn/tjsj/zxfb/201502/t20150226_685799.html.
3. Y. Zhang, Transient Temperature Field of Surrounding Rock of the High Geothermal Roadway and its Heat Control Mechanism by Heat Insulation, China University of Mining and Technology Press, Xuzhou, China, 2013.
4. Z.-J. Wan, Y. Zhang, J.-Y. Cheng, C.-B. Zhou, B. Gu, and P. Zhou, "Mine geothermal and heat hazard prevention and control in China," Disaster Advances, vol. 6, supplement 5, pp. 86–94, 2013.
5. H.-P. Xie, H.-W. Zhou, D.-J. Xue, H.-W. Wang, Z. Ru, and F. Gao, "Research and consideration on deep coal mining and critical mining depth," China Coal Society, vol. 37, no. 4, pp. 535–542, 2012.
6. M.-C. He and P.-Y. Guo, "Deep rock mass thermodynamic effect and temperature control measures,"Chinese Journal of Rock Mechanics and Engineering, vol. 32, no. 12, pp. 2377–2393, 2013.
7. X.-B. Li, T.-B. Yin, Z.-L. Zhou, L. Hong, and K. Gao, "Study of dynamic properties of siltstone under coupling effects of temperature and pressure," International Journal of Rock Mechanics and Mining Sciences, vol. 29, no. 12, pp. 2377–2384, 2010.
8. N. A. Al-Shayea, K. Khan, and S. N. Abduljauwad, "Effects of confining pressure and temperature on mixed-mode (I-II) fracture toughness of a limestone rock," International Journal of Rock Mechanics and Mining Sciences, vol. 37, no. 4, pp. 629–643, 2000.
9. Z.-J. Wan, Y.-S. Zhao, F.-K. Dong, Z.-J. Feng, N. Zhang, and J.-W. Wu, "Experimental study on mechanical characteristics of granite under high temperatures and triaxial stresses," Chinese Journal of Rock Mechanics and Engineering, vol. 27, no. 1, pp. 72–77, 2008.
10. J. Knechtel, "Thermal hazard prevention in longwalls run under extreme geothermal conditions,"Archives of Mining Sciences, vol. 56, no. 2, pp. 265–280, 2011.
11. S. Wasilewski and M. Branny, "A preliminary study of the unsteady states of the ventilation parameters at the longwall face during the shearer operation," in Proceedings of the 12th US/North American Mine Ventilation Symposium, Reno, Nev, USA, June 2008.
12. T.-H. Zhu, "Study on influence law of ventilation time to temperature of surrounding rock," Safety in Coal Mines, vol. 424, no. 2, pp. 10–13, 2010.
13. Q. Wu, Y.-P. Qin, L. Guo, and K. Tan, "Analysis of unsteady temperature field in wall rock by finite element method," Journal of Liaoning University of Technology, vol. 21, no. 5, pp. 604–607, 2002.
14. S.-G. Zhang, "Numerical simulation analysis of temperature field in surrounding rock of deeply buried roadways," Science Technology and Engineering, vol. 6, no. 14, pp. 2194–2196, 2006.
15. Y. Zhang, Z.-J. Wan, C.-B. Zhou, et al., "Spatial-temporal distribution characteristics of surrounding rock temperature in high geothermal roadway," Disaster Advances, vol. 6, no. 5, pp. 103–109, 2013.
16. X.-L. Fan and X.-B. Zhang, "Analysis on influence of temperature distribution in strata rock and difference method," Coal Engineering, no. 11, pp. 82–84, 2011.
17. J.-L. Gao and M. Yang, "Analysis of the factors influencing temperature distribution of surrounding rock and cooled zone radius," China Safety Science Journal, vol. 15, no. 2, pp. 73–76, 2005.
18. Y. Zhang, Z. Wan, C. Zhou, and J. Cheng, "Modeling experimental method of unsteady heat conduction in surrounding rock of roadways or tunnels," Journal of Mining and Safety Engineering, vol. 31, no. 3, pp. 441–446, 2014.

CHAPTER 10

FLEXURAL AND THERMAL PROPERTIES OF NOVEL ENERGY CONSERVATION SLOTTED REINFORCED CONCRETE BEAMS

Gao Ma[1], Liang Huang[1], Libo Yan[2,3], Hui Wang[1], and Peng Yin[1]

[1]College of Civil Engineering, Hunan University, Changsha 410082, China

[2]Department of Organic and Wood-Based Construction Materials, Technical University of Braunschweig, Hopfengarten 20, 38102 Braunschweig, Germany

[3]Centre for Light and Environmentally-Friendly Structures, Fraunhofer Wilhelm-Klauditz-Institut WKI, Bienroder

ABSTRACT

Conventional solid reinforced concrete (RC) beams were modified to slotted beams for consideration as thermal insulation structural components. The slotted beam consisted of an outer and an inner beam, respectively, with a slot located near the middle of the beam along its width direction for filling thermal insulation material. Flexural and thermal behavior of the slotted beams were investigated. Three RC reference solid beams and six slotted beams were fabricated and tested under four-point bending tests. The test results indicated that the failure mode of both slotted beams and the solid beams was flexural failure. However, the damage process of the slotted beams was different from that of the solid beams at the final loading stage. The moment curvature analysis indicated that the tensile reinforcement ratio of the outer and inner beams had an important effect on the flexural behavior, especially the ductility of the slotted beams. Thermal study indicated that the heat transfer coefficient of the slotted beam was greatly reduced and the thermal inertia factor increased a lot, compared with the solid beam. In addition, FE simulation results showed that a new frame structure using slotted beams exhibited obvious and attractive thermal insulation property.

INTRODUCTION

Building energy conservation is one of the most important energy-saving policies in developing and developed countries [1]. For a building, thermal bridge refers to a component with a relatively higher heart transfer than the surrounding parts, which leads to a significant reduction in the thermal insulation of the building. Thermal bridges generally exist in building structures [2, 3], such as frame structures, shear wall structures, and masonry structures. For a reinforced concrete (RC) frame structure, the beams and columns all contribute to thermal bridges. Generation and characteristics of thermal bridge are related to the material properties of the structure and the layout of the structure. In a building, its walls, windows, beams, and columns are the main components to generate thermal bridges. If these components can be designed to have an enhanced thermal insulation and reduced heat conduction performance, the issue of thermal bridges in building structures can be solved.

Citation: Gao Ma, Liang Huang, Libo Yan, Hui Wang, and Peng Yin, "Flexural and Thermal Properties of Novel Energy Conservation Slotted Reinforced Concrete Beams," Advances in Materials Science and Engineering, vol. 2016, Article ID 4642810, 14 pages, 2016. doi:10.1155/2016/4642810.

In general, the popular external wall thermal insulation can be classified into exterior insulation and inner insulation of the external wall. Inner insulation of external wall is a technique which uses thermal insulating layer on the inner surface of the external wall of a structure. Because the insulating layer hides inside the walls, this technique reduces the usable area of the structure, affects the redecoration, and generates mildew and moisture condensation in the walls. To date, the technique of exterior insulation of the external wall is still not so mature; for instance, the construction is difficult and the energy-saving investment is high. Meanwhile, the outside insulation layer is under the risk of fire disaster, is easy to drop off and gets seepage in rainy weather; it is difficult to paste ceramic tile or hang marble on the thermal insulating layer and the durability of the exterior insulation system is reduced because it is directly exposed to the outdoor atmosphere. To solve these problems, self-thermal insulation building blocks such as thermal insulation concrete block [4], hollow block [5], and vacuum insulation window [6–8] have been developed and adopted in practical application.

For RC frame structures, the use of self-thermal insulation building blocks and vacuum insulation windows prevents the walls and the windows from generation of thermal bridges, and hence thermal bridges mainly exist in beams and columns. This paper presents a new type of energy conservation slotted beam-wall system, as shown in Figure 1. Conventional reinforced concrete (RC) beams are modified into slotted beams, and the gap between the outer and inner beams is used to fill with thermal insulation materials. Self-thermal insulation wall is constructed on the slotted beams. It is obvious that this type of slotted beam-wall structure possesses a favourable thermal insulation property, because the thermal bridges at the beams and columns can be reduced effectively. Meanwhile, this new energy conservation wall system is convenient in construction.

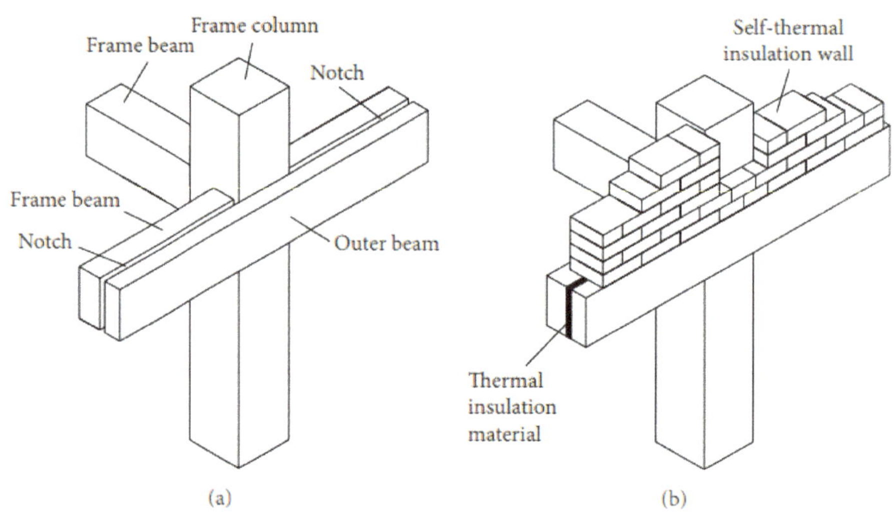

Figure 1: Sketch of energy conservation slotted beam-wall system: (a) constructional details of the slotted beam and (b) the view after self-thermal insulation wall constructed.

In this paper, the feasibility of slotted RC beams as thermal insulation structural components was investigated. Three reference solid beams and six slotted beams were fabricated and tested under four-point bending test. The flexural behavior of the slotted beams was investigated in detail, including failure mode, load-deflection response, and moment curvature analysis. The effects of reinforcement ratio and width ratio of the inner beam to the outer beam on the flexural behavior of slotted beams were investigated, and then a design method for slotted beam was proposed. The thermal properties of the slotted and solid RC beams were analysed and compared. In addition, the thermal performance of the slotted beam-wall system was investigated through finite element method (FEM) analysis.

EXPERIMENTAL

Fabrication of Beams

In order to study the flexural behavior of the slotted beams, a total of 9 beams were designed and constructed, including 6 slotted beams and 3 reference solid beams. As displayed in Figure 2(a), the width of the slot is 30 mm and the distance from the slot to the left/right end is 150 mm. The total width of the outer and inner beam of a slotted beam is 250 mm, which is equal to the width of the solid beam. The depth and the length of the slotted beams are identical to those of the reference solid beams, with a depth of 250 mm and a length of 3000 mm. The inner and the outer beam were designated as "beam A" and "beam B", respectively (Figure 2(a)).

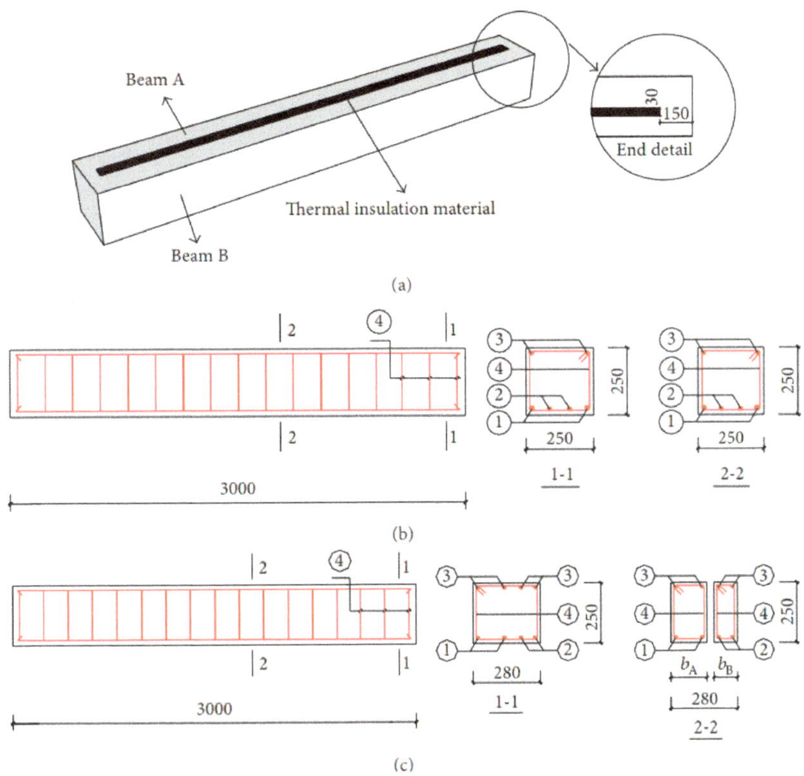

Figure 2: Specimen dimensions and reinforcement: (a) sketch of a slotted beam; (b) solid RC beam; and (c) slotted beam (units: mm).

The reinforcement details of the tested beams are also shown in Figure 2. Detailed descriptions of the specimens are given in Table 1. Constructional reinforcement was used for the compressive region of the beam. The following aspects were considered in the tests: the reinforcement ratio of beam A and beam B and the width ratio (b_A/b_B). The amount of the tensile reinforcement, the compressive reinforcement, and the shear reinforcement of the slotted beam were designed to be close to those of the corresponding reference solid beam, in order to facilitate comparison of the results. In Table 1, SB denotes the slotted beams and RB denotes the reference solid beams. The number behind SB denotes the section type, reflecting the width ratio ((b_A/b_B)). The diameter of the longitudinal reinforcement rebar in beam B is given at the last part of the specimen denomination. For example, SB1-14 represents a slotted beam with section type 1, and the diameter of the rebar in beam B is 14 mm. The reference solid beam for SB1-14 and SB2-14 is RB-14.

Table 1: Detailed descriptions of the specimens.

Specimen identifier	f_c (MPa)	Section $b \times h$ (mm × mm)	Reinforcement details (unit: mm) ①	②	③	④	Tensile reinforcement ratio ρ (%)		Shear reinforcement ratio ρ_v (%)		
RB-10	31.0	250 × 250	2B16	2B10	2B8	A8@100	1.0		0.4		
RB-12	28.8	250 × 250	2B16	2B12	2B8	A8@100	1.1		0.4		
RB-14	29.3	250 × 250	2B16	2B14	2B8	A8@100	1.3		0.4		
		$(b_A + b_B) \times 250$					Beam A ρ_A	Beam B ρ_B	Beam A $\rho_{v,A}$	Beam B $\rho_{v,B}$	Width ratio b_A/b_B [(1)]
SB1-14	29.0	(170 + 80) × 250	2B16	2B14	2B6	A6@100	1.1	1.8	0.3	0.7	2.1
SB2-10	29.7	(150 + 100) × 250	2B16	2B10	2B6	A6@100	1.2	0.7	0.4	0.6	1.5
SB3-12	31.3	(130 + 120) × 250	2B16	2B12	2B6	A6@100	1.4	0.9	0.4	0.5	1.1
SB1-12	29.7	(170 + 80) × 250	2B16	2B12	2B6	A6@100	1.1	1.3	0.3	0.7	2.1
SB2-14	29.0	(150 + 100) × 250	2B16	2B14	2B6	A6@100	1.2	1.4	0.4	0.6	1.5
SB3-10	28.9	(130 + 120) × 250	2B16	2B10	2B6	A6@100	1.4	0.6	0.4	0.5	1.1

Note: [(1)] the width ratio for slotted beam is the ratio of the width of beam A to that of beam B.

The mix proportion of concrete is listed in Table 2. The concrete compressive strength of each specimen was determined from the compression test on 150 × 150 × 300 mm standard concrete prism. Deformed steel was used for longitudinal reinforcement and plain steel was used for hoop reinforcement. Mechanical properties of the rebars are given in Table 3.

Table 2: Concrete mix proportion.

Component	Water	Ordinary Portland cement	Sand	Coarse aggregate
Quantity (kg/m^3)	173	392	487	1249

Table 3: Mechanical properties of the reinforcement rebars.

Material type	Yield strength (MPa)	Ultimate strength (MPa)
B10	382	536
B12	376	556
B14	379	543
B16	385	549
A6	346	464
A8	352	478

Experimental Program

Figure 3(a) shows a slotted beam under four-point bending test. The locations of the two concentrated loads and the supports are given in the schematic view in Figure 3(b). Dial indicators with an accuracy of 0.01 mm were fixed at the bottom of the beams to record the deflection responses, as shown in Figure 3(a). For slotted beams, two dial indicators were used for each position to record the deflection of beam A and beam B, respectively. Dial indicators were also fixed at the left and right ends of each beam to record the displacement of the beam ends. Crack width observation instrument with an accuracy of 0.02 mm was carefully used to measure the maximum crack width of the beams. The spacing between the cracks and the crack number were also measured and marked at load intervals. Strain gauges on the tensile rebars and concrete surface at the middle span of the beams were recorded. During the test, the loads corresponding to the appearance of the first tensile crack and tensile rebar yielding were also recorded. The four-point bending test was conducted according to the standard for test method of concrete structures in China [9].

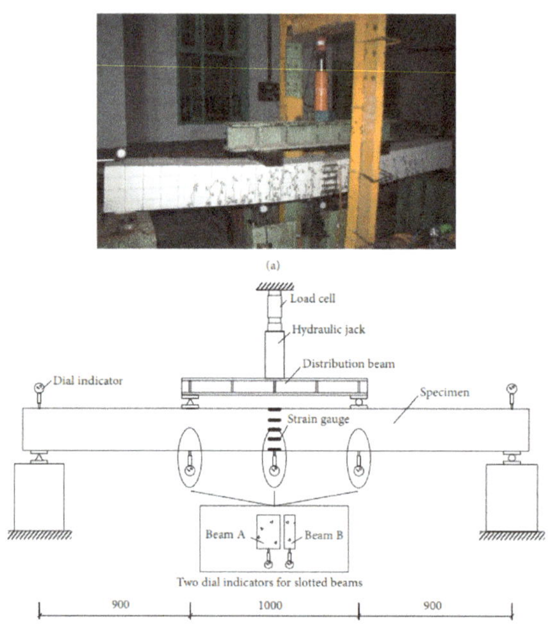

Figure 3: Four-point bending test system of the RC beams: (a) photo of the test devices; (b) sketch of the test devices (units: mm).

FOUR-POINT BENDING TEST RESULTS AND ANALYSIS

Crack and Damage Patterns of the Solid and Slotted Beams

The damage patterns of the solid beams and slotted beams are shown in Figure 4. For the solid beams, the first crack was tensile crack and appeared in the flexural zone near the middle span of the beam. With an increase of the load, new tensile cracks appeared and the former tensile cracks developed into the upper part of the beam. Shear cracks were observed near the ends of the beams at the late loading stage, because the shear force and moment were both significant in this region. Concrete crush was observed on the compressive region of the middle span of the beam at the final loading stage. Therefore, it can be concluded that the final failure mode of the solid beams was the flexural failure.

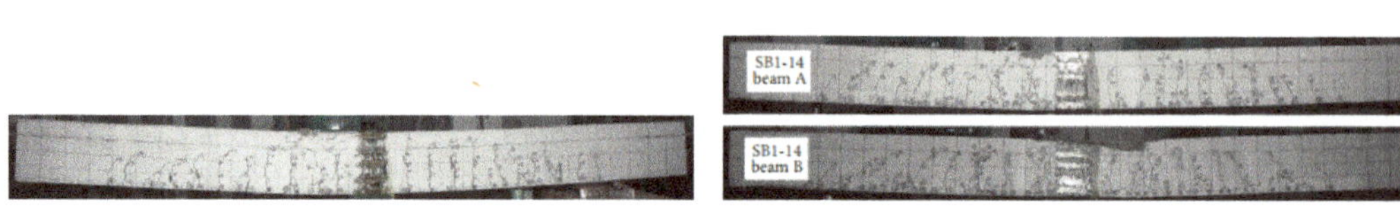

Figure 4: Damage patterns of the beams: (a) RB-14 and (b) SB1-14.

For the slotted beams, the first tensile crack appeared on the surface of the middle span of beam A and beam B, respectively. Then, the cracks in both beam A and beam B developed on the surfaces of the beams with the increase of the load. At the final loading stage, concrete crushing began to appear in the compression region at the middle span of beam A and beam B. The mechanical behavior of beam A and mechanical behavior of beam B of a slotted beam were different from each other, due to the difference in the section dimension and reinforcement ratio. Since the deflections of beam A and beam B at the middle span were almost similar to each other caused by the loading scheme used in the test, the concrete crushing at the middle span of beam A and beam B did not appear simultaneously, resulting in the different ductility behavior of beam A and beam B. The failure process of the two parts of the slotted beams observed during the final loading stage demonstrated this inference. More in-depth analysis concerning the ductility behavior of the two parts of the slotted beams will be given in Section 3.3. Though the mechanical behavior of beam A and beam B was different, severe concrete crushing was observed at the middle span of both beam A and beam B at the failure stage, indicating that the final failure model of the slotted beams was flexural failure. Therefore, it can be concluded that the damage process of the slotted beams exhibited a certain difference from that of the solid beams.

Load-Deflection Curve Analysis

The flexural load-deflection curves of the solid RC beams are given in Figure 5. These curves can be divided into three stages: precrack stage, preyield stage, and postyield stage. The initial linear portion of the curves shows that the stiffness of the beams remained constant prior to the cracking load. The initial cracking point of the three solid beams were almost identical, because the section dimension and the concrete strength of the three beams were almost the same. After the cracking load, the stiffness of the solid beams decreased gradually, as displayed in Figure 5. The yield strength and load capacity increased with the increase in the tensile reinforcement ratio. Because all the three beams were under-reinforced beams, with tensile reinforcement ratio of 1.0%, 1.1%, and 1.3%, respectively, obvious ductility behavior can be observed from the load-deflection curves. The mechanical properties of the solid beam are given in Table 4, where the yielding load was determined when the strain gauges pasted on tensile steel reached yielding strain; the ultimate deflection corresponded to the displacement when vertical load decreased to 85% of the peak load capacity. In Table 4, it can be found that the ductility ratio of the solid beams decreased with the increase in the tensile reinforcement ratio.

Table 4: Test results of the specimens.

Specimen identifier	Yielding deflection Δ_y (mm)	Yielding load (kN)	Peak load capacity (kN)	Ultimate deflection Δ_u (mm)	Ductility ratio μ_Δ (1)
RB-10	11.0	98.6	113.0	69.6	6.3
RB-12	11.9	108.0	120.6	67.1	5.6
RB-14	12.3	128.6	141.0	60.7	5.0
SB1-14	12.3	129.5	132.9	51.2	4.2
SB2-10	14.9	106.7	109.9	56.1	3.8
SB3-12	12.2	108.6	120.0	52.6	4.3
SB1-12	12.6	113.8	118.9	64.4	5.1
SB2-14	12.4	129.5	135.0	61.2	4.9
SB3-10	14.8	96.6	103.8	46.8	3.2

Note: (1) the ductility of the beam was calculated by Δ_u/Δ_y.

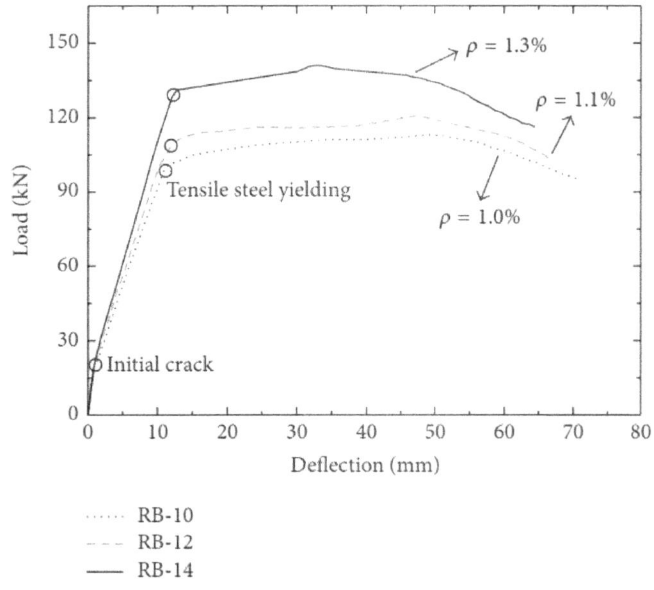

Figure 5: Deflection curves of the solid beams.

It was difficult to fix load cells at the two fulcrums of the steel distribution beam during the four-point bending test, so the concentrated force applied on the two parts of the slotted beams was not measured during the tests. Because both beam A and beam B entered into nonlinear stage after cracking point, it was improper to calculate the actual load applied on each part of the two beams based on the initial elastic cross-section stiffness of each beam. For a slotted beam, the dial indicators showed that the deflections of beam A and beam B were almost the same. Therefore, the total load applied on the top of the distribution beam and the average readings of the dial indicators were used to depict the load-deflection curves of the slotted beams.

Figure 6 shows the deflection curves of the slotted beams and the corresponding solid beams. Before the deflection curves entered into the descending branch, the load-deflection curves of the slotted RC beams were close to that of the corresponding solid RC beam, indicating that the cracking load and yielding load of the slotted beams were almost similar to those of the corresponding solid beam. For SB2-10, the ductility of the load-deflection curves of the slotted beams was lower than that of the corresponding solid RC beam RB-10, and the ductility of SB3-10 became obviously lower than that of SB2-10. Obvious lower ductility was also observed for the slotted beams SB3-12 and SB1-14, comparing with their corresponding solid RC beam. However, the load-deflection curves of SB1-12 and SB2-14 were close to that of the corresponding solid beam and it can be found that the difference between the reinforcement ratio of beam A and beam B of SB1-12/SB2-14 was obviously smaller than that of the other slotted beams. This experimental phenomena indicated that the reinforcement ratio of beam A and beam B had an obvious effect on the flexural behavior of the slotted beams. Detailed analysis for this phenomena will be given in Section 3.3.

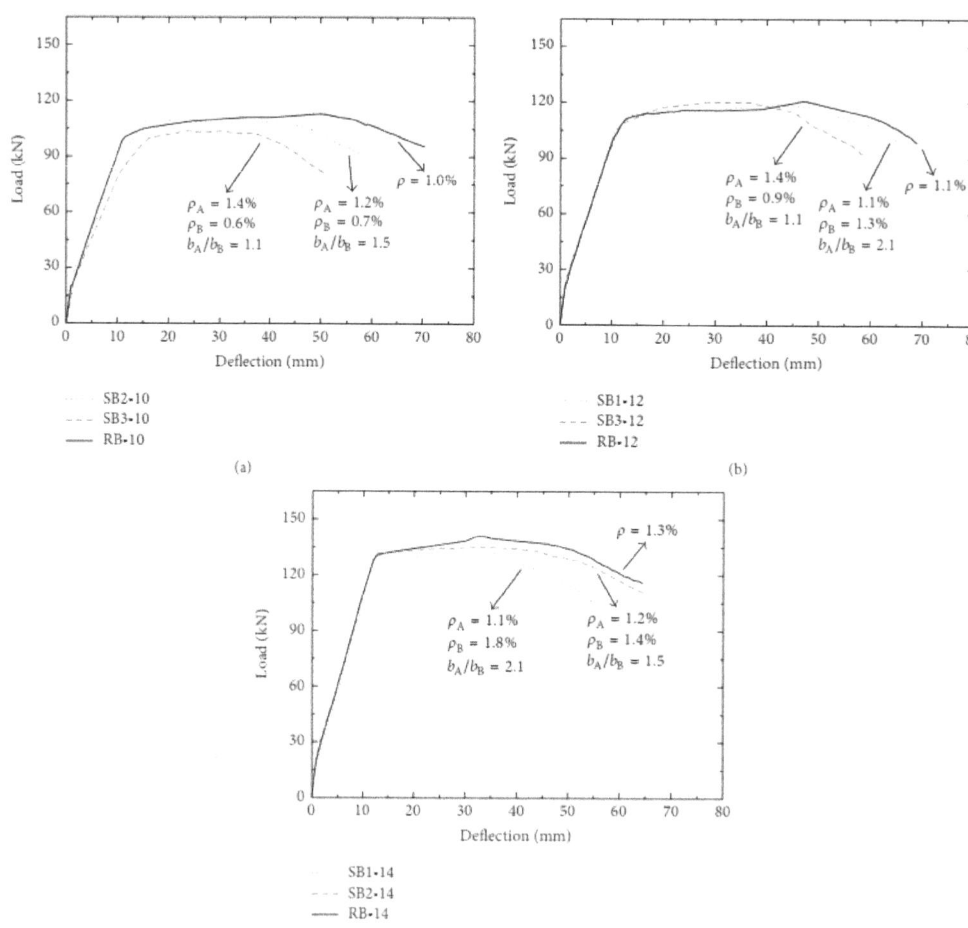

Figure 6: Load-deflection curves of the slotted beams: (a) SB2-10 and SB3-10; (b) SB1-12 and SB3-12; and (c) SB1-14 and SB2-14.

Moment Curvature Analysis

The moment curvature relationship of a beam section can be calculated by using the information of the beam section size, the reinforcement, and the stress-strain curves of the material. In the calculation, the stress-strain curves of the steel rebars obtained from the axial tensile tests were used, and the stress-strain curves of the concrete followed the modified Kent-Park model [10, 11]. The calculation process of moment curvature relationship of a beam section was described in detail in relevant reference [12] and hence is omitted here. The moment curvature curve of all the solid beams is shown in Figure 7, and the values corresponding to the yield point, peak point, and failure point are given in Table 5. The results show that the moment capacity of the beam section increased with the increase in the reinforcement ratio; however, the curvature ductility exhibited a decrease trend, because increased tensile reinforcement ratio can lead to an increase of the height of the compressive region (resulting in decreased curvature).

Table 5: Characteristic points on the calculated moment curvature curves of the specimens.

Specimen identifier	Yielding curvature ϕ_y (rad)	Yielding moment (kN·m)	Peak moment capacity (kN·m)	Ultimate curvature ϕ_u (rad)	Ductility ratio μ_ϕ (1)
RB-10	0.012	41.6	48.1	0.158	12.7
RB-12	0.013	45.8	52.6	0.139	11.0
RB-14	0.014	52.0	58.0	0.122	8.9
SB1-14	0.014	51.5	57.3	0.112	8.0
SB2-10	0.013	41.1	47.5	0.136	10.6
SB3-12	0.014	46.3	52.0	0.122	8.8
SB1-12	0.013	46.0	52.9	0.141	11.2
SB2-14	0.014	52.1	58.1	0.123	8.9
SB3-10	0.013	41.3	46.7	0.120	9.0

Note: (1) the curvature ductility of the beam section was calculated by ϕ_u/ϕ_y.

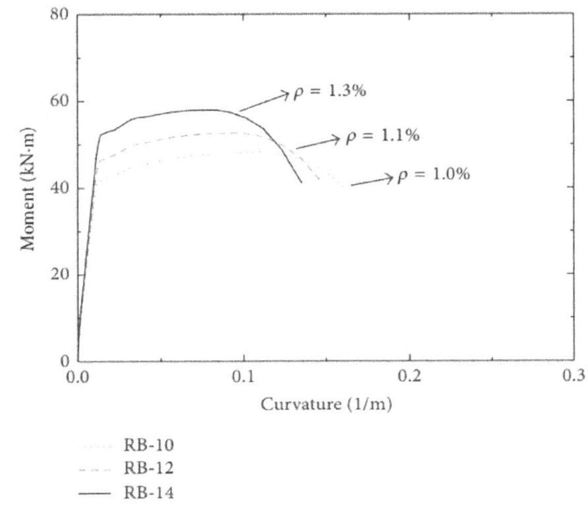

Figure 7: Moment curvature curves of the solid beams.

The deflection curves of beam A and beam B for a slotted beam during the tests were almost the same, based on the readings of the dial indicators fixed on the bottom surfaces of beam A and beam B. Therefore, it can be inferred that the curvatures of the section at the middle span of the two beams are almost identical during the tests, based on the deflection theory of beam. Hence, the moment curvature curve of the entire slotted beam can be derived through superimposing the moment of beam A and beam B at a certain curvature. The moment curvature curves of the two parts of the slotted beams are shown in Figure 8, where the moment curvature curves of the corresponding entire slotted beam are also given for the purpose of comparison.

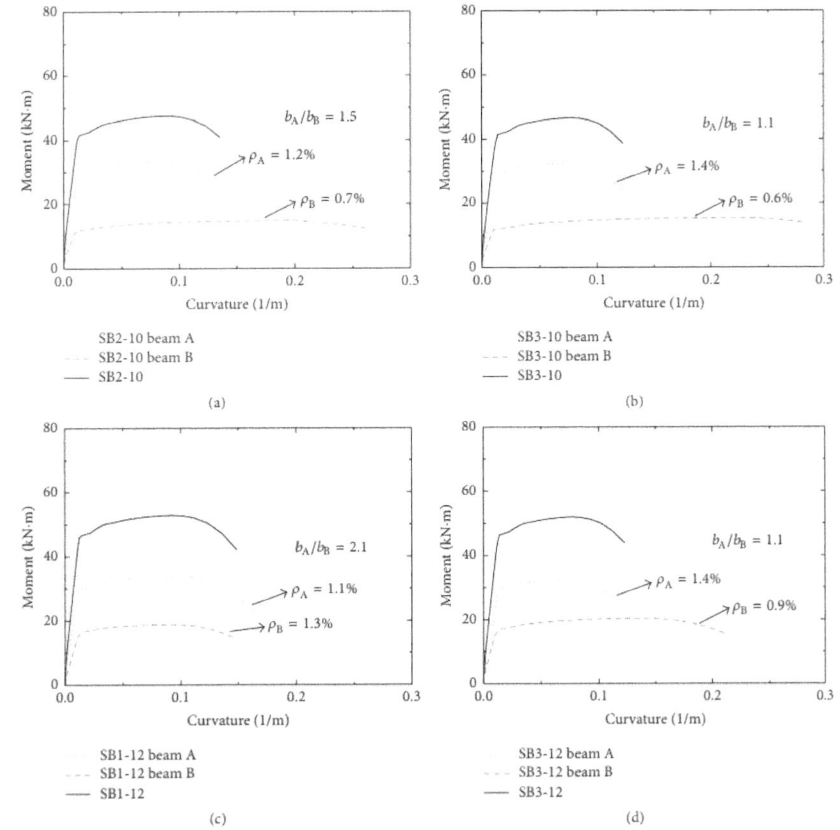

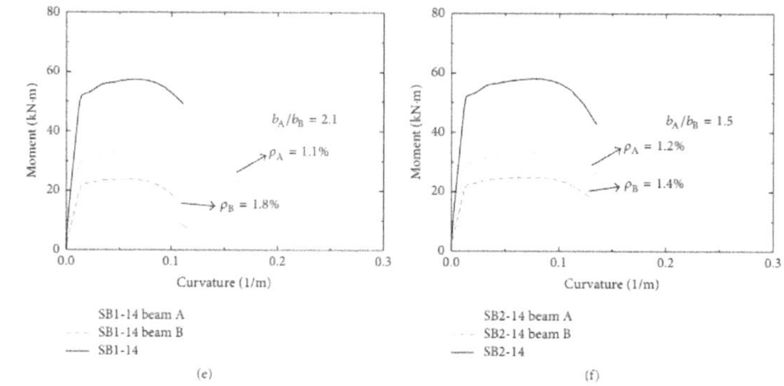

Figure 8: Moment curvature curves of the slotted beams: (a) SB2-10; (b) SB3-10; (c) SB1-12; (d) SB3-12; (e) SB1-14; and (f) SB2-14.

For SB1-12, the ductility of the moment curvature curves of beam A and beam B was almost the same, and the tensile reinforcement ratio of beam A and beam B of SB2-14 was 1.1% and 1.3%, respectively, similar to each other. Similar phenomenon was also found for SB2-14. The similar reinforcement ratio resulted in the fact that the heights of the compression region of beam A and beam B at the concrete crush point were almost equivalent to each other [12]. Hence, the ultimate curvature of beam A and beam B of SB1-12/SB2-14 was nearly the same, because the height of the section was identical for all beams.

For SB2-10, SB3-10, and SB3-12, the ductility of the moment curvature curves of beam B was obviously greater than that of the corresponding beam A, due to the fact that the reinforcement ratio of beam B was much less than that of the corresponding beam A. For SB1-14, the ductility of beam B for SB1-14 was smaller than that of beam A, because the reinforcement ratio of beam B was larger than that of beam A. In Figures 8(a), 8(b), 8(d), and 8(e), it can be found that the ductility of the entire slotted beam was determined by the beam, of which the ductility was worse than the other part. Therefore, the ductility of both beam A and beam B should be considered and designed carefully in the application of slotted beams in practical engineering.

It should be noted that the width ratio b_A/b_B of SB1-12 was 2.1, which was the largest width ratio in the tests. However, the ductility of SB1-12 and RB-12 was nearly the same. For SB3-12, the width ratio of beam A and beam B was 1.1, which was the smallest width ratio in the tests. However, the ductility of SB3-12 and RB-12 was obviously different. Therefore, it can be concluded that the reinforcement ratio rather than the width ratio of beam A to beam B is the key factor affecting the ductility behavior of the slotted beams.

The moment curvature curves of the slotted beams and corresponding solid beams are shown in Figure 9. The moment curvature curves of SB1-12 and RB-12 were almost coincident, because the reinforcement ratio of RB-12 and the two parted beams of SB1-12 were almost the same. Similar phenomenon was also found for SB2-14 and RB-14. In Figure 9(a), it can be clearly found that the larger the difference between ρ_A and ρ_B, the worse the ductility of the slotted beam. Therefore, the two parted beams of the slotted beam should be designed to pose similar tensile reinforcement ratio, in order to achieve a similar ductility capacity of the corresponding solid beam.

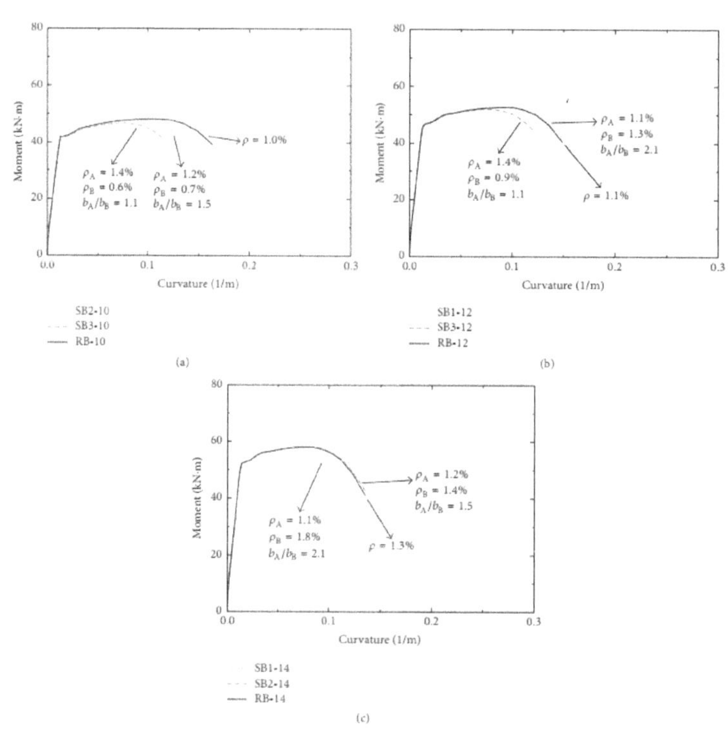

Figure 9: Comparison of the moment curvature curves of the slotted beams and corresponding solid beams: (a) SB2-10, SB3-10, and RB-10; (b) SB1-12, SB3-12, and RB-12; and (c) SB1-14, SB2-14, and RB-14.

Load Capacity Analysis

The flexural capacities of RC beams can be determined when the geometry, the reinforcement, and the strength of the material are known. The flexural capacities of the solid and slotted beams, calculated according to the concrete code of China [13], are given in Table 6, where the experimental results are also given. In Table 6, M_{cap}^{cal} is yield capacity of the beams obtained from experiments and M_{cap}^{cal} is theoretically calculated flexural capacity of the beams. Because the yield strength of the steel rebar is used to calculate the flexural capacity of the beam in the concrete code, the yield capacities of the beams obtained from the tests were used for the comparison with the theoretically calculated values.

Table 6: Yield capacities of the specimens.

Specimen identifier	M_{cap}^{exp} (kN·m)		M_{cap}^{cal} (kN)		$M_{cap}^{exp}/M_{cap}^{cal}$
RB-10	44.4		40.8		1.09
RB-12	48.6		45.6		1.07
RB-14	57.9		51.6		1.12
		Beam A	Beam B	Beam A + beam B	
SE1-14	58.3	29.4	22.2	51.6	1.13
SE2-10	49.5	29.4	11.4	40.8	1.21
SE3-12	48.5	29.4	16.2	45.6	1.06
SE1-12	51.2	29.4	16.2	45.6	1.12
SE2-14	55.9	29.4	22.2	51.6	1.08
SE3-10	44.6	29.4	11.4	40.8	1.09

Table 6 shows that the calculated flexural capacity M_{cap}^{cal} was close to the value obtained from the tests for both the solid and slotted beams, indicating that the method used in the code was effective in predicting the flexural capacity of these two types of beams. Meanwhile, the calculated flexural capacities were conservative compared with the test results. It was found that the sum of the flexural capacities of beam A and beam B of the slotted beams was equal to that of the corresponding solid beam. This observation was coherent with the phenomena shown in Figure 6, where it was found that the load-deflection curves of the solid beam and slotted beam were close to each other before the descending branch. However, it should be noted that the slotted beam exhibited decreased ductility capacity compared with the corresponding solid beam, as discussed in Section 3.2.

STRUCTURAL DESIGN ADVICES FOR THE SLOTTED RC BEAMS

Based on the above experimental results and analysis, several design advices can be made for the slotted RC beams, in order to obtain favourable mechanical behavior:

(1) The target failure mode of the slotted RC beam should be typical flexural failure mode, in order to ensure sufficient ductility. Therefore, sufficient tensile reinforcement is needed at the bottom of the beam. Meanwhile, sufficient stirrup reinforcement for shear is also needed in order to prevent shear failure.

(2) The ductility of beam A and ductility of beam B of a slotted beam should be close to each other in order to insure that these two beams underwent similar damage process under flexural load and reach the final failure nearly at the same time. In order to achieve the above flexural behavior of a slotted beam, the reinforcement ratios of beam A and beam B should be close to each other.

(3) In addition, the concrete strength, the span limitation, and the ratio of beam width to beam height should meet the current specification requirement for reinforced beam in the code.

THERMAL INSULATION BEHAVIOR OF THE SLOTTED BEAMS

Thermal performance is an important parameter to evaluate the feasibility of the slotted beam as a thermal insulation structural component. The heat transfer coefficient is a frequently used index to evaluate the thermal properties of materials. This index is influenced by the thermal conductivity and thickness of the materials through which heat is transferred. The heat transfer coefficient K of the slotted beam can be calculated using the following equations based on the thermal deign code for civil building in China [14]:

$$R = \frac{\delta_A}{\lambda_c} + \frac{\delta_{insul}}{\lambda_{insul}} + \frac{\delta_B}{\lambda_c},$$

$$R_0 = R_{in} + R + R_{out},$$

$$K = \frac{1}{R_0}, \tag{1}$$

where R is sum of thermal resistance values of all the material layers for the slotted beam; δ_A, δ_{insul}, and δ_B are thickness of beam A, the thermal insulation layer, and beam B, respectively; λ_c and λ_{insul} is thermal conductivity of the concrete and the thermal insulation layer, respectively; R_0 is thermal resistance value of the slotted beam; R_{in} and R_{out} is thermal resistance of the inner and outer surface against heat convection, respectively, and are set to be 0.11m^2. K/W and 0.04 m^2. K/W , respectively, based on the thermal design code for civil building in China [14]; and K is heat transfer coefficient of the beam. The larger the coefficient is, the easier the heat is transferred through the beam. For the solid beam, (1) is changed to $R = \delta_c/\lambda_c$ and the calculation process becomes simpler.

Thermal inertia index, as a dimensionless quantity, represents the resistance of the material to variations in temperature. The thermal inertia index D of the slotted beam can be computed using the following equation [14]:

$$D = \frac{\delta_A}{\lambda_c} S_c + \frac{\delta_{insul}}{\lambda_{insul}} S_{insul} + \frac{\delta_B}{\lambda_c} S_c, \tag{2}$$

where S_c and S_{insul} are thermal storage coefficient of the concrete and the thermal insulation layer. Materials with high thermal inertia show small changes in temperature under the effect of outer environmental temperature, showing good heat insulation effect. From (1), it is easy to find that, for the slotted beam with a certain sum width of beam A and beam B, the width ratio of beam A to beam B has no effect on the thermal resistance value R and only δ_{insul} and λ_{insul} of the thermal insulation layer have effect on R. Similarly, only δ_{insul}, λ_{insul}, and S_{insul} have effect on the thermal inertia index of the slotted beam when the sum width of beam A and beam B is determined.

Rigid extruded polystyrene foam board, also called XPS, can provide excellent insulation against both cold and heat and has the following advantages: high corrosion resistance, moisture proof, light weight, and antiaging capacity. Hence, XPS has been widely used as thermal insulation material in practical engineering. In this paper, XPS was used to fill the gap in the slotted beams. In addition to XPS, it should be noted that other insulation materials can also be used as the thermal insulation layer in the slotted beam. The values of λ and S for reinforced concrete and XPS recommended in the Chinese code [14, 15] are given in Table 7. The thermal calculation result of the beams is given in Table 8. It can be seen that the heat transfer coefficient of the slotted beam decreases by 77.4% compared with the solid RC beam. The thermal inertia factor of the slotted beam is 2.83, higher than that of the solid beam, 2.47. These data indicate that the thermal bridge of the slotted beam is effectively mitigated and the thermal performance of the slotted beam becomes much better than that of the conventional solid RC beam.

Table 7: Thermophysical parameters of the materials.

Material type	Thermal conductivity λ (W/(m·K))	Thermal storage coefficient S (W/(m·K))
Reinforced concrete	1.74	17.2
XPS	0.03	0.36

Table 8: Thermal calculation results of the beams.

Beam type	Thermal resistance R (m²·K/W)	Heat transfer coefficient K (W/(m²·K))	Thermal inertia factor D
Solid beam	0.29	3.45	2.47
Slotted beam	1.29	0.78	2.83

THERMAL INSULATION BEHAVIOR OF THE SLOTTED BEAM-WALL SYSTEM

FEM Model Building

A FEM software, COMSOL Multiphysics [16], which has passed the model tests required by the European standard EN 10211: 2007 for thermal bridges in building constructions [17] and was found to be suitable for simulating the heat transport through buildings [18], was used to simulate the thermal insulation behavior of the frame structure using the slotted beam-wall system. The thermal insulation behavior of this new type of structure in winter was considered in the FEM model and a normal frame structure FEM model was also built for comparison purpose. The temperatures inside and outside the room are set to 18°C and 0°C, respectively, according to the design standard for energy efficiency of residential buildings in hot summer and cold winter zone in China [19]. These temperatures are typical for the buildings using heating in the south area of China in winter. The plan and elevation layout are always regular for residential frame buildings; therefore only one unit of the frame was modeled in COMSOL due to the symmetrical structure of the frame structure, for the purpose of conciseness. The dimension information of the frame FEM model is shown in Figure 10, where the beam span, the floor height, and the size of columns and beams are typical for 10-story height residential frame buildings in China. The exterior vacuum insulation windows, beam perpendicular to the exterior wall, and the floor slabs were not modeled, in order to make the comparison of the thermal properties between the normal frame and frame using the slotted beam-wall system more specific. The boundary condition of the four sides of the model was thermal insulation, due to the symmetry of the model. 3D steady state thermal bridge was considered in COMSOL analysis, meaning the field variables are independent of time. More detailed modeling and analysis process can be referred to in COMSOL user manual. Thermophysical parameters of the materials used in the FEM model are given in Table 9 and it is assumed that these thermal properties are independent of the temperature during the FEM analysis.

Table 9: Thermophysical parameters of the materials.

Material type	Density (kg/m³)	Thermal conductivity (W/(m·K))	Thermal storage coefficient (W/(m²·K))	Heat capacity (J/(kg·K))
Reinforced concrete	2500	1.74	17.20	920
XPS	30	0.03	0.36	1380
Aerated concrete wall	500	0.19	2.81	1050

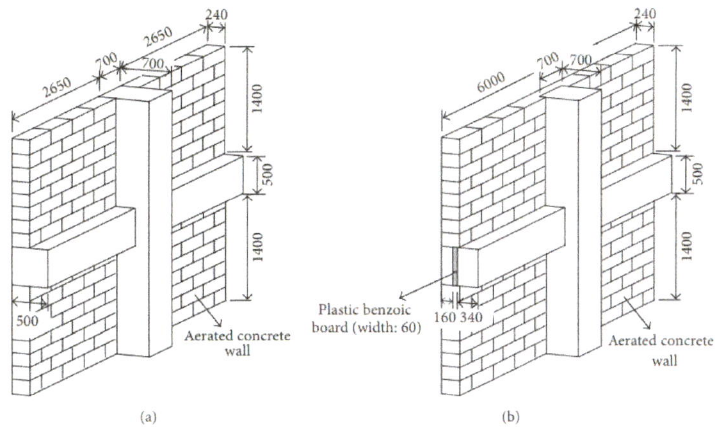

Figure 10: Geometry of the unit frame built in COMSOL Multiphysics: (a) normal frame and (b) frame using the slotted beam-wall system (units: mm).

Simulation Result and Discussion

The temperature distribution on the surfaces of the FEM models is shown in Figure 11. It is obvious that the beams and columns constitute the thermal bridges for the normal frame in Figure 11(a). However, the thermal bridges have been controlled after the slotted beam-wall system have been utilized, as shown in Figure 11(b), where the location of beam-column joint becomes the thermal bridge. In fact, this thermal bridge can be reduced in practical engineering because it is easy to carry out heat insulation treatment on this small area.

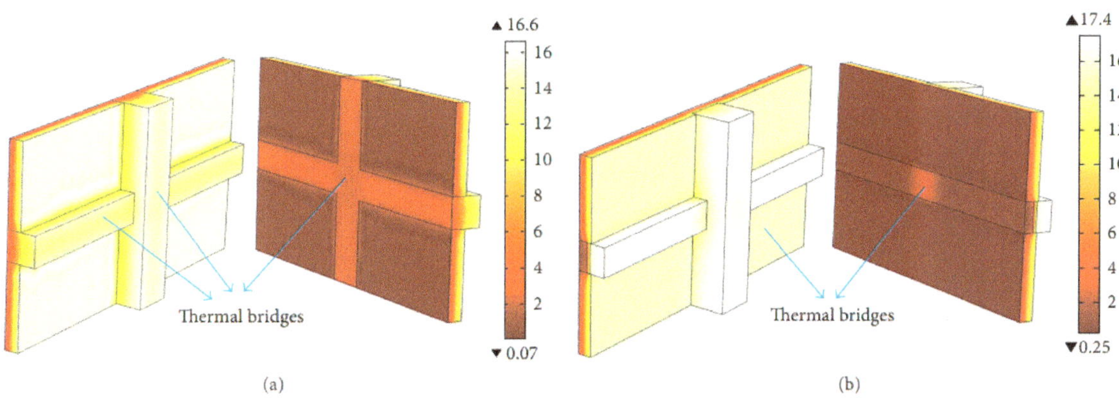

Figure 11: Surface temperature distribution of the FEM models (units: °C): (a) normal frame and (b) frame using the slotted beam-wall system.

The results of the calculated heat flux are given in Table 10. The heat flux in the normal frame and frame using the slotted beam-wall system was 460.3 W and 287.1 W, respectively. The heat flux through the beams and columns in the normal frame structure was 236.2 W (51.3% of the total heat flux of the normal structure), indicating that the thermal bridge effect was obvious in the normal frame structure. However, the heat flux through the beams and columns in the frame using the slotted beam-wall system was 85.5 W (29.8% of the total heat flux of this new type structure). The thermal bridge effect has been controlled in the new frame structure through using the slotted beam-wall system. In addition, though the area of the walls of this new type structure was larger than that of the normal frame, the heat flux on the walls of this new type structure was 22.4 W less than that of the normal frame.

Table 10: Comparison of the FEM simulation results.

Structure type	Heat transfer area (m^2)	Total heat flux (W)	Heat flux on walls (W)	Heat flux on columns (W)	Heat flux on beams (W)
Normal RC frame	19.8	460.3	224.1	102.2	134.0
RC frame using slotted beams	19.8	287.1	201.7	49.4	36.1

CONCLUSIONS

In this study, a novel slotted beam-wall system was proposed for thermal insulation of RC frame buildings. Four-point bending test was carried out to investigate the flexural behavior of the slotted beams. The flexural behavior of the slotted beams was investigated in detail, including failure mode, load-deflection curves, and moment curvature analysis. Thermal insulation behavior of the slotted beam was analyzed through theoretical calculation. Finally, the thermal behavior of the frame using the slotted beam-wall system was simulated through FEM method. The following conclusions can be drawn:

(1) Though the final failure mode of the slotted beams was flexural failure mode, the failure process of the slotted beams was different from that of the solid beams at the final loading stage.

(2) Before the final failure stage, the load-deflection curve of the slotted beam was close to that of its corresponding solid beam. The tensile reinforcement ratio rather than the width ratio of the two parted beams had an important effect on the flexural behavior of the slotted beams. The tensile reinforcement ratios of the two parted beams of a slotted beam should be close to each other, in order to obtain similar ductility behavior of their corresponding solid beam.

(3) The slotted beams exhibited attractive thermal insulation properties. The heat transfer coefficient of the slotted beam was greatly reduced and the thermal inertia factor increased, compared with the corresponding solid beams. Hence, the slotted beam has a great potential for the practical thermal insulation application of RC frame buildings.

(4) The FEM simulation results showed that the frame structure using the slotted beam-wall system exhibited obvious and attractive thermal insulation property, compared with the normal frame structure. It is believed that this new type of structure can become an effective measure to solve the thermal insulation problem.

Competing Interests

The authors declare that they have no competing interests.

ACKNOWLEDGMENTS

This research was sponsored by the Science and Technology Plan of Changsha, China (Grant no. CSCG-HNTF-GK20130576), the National Natural Science Foundation of China (Grant no. 51408211), the Natural Science Foundation of Hunan Province, China (Grant no. 2015JJ3032), the Fundamental Research Funds for the Central Universities through the project of Young Teacher Growth of Hunan University (Grant no. 531107040799), and the Open Fund of Hunan Province Engineering Laboratory of Bridge Structure (Changsha University of Science & Technology, Grant no. 14KD02).

REFERENCES

1. J. Iwaro and A. Mwasha, "A review of building energy regulation and policy for energy conservation in developing countries," Energy Policy, vol. 38, no. 12, pp. 7744–7755, 2010.
2. B. P. Jelle, "Traditional, state-of-the-art and future thermal building insulation materials and solutions—properties, requirements and possibilities," Energy and Buildings, vol. 43, no. 10, pp. 2549–2563, 2011.
3. T. G. Theodosiou and A. M. Papadopoulos, "The impact of thermal bridges on the energy demand of buildings with double brick wall constructions," Energy and Buildings, vol. 40, no. 11, pp. 2083–2089, 2008.
4. K. S. Al-Jabri, A. W. Hago, A. S. Al-Nuaimi, and A. H. Al-Saidy, "Concrete blocks for thermal insulation in hot climate," Cement and Concrete Research, vol. 35, no. 8, pp. 1472–1479, 2005.
5. S. Kumar, "Fly ash-lime-phosphogypsum hollow blocks for walls and partitions," Building and Environment, vol. 38, no. 2, pp. 291–295, 2003.
6. J. Wang, P. C. Eames, J. F. Zhao, T. Hyde, and Y. Fang, "Stresses in vacuum glazing fabricated at low temperature," Solar Energy Materials and Solar Cells, vol. 91, no. 4, pp. 290–303, 2007.
7. N. Ng, R. E. Collins, and L. So, "Thermal conductance measurement on vacuum glazing," International Journal of Heat and Mass Transfer, vol. 49, no. 25-26, pp. 4877–4885, 2006.
8. N. Ng, R. E. Collins, and L. So, "Characterization of the thermal insulating properties of vacuum glazing," Materials Science and Engineering B, vol. 138, no. 2, pp. 128–134, 2007.
9. Code of China, Standard for Test Method of Concrete Structures (GB/T 50152-2012), China Architecture & Building Press, Beijing, China, 2012 (Chinese).
10. D. C. Kent and R. Park, "Inelastic behaviour of reinforced concrete members with cyclic loading,"Bulletin of the New Zealand Society for Earthquake Engineering, vol. 4, no. 1, pp. 108–125, 1971.
11. B. D. Scott, R. Park, and M. J. N. Priestley, "STRESS-strain behavior of concrete confined by overlapping hoops at low and high strain rates," Journal of the American Concrete Institute, vol. 79, no. 1, pp. 13–27, 1982.
12. J. K. Wight and J. G. MacGregor, Reinforced Concrete: Mechanics and Design, Pearson Education, 6th edition, 2011.
13. Code of China, Code for Design of Concrete Structures (GB 50010-2010), China Architecture & Building Press, Beijing, China, 2010 (Chinese).
14. Code of China, Thermal Design Code for Civil Building (GB50176-93), China Planning Press, Beijing, China, 1993 (Chinese).
15. Code of China, "Rigid extruded polystyrene foam board for thermal insulation (XPS)," GB/T 10801.2-2002, Standards Press of China, Beijing, China, 2002 (Chinese).
16. COMSOL, Multiphysics User's Guide, Comsol AB, Stockholm, Sweden, 2010.
17. European Committee for Standardization. EN ISO 10211, "Thermal bridges in building construction—heat flows and surface temperatures—detailed calculations," ISO 10211:2007, 2007.
18. A. W. M. van Schijndel, "Multiphysics modeling of building physical constructions," Building Simulation, vol. 4, no. 1, pp. 49–60, 2011.
19. Code of China, "Design standard for energy efficiency of residential buildings in hot summer and cold winter zone," JGJ 134-2010, China Architecture & Building Press, Beijing, China, 2010 (Chinese).

CHAPTER 11 RESEARCH ON THE HEAT DISSIPATION CHARACTERISTICS OF LITHIUM BATTERY SPATIAL LAYOUT IN AN AUV

Zhaoyong Mao and Shaokun Yan

School of Marine Science and Technology, Northwestern Polytechnical University, Xi'an 710072, China

ABSTRACT

To meet the power demand requirements of autonomous underwater vehicles (AUVs), the power supply is generally composed of a large number of high-energy lithium battery groups. The lithium battery heat dissipation properties not only affect the underwater vehicle performance but also bring some security risks. Based on the widespread application of lithium batteries, lithium batteries in an AUV are taken as an example to investigate the heat dissipation characteristics of the lithium battery spatial layout in an AUV. With the aim of increasing the safety of lithium batteries, a model is developed for the heat transfer process based on the energy conservation equation, and the battery heat dissipation characteristics of the spatial layout are analyzed. The results indicate that the most suitable distance between the cells and the cross arrangement is better than the sequence arrangement in terms of cooling characteristics. The temperature gradient and the temperature change inside the cabin with time are primarily affected by the navigation speed, but they have little relationship with the environmental temperature.

INTRODUCTION

As autonomous underwater vehicles (AUVs) are developing toward the direction of long ranges and high speeds, increasingly more power is urgently needed to support navigation. Because electrochemical reactions occurring within lithium-ion batteries will generate heat, the battery compartment of autonomous underwater vehicles works for a long time on large-scale integrated lithium-ion battery packs in a confined space, and thus security and reliability problems will exist. In [1], the heat can be divided into two parts. On the one hand, heat accumulation occurs in the battery cabin because the heat from the battery pack cannot dissipate in a timely manner. On the other hand, the battery pack radiating heat unevenly will cause a local temperature difference, which results in uneven performance between the batteries and ultimately affects the overall performance of the batteries.

At present, domestic and foreign scholars have been focusing on the security problem of AUVs using a lithium battery group to conduct related studies. In [2–7], a study on the control strategy of the lithium battery thermal balance and

Citation: Zhaoyong Mao and Shaokun Yan, "Research on the Heat Dissipation Characteristics of Lithium Battery Spatial Layout in an AUV," Discrete Dynamics in Nature and Society, vol. 2016, Article ID 6269539, 15 pages, 2016. doi:10.1155/2016/6269539.

thermal control system that was designed for the inconsistent influence of the battery power service life was performed. In [8–10], the fluid-solid coupled heat transfer method was used to establish a mathematical physical model of the AUV battery cabin interior heat conduction process with respect to the AUV battery cooling problem. Moreover, the battery discharge current and battery frame thermal conductivity of navigation devices were also analyzed in [11], in which the finite element analysis software ANSYS was used to analyze the lithium battery group temperature field of AUVs and discuss the impacts of different discharge time and boundary conditions on the battery temperature field. In [12], AUV battery cabin steady-state thermal modeling analysis was performed according to the heat transfer process of the key point passive thermal control design of the structure. For the battery thermal aspects in research works, the main focus is in the field of electric power for new energy vehicles. In [13], a model of a large-scale battery pack was established for investigating battery pack heat dissipation; it primarily concentrated on the field of electric power for new energy vehicles. In [14, 15], a model for predicting the performance of lithium batteries was established for electric vehicles, and the effect of different groups on the performance of the battery was analyzed in the same cooling mode with 9 single-cell batteries as the battery pack. In addition, using forced air cooling and phase-change materials, the cooling performance of the power car battery pack was analyzed based on the computational fluid dynamics method in [16, 17]. An appropriate battery model is necessary for the proper design and operation of battery systems using BMS. Several modeling approaches are available: empirical models, statistical models, and electrical models [18, 19]. In [20], the local heat generation in a single-stack lithium-ion battery cell was investigated as a function of the C-rate and state of charge (SOC). In [21], a combined model was used to study heat generation and heat dissipation as well as their influences on the temperatures of the battery pack with and without a fan under constant current discharge and variable current discharge based on electric vehicle (EV) driving cycles.

The existing studies primarily focus on battery thermal balance control system design. Regarding investigations of the AUV battery cooling layout, analyses have been performed only for navigation within the battery compartment temperature field, but little connections exist with structural layout studies on the thermal performance of the battery pack. In addition, compared to electric power vehicles, the battery cabin of AUVs is a closed compact space, and the use of common cooling methods, such as cold wind and solvent cooling, is limited. The battery pack heat conduction can only be achieved through the battery shell body and seawater, and physical problems involved are how to realize battery cooling through air flow engendered by local temperature variations inside the battery cabin and the heat conduction structure.

The main contribution of this paper is twofold: (i) we analyze the heat exchange process of the vehicle battery pack and establish the natural convection and heat transfer model for the confined space of the battery compartment and (ii) we investigate the heat transfer characteristics of lithium batteries in different spatial distributions.

MODELING OF THE AUV LITHIUM BATTERY SPACE WITH EXTERNAL COOLING

According to the internal structure of the AUV battery cabin and theoretical knowledge of heat transfer, the heat transferred from the battery to the outside seawater can be summarized as the following three aspects of heat conduction. The first part of the heat conduction includes the heat generated by the battery pack and the heat exchange process between the battery cabin and the housing wall. The second part of the heat conduction process is between the cabin housing wall and the outer wall of the housing. Finally, the third part of the heat conduction is battery heat exchange between the outer wall of the cabin housing and seawater. The procedure is shown in Figure 1.

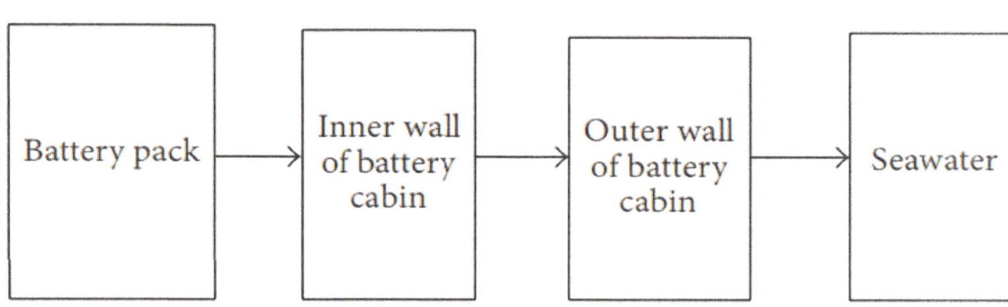

Figure 1: Schematic diagram of heat dissipation and conduction of battery cabin.

To facilitate analysis of the temperature distribution in the battery cabin under different working conditions, the heat transfer process in the battery cabin was hypothesized and simplified as follows:

(1) The ends of the battery cabin and inner battery pack are insulated.

(2) The temperature distributions inside the battery cabin and battery pack only change in the radial direction and remain essentially unchanged in the axial direction.

(3) When the battery cabin works, the thermal parameters do not change over time.

Based on the above analysis, the heat dissipation model of the confined space of the battery cabin is equivalent to the problems of constant properties, inner heat source, and three-dimensional unsteady heat transfer.

Battery inside the Module Heat Analysis

The internal heat transfer process in a lithium battery can be simplified into a regular physical, three-dimensional unsteady heat transfer process within the heat source. For this reason, the internal lithium/thionyl chloride battery energy equation can be expressed as

$$\mathrm{Bi} = \frac{r_b h_b}{\lambda_b},$$

$$\lambda_b \frac{\partial^2 t}{\partial r^2} + \frac{\lambda_b}{r}\frac{\partial t}{\partial r} + \dot{Q}_b = \rho_b c_{pb} \frac{\partial t}{\partial \tau}. \quad (1)$$

Boundary conditions are

$$\tau = 0,$$

$$t = T_w,$$

$$r = 0,$$

$$\frac{\partial t}{\partial r} = 0,$$

$$r = r_b,$$

$$-\lambda_b \frac{\partial t}{\partial r} = h_a (t - t_a), \quad (2)$$

where $\dot{Q}_b$ is the heat generation rate in the overall battery (W/m^3), λ_b is the thermal conductivity of the battery (W/(m·k)), ρ_b is density of the battery (kg/m^3), Bi is the Biot number, C_{pb} is the specific heat capacity of the battery (J/(kg·K)), and r_b is the diameter of the battery .

When the Biot number of the battery is less than 0.1 under the natural convection environment, the internal temperature of the battery can be considered to be approximately evenly distributed. According to the Bernardi hypothesis, the heat generation rate in a single battery is constant, which can be approximately expressed as follows:

$$Q = \frac{I_b}{V_b}\left(E_{oc} - E - t\frac{\partial t}{\partial r}\right) \approx \frac{I_b^2 R_b}{V_b}, \quad (3)$$

where V_b is the volume of the single battery (m^3), E_{oc} is the open circuit voltage of the battery (V), E is the discharge voltage of the battery, R_b is the internal resistance of the battery (Ω), r_b is the radius of the battery (m), and I_b is the current of the single battery.

The heat convection of single cells occurs primarily through air convection and radiative heat transfer according to the ideal gas equation:

$$Pv = \frac{mRt_a}{M_a}. \quad (4)$$

The air density is

$$\rho_a = \frac{M_a P}{Rt_a}. \quad (5)$$

As shown by the above equation, a change in temperature can cause a change in air density in the battery cabin, and natural convection forms under the effect of gravity. Ignoring the effects of volume force and viscous force, the momentum conservation equation of air in the battery cabin can be expressed as follows:

$$\frac{\partial(\rho_a u)}{\partial \tau} + \mathrm{div}(\rho_a u\vec{v}) = \mathrm{div}(\eta_a \,\mathrm{grad}\, u) - \frac{\partial p}{\partial x},$$

$$\frac{\partial(\rho_a v)}{\partial \tau} + \mathrm{div}(\rho_a v\vec{v}) = \mathrm{div}(\eta_a \,\mathrm{grad}\, v) - \frac{\partial p}{\partial y} - \rho_a g, \quad (6)$$

where η_a is the kinematic viscosity of air (Pa·s), g is gravitational acceleration (9.8 m/s^2), Ma is the molar mass of air, P is air pressure (Pa), and t_a is air temperature (K).

The integral equation of the conservation of energy in the form of the cabin battery operation equation can be expressed as follows:

$$Q = Q_{out} + Q_{uns},$$

$$Q = \int_{V_b} Q_b dV, \quad (7)$$

where Q is the total heat generated by the battery pack (J), Q_{out} is the heat dissipating from the battery cabin to the outside (J), and Q_{uns} is the heat absorbed by the battery cabin (J).

Heat transfers between the battery and vehicle wall primarily through natural air convection, and the heat generated by the battery part dissipates into the environment via the shell. The other part of the heat is absorbed by the shell of the vehicle, which causes the battery cabin temperature to increase. The purpose of this study is aimed at increasing the proportion of Q_{out} and lowering the proportion of Q_{uns}, thereby reducing the temperature of the battery module:

$$Q_{out} = \int_{A_c} h_a (t - t_{ci})\, dS + \int_{A_k} h_w (t - t_h)\, dS,$$

$$Q_{uns} = \frac{\partial}{\partial \tau}\int_{V_b} \rho_b c_{pb} t\, dV + \frac{\partial}{\partial \tau}\int_{V_a} \rho_a c_{pa} t\, dV. \quad (8)$$

Boundary conditions are

$$\tau = 0,$$

$$t = T_w,$$

$$r = r_c,$$

$$-\lambda_w \frac{\partial t}{\partial \tau} = h_a (t - t_{ci}), \quad (9)$$

where A_c is the heat transfer area of the inner wall of the vehicle (m^2), V_a is the volume of air inside the battery cabin (), ρ_a is the air density (kg/m^3), c_{pa} is the specific heat capacity of air (J/(kg·K)), λ_w is the heat conductivity of the vehicle housing (W/(m·K)), h_a is the convective heat transfer coefficient of air in the battery cabin (W/(m^2·k)), t_h is the seawater temperature (°C), and t_{ci} is the initial temperature in the battery compartment (°C).

Heat Exchange between the Inner and Outer Walls of the Battery Cabin

Heat transfer from the inner wall to the outer wall of the battery cabin can be considered as heat conduction of a cylindrical wall, which can be expressed as follows:

$$\phi = A_{ci}k_c\left(t_{ci} - t_{co}\right),$$

$$t_{ci} = t_{co} + \frac{\phi}{A_{ci}k_c}, \tag{10}$$

where ϕ is the total heat through the bulkhead and the water exchange (J), A_{ci} is the heat transfer area of the inner wall (m^2), k_c is the equivalent heat conductivity of the battery wall (W/(m·k)), and t_{ci} is the temperature of the battery inner wall (°C).

Heat Exchange between the Outer Wall of the Vehicle Housing and Seawater

During underwater movement, forced convection heat transfer occurs between the outer wall of the battery cabin and seawater, which can be expressed as follows:

$$\phi = A_{co}h_{cf}\left(t_{co} - T_w\right),$$

$$A_{co} = 2\pi r_{co}l_c,$$

$$t_{co} = \frac{\phi}{2\pi r_{co}l_c h_{cf}} + T_w, \tag{11}$$

where ϕ is the heat exchanged between the battery cabin and seawater (W), A_{co} is the heat transfer area of the inner wall (m^2), r_{co} is the outer diameter of the vehicle housing (m), t_{co} is the outer wall temperature (°C), T_w is the seawater temperature (°C), and h_{cf} is the forced convection heat transfer coefficient between the outer wall and seawater (W/(m^2·K)).

The forced convection heat transfer coefficient between the outer wall and seawater is related to the sailing speed, which can be determined by the Reynolds number and Nusselt number of convective heat transfer between the outer wall and seawater:

$$h_{cf} = \mathrm{Nu}_{cf}\frac{\lambda_f}{l_c},$$

$$\mathrm{Nu}_{cf} = \left(0.037\mathrm{Re}^{4/5} - 871\right)\mathrm{Pr}^{1/3},$$

$$\mathrm{Re} = \frac{u_{cf}l_c}{v_f}. \tag{12}$$

Without consideration of seawater flow,

$$u_{cf} = v, \tag{13}$$

where λ_f is the heat conductivity of seawater (W/(m·K)), Nu_{cf} is the Nusselt number of convective heat transfer between the vehicle housing and seawater, Re is the Reynolds number of convective heat transfer between the vehicle housing and seawater, Pr is the Prandtl number of seawater, v_f is the kinematic viscosity of seawater (m^2/s), u_{cf} is the relative velocity (m/s) between seawater and the vehicle, v is the barycentric velocity of the vehicle (m/s), and l_c is the characteristic length of the battery cabin (m).

ANALYSIS OF THE INFLUENCE OF SPACE AND ARRANGEMENT ON THE HEAT DISSIPATION CHARACTERISTICS OF THE BATTERY PACK

Based on the lithium-ion battery pack for underwater space external thermal model and because the AUV battery compartment is a closed and compact space, the distance between batteries and combining types for the distribution of the temperature gradient inside the battery compartment have a great impact. This paper selects a winding-type lithium/thionyl chloride battery named 18650 as an example to analyze cells with different spaces and different permutations. Numerical parameters related to a single 18650 battery are shown in Table 1.

Table 1: Parameters of 18650-type battery.

Parameter	Value
Diameter [m]	0.018
Length [m]	0.065
Weight [kg]	0.048
Internal resistance [Ω]	0.03–0.06
Density [$kg \cdot m^{-3}$]	2900
Specific heat capacity [$J \cdot kg^{-1} \cdot K^{-1}$]	1000
Equivalent heat conductivity [$W \cdot m^{-1} \cdot K^{-1}$]	3
Nominal voltage [V]	3.6
Rated capacity [Ah]	2.5

Influence of Battery Spacing on Temperature Field of Battery Pack

The domain is described using triangular elements, with a total number of approximately 20,000. Grids closest to the profiles of the batteries were refined with triangular boundary elements to describe the boundary flow with sufficient precision. The distance between two adjacent cells is the same, and the distance between the boundary and batteries remains constant. The distances between the batteries change continually under the premise of shape and constant number of batteries.

With 5 single 18650 batteries as the objects, the space between the batteries (L) is studied. The temperature of the battery compartment is determined when the values of L are $0.1d$, $0.2d$, $0.3d$, $0.4d$, $0.5d$, $0.6d$, $0.7d$, $0.8d$, $0.9d$, and $1.0d$ (where d is the diameter of the battery). The temperature distributions are shown in Figure 2.

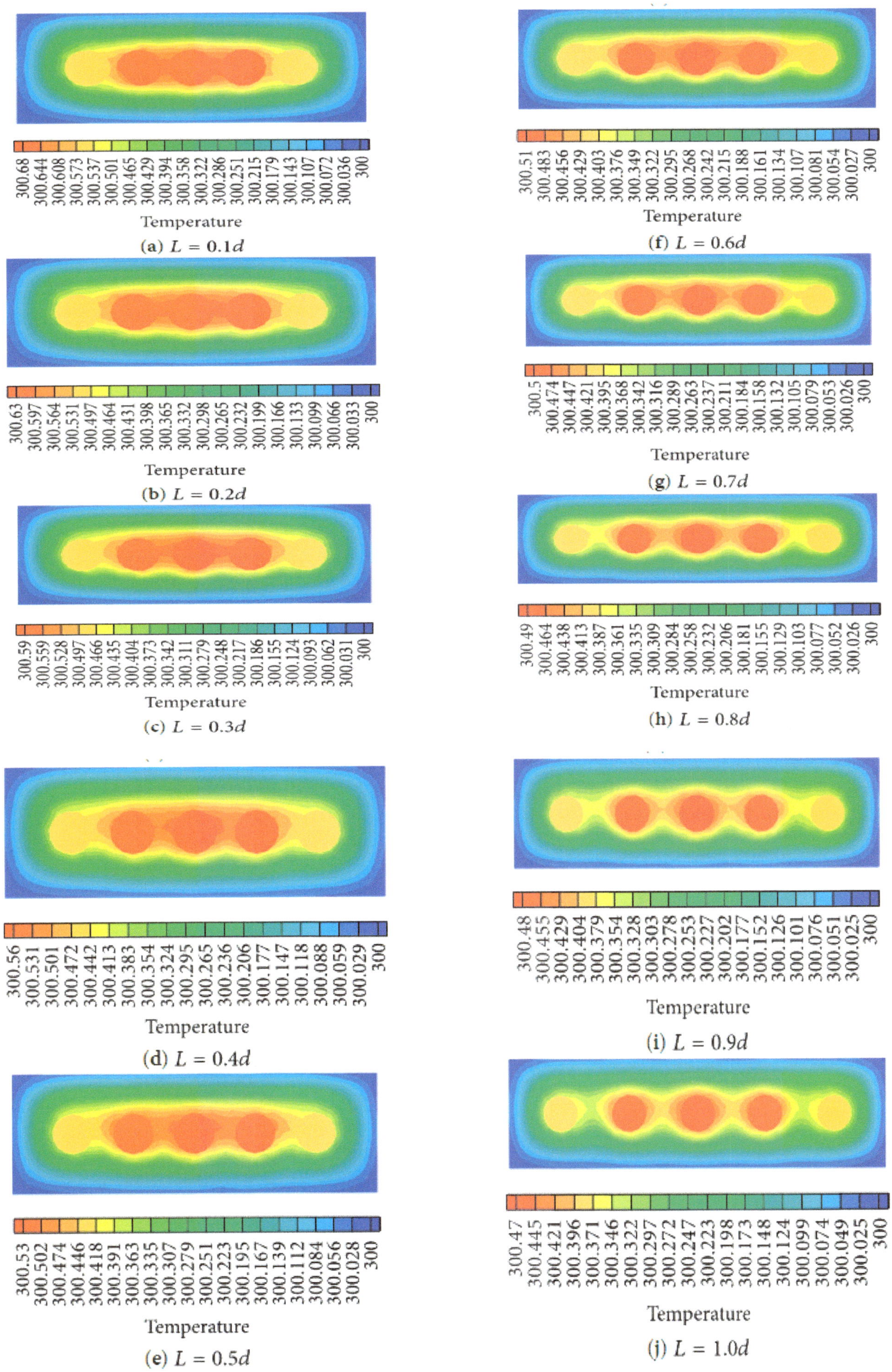

Figure 2: Temperature distribution inside the battery compartment under different spacings between cells.

Table 2 presents the temperature differences inside the battery compartment when the batteries are located at different distances between cells. The variations in the temperature differences in the battery compartment with varying cell distances are shown in Figure 3.

Table 2: Changes in temperature differences with varying cell spacings.

Cell spacing (m)	0.1*d*	0.2*d*	0.3*d*	0.4*d*	0.5*d*	0.6*d*	0.7*d*	0.8*d*	0.9*d*	1.0*d*
Temperature difference value (°C)	0.68	0.63	0.59	0.56	0.53	0.51	0.50	0.49	0.48	0.47

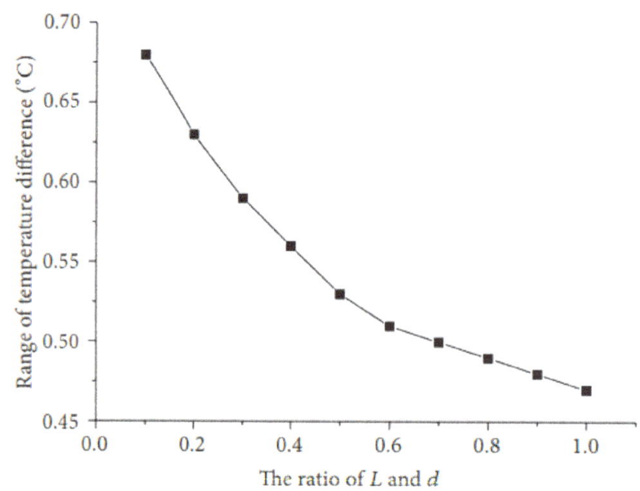

Figure 3: Curve of the temperature difference variations with the distance between the batteries.

Table 2 and Figure 3 show that as the distance between the batteries increases, the internal temperature gradient gradually decreases. The temperature change inside the battery compartment is stable, and the temperature decreases 0.01°C when the distance between batteries increases, 0.1d from L=0.6d. Therefore, the best distance between batteries is 0.6d.

Influence of Battery Arrangement on Temperature Field of Battery Pack

Under the premise of the most suitable distance (L=0.6d) of the battery inside the model, 15 single 18650 batteries are divided into three parts as objects, in which each is composed of 5 batteries and the distance between the batteries is 0.6d. In this paper, we investigate the influence of the sequential arrangement and the cross arrangement of the batteries on the temperature field of the battery pack. The first row and the third row of the battery pack move to the left at 0.3d, and the second row moves to the right at 0.3d. The temperature field distribution in the battery pack is shown in Figure 4.

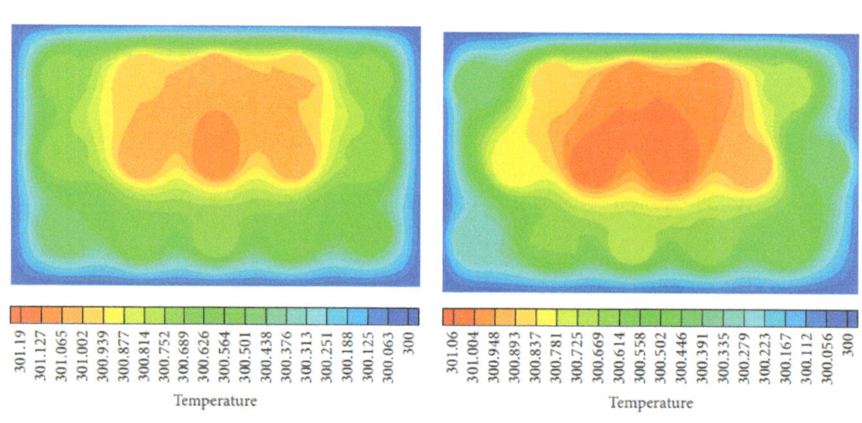

Figure 4: Simulation results of sequential arrangement and cross arrangement.

Figure 4 shows that the temperature difference between the battery packs is 1.19°C when in the sequential arrangement and that the temperature difference between the battery packs is 1.06°C when in the cross arrangement. Comparing the two sets of data reveals that the cross arrangement is better than the sequential arrangement in terms of the temperature gradient.

UNDERWATER VEHICLE BATTERY PACK SPATIAL LAYOUT THERMAL SIMULATION ANALYSIS

On the basis of the previous discussion, take the batteries that are in the cross arrangement. Additionally, select the distance between batteries as 0.6d; the project team developed a certain type of AUV lithium battery for the study (Figure 5). Its battery spatial structure and layout design are studied. Moreover, simulation of the temperature distribution in the battery compartment of an AUV at different speeds and different sea temperature is performed to investigate the effects of vehicle speed and water temperature on the cabin temperature distribution.

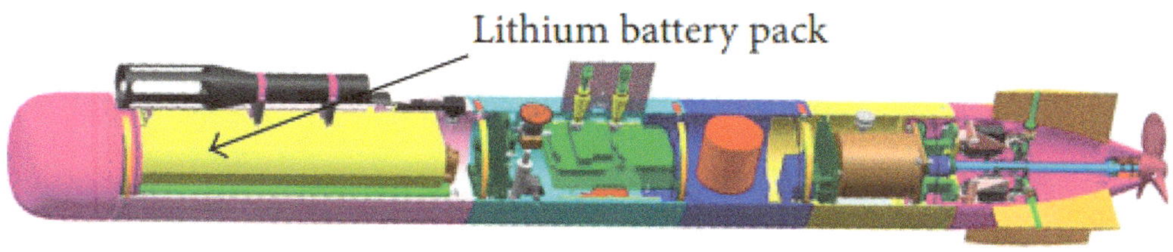

Figure 5: Certain structural layout AUV.

Underwater Vehicle Battery Pack Spatial Structure Layout

Certain underwater vehicle powertrain design specifications are shown in Table 3. According to energy estimates, the power battery requires at least 189 batteries when 18650 lithium/thionyl chloride batteries are used.

Table 3: Certain underwater vehicle dynamic system indicators.

Parameter	Value
Speed	4 kn
Flight	70 km
Power	180 W
Operating voltage	21 V~30 V

The formula for calculating the number of batteries is as follows:

$$N = \frac{sP}{vVC}, \quad (14)$$

where N is the number of batteries required, s is the flight (km), P is the power to the AUV (W), v is the speed of the AUV (kn), V is the nominal voltage of the 18650 battery (V), and C is the battery-rated capacitance of the 18650 battery (Ah).

The battery pack sets 189 batteries into 7 series groups, with each group including 27 parallel batteries. The battery pack was installed in the battery cabin with a diameter of 200 mm to provide an operating voltage of 21 to 30 V. See Figure 6 for the structure.

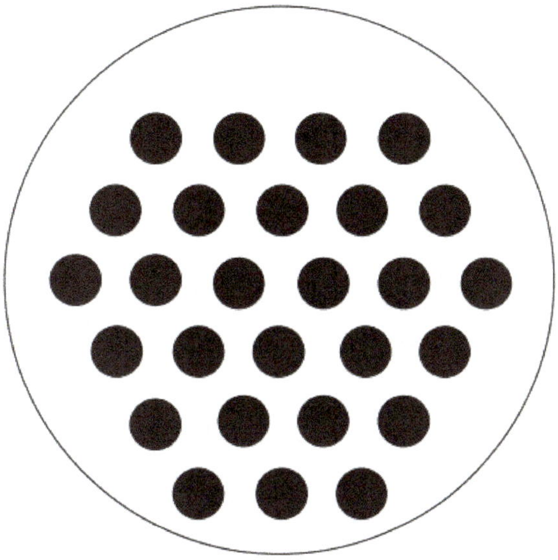

Figure 6: Schematic layout of the battery pack.

AUV Lithium Battery Thermal Simulation Analysis

According to the analysis assumptions, the heat dissipation model of the confined space of the battery cabin is equivalent to the problems of constant properties, inner heat source, and two-dimensional unsteady heat transfer. A cross section of the battery cabin was taken as the calculation area, and the preprocessing program of ANSYS was used to construct the finite element analysis model through cell type selection, material parameters decision, geometric modeling, and cell generation.

The domain is described with triangular elements, with a total number of approximately 72,000. The total number of nodes is approximately 7300, and some grids closest to the profiles of the batteries and navigation shell were refined with triangular boundary elements to describe the boundary flow with sufficient precision. The grid of the model was divided as shown in Figure 7.

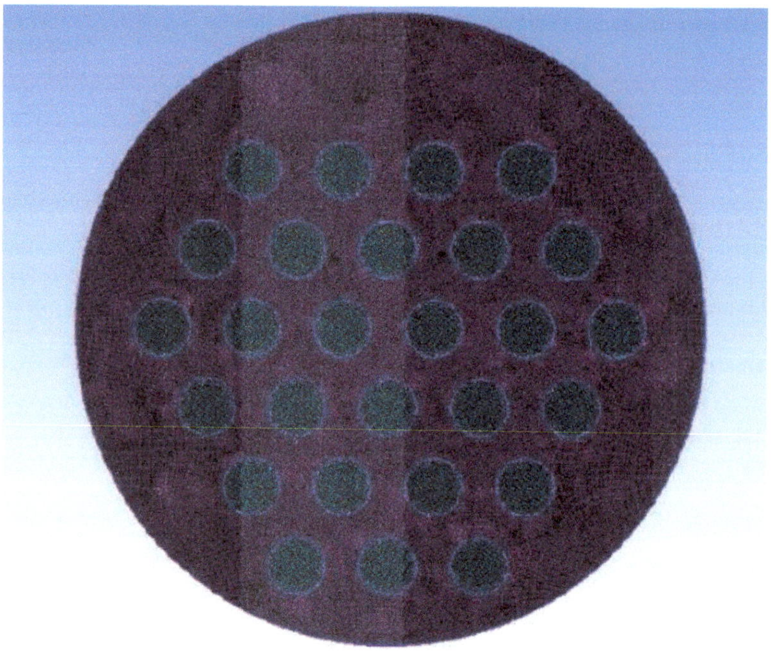

Figure 7: Underwater vehicle battery compartment mesh.

The Effect of Navigating Speed on Temperature inside the Battery Compartment Changing over Time

Batteries for underwater cabins have different heats per unit of time when sailing at different speeds. After analysis, the direction in which the vehicle experiences resistance (the direction of the center of gravity speed) is opposite to the direction of navigation. Namely, the speed line is in the opposite direction of the -axis. Dimensionless coefficients can be expressed as

$$X = c_x \frac{1}{2} \rho v^2 S_c. \tag{15}$$

Vehicle power can be expressed as follows:

$$P = Xv = c_x \frac{1}{2} \rho S v^3, \tag{16}$$

where X is the resistance to the vehicle, c_x is the resistance coefficient, ρ is the seawater density, and S is the maximum cross-sectional area of the vehicle.

Therefore, the current through a single battery is

$$I_b = \frac{P_b}{E_b} = \frac{P}{N \cdot E_b} = \frac{c_x r S v^3}{2 \cdot N \cdot E_b}, \tag{17}$$

where P_b is the output power of a single battery (W) and E_b is its nominal voltage (V). N is the total number of battery cells in the body.

It can be seen that the heat production of a single battery is associated with the vehicle speed. Select vehicle speeds of 4 kn, 5 kn, and 6 kn. The performance parameters of batteries at different speeds are shown in Table 4.

Table 4: Battery performance parameters at different speeds.

Speed/kn	Single battery current/A	Surface heat transfer coefficient (W/(m^2·K))	Heat rate Q (W/m^3)
4	0.27	1564	172
5	0.52	2854	655
6	0.90	4094	1958

After the batteries discharge for 10 hours, the simulation analysis shows that the temperature distribution in the battery cabin is as shown in Figure 4 when the seawater temperature is 15°C and the sailing speeds are 4 kn, 5 kn, and 6 kn. Moreover, the curve of the change in maximum temperature over time is presented in Figure 8.

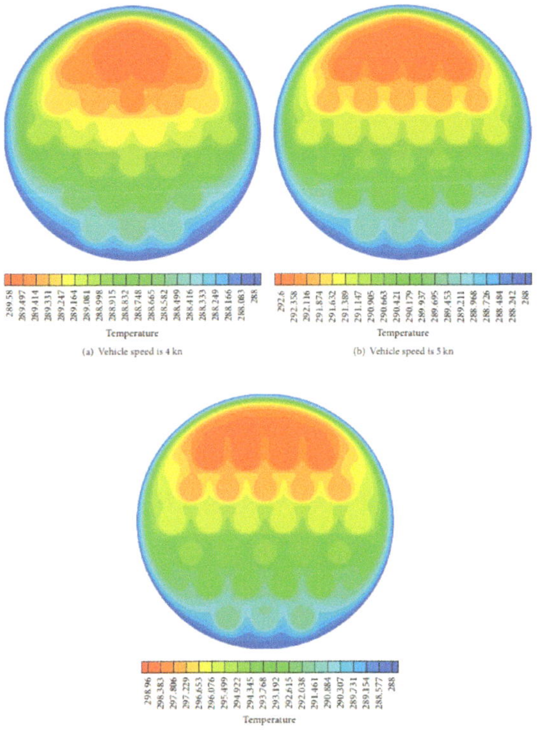

Figure 8: Battery compartment internal temperature profile at different speeds.

After continuous discharging for 10 h, the curve of the maximum temperature of the battery compartment over time is as shown in Figure 9.

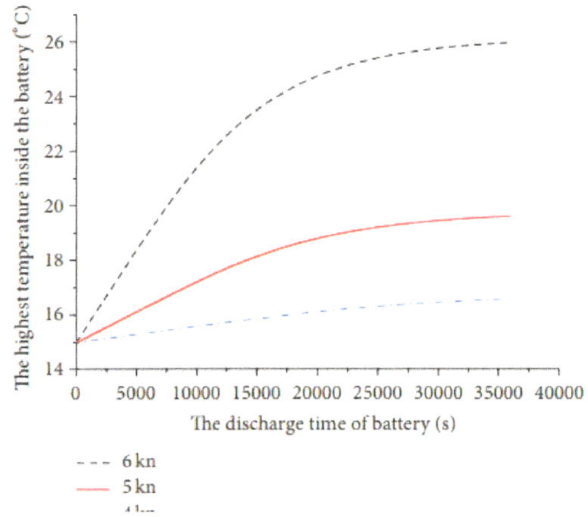

Figure 9: The maximum temperature changes over time.

As shown in Figures 8 and 9, the seawater temperature is 15°C and the discharge time is 10 h. When the sailing speed is 4 kn, the temperature difference is 1.58°C; when the speed is 5 kn, the temperature difference is 4.60°C; and when the speed is 6 kn, the temperature difference is 10.96°C.

Therefore, as the vehicle speed increases, the maximum temperature inside the battery compartment increases and temperature differences increase accordingly. The reasons for this phenomenon are summarized as follows: as the underwater vehicle speed increases, the heat production rate is greater, and more heat is generated per unit time. Because the surface heat transfer coefficient is small, the battery heat generation per unit of time is less than the shell heat distribution per unit of time, resulting in heat concentration and the maximum temperature increasing. In addition, the minimum temperature is always the environmental temperature, which remains unchanged, and the battery compartment temperature increases as the vehicle speed increases.

The Effects of Different Environmental Temperatures on the Battery inside Cabin Temperature versus Time Relationship

For accuracy, select navigation speeds of 4 kn, 5 kn, and 6 kn to study the effects of seawater temperature on the temperature distribution inside the battery cabin when the water temperatures are 10°C, 15°C, and 20°C, respectively.

After the batteries continuously discharge for 10 hours, the simulation analysis shows that when the vehicle cruising speed is 4 kn, the seawater temperatures are 10°C, 15°C, and 20°C, and the battery compartment temperature distribution is shown in Figure 10.

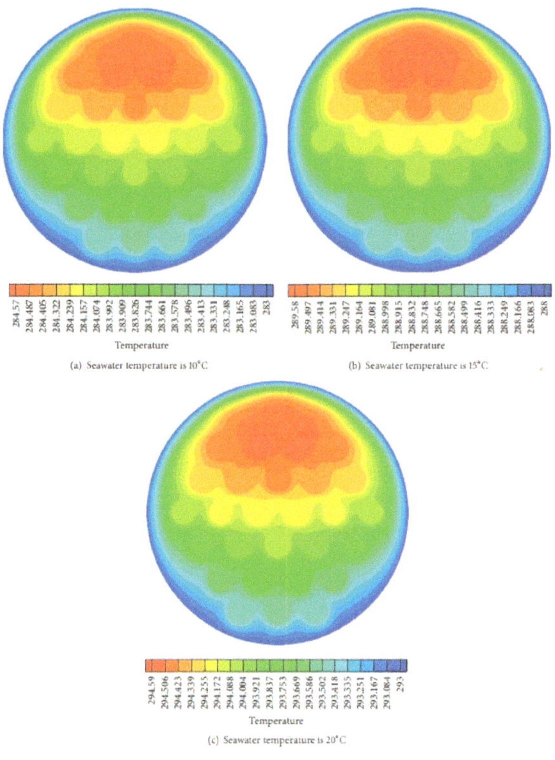

Figure 10: The battery compartment temperature distribution when the speed is 4 kn.

After continuously discharging for 10 h, the battery compartment maximum temperature curve over time is as shown in Figure 11. As shown in Figures 10 and 11, after the underwater vehicle sails at a speed of 4 kn and continuously works for 10 h, the battery compartment internal temperature difference is 1.57°C when the water temperature is 10°C; the battery compartment internal temperature difference is 1.58°C when the water temperature is 15°C; and the battery compartment internal temperature difference is 1.59°C when the water temperature is 20°C.

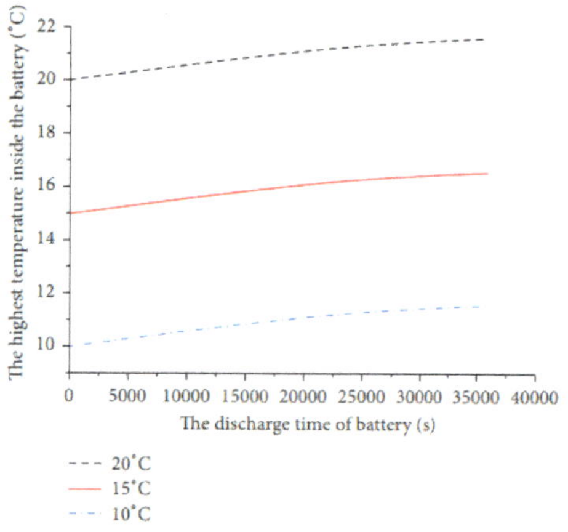

Figure 11: The battery compartment maximum temperatures over time when the speed is 4 kn.

After the batteries continuously discharge for 10 hours, the simulation analysis shows that when the vehicle cruising speed is 5 kn, the seawater temperatures are 10°C, 15°C, and 20°C, and the battery compartment temperature distribution is shown in Figure 12.

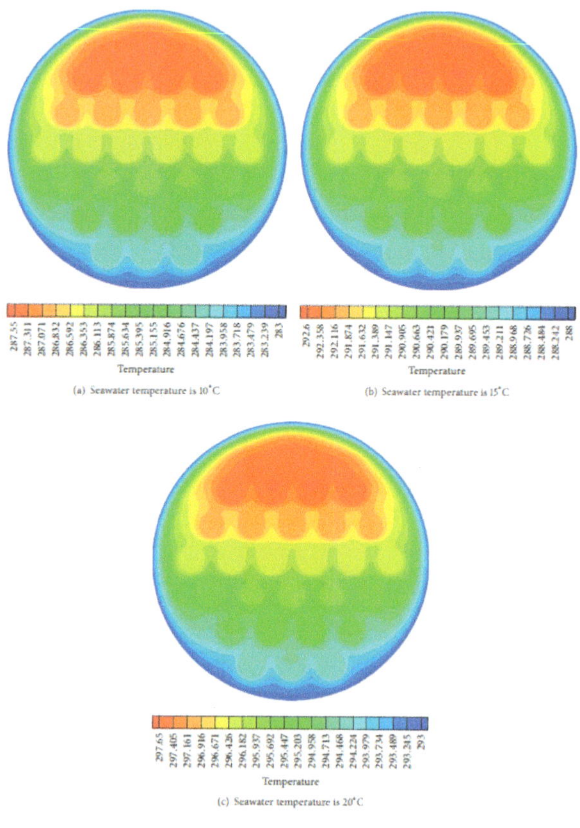

Figure 12: The battery compartment temperature distribution when the speed is 5 kn.

After continuously discharging for 10 h, the battery compartment maximum temperature curve over time is as shown in Figure 13. As shown in Figures 12 and 13, after the underwater vehicle sails at a speed of 5 kn and continuously works for 10 h, the battery compartment internal temperature difference is 4.55°C when the water temperature is 10°C; the battery compartment internal temperature difference is 4.60°C when the water temperature is 15°C; and the battery compartment internal temperature difference is 4.65°C when the water temperature is 20°C.

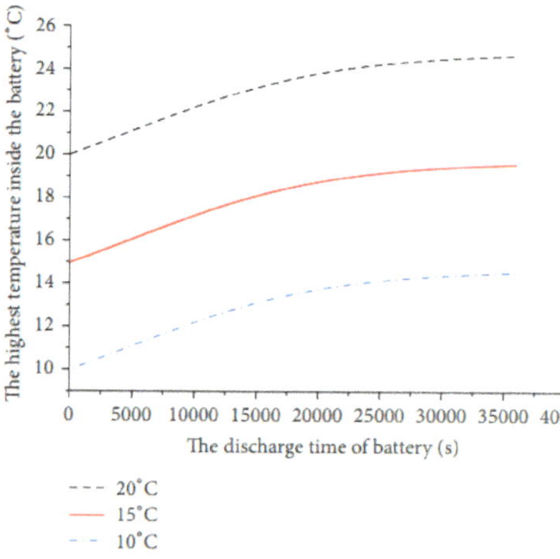

Figure 13: The battery compartment maximum temperatures over time when the speed is 5 kn.

After the batteries continuously discharge for 10 hours, the simulation analysis shows that when the vehicle cruising speed is 6 kn, the seawater temperatures are 10°C, 15°C, and 20°C, and the battery compartment temperature distribution is shown in Figure 14.

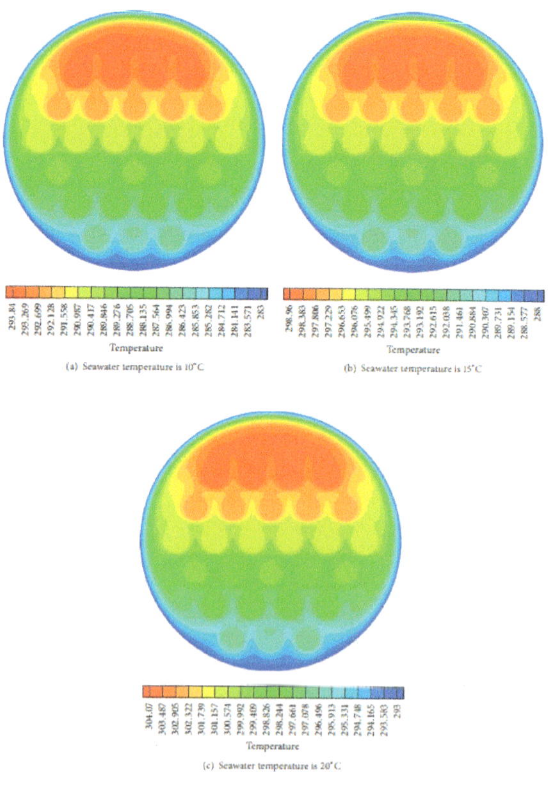

Figure 14: The battery compartment temperature distribution when the speed is 6 kn.

After continuously discharging for 10 h, the maximum temperature inside the battery compartment over time is as shown in Figure 15. As shown in Figures 14 and 15, after the underwater vehicle sails at a speed of 6 kn and continuously works for 10 h, the battery compartment internal temperature difference is 10.84°C when the water temperature is 10°C; the battery compartment internal temperature difference is 10.96°C when the water temperature is 15°C; and the battery compartment internal temperature difference is 11.07°C when the water temperature is 20°C.

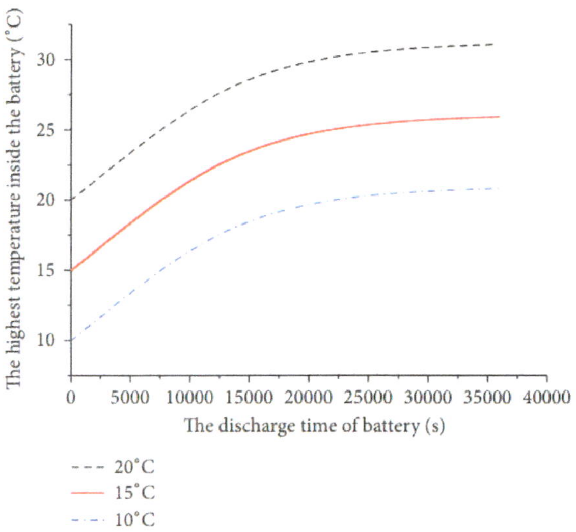

Figure 15: The battery compartment maximum temperatures over time when the speed is 6 kn.

To summarize, the water temperature has essentially no effect on the temperature difference inside the battery compartments. The reasons for this phenomenon are summarized as follows: the heat generation rate of batteries within unit time and the forced convection heat transfer coefficient between the outer wall of the vehicle and seawater are unchanged and almost equivalent when the sailing speed is constant. When the seawater temperature increases, the overall temperature in the battery cabin increases, but the temperature difference is basically unchanged.

CONCLUSION

In this paper, using theoretical analysis combined with the actual situation and the finite element software ANSYS, we establish the space of lithium batteries for an underwater thermal model. Then, we study the spatial layout of thermal characteristics, and the conclusions are as follows:(1)A correlation exists between the underwater battery compartment temperature and the distance between the batteries. As the distance between the batteries increases, the temperature gradient gradually changes, and when it reaches a certain threshold, the gradient gradually stabilizes.(2)Different permutations and combinations between the batteries have effects on the temperature gradient. The cross arrangement is better than the sequential arrangement in terms of the temperature gradient.(3)The sailing speed affects the change in the temperature gradient and the maximum temperature inside the battery cabin over time. The greater the speed and the battery discharge current, the more the heat that the batteries generate. Then, the temperature increases faster, and less time is required to achieve the steady state.(4)The seawater temperature rise may cause an overall temperature increase in the battery cabin. However, it has almost no impact on the temperature gradient and uniformity.

Competing Interests

The authors declare that they have no competing interests.

ACKNOWLEDGMENTS

This work was supported by the National Natural Science Foundation of China (NSFC) under Grant 51509205 and by the Shaanxi Provincial Natural Science Foundation of China 2015JQ5136.

REFERENCES

1. J. Wang and Y. Hu, "Lithium-ion battery temperature field analysis," Power Technology, vol. 32, no. 2, pp. 120–121, 2008.
2. S. Ogura, I. Kawama, T. Sakurai et al., "Development of oil filled pressure compensated lithium-ion secondary battery for DSV shinkai 6500," in Proceedings of the IEEE MTTS/IEEE TECHNO (OCEANS '04), vol. 3, pp. 1720–1726, Kobe, Japan, November 2004.
3. T. Stuart and W. Zhu, "Fast equalization for large lithium ion batteries," IEEE Aerospace and Electronic Systems Magazine, vol. 24, no. 7, pp. 27–31, 2009.
4. T. A. Stuart and W. Zhu, "A targeted equalizer for lithium ion battery packs," in Proceedings of the 5th IEEE Vehicle Power and Propulsion Conference (VPPC '09), pp. 175–180, IEEE, Dearborn, Mich, USA, September 2009.
5. T. Hyakudome, H. Yoshida, S. Ishibashi, T. Sawa, and M. Nakamura, "Development of advanced lithium-ion battery for underwater vehicle," in Proceedings of the IEEE Symposium on Underwater Technology (UT) and Workshop on Scientific Use of Submarine Cables and Related Technologies (SSC '11), pp. 1–4, IEEE, Tokyo, Japan, 2011.
6. I. Masmitjà Rusinyol, J. González, G. Masmitjà, S. Gomáriz, and J. del-Río-Fernández, "Power system of the Guanay II AUV," Acta IMEKO, vol. 4, no. 1, pp. 35–43, 2015.
7. I. Shnaps and E. Rimon, "On-line coverage of planar environments by a battery powered autonomous mobile robot," in Algorithmic Foundations of Robotics XI: Selected Contributions of the Eleventh International Workshop on the Algorithmic Foundations of Robotics, vol. 107 of Springer Tracts in Advanced Robotics, pp. 571–589, Springer, Berlin, Germany, 2015.
8. Y. Wang and Y. Hu, "Battery rack with aluminum injection improvement of battery cabin cooling effect,"Northwestern Polytechnical University, vol. 30, no. 3, pp. 428–434, 2012.
9. Y. Wang and Y. Hu, Affect the Thermal Conductivity of the Battery Holder Auv Battery Inside Cabin Fever and Heat Transfer Processes, vol. 5, Shanghai Jiaotong University, 2012.
10. Y. Wang, Y. Hu, S. Meng, and P. Wang, "A certain type of underwater vehicle battery cabin thermal process CFD analysis," Defence Technology, vol. 33, no. 4, pp. 476–482, 2012.
11. B. Ji, X. Song, W. Cao, and V. Pickert, "Active temperature control of Li-ion batteries in electric vehicles," in Proceedings of the IET Hybrid and Electric Vehicles Conference (HEVC '13), pp. 1–5, London, UK, November 2013.
12. M. Guo and J. Chen, "Passive thermal control structure of the program underwater vehicle lithium battery

pack," Mine Warfare and Ship Protection, vol. 18, no. 4, pp. 15r–20r, 2010.

13. D. Bernardi, E. Pawlikowski, and J. Newman, "A general energy balance for battery systems," Journal of the Electrochemical Society, vol. 132, no. 1, pp. 5–12, 1985.
14. S. Bhide and T. Shim, "Development of improved Li-ion battery model incorporating thermal and rate factor effects," in Proceedings of the 5th IEEE Vehicle Power and Propulsion Conference (VPPC '09), pp. 544–550, Dearborn, Mich, USA, September 2009.
15. S. Bhide and T. Shim, "Novel predictive electric Li-ion battery model incorporating thermal and rate factor effects," IEEE Transactions on Vehicular Technology, vol. 60, no. 3, pp. 819–829, 2011.
16. D. Chen, J. Jiang, Y. Duan, Z. Wang, and F. Wen, "Fluid and thermal analysis of power li-ion battery pack and experimental verification," in Proceedings of the 2013 International Conference on Electrical and Information Technologies for Rail Transportation (EITRT2013)-Volume II, vol. 288 of Lecture Notes in Electrical Engineering, pp. 161–170, Springer, Berlin, Germany, 2014.
17. N. Javani, I. Dincer, G. F. Naterer, and B. S. Yilbas, "Heat transfer and thermal management with PCMs in a Li-ion battery cell for electric vehicles," International Journal of Heat and Mass Transfer, vol. 72, pp. 690–703, 2014.
18. Y. Wang, C. Zhang, and Z. Chen, "A method for state-of-charge estimation of $LiFePO_4$ batteries at dynamic currents and temperatures using particle filter," Journal of Power Sources, vol. 279, pp. 306–311, 2015.
19. E. Samadani, S. Farhad, W. Scott et al., "Empirical modeling of lithium-ion batteries based on electrochemical impedance spectroscopy tests," Electrochimica Acta, vol. 160, pp. 169–177, 2015.
20. C. Heubner, M. Schneider, C. Lämmel, U. Langklotz, and A. Michaelis, "In-operando temperature measurement across the interfaces of a lithium-ion battery cell," Electrochimica Acta, vol. 113, pp. 730–734, 2013.
21. P. Amiribavandpour, W. Shen, D. Mu, and A. Kapoor, "An improved theoretical electrochemical-thermal modelling of lithium-ion battery packs in electric vehicles," Journal of Power Sources, vol. 284, pp. 328–338, 2015.

CHAPTER 12 STRUCTURAL OPTIMIZATION OF A WEARABLE DEEP BODY THERMOMETER: FROM THEORETICAL SIMULATION TO EXPERIMENTAL VERIFICATION

Ming Huang[1], Toshiyo Tamura[2], Zunyi Tang[2], Wenxi Chen[3], and Shigehiko Kanaya[1]

[1]Graduate School of Information Science, Nara Institute of Science and Technology, Ikoma 6300192, Japan

[2]Department of Biological Engineering, Osaka Electro-Communication University, Shijonawate, Japan

[3]Biomedical Information Technology Laboratory, The University of Aizu, Tsuruga, Aizuwakamatsu, Japan

ABSTRACT

Deep body temperature (DBT) has yet to be measured continuously in everyday life, even though it is useful in physiological monitoring and chronobiology studies. We tried to address this issue by developing a transcutaneous thermometer based on the dual-heat-flux method (DHFM) invoking the principle of heat transfer, for which measurement error was mitigated by elaborate design. First, a structural modification based on the original design of the DHFM was implemented by the finite element method. Based on the results of the simulations, prototypes were then implemented and tested with an experimental system that mimicked the thermometer being applied to skin. The simulation phase proposed the adoption of an aluminum cover to boost measurement accuracy and suggested that thermometers of different height be chosen according to specified requirements. The results of the mock-up experiments support the modification put forward in the simulation phase: the standard type (15 mm in height) achieved the accuracy with error below 0.3°C while the thin type (9 mm in height) attained accuracy with error less than 0.5°C under normal ambient temperature ranging from 20 to 30°C. Even though the design should also be examined in vivo, it is believed that this study is an important step in developing a practical noninvasive deep body thermometer.

INTRODUCTION

Deep body temperature (DBT) is one of the vital signs for human beings and is typically referred to as the temperature of the natural cavities, for example, the abdomen and the thorax. Strictly speaking, DBT can only be measured by invasive methods such as catheter insertion into rectum [1].

However, noninvasive methods are more acceptable and are therefore more widely used. They can provide approximation to the DBT, which is the temperature at a certain depth under the skin reflecting the real DBT. One good alternative to the invasive methods is the zero-heat-flux method [2], which was improved by Togawa's group [3] and implemented in the CoreTemp medical device (Terumo, Tokyo, Japan). It showed good agreement with distal esophageal temperature [4] and pulmonary blood temperature [5]. Developed with an inlaid heater, this device is considered to be stable (robust to changes

Citation: Ming Huang, Toshiyo Tamura, Zunyi Tang, Wenxi Chen, and Shigehiko Kanaya, "Structural Optimization of a Wearable Deep Body Thermometer: From Theoretical Simulation to Experimental Verification," Journal of Sensors, vol. 2016, Article ID 4828093, 7 pages, 2016. doi:10.1155/2016/4828093.

in the ambient environment) and sensitive [5]. On the other hand, the heater consumes considerable power, which makes this kind of device only applicable indoors, mainly in hospitals during therapy.

However, continuous DBT measurement is necessary in situations such as heat strain monitoring [6] and female menstrual cycle management [7] and in the treatment of sleep disorders [8] for the estimation of biorhythms. These needs could be met by wearable devices. We suggest that the display and data accumulating functions could be allocated to a smartphone/watch, with the data further analyzed and modeled on a server. In the initial stage of signal extraction, a noninvasive thermometer with low power consumption is indispensable.

More often, it is the temperature fluctuations rather than absolute temperature readings that interest us in the healthcare domain. However, the changes in readings by the devices should be caused by actual changes in the physiological state and not by some external influence. From this viewpoint, an ear-inserted thermometer is not a good choice because its measurements are evidently influenced by the ambient environment. The infrared tympanic thermometer is able to reflect internal change but is sensitive to its positioning.

The mechanism of the thermometer should also be universal; it should be applicable to different people. More specifically, even in the same person, the thermal conductivity of the skin might differ temporally and spatially. A thermometer based on the exact value of the personal physiological condition [9] should be further generalized.

The above concerns can be addressed properly by the dual-heat-flux method (DHFM) [10], the mechanism of which is depicted in Figure 1. The DHFM uses at least four temperature sensors to calculate the DBT, based on the idealized situation that heat flows from the human body into the thermometer longitudinally. This assumption would bring about error in the temperature calculation; however, it requires little information from the user. Moreover, the idealized situation can be approximated by dimensional or thermophysical modification [11].

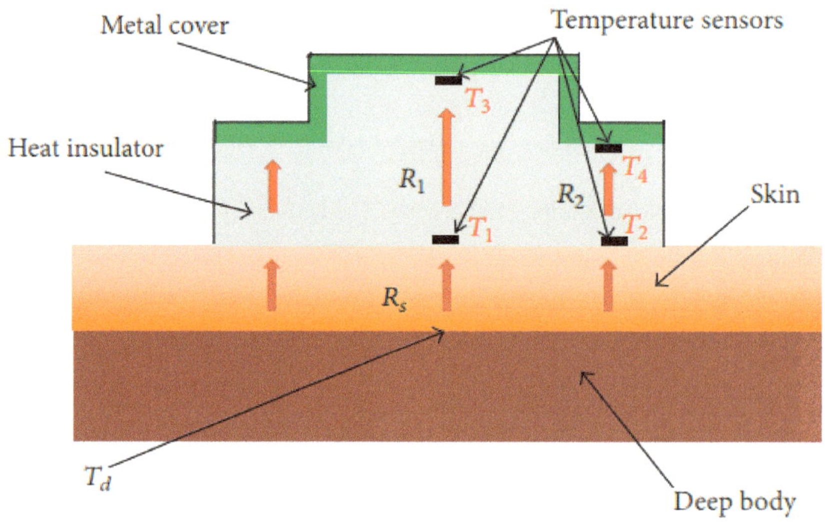

Figure 1: Illustration of dual-heat-flux method. The heat flows from deep body into the thermometer longitudinally and thus the DBT can be calculated with the four inlaid sensors ($T_1 - T_4$) theoretically.

The criterion of accuracy for clinical use is 0.1°C conventionally, which is not readily met by noninvasive methods [9, 11]. Hence, we considered 0.5°C as the margin for acceptable accuracy in this study. However, for the original design, this level of accuracy is achieved at the cost of a device that is too large to wear or an ultrathin design that is difficult to implement [11]. On the basis of the above understanding, this study aimed to improve the structural design to achieve this level of accuracy without compromising its small size. To reach this target, we first designed and examined the structural modification based on the finite element method (FEM). Then, the validated new structures were fabricated and their capabilities were examined by mock-up experiments.

MATERIALS AND METHOD

About DHFM

At least four inlaid temperature sensors are necessary for a thermometer based on DHFM. They are $T_1 - T_4$ in Figure 1. On account of temperature gradient, heat flow arises and flow through the skin layer into the thermometer as shown by red arrows. T_d is the general DBT beneath the skin and subcutaneous layer. Assuming that the heat flow from the deep body into the thermometer remains the same, Fourier's law applies, giving

$$\frac{(T_d - T_1)}{R_s} = \frac{(T_1 - T_3)}{R_1},$$
$$\frac{(T_d - T_2)}{R_s} = \frac{(T_2 - T_4)}{R_2}. \quad (1)$$

T_1 and T_2 are skin temperatures measured by two cutaneous temperature sensors inside the thermometer. T_3 and T_4 are the temperatures measured by the other two sensors. R_s is the heat resistance of the skin, with R_1 and R_2 being the heat resistance of the two heat paths inside the thermometer. According to (1), T_d then can be expressed as [10]

$$T_d = T_1 + \frac{(T_1 - T_2)(T_1 - T_3)}{k(T_2 - T_4) - (T_1 - T_2)}. \quad (2)$$

Hence, $k=R_1/R_2$ and it can be represented as the ratio of the heights of the two concentric cylinders used in the thermometer fabrication.

Generally speaking, DBT is strictly regulated by hypothalamus and it changes tardily in accordance to biorhythm. These characteristics make the measurement with DHFM applicable. For such a passive method with heat insulator as the substrate material, it takes time for initial response.

Simulation Based on FEM

Bioheat transfer involves blood perfusion and metabolic processes. It can be well described by the Pennes equation [12]:

$$\rho c_p \frac{\partial T(\mathbf{X},t)}{\partial t} = \nabla \cdot [k(\mathbf{X}) \nabla T(\mathbf{X},t)] + \omega_b \rho_b c_b (T_b - T(\mathbf{X},t)) + q_m(\mathbf{X},t) \quad \mathbf{X} \in \Omega_T. \quad (3)$$

The parameters T, ρ, c_p, and k are the local temperature, the density (kg/m³), the specific heat (J/kg·°C), and the thermal conductivity (W/m·°C) of the local tissue, respectively. T_b, ω_b, ρ_b, and c_b are the blood temperature, blood perfusion rate (m³/m³·s), blood density (kg/m³), and the specific heat of blood (J/kg·°C), respectively.

The second and third terms on the right denote the heating effect of blood perfusion and metabolism. However, these two terms were not considered in this model because this kind of evenly distributed heat source would alter the actual value of the measurement somewhat but would not invalidate the qualitative relations from the simulation. Further, because the DBT should vary in a quasistatic manner, the temporal change was not considered.

Because we assumed that the thermometer and skin were covered by suitable clothing, heat convection was neglected. However, radiation was inevitable; therefore, the boundary conditions of the thermometer and its surrounding skin could be described by the Stefan-Boltzmann law [13]:

$$-\vec{\mathbf{n}} \cdot [-k(\mathbf{X}) \nabla T(\mathbf{X},t)] = \varepsilon\sigma \left(T_{\text{amb}}^4 - T_{\text{boun}}^4(\mathbf{X},t)\right), \quad X \in \Gamma, \quad (4)$$

where T_{amb} is the ambient temperature, T_{boun} the local temperature value at the boundary, σ the Stefan-Boltzmann constant, and ε the emissivity of the boundary.

In the simulation phase, we sought to improve the thermometer's accuracy by structural modification, while reserving the wearability of the thermometer. Therefore, an additional component, a peripheral aluminum ring (PAR), was introduced in view of the high emissivity of the insulator component. There were two kinds of thermometers in this simulation, with and without the PAR, as shown in Figures 2(a) and 2(b).

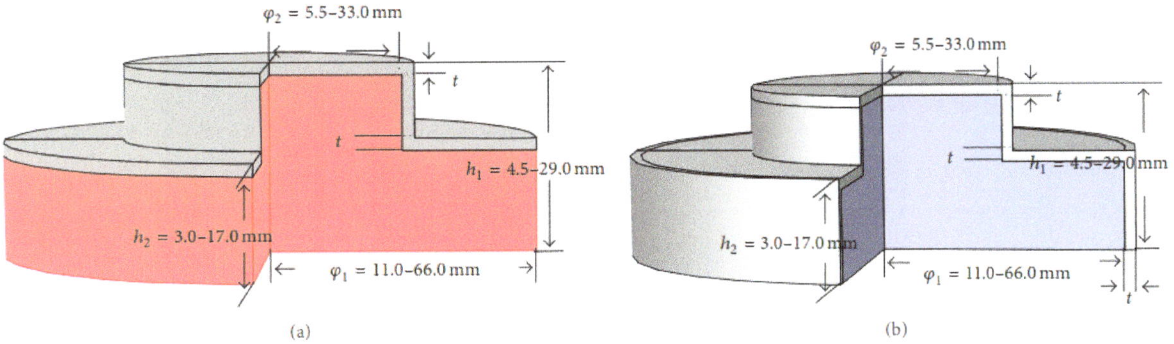

Figure 2: Sectional view of two models. The original structure is without PAR (a), while the modified structure with PAR is shown by (b).

As shown in Figure 2, we considered four combinations of heights, namely, (29.0, 17.0), (15.0, 9.0), (8.0, 5.0), and (4.5, 3.0), for (h_1, h_2). Similarly, we considered six combinations of radii, namely, (66.0, 33.0), (55.0, 27.5), (44.0, 22.0), (33.0, 16.5), (22.0, 11.0), and (11.0, 5.5), for (φ_1, φ_2), respectively. (We also labeled these configurations in terms of their h_1 and φ_1 values.) In fact, because a radius larger than 50.0 mm is not suited for wearable application, we introduced 66.0 mm and 55.0 mm sizes just for the whole picture of the relation between radius and accuracy. In total, there were 24 dimensional combinations, all of which were simulated by our models. In Figure 2(b), the PAR made of aluminum was represented in silver with t=1.0 mm, a size that will not bring about distinct change in volume and weight. The ambient temperature of the simulation was T_{amb}=25.0°C and T_d=37.0°C. The thickness of the skin and subcutaneous layer was 10.0 mm.

In this model, the rubber was used as insulator and the aluminum was used as the metal. Necessary physical parameters are tabulated in Table 1. The models were built, simulated, and analyzed with LiveLink for MATLAB based on COMSOL Multiphysics 4.3a (COMSOL Inc., Stockholm, Sweden). Unconstructed mesh was generated by COMSOL adapting to the current physics and geometries. The numbers of elements were different across the models (e.g., 57544 for standard type model); however, the minimum element quality (q) was kept bigger than 0.1. q has a considerable impact on accuracy of the solution, and, for 3D model, the mesh quality should not affect the solution's quality if q>0.1.

Table 1: Thermal properties of various materials.

Component	Conductivity (W/m·°C)	Density (kg/m^3)	Specific heat (J/kg·°C)	Emissivity
Skin	0.17	1100	3500	0.98
Rubber	0.06	180	2010	0.95
Aluminum	400	8700	385	0.05

*The properties used here were cited from the materials library of COMSOL Multiphysics.

In practical implementation, the values of the parameters or even the specified materials may be different. The lower the conductivity of the insulator, the higher the accuracy at the cost of a longer initial response in general. The simulation here serves as an examination of the new design.

Fabrication of Prototypes

Based on the results of the simulation, we fabricated two prototypes of different heights, h_1=9.0 mm (thin type) and h_1=15.0 mm (standard type), as shown in Figure 3. Both prototypes were of the same radius, $\varphi_1 = 22.0$ mm. The prototypes consisted of two parts: the probe to be applied to the skin for measurement and the processing circuit. As designed in the simulation phase, the probe consists of a heat resistor made of chloroprene rubber and metal components made of aluminum 2017 alloy. Miniature digital temperature sensors LM73 (11–14 bits; Texas Instruments, Dallas, TX) were used as the four inlaid sensors and communicated with the central board with I^2C.

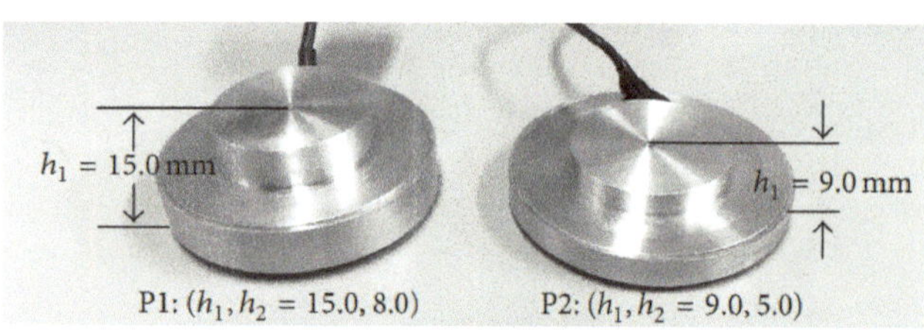

Figure 3: Prototypes of thermometers based on DHFM. Line-up of the standard type and thin type is shown.

The probe of the thermometer connects to the main processing board by USB cable (Mini to Standard plug). The main processing board consists mainly of the controlling unit (ATmega164A, Atmel RISC Microcontroller; Atmel, San Jose, CA), a memory unit with 8 Mb serial flash memory to store the infradian data, and a battery unit to supply three days of energy for the whole local system. Data stored locally can be retrieved with a specialized program run on the PC as a CSV file.

Mock-Up Experiment

The structures suggested by the simulation phase should be validated by mock-up experiments before being applied in practical measurement. A standard experimental system [10, 14] was used to examine the thermometer's capability. An illustration of the system is shown in Figure 4. The thermometer was placed on a natural rubber sheet (sheets of 2.0, 4.0, 6.0, 8.0, and 10.0 mm in thickness), which was laid inside a copper vat. The vat floated on temperature-controlled water. In view of the excellent heat conductivity of copper, the boundary between the rubber sheet and copper vat was considered isothermal. Silicon grease was applied to the boundary between the probe and the rubber sheet to minimize thermal resistance. The temperature of the water inside a thermostatic water bath (FR-004, ±0.1°C; TGK, Tokyo, Japan) was used to mimic T_d, while the natural rubber sheet was used to mimic human skin, based on the physical similarities of these two materials.

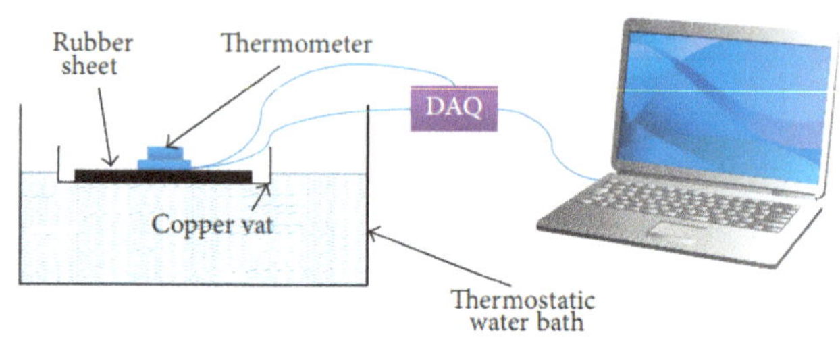

Figure 4: Illustration of the mock-up experiment system. The temperature of the water was used to mimic the DBT, while the natural rubber was used to mimic the skin layer.

The prototypes were tested in three conditions: without PAR (N), with PAR (A), and without PAR but with sponge cover (S). Through this phase, we hoped to find out whether the new component (i.e., the PAR) would influence the measurements as predicted by the simulation. Condition S was adopted to confirm the effect of the sponge in measuring accuracy; although it is obstructive in practical measurement, it is reportedly indispensable [10].

The range of normal T_d falls in between 36.0°C and 38.0°C, while normal indoor T_{amb} is 20.0–30.0°C, which may influence the measurements. Thus, we further tested the prototypes in the following three combinations: Combination

Combination 1: $T_{amb} = 20.0^{\circ}C, T_d = 36.0^{\circ}C.$

Combination 2: $T_{amb} = 25.0^{\circ}C, T_d = 37.0^{\circ}C.$

Combination 3: $T_{amb} = 30.0^{\circ}C, T_d = 38.0^{\circ}C.$

For each combination, experiments were conducted three times and for each condition of the thermometer (N, A, and S) 30 minutes was given for the establishment of heat equilibrium inside the probes. The two probes (thin and standard types) were tested in rotation to remove the heat stored in the previous trial.

RESULTS

Results of Simulation

In this phase, while the ambient temperature T_{amb} and the DBT T_d did not change, the specific combination of these two parameters is sufficient to reveal the relations between the physical structure and the accuracy [11].

For thermometers both with and without PAR, all 24 thermometers defined by the combinations of 4 different heights and 6 different radii were considered and modeled. The measurements from each thermometer are summarized in Figure 5(a) (without PAR) and Figure 5(b) (with PAR). Markers in one interpolating curve indicate the measuring values with thermometers of the same height.

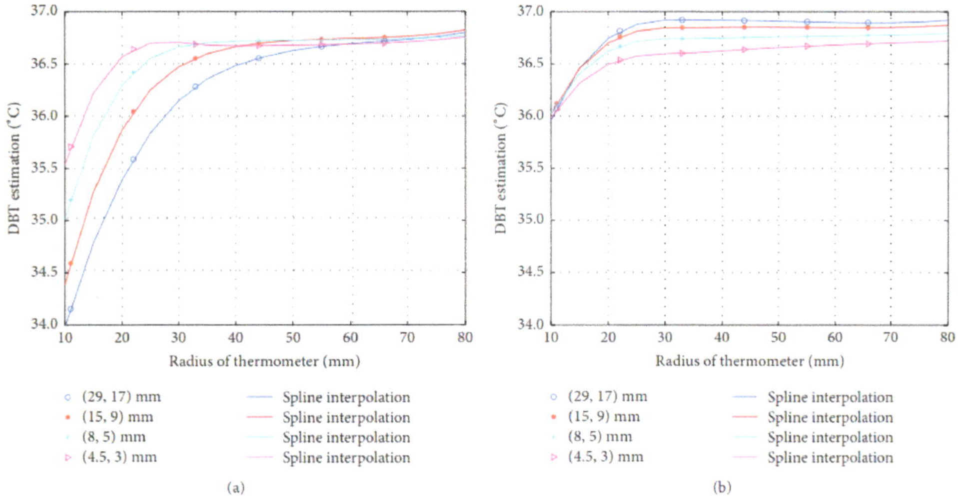

Figure 5: Results of simulation based on FEM. (a) and (b) show the measurements of thermometers in condition N and condition A, respectively.

From the results, we can see that the effects of the dimensions were different. For the thermometer without PAR, generally, the accuracy was proportional to the radius, but inversely proportional to the height. A change of dimensions influences the measuring accuracy greatly when the radius is less than 40 mm. Furthermore, the radius is more effective than the height. A twofold increase in radius benefits the accuracy much more significantly than does a half-size decrease in height.

For the thermometers with PAR, the accuracy was proportional to height and radius. It is easier for them to attain much greater accuracy with a radius greater than 20 mm. Moreover, the height is definitive in these thermometers, as no distinct improvement can be seen by changing the radius of thermometers of the same height, with a radius larger than 30 mm.

Results of the Mock-Up Experiment

According to the simulation results, with both standard and thin types, we can attain the acceptable margin of error (<0.5°C) that is hard to attain in the designs without PAR.

In this phase, the combinations of experimental conditions (Combination 1, Combination 2, and Combination 3) were all tested and Combination 2 was the same as in the simulation phase. For both thermometers, all three conditions (A, N, and S) were tested three times each.

The results for these tests are shown in Figure 6, where the results were concluded based on the type of thermometer. In both A and N conditions, the quantitative relations conform to the results of the simulation. That is, thermometers with PAR attain better accuracy than their condition N counterparts. In condition A, a thicker design gives good accuracy at the expense of a larger volume. In condition S, the measurements showed the best accuracy for both prototypes.

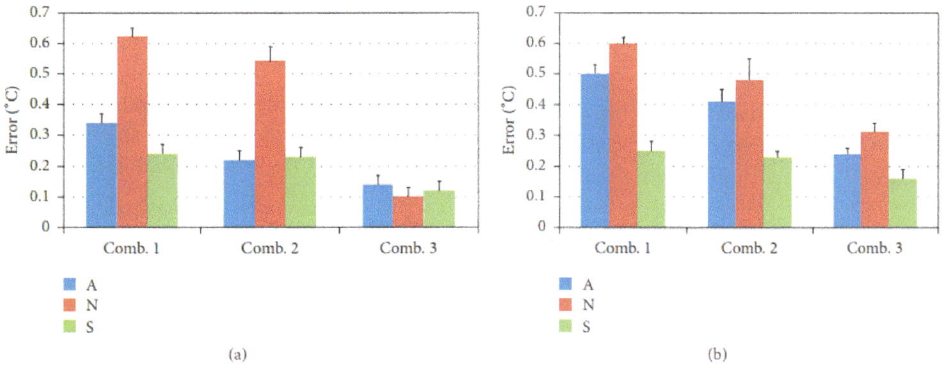

Figure 6: Results of mock-up experiments for both thermometers. Results of standard type were summarized in (a) and thin type in (b).

In condition N, the spans of the error were about 0.5 and 0.3°C for standard and thin type, respectively. By contrast, the spans were about 0.2°C for both types in condition A. This suggests that the PAR takes effect in resisting the influence of the ambient environment.

DISCUSSION

The Practical Use

As we have introduced, the method is applicable in measuring the DBT that change at a slow manner. For extreme situation such as anesthesia or heatstroke, time delay happens. In a simulation whose results were not shown here, the standard type was used to monitor the abrupt change of DBT at a rate of 0.2°C/minute lasting for 10 minutes. The thermometer could indicate the change in about 4 minutes and became stable in 20 minutes after the change. For such situation, the thin type, whose reading will become stable in about 10 minutes, might be suited to trace the change.

If the thermometer is applied to the upper torso, for example, the thorax, the thermometer may come into contact with clothes from time to time. The physically thin layer and low conductivity of clothes would hardly change the resistance of the thermometer. However, the relatively high emissivity (0.75–0.90) [13] may be influential. Assuming that the thermometer comes into contact with the clothes made of cotton of 0.5 mm thickness for an hour, the transient study of simulation was carried out. Its result showed that the measurement of the sensor changed (decreased) after about half an hour by 0.1°C, and then the measurement became stable gradually. In a practice, contact between the thermometer and the clothes is often erratic. In consideration of the slow and subtle change of the measurement, we considered the influence of clothes to be minor in common situation.

The Simulation

The adoption of a PAR was based on the physical characteristics of metals, which usually have a much lower emissivity than heat insulators (e.g., rubber), and, at the same time, the radioactive energy exchange is the main reason for the horizontal heat flow distorting the practical situation from the theoretical assumption of DHFM. With a PAR, the influence of the ambient environment may be mitigated remarkably because this thermometer is designed to be worn on the torso with suitable clothing, which will shield the thermometer from the influence of airflow from concerted movement. It is also the reason why the convection was neglected in the boundary condition of the simulation.

In the simulation phase, only one pair of T_{amb} and T_d was used, and the quantitative conclusions were extended to other conditions. This extension is plausible because these two factors affect the readings of the individual thermometer; however, the qualitative relations remain valid.

By comparing the results with and without PAR, distinctly different patterns of the dimensional effects (height and radius) could be seen. Further, improvements in accuracy were attained for thermometers of the same size.

In the previous study on DHFM-based thermometers [11], the radius was decisive because a larger radius means a larger area covering the skin; therefore, the heat dissipated on boundaries could be greatly compensated for by the heat from the surrounding skin. In other words, the situation will approximate the theoretical assumption of DHFM. However, with PAR, the influence of radius becomes minor, which means that less heat is demanded from the skin, resulting from less radioactive heat dissipation.

In the simulation phase, multiple sizes (height and radius) were tested, but only the 22.0 mm radius and the 15.0 and 9.0 mm heights were adopted in the fabrication stage; this limitation was imposed to retain only compact wearable sizes. An increase in radius to more than 30 mm gives minor improvement but will bring about difficulties in installation and an obstructive feeling.

The Mock-Up Experiment

Only 3 combinations of the ambient temperature and the DBT were investigated here. For ambient temperature, lower value results in lower accuracy due to radiation, while the relatively higher DBT would contribute to a more accurate result [11]. Therefore, Combination 1 and Combination 3 can be used to confirm the span of error under moderate environment.

In the experimental phase, condition S was adopted to confirm the influence of the sponge in measuring accuracy. Its positive influence on measuring accuracy may be because of the heat-insulating effect shielding the thermometer and the surrounding surface of skin from the ambient environment. However, it is actually obstructive in practical measurement and may not be suitable for wearable use.

As previously mentioned, at the mock-up experiment stage, rubber sheets of different thicknesses were tested. As we expected, the thicker the sheet, the larger the measuring error. The two prototypes were able to measure the DBT (water temperature) accurately until the thickness was increased to 8.0 mm. The error became larger than 0.5°C for both thermometers when the sheet was increased to 10.0 mm in thickness. The main reason may lie in the abrupt decrease of the reading of sensor T_1. It may suggest that the experimental system is unable to deliver enough heat to the surface of the rubber sheet because of the lower heat conductivity of the rubber sheet [12] compared with skin. From this point of view, it is still possible for a DHFM-based thermometer to monitor the temperature about 10 mm under the skin. Further, to measure sites such as the forehead or the pit of the stomach with thin skin layer and bone immediately beneath [15], the measuring depth may be extended. Simulation or experiments on the measuring depth for these positions will be interesting.

For this kind of thermometer, which depends on a heat source from the human body, the resistance to the influence of ambient change is crucial. Judging from the results of these experiments, the error spans that came with the change of T_{amb} were shortened considerably with this PAR.

The theory of this method suggests that k here can be represented by the lengths of both heat paths. In practice, k can be optimized experimentally.

With the measurements in this study, we also tried to adjust k, denoted by K_A hereafter. K_A for both prototypes was different and K_A was set as 1.3 for the standard type and 1.1 for the thin type. For all the measurements, the accuracies were improved in varying degrees. What is more, the standard type in condition S achieves the accuracy level within ±0.1°C. In the coming in vivo experiment to verify the thermometers' accuracies, adjustment of k seems beneficial in attaining the best accuracies.

CONCLUSIONS

This was a preliminary study on the capability of a new kind of noninvasive DBT thermometer based on DHFM. The simulation study based on FEM brought this theory to a practical stage by proposing a new component, the PAR, to improve the accuracy and stability.

The designs were then implemented and validated through mock-up experiments. The measurement errors were mitigated to a level of less than 0.5°C for both designs. The standard type of 15 mm in height had better accuracy than the thin type, as the simulation predicted. Even though further in vivo experiments are necessary for this kind of thermometer, we believe that it can serve as an alternative to the heat-generating noninvasive DBT thermometer and is suited for wearable modality.

Conflict of Interests

The authors declare that there is no conflict of interests regarding the publication of this paper.

ACKNOWLEDGMENTS

This study was supported by the Keihanna Science City Healthcare Project of the Ministry of Education, Culture, Sports, Science and Technology, Japan, and by Tateisi Science and Technology Foundation, Japan. The authors wish to thank Professor Yoshida Masaki for his kind support throughout the experiment stage.

REFERENCES

1. P. Kyriacou, Biomedical Sensors, Momentum Press, London, UK, 2010.
2. R. H. Fox and A. J. Solman, "A new technique for monitoring the deep body temperature in man from the intact skin surface," Journal of Physiology, vol. 212, no. 2, pp. 8–10, 1971.
3. T. Tamura, T. Nemoto, and T. Togawa, "A zero-heat-flow transducer for monitoring perfusion blood temperature," IEEE Transactions on Biomedical Engineering, vol. 11, pp. 644–646, 1979.
4. L. P. J. Teunissen, J. Klewer, A. De Haan, J. J. De Koning, and H. A. M. Daanen, "Non-invasive continuous core temperature measurement by zero heat flux," Physiological Measurement, vol. 32, no. 5, pp. 559–570, 2011.
5. M. Yamakage and A. Namiki, "Deep temperature monitoring using a zero-heat-flow method," Journal of Anesthesia, vol. 17, no. 2, pp. 108–115, 2003.
6. D. S. Moran, A. Shitzer, and K. B. Pandolf, "A physiological strain index to evaluate heat stress," The American Journal of Physiology—Regulatory Integrative and Comparative Physiology, vol. 275, no. 1, pp. R129–R134, 1998.
7. M. D. Coyne, C. M. Kesick, T. J. Doherty, M. A. Kolka, and L. A. Stephenson, "Circadian rhythm changes in core temperature over the menstrual cycle: method for noninvasive monitoring," American Journal of Physiology—Regulatory Integrative and Comparative Physiology, vol. 279, no. 4, pp. R1316–R1320, 2000.
8. B. Bjorvatn and S. Pallesen, "A practical approach to circadian rhythm sleep disorders," Sleep Medicine Reviews, vol. 13, no. 1, pp. 47–60, 2009.
9. H.-C. Gunga, M. Sandsund, R. E. Reinertsen, F. Sattler, and J. Koch, "A non-invasive device to continuously determine heat strain in humans," Journal of Thermal Biology, vol. 33, no. 5, pp. 297–307, 2008.
10. K.-I. Kitamura, X. Zhu, W. Chen, and T. Nemoto, "Development of a new method for the noninvasive measurement of deep body temperature without a heater," Medical Engineering & Physics, vol. 32, no. 1, pp. 1–6, 2010.
11. M. Huang, T. Tamura, W. Chen, and S. Kanaya, "Evaluation of structural and thermophysical effects on the measurement accuracy of deep body thermometers based on dual-heat-flux method," Journal of Thermal Biology, vol. 47, pp. 26–31, 2015.
12. H. H. Pennes, "Analysis of tissue and arterial blood temperatures in the resting human forearm," Journal of Applied Physiology, vol. 1, no. 2, pp. 93–122, 1948.
13. F. P. Incropera, D. P. DeWitt, T. L. Bergman, and A. S. Lavine, Fundamentals of Heat and Mass Transfer, John Wiley & Sons, 6th edition, 2006.
14. T. Nemoto and T. Togawa, "Improved probe for a deep body thermometer," Medical and Biological Engineering and Computing, vol. 26, no. 4, pp. 456–459, 1988.
15. R. Y. Ha, K. Nojima, W. P. Adams Jr., and S. A. Brown, "Analysis of facial skin thickness: defining the relative thickness index," Plastic and Reconstructive Surgery, vol. 115, no. 6, pp. 1769–1773, 2005.

CHAPTER 13

EXPERIMENTAL RESEARCH ON THE THERMAL PERFORMANCE OF COMPOSITE PCM HOLLOW BLOCK WALLS AND VALIDATION OF PHASE TRANSITION HEAT TRANSFER MODELS

Yuan Zhang[1], Sunqi Zhuang[2], Qian Wang[1], and Jiapeng He[2]

[1]School of Energy and Power Engineering, Jiangsu University, 301 Xuefu Road, Zhenjiang, Jiangsu 212013, China

[2]College of Urban Construction, Nanjing Tech University, Nanjing, China

ABSTRACT

A type of concrete hollow block with typical structure and a common phase change material (PCM) were adopted. The PCM was filled into the hollow blocks by which the multiform composite PCM hollow blocks were made. The temperature-changing hot chamber method was used to test the thermal performance of block walls. The enthalpy method and the effective heat capacity method were used to calculate the heat transfer process. The results of the two methods can both reach the reasonable agreement with the experimental data. The unsteady-state thermal performance of the PCM hollow block walls is markedly higher than that of the wall without PCM. Furthermore, if the temperature of the PCM in the wall does not exceed its phase transition temperature range, the PCM wall can reach high thermal performance.

INTRODUCTION

As the building energy consumption increases, the energy-saving technology in buildings has become one of the hot fields in the world; meanwhile, people's demand on the indoor thermal comfort also increases continuously. One of the traditional ways to solve this problem is to introduce thermal insulation materials into building envelope to enhance the thermal performance of wall. Traditional thermal insulation materials have low thermal conductivities which can effectively reduce the heat flow through the building envelope; however, they are mostly organic thermal insulation materials with low specific heat capacity. This can probably result in the indoor temperature fluctuation with large amplitude and frequency, which reduces the indoor thermal comfort degree, gives rise to the frequent open and close of air-conditioning device, and decreases the opportunity of device operating under the optimum condition, which increases the energy consumption of device. In addition, the fire safety hidden trouble is accompanied, and the thermal insulation material does not have a lifetime as long as the building itself (the general service life of the organic thermal insulation material is 10 to 20 years). Therefore, people start to find new and better methods to save energy.

Citation: Yuan Zhang, Sunqi Zhuang, Qian Wang, and Jiapeng He, "Experimental Research on the Thermal Performance of Composite PCM Hollow Block Walls and Validation of Phase Transition Heat Transfer Models," Advances in Materials Science and Engineering, vol. 2016, Article ID 6359414, 15 pages, 2016. doi:10.1155/2016/6359414.

It is one of the potential ways to use phase change material (PCM) into building envelope. The proper use of the latent heat of phase transition effect can substantially increase the thermal capacity of building envelope. By doing this, the heat flow into building envelope in daytime is conserved; when the outdoor air's temperature falls to a low value in night time, the PCM releases heat and then makes the conserved heat back to the outside. In this way, the heat entering into the interior of the room throughout a day will be remarkably reduced; furthermore, the peak of heat entering building envelope in daytime will enter into the interior of the room at dusk or night with a lower peak value and larger time lag because of the decrement and delay effects. At this time, the indoor cooling load is reduced; also, the electric charge is in the valley time period, by which the indoor thermal comfort degree is improved and the money is saved.

At present, a lot of researchers studied the thermal performance of the building envelope outfitted with PCM layer. Aelenei used a new structure of wall with photovoltaic board and PCM board and calculated and tested the application effect under the climate of winter. The results indicated that the electrical energy was saved by 10% and the efficiency of the whole system reached 20% by using this type of wall [1]. Ahmed et al. introduced the PCM (paraffin) into the wall of the truck box which was used for the refrigeration of things. The test results demonstrated that the peak heat flow rate entering the truck box was averagely reduced by 29.1% when all the walls were outfitted with PCM; if one wall of the walls was outfitted with PCM, the heat flow rate can be decreased for 11.3%–43.8%; the heat flow rate that entered the truck box every day can be reduced for 16.3% [2]. Alqallaf and Alawadhi analyzed the thermal performance of a concrete roof deck outfitted with PCM. The test results indicated that the heat gain became lower with the increasing amount of PCM; furthermore, the heat gain was influenced by the phase transition temperature of the PCM [3]. Bontemps et al. applied PCM into the wall of an experimental box with twin-box structure. The experimental box was placed under the environment of the local typical outdoor climate. The test results demonstrated that the temperature fluctuation inside the test box with PCM was significantly lower than that of the box without PCM [4]. Borreguero et al. tested the thermal performance of a gypsum board outfitted with PCM. The result showed that the thermal storage capacity of the gypsum board was strengthened by the increasing amount of PCM; meanwhile, the temperature fluctuation inside the wall was reduced; in addition, containing the same thermal performance, by increasing 5% (mass fraction) amount of PCM inside the board, the thickness of the board can be decreased for 8.5% [5]. Cabeza et al. built a concrete room with actual size. The south and west walls and the roof were outfitted with 5% PCM, and the room was exposed to the natural environment. The results demonstrated that the thermal inertia of the wall with PCM was remarkably higher than the wall without PCM, and the interior surface temperature of the PCM wall was lower (reducing for 0 to 2°C); the peak heat was shifted for about two hours by the PCM wall [6]. Castell et al. built the house using two types of traditional perforated bricks with and without PCM. The houses were tested under the typical climate of Spain. The results showed that the PCM made the indoor peak temperature reduce for about 0.9°C and 0.73°C, and the energy consumption reduced for 15% and 17% [7].

It can be seen from the above research that the building envelope outfitted with PCM can reach better thermal performance. At present, the study of the phase transition heat transfer model is mostly around the one-dimensional models [4, 5, 8]. However, it is easy to leak and be damaged and occupies the thickness of wall, if the PCM exists in building envelope with the form of one layer. The hollow brick wall and hollow block wall are widely used in some parts of world for energy conservation now. If the PCM is properly put into the holes of the hollow bricks and hollow blocks, the construction procedure will be simplified and the negative influences will be avoided with maintaining high thermal performance.

On the field of multidimensional heat transfer, Antoniadis et al., using Comsol Multiphysics software [9] and two-dimensional finite element method, analyzed the impacts of the hole rows, hole thickness, and the staggered or unstaggered arrangement of hole rows on the thermal performance of hollow brick, taking the effective thermal conductivity as the evaluation index [10]. Arendt, using two-dimensional energy equation, studied the impacts of hollow ratio on the steady-state and unsteady-state thermal parameters of hollow bricks, with the evaluation indexes of time lag, decrement factor, and equivalent thermal conductivity [11]. Li et al. studied the heat transfer process of the hollow bricks with different structures using the Tri-diagonal Matrix (TDMA) method and Alternating Direction Implicit (ADI) method [12, 13]. The computational efficiency and accuracy of the methods are both superior to or equivalent to the Gaussian elimination method [14], which are appropriate for the solution of multidimensional heat transfer problems of the nonhomogeneous materials such as hollow blocks [15, 16]. Mackerle et al. analyzed the thermal performance of a concrete roof outfitted with PCM using three-dimensional heat conduction model and finite element method. In the mathematical model, the specific heat of PCM within the phase transition temperature range was equivalent to a certain value. The discrete equations were solved by preconditional generalized minimum residual (PGMR) [17] solver. This concrete roof had some cylindrical holes inside, which were filled with PCM (A39). The numerical calculation and experimental test were performed on the roofs with and without PCM, respectively. The results demonstrated that the maximum errors between the calculation results and the experimental data were 7.25% and 7.4%, and there was reasonable agreement between calculation results obtained by the PCM mathematical model and the experimental data [3]. Carbonari et al. developed a two-dimensional phase transition heat transfer model. He used the model and finite element method calculated the heat transfer process of a type of wall board with PCM layer. After the comparison between the calculation results and the experimental data, it was found that a good agreement is achieved [18]. It can be known that, at present, the study on the multidimensional heat transfer process of the constant-property material wall is relatively sufficient but that of the composite PCM wall is insufficient.

In consideration of the reasons above, a type of concrete hollow blocks in typical structure and a commonly used phase change material (PCM) were adopted and the multiform composite PCM hollow blocks were produced in this paper. The PCM was put into different positions of hollow blocks. The temperature-changing hot chamber method was used to test the thermal performance of the hollow block walls with or without PCM under different temperature conditions; meanwhile, the two-dimensional enthalpy model and effective heat capacity model were used to calculate the phase transition heat transfer process under the conditions of experiments. The results of the models were validated by the experimental data, and the thermal performances of the hollow block walls with and without PCM and the walls with the PCM at different positions were compared. The difference on the thermal performance of each wall was analyzed, which provides the data reference and guide for the engineering application of composite PCM hollow block wall.

MATHEMATICAL MODELS

Assumptions

To simplify the heat transfer model, the following assumptions were made: (1) all the materials were considered to be thermally homogeneous and isotropic; (2) because the heat transfer process was mainly heat conduction, the thermal conductivity of the liquid PCM was considered to be not changed; (3) because the specific volume ratio of the solid state to that of the liquid state was less than 1.15, the volume change of the PCM during the phase change process was ignored; (4) the supercooling effect during the freezing process and the natural convection effect were ignored (the used PCM in the experiment and calculation was the composite PCM which was the mixture of the PCM and gypsum in a certain ratio; the natural convection effect was effectively restrained); (5) the equivalent thermal conductivity was applied for the air cavities inside the blocks [19]; (6) because the net heat flow was 0 on the vertical direction of wall, the heat transfer in the computational domain was simplified to be two-dimensional heat transfer, and the total heat flow was on the direction of thickness (the boundaries of other directions were adiabatic, where the second type thermal boundary condition was applied).

For point (4), there are some explanations which are worthy to state. According to the formula, the Grashof and Prandtl numbers can be calculated as $\mathrm{Gr} = g\alpha\Delta t l^3/\nu^2$ and Pr=v/a. After the calculations, for the side holes with air in the experimental hollow blocks, the average Gr is calculated as 0.0167, and the 0.3998 for the middle holes with air. The average Pr for the side holes with air and the middle holes with air is around 0.699. The average Gr for the side holes with air and the middle holes with air is relatively small, which demonstrates that the buoyancy force is small, relative to the viscous force. Meanwhile, the average Pr is relatively large, which indicates the obvious influence of viscidity during the heat transfer process. Therefore, it can be seen that the convection effect in the holes with air is unapparent and the main heat transfer approach is heat conduction. For the holes with PCM, the used PCM in the experiment and calculation was the composite PCM which was the mixture of the PCM and gypsum in a certain ratio. Then the liquid PCM (if it has melted) exists in the microvoid of the solid gypsum. Because of these, the natural convection effect was effectively restrained, the value of the Gr for the holes with PCM becomes very small, and the Pr is the same as the status of the holes with air. With the situations above, we assumed the natural convection to be negligible and we considered that it will not produce large errors in the calculations.

For point (6), it is worthy to explain that, according to the instruction book of the device, the heat loss of the chambers and the wall frames is less than 5%. Furthermore, the thermal conductivity of the insulation material inside the frames of chambers (extruded polystyrene) is low enough (0.029 W/m·K) and the thickness of the frames is large enough (larger than 0.15 m). Therefore, we consider that the heat losses through the chambers and the wall frames have little impact on the accuracy of the experimental data.

Governing Equation

In this research, the traditional two-dimensional heat transfer equation was used to solve the problem of the hollow block wall without PCM [20], and the two-dimensional enthalpy model and effective heat capacity model were adopted for the hollow block wall with PCM.

Two-Dimensional Enthalpy Model

One has

$$\rho\frac{\partial H}{\partial \tau} = \lambda\left(\frac{\partial^2 T}{\partial x^2} + \frac{\partial^2 T}{\partial y^2}\right), \tag{1}$$

where ρ, λ, and H were the density, thermal conductivity, and enthalpy of PCM, respectively. The thickness and width directions were represented by x and y, and the temperature at any point in time τ was T. The enthalpy of PCM was calculated by

$$H = \begin{cases} c_{p,s}\cdot T & T < T_m - \varepsilon \\ c_{p,s}\cdot T_m + \dfrac{L}{2\varepsilon}\cdot[T-(T_m-\varepsilon)] & T_m - \varepsilon \le T \le T_m + \varepsilon \\ c_{p,s}\cdot T_m + L + c_{p,l}\cdot[T-(T_m+\varepsilon)] & T > T_m + \varepsilon, \end{cases} \tag{2}$$

where $c_{p,s}$ and $c_{p,l}$ were the specific heat of the PCM in its solid and liquid states, respectively, T_m was the mean temperature of the phase transition temperature range. The phase change radius ε represented the half range of the phase change temperature range and L was the latent heat during phase change. The temperature range of phase transition was from $T_m-\varepsilon$ to $T_m+\varepsilon$. When $T < T_m-\varepsilon$ or when $T > T_m+\varepsilon$, the value of $c_{p,s}$ was equal to the value of $c_{p,l}$ (=2.919kJ·kg^{-1}·K^{-1}); when $T_m - \varepsilon \le T \le T_m + \varepsilon$, the value of $c_{p,s}$ was equal to $c_{p,\mathrm{equ}} = L/2\varepsilon$, where $c_{p,\mathrm{equ}}$ was the equivalent specific heat of the PCM during phase change.

Two-Dimensional Effective Heat Capacity Model

One has

$$\rho c_{\mathrm{eff}}\frac{\partial T}{\partial \tau} = \lambda\left(\frac{\partial^2 T}{\partial x^2} + \frac{\partial^2 T}{\partial y^2}\right), \tag{3}$$

$$c_{\mathrm{eff}} = \frac{dH/d\tau}{(m\cdot(dT/d\tau))}, \tag{4}$$

where ρ and λ represented the density and the thermal conductivity of the PCM, respectively. c_{eff} was the effective heat capacity of the PCM, which was calculated using (4). The parameters x and y were the thickness and width directions, and T was the temperature of the specimen at a point in time τ ; dh/d τ and m were the heat flow rate across the PCM and the mass of the PCM. The heating or cooling rate of the PCM was given by dh/d τ .

Initial and Boundary Conditions

Because the parameters were tested under the periodical thermal boundary conditions in the experiments, four temperature periods were performed in each experiment to eliminate the impact of the initial condition on the testing results. The initial temperature was shown as follows:

$$T(x, y, \tau)|_{\tau=0} = T_{\text{init}}, \quad (5)$$

where $T(x, y, \tau)$ was the temperature at point (x, y) at time τ . T_{init} was the initial temperature.

The third type thermal boundary condition was applied when x=1 (exterior surface on the thickness direction) and x=N_x (interior surface on the thickness direction):

$$x = 1{:}\ \rho c \frac{\partial T}{\partial \tau} dx\, dy = \lambda \left(\frac{\partial T}{\partial x} \right)\Big|_{P-E} dy + h_a \left(T_{fa} - T_P \right) dy + \lambda \left(\frac{\partial T}{\partial y} \right)\Big|_{P-N} dx - \lambda \left(\frac{\partial T}{\partial y} \right)\Big|_{S-P} dx$$

$$x = N_x{:}\ \rho c \frac{\partial T}{\partial \tau} dx\, dy = -\lambda \left(\frac{\partial T}{\partial x} \right)\Big|_{W'-P'} dy + h_b \left(T_{fb} - T_{P'} \right) dy + \lambda \left(\frac{\partial T}{\partial y} \right)\Big|_{P'-N'} dx - \lambda \left(\frac{\partial T}{\partial y} \right)\Big|_{S'-P'} dx, \quad (6)$$

where N_x was the total number of the meshes on the x direction; point p and p' were the boundary meshes when x=1 and x = N_x; E, N, and S were the points right to, above, and under point P; W', N', and S' were the points left to, above, and under point P'; h_a, T_{fa}, h_b, and T_{fb} were the convective heat transfer coefficients and boundary air temperatures when was equal to 1 and , respectively. Each item of the two equations took the positive value when heat flowed into the point p (or p') and negative value when heat flowed out of the point p (or p').

The boundaries on the direction of y (width direction, when y=1 and y=N_y, N_y was the total number of the meshes on the direction) were adiabatic (the second type thermal boundary condition). The equation is

$$-\lambda \frac{\partial T}{\partial y}\Big|_{y=1} = -\lambda \frac{\partial T}{\partial y}\Big|_{y=N_y} = f(\tau),$$

where f(τ) was the heat flow function in which the independent variable was time τ , f(τ).

Numerical Method

The point iteration method was used for the two-dimensional numerical calculation in this work. The governing equations and boundary conditions were discretized by Finite Difference Method (FDM) [21, 22]. The sizes of the mesh and time steps were 0.005 m and 2 s, which guarantees that the results were independent of the used mesh and the time step sizes. Central differences were applied in space step while a fully implicit finite difference scheme and forward differences were applied in time step. The Gauss-Seidel scheme was employed in iteration. The convergence criterion is shown as in (8). The calculation program was developed in MATLAB environment [23]:

$$\max |T(x, y, \tau) - T(x, y, \tau + 1)| \le 10^{-10}, \quad (8)$$

where T(x, y, τ) and T(x, y, τ +1) were the temperatures of the point (x+y) at the time of τ and τ +1, respectively.

EXPERIMENTS AND MODELS VALIDATION OF COMPOSITE PCM HOLLOW BLOCK WALLS

PCM and Composite PCM Hollow Blocks

Figure 1 is the schematic diagram of the four blocks used in experiments (the shaded part in the figures is PCM), they are the hollow block without PCM (Block 1), the block with a hole row of PCM which is near the left surface (Block 2, the left side of the block wall was disposed as the outdoor side, and the right side was disposed as the indoor side), the block with a hole row of PCM which is in the middle of the block (Block 3), and the block with a hole row of PCM which is near the right surface (Block 4), respectively. With filling PCM into the different rows of blocks, and taking the block without PCM as a reference, the thermal performance of composite PCM hollow block walls was analyzed. The PCM used in the experiments is Capric Acid. The thermal-physical parameters are shown in Table 1 [24, 25].

Table 1: Thermal-physical parameters of PCM (Capric Acid).

Density ρ ($kg{\cdot}m^{-3}$)	Sensible heat c ($J{\cdot}kg^{-1}{\cdot}K^{-1}$)	Latent heat L ($J{\cdot}kg^{-1}$)	Thermal conductivity λ ($W{\cdot}m^{-1}{\cdot}K^{-1}$)	Melting peak temperature range (°C)	Freezing peak temperature range (°C)
886 (liquid) 1004 (solid)	2919	138040 (melting) 150970 (freezing)	0.152	31.4~34.5	24.6~27.1

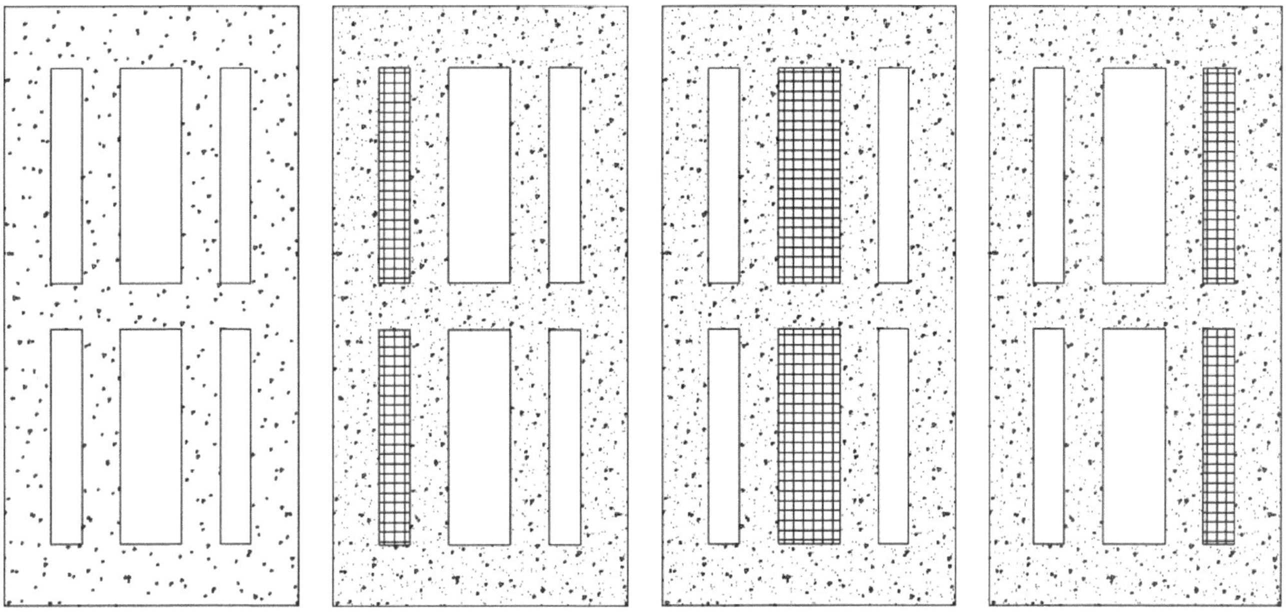

(a) Block 1 (b) Block 2 (c) Block 3 (d) Block 4

Figure 1: Schematic diagram of the PCM's positions in the blocks.

After the sealing and screeding on the undersurfaces of the corresponding holes of blocks with cement slurry sitting method (CSSM), the mixture of the PCM and gypsum was filled into different rows of holes in the blocks, as shown in Figure 2. Because the holes in the middle part of blocks were thicker than the side holes, the amount of PCM in each row of holes should be equal with maintaining the comparability; therefore, the material in each row of holes was the mixture of the PCM and gypsum in different proportions, and the mass of the PCM in each hole was 0.314 kg. In addition, four thermocouples were pasted on the surfaces of the middle holes to test the varied temperatures on the surfaces of the PCM layer in Block 3 wall. In the calculations, the compound specific heat was used to calculate the heat transfer of the composite PCM, which was obtained by weighting the mass ratio of the PCM and gypsum.

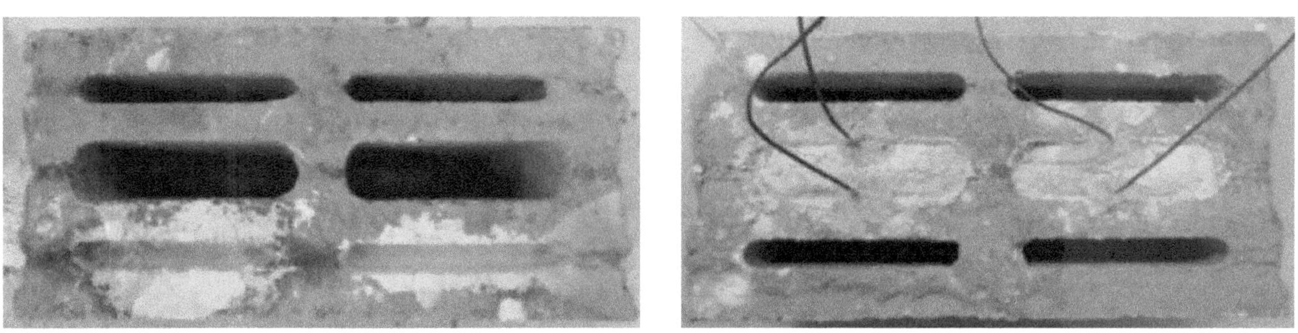

(a) PCM in the side holes (Block 2 and Block 4) (b) PCM in the middle holes (Block 3)

Figure 2: PCM blocks.

Experimental Methodology and Conditions

The temperature-changing hot chamber method [20] was used in this research. The experimental device is in Figure 3. Both the left chamber and the right chamber can stabilize the temperatures inside and change them dynamically in the experiment. In order to maintain the adiabatic condition around the experimental wall, the frames of the specimen bracket and the two chambers were treated to be approximately adiabatic. In the experiments, the temperatures on the two surfaces of wall were tested by the uniformly distributed thermocouples (nine thermocouples for each surface of the wall) and recorded by data logger.

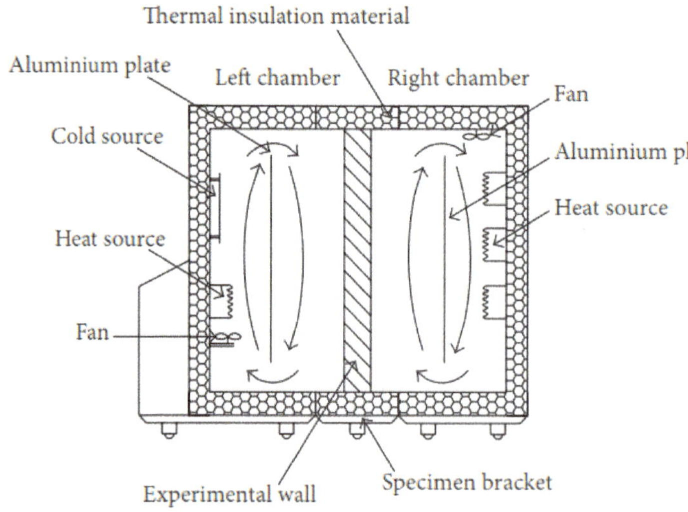

Figure 3: Experimental device.

Each block wall was tested under the common four boundary temperature conditions shown in Table 2; besides, for further exploration of the composite PCM blocks, the walls of Block 2, Block 3, and Block 4 were tested under the additional four boundary temperature conditions, respectively. In the right chamber of the experiments, the adiabatic condition was maintained around the frames but the temperature was not controlled; therefore, the temperature in the right chamber was influenced by the varied temperature of the right surface. Before the experiments, by multiple tests, the convective heat transfer coefficients on the left and right surfaces of wall were nearly stabilized at the values around 20.0 W/m^2·K and 11.0 W/m^2·K.

Table 2: Boundary temperature ranges of each experiment.

Experiment number	Left side air's temperature ranges (°C)
Common boundary temperature ranges	
Temperature range 1	19.9~43.2
Temperature range 2	1.2~67.2
Temperature range 3	3.9~72.6
Temperature range 4	1.0~83.4
Additional boundary temperature ranges for Block 2	
Temperature range 5	27.0~34.5
Temperature range 6	23.5~36.9
Temperature range 7	20.9~40.4
Temperature range 8	17.5~42.7
Additional boundary temperature ranges for Block 3 and Block 4	
Temperature range 5	17.6~42.4
Temperature range 6	11.6~48.0
Temperature range 7	10.7~54.1
Temperature range 8	11.9~60.3

The Uncertainty of the Experimental Data

The main devices used in the experiment are left and right chambers, temperature controller, data logger, thermocouples, and cooling and heating units. The temperature amplitude inside the left and right chambers is equal to or lower than 0.5°C, and the heat loss of the chambers is lower than 5%. Because the tested data are transient varied temperatures, the devices that influence the accuracy of the experimental data are the thermocouples. These thermocouples are all T-type thermocouples. The range of measurement is −50°C to 100°C with the maximum uncertainty of 0.5°C. Therefore, the maximum uncertainty of the experimental data is 0.5°C.

Evaluation Indexes

To analyze the calculation accuracy, the standard errors and maximum absolute errors between the calculation curves and experimental curves were calculated using (9). The standard error reflects the average extent of the calculation curves deviating from the experimental curves, while the maximum absolute error reports the maximum deviation between the two curves:

$$\sigma = \sqrt{\frac{\sum_{i=1}^{n} \left(N_i - N_{\mathrm{sta},i}\right)^2}{n}},$$

$$\sigma' = \max\left(\left|N_1 - N_{\mathrm{sta},1}\right|, \left|N_2 - N_{\mathrm{sta},2}\right| \cdots \left|N_n - N_{\mathrm{sta},n}\right|\right), \tag{9}$$

where σ and σ' are standard error and maximum absolute error, respectively, n is the amount of data, i is the sequence number of data, N_i is the ith datum, and $N_{\mathrm{sta},i}$ is the ith standard datum.

Models Validation and Experimental Analysis

Figures 4–7 are the curves of the surface temperatures (including the experimental data and calculation results), decrement factors, time lags, standard errors, and maximum absolute errors of the blocks in Figure 1under the boundary temperature conditions in Table 2. The values of the temperature decrements, decrement factors, time lags, standard errors, and maximum absolute errors of each block wall are shown in Table 3.

Table 3: Temperature decrements, decrement factors, time lags, standard errors, and maximum absolute errors of each block wall.

Indexes	Temperature decrements (°C)	Decrement factors	Time lags (h)	Standard errors (°C)		Maximum absolute errors (°C)	
				Left surface Eff/Ent	Right surface Eff/Ent	Left surface Eff/Ent	Right surface Eff/Ent
Block walls and temperature ranges				Experimental values			
Block 1 wall							
Temperature range 1	17.5	4.02	6.17	0.50	0.28	0.90	0.54
Temperature range 2	49.9	4.10	5.42	0.62	0.69	1.29	1.23
Temperature range 3	51.6	4.07	5.58	0.60	0.72	1.14	1.38
Temperature range 4	61.0	3.85	5.17	1.04	0.77	1.79	1.25
Block 2 wall							
Temperature range 1	19.0	5.30	7.25	0.62/0.76	0.24/0.27	0.83/1.40	0.46/0.48
Temperature range 2	51.2	5.43	6.17	0.36/0.33	0.67/0.47	0.72/2.64	1.25/0.99
Temperature range 3	51.5	5.17	6.08	0.63/1.41	0.60/0.46	0.98/2.91	1.02/0.87
Temperature range 4	65.4	5.18	5.92	0.52/1.69	0.70/0.63	1.12/3.52	1.34/1.46
Temperature range 5	6.0	5.47	7.33	0.19/0.13	0.12/0.16	0.39/0.32	0.20/0.26
Temperature range 6	10.9	5.65	6.92	0.24/0.39	0.19/0.28	0.54/0.76	0.37/0.48
Temperature range 7	15.9	5.62	6.50	0.44/0.30	0.24/0.26	0.78/0.55	0.45/0.49
Temperature range 8	20.7	5.72	7.25	0.56/0.68	0.31/0.37	1.18/1.07	0.55/0.62
Block 3 wall							
Temperature range 1	19.7	6.82	7.67	0.09/0.13	0.24/0.30	0.19/0.28	0.45/0.53
Temperature range 2	53.0	6.47	7.17	0.70/0.98	0.77/0.44	1.42/1.80	1.10/0.99
Temperature range 3	56.2	6.43	6.83	0.72/0.85	0.70/0.50	1.47/1.63	1.24/1.05
Temperature range 4	67.2	5.63	5.08	0.71/0.99	0.94/1.43	1.60/2.03	2.18/3.15
Temperature range 5	20.5	6.61	6.67	0.40/0.44	0.20/0.25	0.66/0.71	0.40/0.45
Temperature range 6	30.4	6.24	5.92	0.80/0.97	0.34/0.47	1.18/1.46	0.68/0.86
Temperature range 7	40.3	6.14	5.5	0.78/1.00	0.47/0.66	1.24/1.71	1.01/1.21
Temperature range 8	49.3	5.96	5.75	0.73/1.01	0.62/0.93	1.20/1.78	1.37/1.55
PCM in the holes of Block 3 wall							
Temperature range 1	—	—	—	1.73/2.18	1.29/1.88	3.30/3.13	2.34/2.66
Temperature range 2	—	—	—	2.20/2.91	1.58/2.25	4.82/4.18	3.43/4.52
Temperature range 3	—	—	—	2.22/2.25	1.74/1.50	4.96/3.61	3.76/2.92
Temperature range 4	—	—	—	2.57/2.89	2.23/2.27	5.30/4.25	4.18/5.47
Temperature range 5	—	—	—	1.44/1.94	0.99/1.50	2.81/2.72	1.96/1.78
Temperature range 6	—	—	—	1.73/2.10	1.04/1.57	3.44/3.44	2.25/2.10
Temperature range 7	—	—	—	1.66/2.52	1.17/1.71	3.47/4.03	2.39/2.83
Temperature range 8	—	—	—	1.73/2.74	1.34/1.99	3.74/4.50	2.58/3.75
Block 4 wall							
Temperature range 1	19.3	5.71	6.33	0.49/0.59	0.25/0.32	1.01/1.13	0.45/0.72
Temperature range 2	49.8	5.45	6.33	1.32/1.38	0.54/0.35	2.42/2.64	1.13/0.61
Temperature range 3	56.1	5.96	6.33	1.31/1.26	0.65/0.26	2.50/2.50	1.24/0.71
Temperature range 4	62.7	4.87	5.92	1.37/1.43	0.50/0.45	2.75/3.04	1.05/0.98
Temperature range 5	19.9	5.45	5.83	0.52/0.49	0.22/0.40	0.94/0.83	0.34/0.69
Temperature range 6	29.5	5.41	5.67	0.54/0.59	0.38/0.58	1.01/1.03	0.64/1.01
Temperature range 7	34.8	5.10	5.67	0.62/0.72	0.44/0.92	1.17/1.25	0.74/1.74
Temperature range 8	38.8	5.12	6.08	0.70/0.78	0.36/0.62	1.34/1.43	0.69/1.32

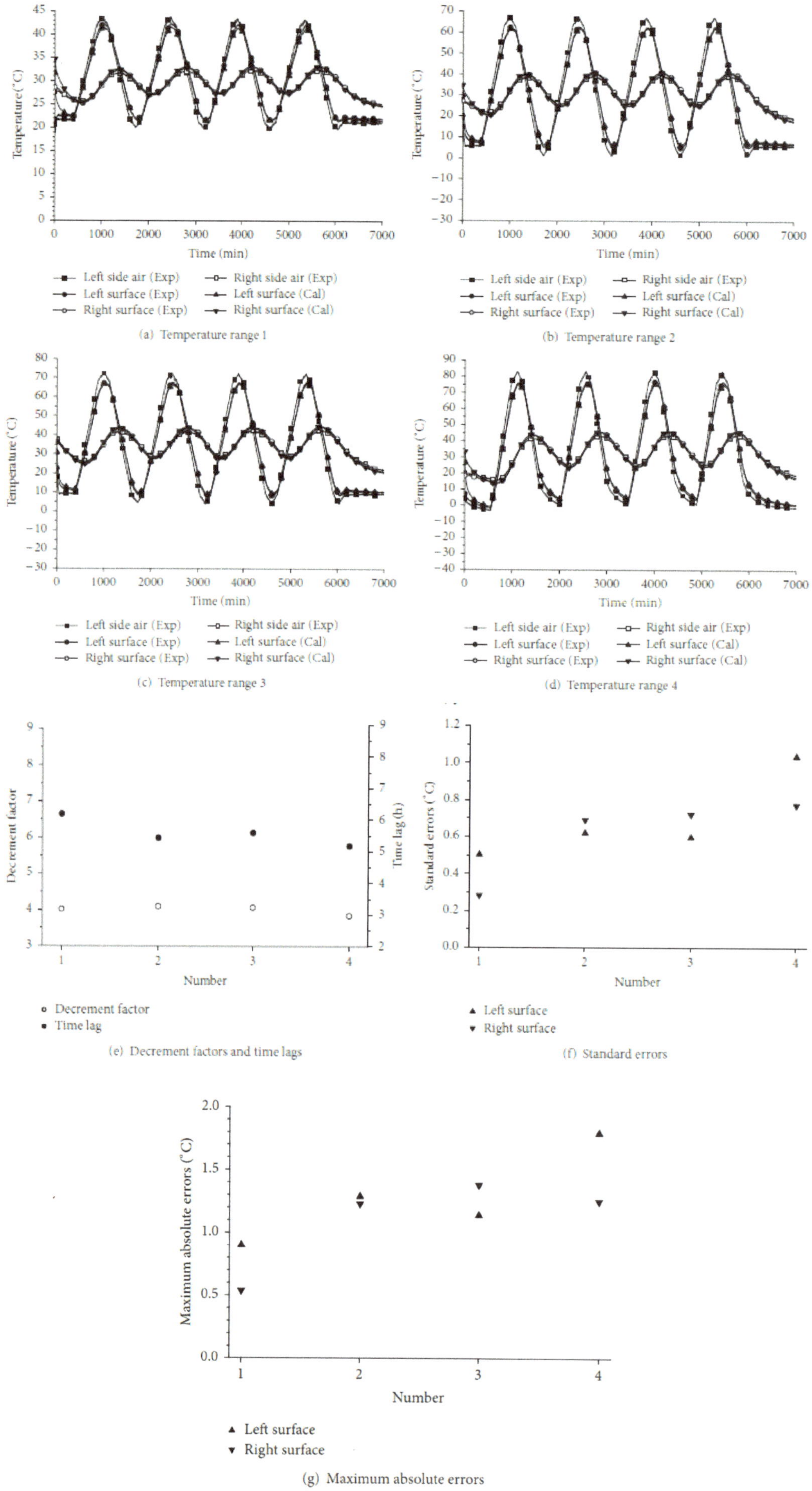

Figure 4: Experimental data of Block 1 wall and calculation errors.

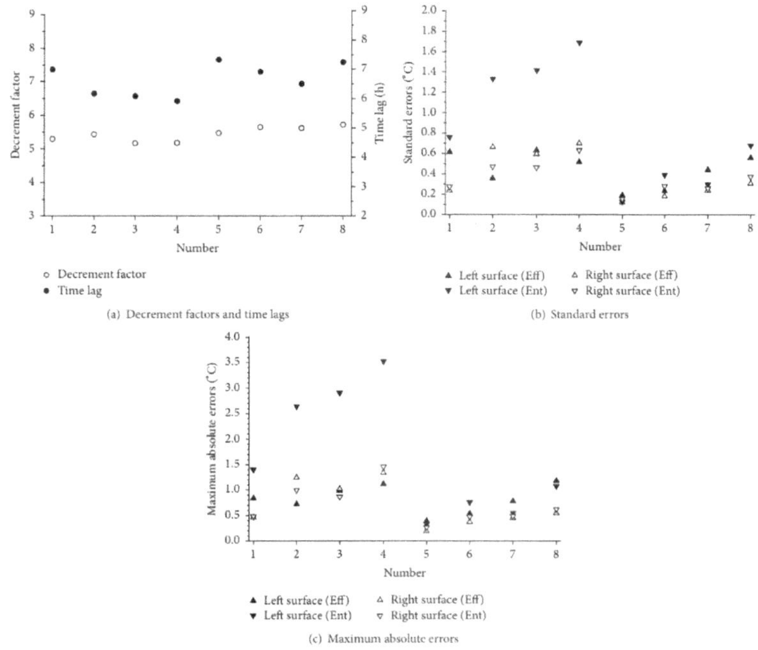

Figure 5: Experimental data of Block 2 wall and calculation errors.

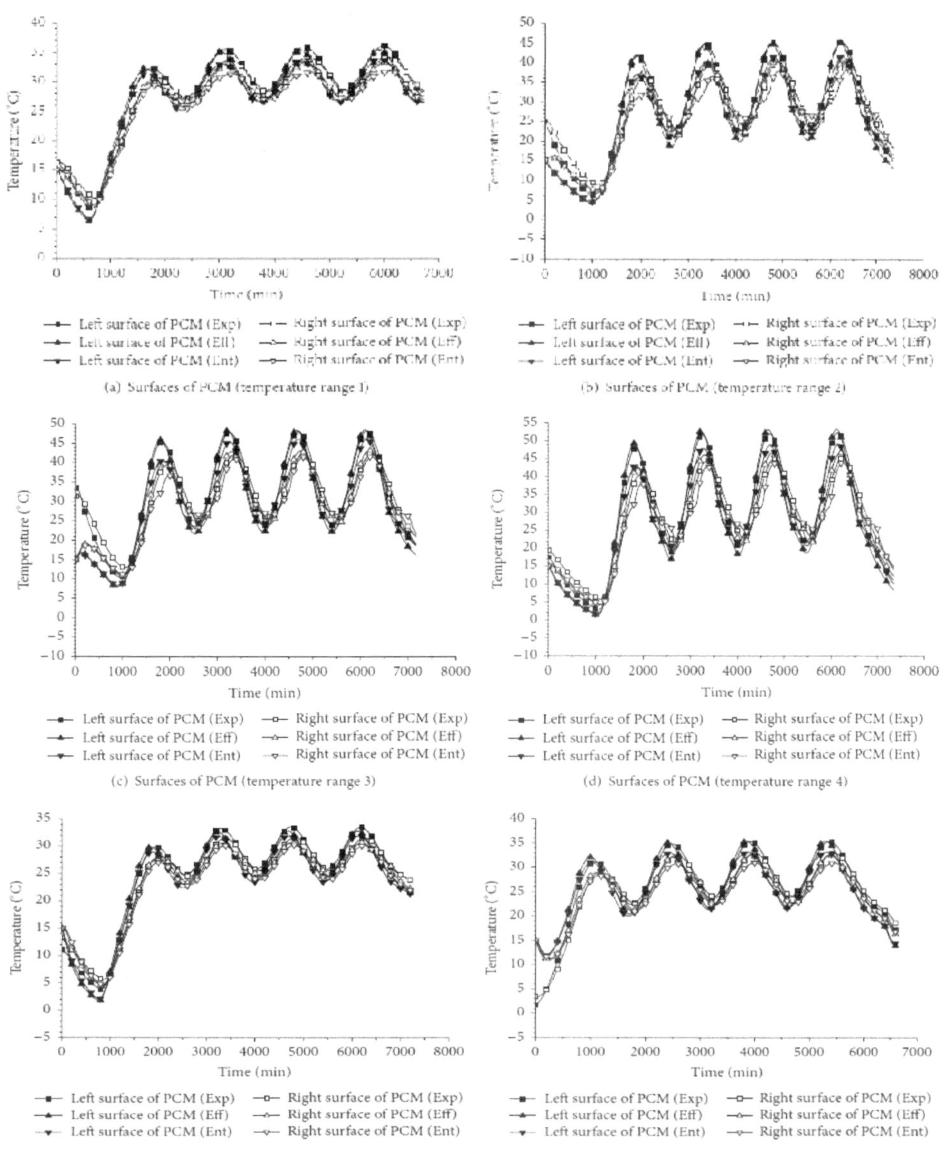

Figure 6: Continued.

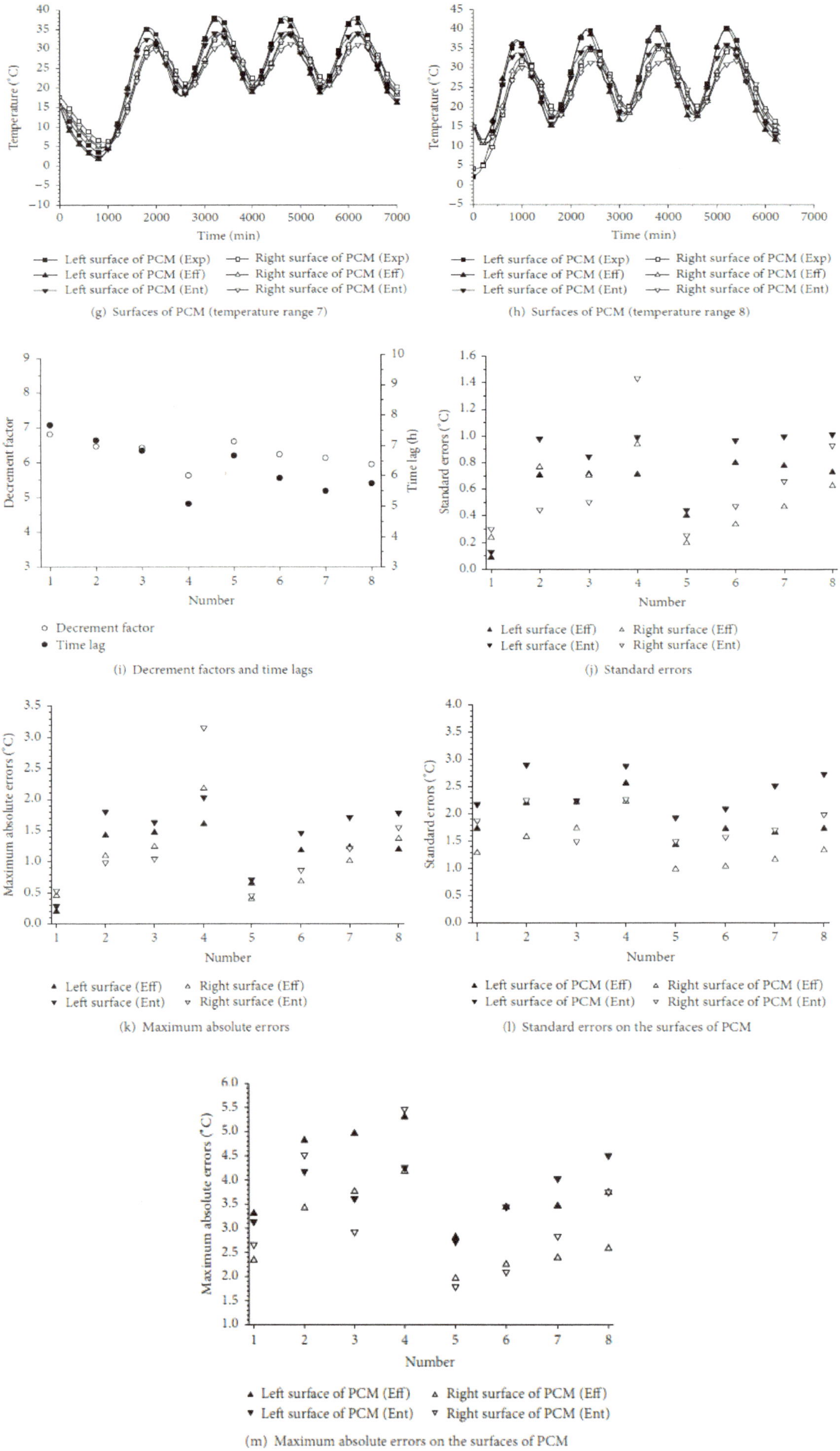

Figure 6: Experimental data of Block 3 wall and calculation errors.

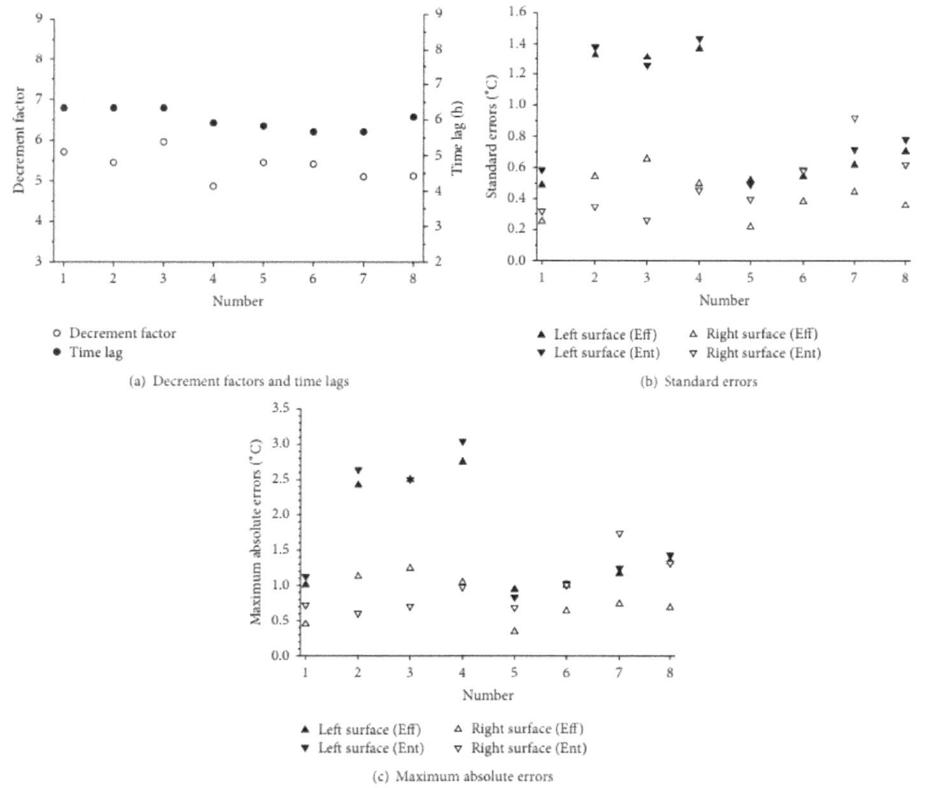

Figure 7: Experimental data of Block 4 wall and calculation errors.

Models Validation

It can be seen that, as a whole, the standard errors and maximum absolute errors of the calculation results increase slowly with the temperature range's width on the left surface of wall. The standard errors of the effective heat capacity model (Eff) and enthalpy model (Ent) are lower than 1.4°C and 1.7°C, and the maximum absolute errors are lower than 2.8°C and 3.6°C, respectively. The maximum errors always exist for the calculation results of the left surface. The reason for this is that the decrement effect of wall makes the temperature fluctuation on the right surface be significantly lower than that on the left surface; as a result, the calculation errors on the right surface reduce, while the temperature on the left surface is slightly influenced by the wall, which makes the errors be relatively larger. Figures 6(l) and 6(m) are the calculation errors of the temperature on the two surfaces of PCM layer in Block 3 wall. It can be seen from the data in Table 3 that the standard errors of the two models are both lower than 2.9°C and mostly in the range from 1.0°C to 2.5°C; the standard errors of the effective heat capacity model (Eff) are lower than that of the enthalpy model (Ent). The errors are produced by these factors: (1) in the experiment, it is impossible to reach the absolute adiabatic condition around the walls, but in the mathematical models, it is the idealized adiabatic condition around the walls; (2) the error inevitably exists in the obtained experimental data; (3) to control the calculation amount, the size of the computational mesh is set as 0.005 m × 0.005 m, so the calculation result is the average value of the area of each mesh; and (4) not all the assumptions in the models and the conditions in the experiment are unified. In spite of these, there is reasonable agreement between the curves of the calculation results and experiment data (the error is lower than 5%); therefore, it is considered that the results of the two models can both meet the requirement for the analysis, and the effective heat capacity model is more accurate than the enthalpy model.

Experimental Analysis

Seen from Figures 4–7 that, for the four experiments under the common boundary temperature ranges (from temperature range 1 to temperature range 4) of each wall, the temperature decrements, decrement factors, and time lags of the block walls with PCM (Block walls 2, 3, and 4) are all significantly larger than that of the block wall without PCM (Block wall 1). The increased amounts of the temperature decrements of Block walls 2, 3, and 4 are from 1.5°C to 5.0°C, and the increased decrement factors are from 1.0 to 2.5 and those for the time lags are from 0.5 h to 3.0 h, compared to Block wall 1. This demonstrates that the phase transition energy storage effect has worked by the use of PCM, which plays an important role on the development of the unsteady-state thermal performance of wall.

Also, the decrement factors and time lags of each wall vary with a slowly declining trend on the whole with increasing width of boundary temperature ranges. The main reasons for this phenomenon can be explained as (1) for the air layers in the blocks, the width addition of the boundary temperature ranges increases the temperature difference between the colder surface and the hotter surface to some extent, which enhances the heat convection and radiation in the holes and makes the thermal performance of wall reduce, and consequently produces the fall of the decrement factors and time lags; (2) for the blocks with PCM, when the temperature of the PCM in the wall has not exceeded the phase transition temperature range (the phase transition of PCM has not totally finished), the PCM wall can reach high unsteady-state thermal performance (Figures 6(a) and 6(e)), but once the temperature of the PCM has exceeded it, as shown in Table 1 (it is apparent in Figures 6(b), 6(c), 6(d), 6(f), 6(g), and 6(h)), the phase transition process has been totally finished: namely, the latent heat of PCM has been used up, and only the sensible heat of the liquid or solid PCM can store the heat; because the sensible heat of PCM is at least one order of magnitude lower than the latent heat, the thermal performance of wall greatly reduce after the phase transition has finished, and the function of PCM gradually reduces with the increased width of the boundary temperature range, which makes the decrement factors and time lags of wall decrease. Under the common boundary temperature ranges (from temperature range 1 to temperature range 4), the temperature decrements, decrement factors, and time lags between Block wall 2 and Block wall 4 have little difference.

For the additional boundary temperature ranges (from temperature range 5 to temperature range 8) of Block walls 2, 3, and 4, the decrement factors and time lags of Block 2 wall are slightly larger than that of Block 4 wall, but the temperature decrements of Block 4 wall are greatly larger than that of Block 2 wall. This phenomenon results from the large difference of the boundary temperature ranges between the two walls: the boundary temperature ranges of Block 4 wall are larger; therefore, in spite of its smaller decrement factors and time lags compared to Block 2 wall, the temperature decrements are large; it should not be simply considered that the thermal performance of Block 4 wall is lower than Block 2 wall. Hence, the decrement factor and the temperature decrement should be both considered as the indexes for evaluating the unsteady-state thermal performance of PCM wall.

For the data of Block 3 wall, under the common boundary temperature ranges (from temperature range 1 to temperature range 4), its temperature decrements, decrement factors, and time lags are higher than Block 2 wall and Block 4 wall with the values of 0°C–5°C, 0.5–1.5, and 0 h–2.5 h, respectively; under the additional boundary temperature ranges (from temperature range 5 to temperature range 8), its unsteady-state thermal performance also apparently improves. It is worthy to point out that, although the mass of the PCM filled into each block is not changed, the thermal performance of Block 3 wall is higher than Block 4 wall under the same boundary condition. The reasons are complicated: one of the important reasons is that the addition of a certain amount of gypsum takes the place of a part of air, which increases the heat capacity and consequently enhances the unsteady-state thermal performance of the Block 3 wall.

It can be concluded from the experimental data that suitable environmental condition for the application of the PCM wall with a certain structure exists; for different boundary temperature conditions, the different optimum structure of PCM wall should be selected; the unsteady-state thermal performance of PCM wall under different environmental conditions is related to the factors such as phase transition temperature, latent heat, PCM's thickness, and position in the wall.

CONCLUSIONS

In this paper, a type of concrete hollow blocks with typical structure and a common used phase change material (PCM) were adopted, and the PCM was filled into different positions of the hollow blocks, and then different types of composite PCM hollow blocks were produced. The temperature-changing hot chamber method was used to test the thermal performance of the hollow block walls with and without PCM under different temperature conditions; meanwhile, the enthalpy model and effective heat capacity model were used to calculate the phase transition heat transfer process under the conditions of experiments. The results of the models were validated by the experimental data, the thermal performances of different hollow block walls were compared, and the reasons for the difference on the thermal performance of each wall were analyzed, which provide the data reference and guide for the engineering application of the composite PCM hollow block wall. After the analysis of this research, it can be concluded that(1)for the dynamic heat transfer processes of the composite PCM hollow block walls with multiple structures, the calculation results of the adopted two-dimensional enthalpy model and effective heat capacity model have reasonable agreement with the experimental data (the errors are lower than 5%); therefore, it is considered that the two mathematical models can both satisfy the analysis requirement on the problem of the unsteady-state phase transition heat transfer, and the effective heat capacity model is more accurate than the enthalpy model;(2)when the phase transition temperature of the PCM in the wall is within the temperature range of the environment, the unsteady-state thermal performance of the hollow block wall with PCM is higher than that of the same wall without PCM (in the research of this paper, the increased amounts of the temperature decrements of the PCM wall are from 1.5°C to 5.0°C, and the increased decrement factors are from 1.0 to 2.5 and those for the time lags are from 0.5 h to 3.0 h);(3)when the temperature of the PCM in the wall has not exceeded the phase transition temperature range, the PCM wall can reach high unsteady-state thermal performance; otherwise, the function of PCM will significantly reduce, which makes the unsteady-state thermal performance of wall reduce;(4) the temperature decrement, decrement factor, and time lag should be considered together as the indexes for evaluating the unsteady-state thermal performance of PCM wall; otherwise, the validity of the evaluation will probably be affected;(5) optimum boundary condition exists for the PCM wall with a certain structure; for different boundary conditions, the PCM wall structure should be properly selected; the unsteady-state thermal performance of the PCM wall is related to the factors such as phase transition temperature, latent heat, thickness, and the position in the wall.

Highlights

(i)The two mathematical models for the phase transition calculation were validated.(ii)The thermal performance of composite PCM hollow block walls was analyzed.(iii)The impact factors on the thermal performance of PCM hollow block wall were analyzed.(iv)The thermal performances of the wall with and without PCM were compared.

Nomenclature

c: Specific heat, $kJ·kg^{-1}·K^{-1}$
c_{eff}: Effective heat capacity of PCM, $kJ·kg^{-1}·K^{-1}$
$c_{p,equ}$: Equivalent specific heat of PCM in the phase transition temperature range, $kJ·kg^{-1}·K^{-1}$
$c_{p,s}$: Specific heat of the PCM in the solid state, $kJ·kg^{-1}·K^{-1}$
$c_{p,l}$: Specific heat of the PCM in the liquid state, $kJ·kg^{-1}·K^{-1}$
$f(\tau)$: Heat flow function as a function of τ, $W·m^{-2}$
H: Enthalpy of PCM ($kJ·kg^{-1}$)
h_a: Convective heat transfer coefficient when $x = 1$, $W·m^{-2}·K^{-1}$
h_b: Convective heat transfer coefficient when $x = N_x$, $W·m^{-2}·K^{-1}$
i: Sequence number of data
L: Latent heat of PCM, $kJ·kg^{-1}$
m: Mass of the PCM, kg
n: Amount of data
N_i: ith datum
$N_{sta,i}$: ith standard datum
N_x: Total number of the meshes on the x direction
N_y: Total number of the meshes on the y direction
T: Temperature, °C
T_{fa}: Boundary air temperature when $x = 1$, °C
T_{fb}: Boundary air temperature when $x = N_x$, °C
T_{init}: Initial temperature, °C
T_m: Mean temperature of the phase transition temperature range, °C

$T(x, y, \tau)$: Temperature of the point (x, y) at time point τ, °C
$T(x, y, \tau + 1)$: Temperature of the point (x, y) at the time of $\tau + 1$, °C
x: Direction of thickness, m
y: Direction of width, m.

Greek Symbols

ρ: Density, $kg·m^{-3}$
τ: Time, s
λ: Thermal conductivity, $W·m^{-1}·K^{-1}$
ε: Half of the phase transition temperature range, °C
σ: Standard error, °C
σ': Maximum absolute error, °C

Subscripts

E: Point right to the point p

init: Initial
N: Point above the point p
N': Point above the point p'
p: Boundary meshes when $x = 1$
p': Boundary meshes when $x = N_x$
S: Point under the point p
S': Point under the point p'
sta: Standard
W': Point left to the point p'.

Conflict of Interests

The authors declare that there is no conflict of interests regarding the publication of this paper.

ACKNOWLEDGMENTS

This work was financially supported by the Natural Science Foundation of China (no. 51508232), the Natural Science Foundation of Jiangsu Province (no. BK20150521), the Jiangsu Province Postdoctoral Science Foundation (no. 1501106B), and the Scientific Research Starting Foundation for Advanced Talents of Jiangsu University (no. 15JDG049).

REFERENCES

1. L. Aelenei, R. Pereira, H. Gonçalves, and A. Athienitis, "Thermal performance of a hybrid BIPV-PCM: modeling, design and experimental investigation," Energy Procedia, vol. 48, pp. 474–483, 2014.
2. M. Ahmed, O. Meade, and M. A. Medina, "Reducing heat transfer across the insulated walls of refrigerated truck trailers by the application of phase change materials," Energy Conversion and Management, vol. 51, no. 3, pp. 383–392, 2010.
3. H. J. Alqallaf and E. M. Alawadhi, "Concrete roof with cylindrical holes containing PCM to reduce the heat gain," Energy and Buildings, vol. 61, pp. 73–80, 2013.
4. A. Bontemps, M. Ahmad, K. Johanns, and H. Sallée, "Experimental and modelling study of twin cells with latent heat storage walls," Energy and Buildings, vol. 43, no. 9, pp. 2456–2461, 2011.
5. A. M. Borreguero, M. Luz Sánchez, J. L. Valverde, M. Carmona, and J. F. Rodríguez, "Thermal testing and numerical simulation of gypsum wallboards incorporated with different PCMs content," Applied Energy, vol. 88, no. 3, pp. 930–937, 2011.
6. L. F. Cabeza, C. Castellón, M. Nogués, M. Medrano, R. Leppers, and O. Zubillaga, "Use of microencapsulated PCM in concrete walls for energy savings," Energy and Buildings, vol. 39, no. 2, pp. 113–119, 2007.
7. A. Castell, I. Martorell, M. Medrano, G. Pérez, and L. F. Cabeza, "Experimental study of using PCM in brick constructive solutions for passive cooling," Energy and Buildings, vol. 42, no. 4, pp. 534–540, 2010.
8. H. Ye, H. He, X. Ge, and B. Xu, "Comparative numerical investigations on the melting process of form-stable phase change material using enthalpy formulation method and effective heat capacity formulation method," Acta Energiae Solaris Sinica, vol. 25, no. 4, pp. 488–491, 2004.
9. COMSOL Multiphysics, COMSOL Multiphysics Engineering Simulation Software, COMSOL Multiphysics, Stockholm, Sweden, 2012.
10. K. D. Antoniadis, M. J. Assael, C. A. Tsiglifisi, and S. K. Mylona, "Improving the design of greek hollow clay bricks," International Journal of Thermophysics, vol. 33, no. 12, pp. 2274–2290, 2012.
11. K. Arendt, M. Krzaczek, and J. Florczuk, "Numerical analysis by FEM and analytical study of the dynamic thermal behavior of hollow bricks with different cavity

concentration," International Journal of Thermal Sciences, vol. 50, no. 8, pp. 1543–1553, 2011.

12. L. P. Li, Z. G. Wu, Y. L. He, G. Lauriat, and W. Q. Tao, "Optimization of the configuration of 290 × 140 × 90 hollow clay bricks with 3-D numerical simulation by finite volume method," Energy and Buildings, vol. 40, no. 10, pp. 1790–1798, 2008.
13. L. P. Li, Z. G. Wu, Z. Y. Li, Y. L. He, and W. Q. Tao, "Numerical thermal optimization of the configuration of multi-holed clay bricks used for constructing building walls by the finite volume method,"International Journal of Heat and Mass Transfer, vol. 51, no. 13-14, pp. 3669–3682, 2008.
14. M. S. El-Genk and C. Gao, "Transient heat conduction during quenching of downward facing copper and stainless steel convex surfaces," Numerical Heat Transfer Part A: Applications, vol. 29, no. 6, pp. 543–573, 1996.
15. C. Gao, "Accuracy of the general Peaceman and Rachford alternating direction implicit method,"International Communications in Heat and Mass Transfer, vol. 23, no. 8, pp. 1193–1199, 1996.
16. C. Gao and Y. Wang, "A general formulation of peaceman and rachford ADI method for the N-dimensional heat diffusion equation," International Communications in Heat and Mass Transfer, vol. 23, no. 6, pp. 845–854, 1996.
17. J. Mackerle, "Finite elements and boundary elements applied in phase change, solidification and melting problems. A bibliography (1996–1998)," Finite Elements in Analysis and Design, vol. 32, no. 3, pp. 203–211, 1999.
18. A. Carbonari, M. De Grassi, C. Di Perna, and P. Principi, "Numerical and experimental analyses of PCM containing sandwich panels for prefabricated walls," Energy and Buildings, vol. 38, no. 5, pp. 472–483, 2006.
19. Ministry of Housing and Urban-Rural Development of the People's Republic of China, Thermal Design Code for Civil Building (GB50176-93), China Planning Press, Beijing, China, 1993.
20. Y. Zhang, K. Du, J. He, L. Yang, Y. Li, and S. Li, "Impact factors analysis on the thermal performance of hollow block wall," Energy and Buildings, vol. 75, pp. 330–341, 2014.
21. S. V. Patankar, Numerical Heat Transfer and Fluid Flow, McGraw-Hill, New York, NY, USA, 1980.
22. W. Q. Tao, Numerical Heat Transfer, Xi'an Jiaotong University Press, Xi'an, China, 2nd edition, 2001.
23. B. Z. Liu, Y. H. Su, and H. L. Zhang, MATLAB 7.0 from Entry to Master, Posts & Telecom Press, Beijing, China, 2010.
24. Y. Zhang, K. Du, J. P. He, L. Yang, and Y. J. Li, "Impact factors analysis of the enthalpy method and the effective heat capacity method on the transient nonlinear heat transfer in phase change materials (PCMs)," Numerical Heat Transfer Part A: Applications, vol. 65, no. 1, pp. 66–83, 2014.
25. Y. Zhang, K. Du, M. A. Medina, and J. He, "An experimental method for validating transient heat transfermathematical models used for phase change materials(PCMs) calculations," Phase Transitions, vol. 87, no. 6, pp. 541–558, 2014.

CHAPTER 14 HEAT TRANSFER CHARACTERISTICS OF CALCINED PETROLEUM COKE IN WASTE HEAT RECOVERY PROCESS

Bin Zheng[1,2], Yongqi Liu[2], Lichen Zou[2], and Ruiyang Li[1]

[1]School of Energy and Power Engineering, University of Shanghai for Science and Technology, Shanghai 200093, China

[2]School of Transportation and Vehicle Engineering, Shandong University of Technology, Zibo 255000, China

ABSTRACT

This paper reports the results of heat transfer characteristics of calcined petroleum coke in waste heat recovery process. The model of heat exchanger was set up. The model has been used to investigate the effects of porosity (0.58 to 0.79), equivalent heat conductivity coefficient (0.9 to 1.1), and equivalent specific heat (0.9 to 1.1). The calculated values of calcined petroleum coke temperature showed good agreement with the corresponding available experimental data. The temperature distribution of calcined petroleum coke, the calcined petroleum coke temperature at heat exchanger outlet, the average heat transfer coefficient, and the heat recovery efficiency were studied. It can also be used in deriving much needed data for heat exchanger designs when employed in industry.

INTRODUCTION

With the rapid development of economy in China, energy consumption has been increasing on a vast scale in recent years. The equivalent coal consumption of China was 3.75 billion tons in 2013 [1]. To solve this problem, it is widely believed that energy recycle is an effective measure to decrease energy consumption. In China's industrial production, waste heat resource is rich and various, such as waste heat from high temperature gas, coke slag, and other solid particles [2]. The researches on waste heat utilization of high temperature gas and liquid are relatively wide, and technology of waste heat recovery is relatively mature. However, the researches and technology of waste heat utilization of high temperature solid are not inadequate.

Calcined petroleum coke is one of the important basic raw materials. It is widely used in producing anode of aluminum electrolytic, graphite electrode, recarburizer, industrial silicon, and other carbon products. The production capacity of calcined petroleum coke in China is the largest in the world, over 70% of which is produced in tank calcined furnaces. When calcined petroleum coke is discharged from tank calcined furnaces, the temperature is 1000°C. The heat of high temperature calcined petroleum coke is about 33.5% of the total heat of calcination process [3]. The waste heat of it has high possibility to be utilized.

Citation: Bin Zheng, Yongqi Liu, Lichen Zou, and Ruiyang Li, "Heat Transfer Characteristics of Calcined Petroleum Coke in Waste Heat Recovery Process," Mathematical Problems in Engineering, vol. 2016, Article ID 2649383, 9 pages, 2016. doi:10.1155/2016/2649383.

Waste heat reutilization of the solid particle has been studied in some field. Cai and Ding [4, 5] summarized the current energy consumption condition of steel industry and distribution of waste heat resource of steel production process. Wang et al. [6] analyzed the thermal equilibrium of molten slag heat recovery system and discussed the relationship between waste heat utilization efficiency and the amount of cooling air. The cooling process of steel slag particles in the moving bed was analyzed. Chen [7] simulated the waste heat recovery system of high temperature slag with fluent software, analyzed the flow field and temperature field of steel slag, and researched the change of temperature of air at outlet and heat recovery efficiency when the wind speed and air temperature at inlet are varied. Liu [8] proposed a new technology of fluidized bed. The waste heat of slag can be efficiently used with this technology. The flow and heat transfer process were simulated in fluidized bed with CFD software. Herz et al. [9] researched the heat transfer coefficient between moving material layer and the wall of bed through experiment. The effects of operational parameters, moving speed of material layer, and depth of buried material on heat transfer coefficient were studied. Atmakidis and Kenig [10] simulated the movement of irregularly arranged particle near the wall with CFD. Sundaresan and Kolar [11] researched the axial heat transfer characteristics in circulating fluidized bed and analyzed surface heat transfer coefficient of several vertical pipes. The results showed that the length and location of the tube affected the heat transfer coefficient. Guo et al. [12] simulated the heat transfer in gas-solid fluidized bed with Euler-Euler method and obtained the effects of Reynolds number and porosity on heat transfer.

The investigations on the reutilization of calcined petroleum coke waste heat are lacking. In this paper, heat transfer characteristics of calcined petroleum coke in waste heat recovery process were studied.

PHYSICAL AND MATHEMATICAL DESCRIPTION OF THE PROBLEM

The waste heat of calcined petroleum coke is recovered by a heat exchanger. The heat exchanger of calcined petroleum coke includes an internal heat exchanger and an external heat exchanger. Its structure is shown in Figure 1. The external heat exchanger consists of tubes and a diaphragm wall, and the ends of tubes are held together by two ring pipes. Similarly the internal heat exchanger also consists of tubes, with the ends of tubes held together by a straight pipe. The calcined petroleum coke flows from top to bottom in the heat exchanger surrounded by tubes and diaphragm walls, and the water flows from bottom to top in the tubes. The temperature of calcined petroleum coke is high; the water temperature is low; so the heat is transferred to water from calcined petroleum coke.

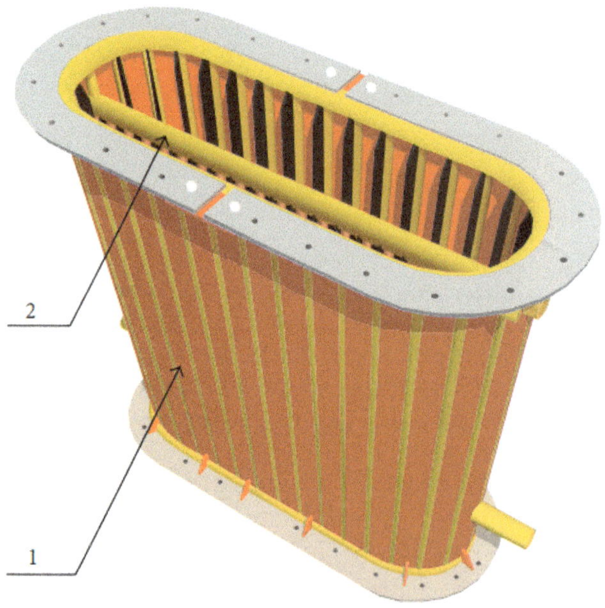

Figure 1: The structure of heat exchanger ((1) the external heat exchanger and (2) the internal heat exchanger).

In this paper, the calculated model of calcined petroleum coke heat exchanger is equal to the experimental heat exchanger model in scale. Considering the structure characteristic of the heat exchanger, the calculation model is appropriately simplified.(1)The lower ring pipe and the upper ring pipe of the internal heat exchanger and the external heat exchanger will not be calculated.(2)The calculated model includes two external heat exchanger tubes, one internal heat exchanger tube, and the region between the tubes and thermal barriers.(3)The tube pitch is equal from top to bottom. Additionally the symmetry plane of calculation model is the plane where the short axis of the oval heat exchanger exists.(4)The calcined petroleum coke particles of high temperature and the gas among them are regarded as homogenous continuum. The physical parameters of calculation model are calculated as the porosity of calcined petroleum coke.(5)Considering the thermal insulation layer outside the heat exchanger, it is assumed that there is no heat loss from the heat exchanger lateral.

According to the simplification above, the model has been set up by Gambit. The lengths of x and y axis are 78 mm and 262 mm. The height of the model is 1550 mm. It is shown as Figure 2.

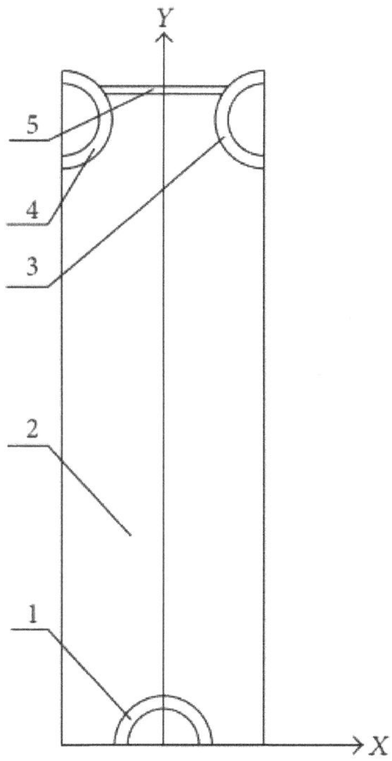

Figure 2: Simplified calculation model ((1) the internal heat exchange tube, (2) the calcined petroleum coke, (3) the 1st external heat exchange tube, (4) the 2nd external heat exchange tube, and (5) the diaphragm wall).

When heat exchanges between the high temperature calcined petroleum coke and water, the mass, momentum, and energy conservation exist. In this paper, the unsteady model is used for calculation. The mass, momentum, and energy conservation equations are shown as follows.

The mass conservation equation is as follows:

$$\frac{\partial \rho}{\partial t} + \frac{\partial (\rho u)}{\partial x} + \frac{\partial (\rho v)}{\partial y} + \frac{\partial (\rho w)}{\partial z} = 0, \quad (1)$$

where ρ is density, kg/m^3, t is time, s, and u, v, and w are the velocity vectors in x, y, and z coordinate directions, m/s.

The momentum conservation equation is as follows:

$$\frac{\partial (\rho u)}{\partial t} + \mathrm{div}(\rho u V) = -\frac{\partial p}{\partial x} + \frac{\partial \tau_{xx}}{\partial x} + \frac{\partial \tau_{yx}}{\partial y} + \frac{\partial \tau_{zx}}{\partial z} + F_x,$$

$$\frac{\partial (\rho v)}{\partial t} + \mathrm{div}(\rho v V) = -\frac{\partial p}{\partial y} + \frac{\partial \tau_{xy}}{\partial x} + \frac{\partial \tau_{yy}}{\partial y} + \frac{\partial \tau_{zy}}{\partial z} + F_y,$$

$$\frac{\partial (\rho w)}{\partial t} + \mathrm{div}(\rho w V) = -\frac{\partial p}{\partial z} + \frac{\partial \tau_{xz}}{\partial x} + \frac{\partial \tau_{yz}}{\partial y} + \frac{\partial \tau_{zz}}{\partial z} + F_z, \quad (2)$$

where p is pressure on the infinitesimal body, Pa, τ_{xx}, τ_{xy} and τ_{xz} are stick stress component of τ on the infinitesimal body caused by molecular viscosity, N/m^2, F_x, F_y, and F_z are body forces, N/m^3, and V is velocity vector quantity, m/s.

The energy conservation equation is as follows:

$$\frac{\partial (\rho T)}{\partial t} + \frac{\partial (\rho T u)}{\partial x} + \frac{\partial (\rho T v)}{\partial y} + \frac{\partial (\rho T w)}{\partial z} = \frac{\partial}{\partial x}\left(\frac{k}{c_p}\frac{\partial T}{\partial x}\right) + \frac{\partial}{\partial y}\left(\frac{k}{c_p}\frac{\partial T}{\partial y}\right) + \frac{\partial}{\partial z}\left(\frac{k}{c_p}\frac{\partial T}{\partial z}\right) + S_T, \quad (3)$$

where c_p is specific heat of water, J/(kg·K), is effective heat transfer coefficient, W/(m·K), T is temperature of fluid, K, and S_T is viscous dissipation term.

The equations of the flow model in the pipe are as follows.

Turbulence intensity is

$$I = 0.16 \times R_e^{-0.125}. \quad (4)$$

Turbulent kinetic energy is

$$k = \frac{3}{2} \times (\bar{u} \times I)^2. \quad (5)$$

Turbulence kinetic energy dissipation rate is

$$\varepsilon = \frac{C_u^{3/4} k^{3/2}}{l}, \quad (6)$$

where I is turbulence intensity, R_e is Reynolds number of fluid, $\bar{u}$ is average fluid turbulent velocity, m/s, C_u is empirical constant, 0.09, and l is characteristic length, m.

According to the simplified model, the heat transfer of the calcined petroleum coke in the heat exchanger mainly includes the heat transfer among calcined petroleum coke particles and the heat transfer between calcined petroleum coke and the wall of the heat exchanger. The heat transfer process in this paper is steady with no inner heat source, so the heat conduction differential equation is as follows:

$$\rho c \frac{\partial T}{\partial \tau} = \frac{\partial}{\partial x}\left(\lambda \frac{\partial T}{\partial x}\right) + \frac{\partial}{\partial y}\left(\lambda \frac{\partial T}{\partial y}\right) + \frac{\partial}{\partial z}\left(\lambda \frac{\partial T}{\partial z}\right), \quad (7)$$

where ρ is density, kg/m^3, c is specific heat, J/(kg·K), is temperature, K, and λ is coefficient of heat conductivity, W/(m·K).

The discrete points within the domain are solved. The implicit format of the unsteady heat conduction equation is used to solve problems in this paper:

$$\frac{T_n^{(i+1)} - T_n^{(i)}}{\Delta \tau} = a \frac{T_{n+1}^{(i+1)} - 2T_n^{(i+1)} + T_{n-1}^{(i+1)}}{\Delta x^2}, \quad (8)$$

where $\Delta\tau$ is time step, s, T is temperature, K, n is the serial number of n node; i is the serial number of i temporal sequence; a is thermal diffusivity, m^2/s, and Δx is distance between two adjacent nodes, m.

The heat transfer equation of calcined petroleum coke and the wall of heat exchanger is as follows:

$$q = k \cdot A \cdot \Delta T_m, \quad (9)$$

where q is heat flux, W, k is heat transmission coefficient, W/(m^2·K), A is heat exchange area, m^2, and ΔT_m is average temperature difference, K.

The calcined petroleum coke velocity, water velocity in the tube, and calcined petroleum coke temperature in the heat exchange inlet are constant. The calcined petroleum coke particles and the gas among them are regarded as homogenous continuum. The physical parameters of calculation model are calculated as the porosity of calcined petroleum coke. The effective heat conductivity coefficient of homogenous continuum is calculated using (10) [13]:

$$\lambda_e = \phi\lambda_f + (1-\phi)\lambda_s, \quad (10)$$

where λ_e is effective thermal conductivity coefficient, W/(m·K), φ is porosity, λ_f is thermal conductivity coefficient of gas, W/(m·K), and λ_s is thermal conductivity coefficient of solid, W/(m·K).

The effective specific heat of homogenous continuum is calculated using

$$C_{pe} = \phi C_{pf} + (1-\phi) C_{ps}, \quad (11)$$

where C_{pe} is effective specific heat, J/(kg·K), ϕ is porosity, C_{pf} is specific heat of gas, J/(kg·K), and C_{ps} is specific heat of solid, J/(kg·K).

When the heat exchanger model is meshed, the internal boundary is refined locally. The quantity and quality of the grid have a great influence on the calculation results. If the grids are much less, it cannot meet the precision request. If the grids are too dense, it will require sufficient computer memory. And the calculating time will increase. So it is critical to confirm the quantity of grids. The test of grid independence has been widely used to solve this problem. There are four schemes for meshing. The quantity of grids is 130442, 215817, 313721, and 467604. When calculation is completed, the temperature of calcined petroleum coke at outlet is used as the evaluation standard. The calculation results are shown as in Table 1.

Table 1: Comparison of the calculation results of different meshing schemes.

	Grid quantity	The maximum temperature of calcined petroleum coke at outlet/K	Relative change/% (relative to the former scheme)
Scheme 1	130442	647.8571	—
Scheme 2	215817	653.0132	0.796
Scheme 3	313721	657.2785	0.653
Scheme 4	467604	662.5237	0.798

Table 1 showed that the relative change of Scheme 2 is within 2% compared to Scheme 1. There will be little influence on calculation results if grids are much dense. Therefore Scheme 2 not only reaches the calculation accuracy but also saves much time. So Scheme 2 was used for calculation.

The velocity of water in tube is 1 m/s, the inlet temperature of water is 300 K, the inlet temperature of calcined petroleum coke is 1073 K, and the velocity of calcined petroleum coke in the heat exchanger is 6×10^{-5}m/s.

The heat recovery efficiency is calculated by

$$\eta = \frac{q_1 \cdot (h_{1out} - h_{1in}) + q_2 \cdot (h_{2out} - h_{2in}) + q_3 \cdot (h_{3out} - h_{3in})}{c \cdot q_4 \cdot T} \times 100\%, \quad (12)$$

where η is heat recovery efficiency, %, q_1 is water flow mass of 1st external heat exchange tube, kg·s^{-1}, h_{1out} is enthalpy value of water in 1st tube outlet, kJ·kg^{-1}, h_{1in} is enthalpy value of water in 1st tube inlet, kJ·kg^{-1}, q_2 is water flow mass of 2nd external heat exchange tube, kg·s^{-1}, h_{2out} is enthalpy value of water in 2nd tube outlet, kJ·kg^{-1}, h_{2in} is enthalpy value of water in 2nd tube inlet, kJ·kg^{-1}, q_3 is water flow mass of internal heat exchange tube, kg·s^{-1}, h_{3out} is enthalpy value of water in internal heat exchange tube outlet, kJ·kg^{-1}, h_{3in} is enthalpy value of water in internal heat exchange tube inlet, kJ·kg^{-1}, c is specific heat capacity, kJ·kg^{-1}·K^{-1}, q_4 is calcined petroleum coke flow mass, kg·s^{-1}, and T is temperature of calcined petroleum coke when it flows into the heat exchanger, K.

The definition of the equivalent heat conductivity coefficient is the ratio of the actual heat conductivity coefficient to the standard heat conductivity coefficient. The standard heat conductivity coefficient is the actual heat conductivity coefficient of the calcined petroleum coke porosity which is 0.7.

The definition of the equivalent specific heat is the ratio of the actual specific heat to the standard specific heat. The standard specific heat is the actual specific heat of the calcined petroleum coke porosity which is 0.7.

EXPERIMENTAL SYSTEM

The experimental system of the waste heat utilization exchanger was built in Weifang Lianxing New Materials Technology Co., Ltd. The heat transfer experiments were carried out by the experimental system (Figure 3). The experimental system is composed of the waste heat exchanger, the calcined petroleum coke supply system, the water circulation system, and the measurement system.

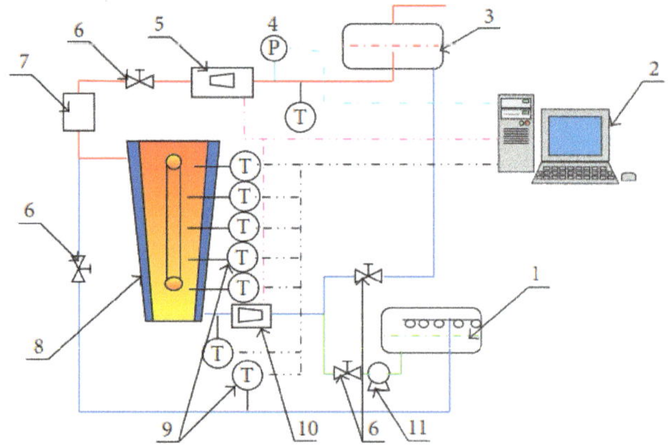

Figure 3: Schematic of experimental system ((1) the cooling pond; (2) the data acquisition system; (3) the drum; (4) the pressure sensor; (5) the two-phase flowmeter; (6) valve; (7) the sight glass; (8) the waste heat exchanger; (9) the temperature sensor; (10) the flowmeter; (11) the pump).

The experimental system is installed in the tank calcined furnace. The high temperature calcined petroleum coke is supplied directly from the tank calcined furnace. The water circulation system is composed of cooling pond, pumps, valves, sight glass, a drum, and so on. There is a WSM-D two-phase flowmeter in the heat exchanger outlet, which is used to measure steam dryness and flow. There is a FLUXUS F601 ultrasonic flowmeter in the heat exchanger inlet, which is used to measure the water flow. Its measuring accuracy is ±0.5%. Its measuring velocity range is from 0.01 m/s to 25 m/s. Its repeatability is 0.15%.

The measurement system is composed of a temperature sensor, a pressure sensor, and a piece of data acquisition instrument. The temperature sensor is used to measure the temperature of water in the heat exchanger inlet, the temperature of calcined petroleum coke in heat exchanger, the temperature of steam in the heat exchanger outlet, and so on. The pressure sensor is used to measure the water pressure and the steam pressure. The data acquisition instrument is used to record the data. The temperature measuring points of calcined petroleum coke in the heat exchanger are shown in Figure 4.

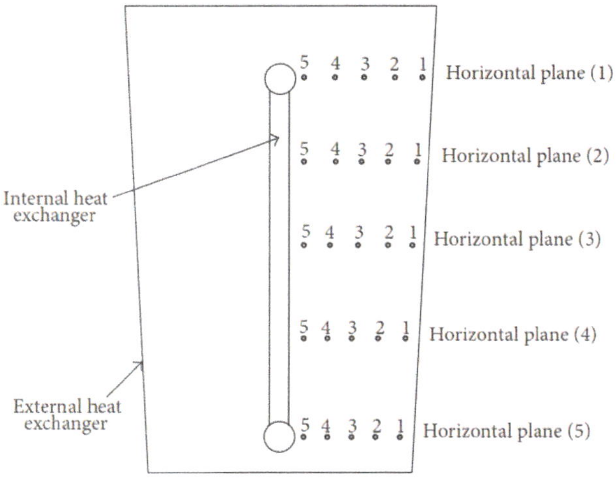

Figure 4: Temperature measuring points of calcined petroleum coke in the waste heat exchanger.

RESULTS AND DISCUSSIONS

Model Validation

The experimental data were used to validate the developed model. Simulations were conducted for the same operational conditions as those employed in the experimental investigation. Figure 5 shows the temperatures of numerical and experimental results in different planes when the porosity is 0.7. The temperature difference of numerical and experimental results is 16 K and the relative error is 1.78%. The results are in good agreement with the experimental value obtained in the same conditions.

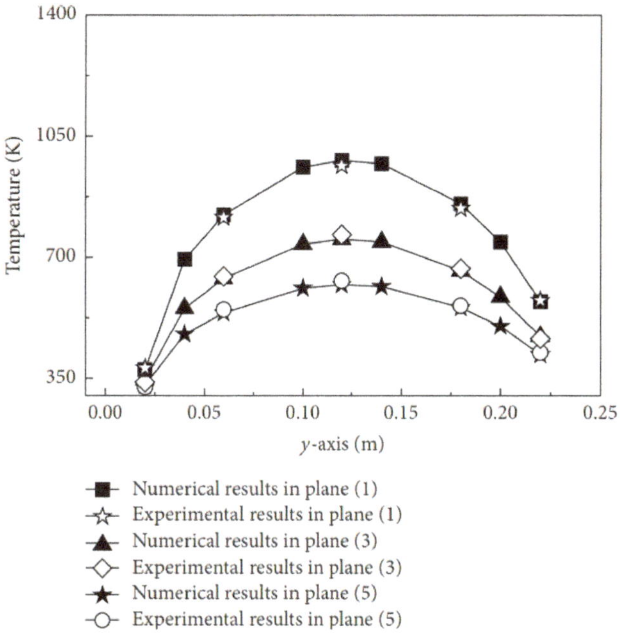

Figure 5: The temperatures of numerical and experimental result.

Temperature Distribution Characteristics in the Heat Exchanger

Figure 6 shows the temperature distribution of calcined petroleum coke in the heat exchanger for the porosity which is 0.7. Figure 6 indicates that the temperature of calcined petroleum coke is low near the wall of the heat exchanger tube and the temperature differences are small in different planes. The temperature difference of the first plane and the fifth plane is only 19 K, while, in the center of the internal and external heat exchangers, the temperature of calcined petroleum coke is high and the temperature differences of different planes are great. The temperature difference of the first plane and the fifth plane is 359 K. The main reason is that the coefficient of the heat conductivity of calcined petroleum coke is very low. The longer the distance of the heat transfer is, the bigger the heat resistance of calcined petroleum coke is and accordingly the higher the temperature is.

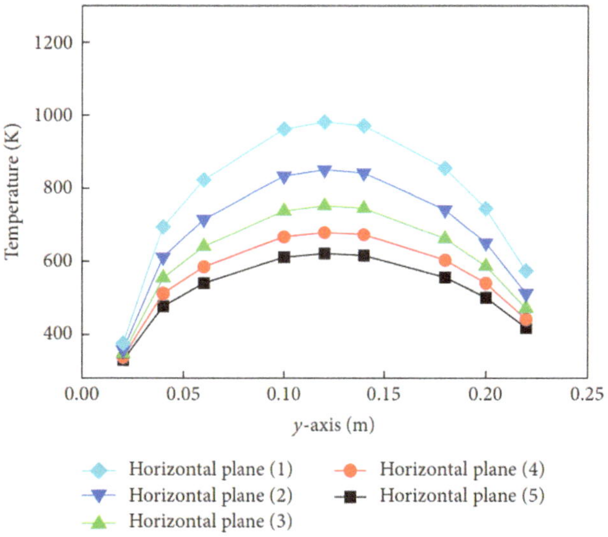

Figure 6: The temperature distribution of calcined petroleum coke.

Effect of Calcined Petroleum Coke Porosity

Figure 7 shows the effects of porosity on calcined petroleum coke temperature at the outlet of the heat exchanger. Figure 7 shows that when the porosity increases from 0.58 to 0.79, the average temperature of calcined petroleum coke decreases from 541 K to 489 K and the maximum temperature decreases from 650 K to 582 K. The main reason is that, with the increase of porosity, the effective specific heat and the effective density decrease and the initial total heat quantity reduces, which make the temperature of calcined petroleum coke decrease at the heat exchanger outlet. However, with the increase of porosity, the effective heat conductivity coefficient decreases. So the ability of the heat conductivity decreases, which makes the temperature of calcined petroleum coke increase at the heat exchanger outlet. But the effect of the initial total heat quantity reduction is a major influence factor, so the calcined petroleum coke temperature decreases with the increase of porosity.

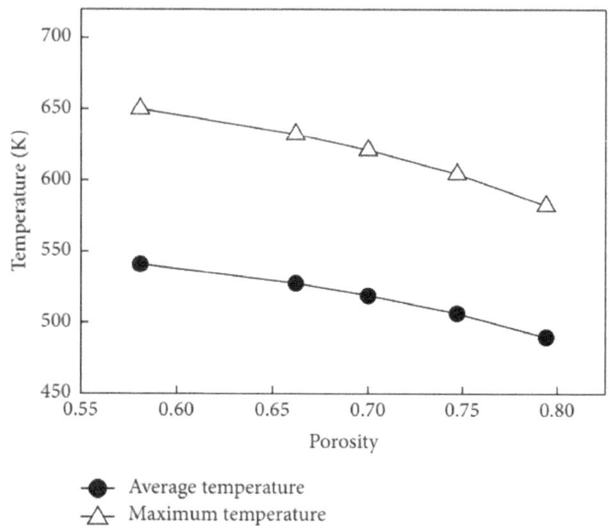

Figure 7: Variations of temperature at the heat exchanger outlet with different calcined petroleum coke velocity porosity.

Figure 8 shows the effects of porosity on average heat transfer coefficient. As can be seen from Figure 8, the average heat transfer coefficient of the internal heat exchanger is higher than the external heat exchanger. When the porosity increases from 0.58 to 0.79, the average heat transfer coefficients of the internal and the external heat exchanger decrease by 6.4 and 5.6 W/(m^2·K), respectively. The reason for this is that, with the increase of porosity, the effective specific heat and effective heat conductivity coefficient decrease and the average heat flux of the heat exchanger tube wall decreases. At the same time, with the decrease of the effective specific heat, the initial total heat quantity reduces while the cooling rate of calcined petroleum coke increases, so the average heat transfer temperature difference decreases. However, the effect of heat exchanger tube average heat flux is a major influence factor, so the average heat transfer coefficient decreases with the increase of porosity.

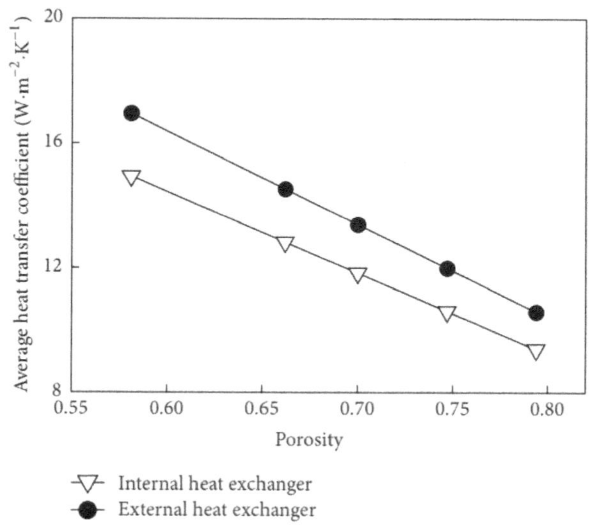

Figure 8: Variations of average heat transfer coefficient with different calcined petroleum coke velocity porosity.

Figure 9 shows the effects of porosity on heat recovery efficiency. Figure 9 shows that the heat recovery efficiency increases by 5.2% when the porosity increases from 0.58 to 0.79. The main reason is that when the porosity increases, the initial total heat quantity and the heat recovery quantity reduce. But the reduce rate of the initial total heat quantity is higher than that of the reduce rate of heat recovery quantity, and thus the heat recovery efficiency increases.

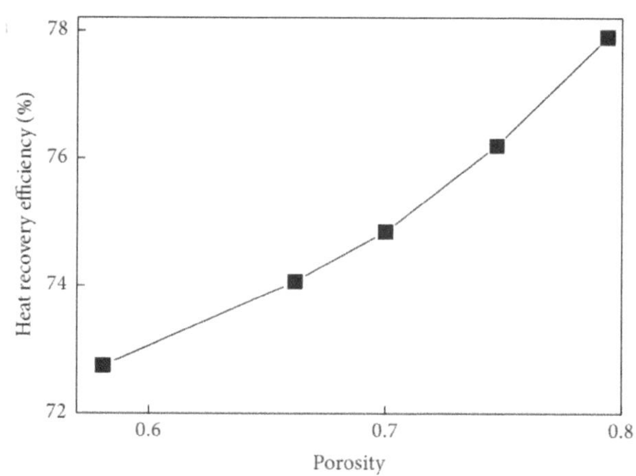

Figure 9: Variations of heat recovery efficiency with different calcined petroleum coke velocity porosity.

Effect of Equivalent Heat Conductivity Coefficient

The normal value of heat conductivity coefficient is the effective heat conductivity coefficient when the calcined petroleum coke porosity is 0.7. The definition of the equivalent heat conductivity coefficient is a ratio of the objective heat conductivity coefficient and the normal value.

Figures 10–12 show the effects of equivalent heat conductivity coefficient at the calcined petroleum coke temperature at the heat exchanger outlet, the average heat transfer coefficient, and the heat recovery efficiency. As can be seen from Figures 10–12, when the equivalent heat conductivity coefficient increases from 0.9 to 1.1, the average temperature reduces by 35 K and the maximum temperature decreases by 50 K, the average heat transfer coefficient of the internal and external heat exchanger increases by 2.8 and 2.6 W/(m^2·K), respectively, and the heat recovery efficiency increases by 4.3%. The reason for this is that, with the increase of the equivalent heat conductivity coefficient, the heat resistance of calcined petroleum coke decreases, the heat transfer quantity in unit time increases, the temperatures at the heat exchanger outlet reduce, and the average heat transfer coefficient and the heat recovery efficiency increase.

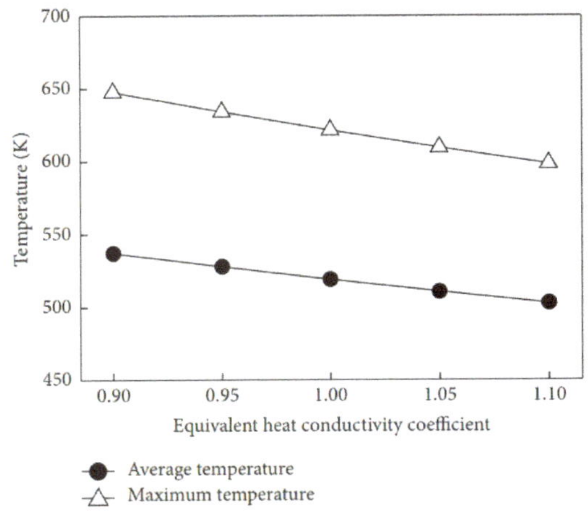

Figure 10: Variations of the temperature at the heat exchanger outlet with different equivalent heat conductivity coefficients.

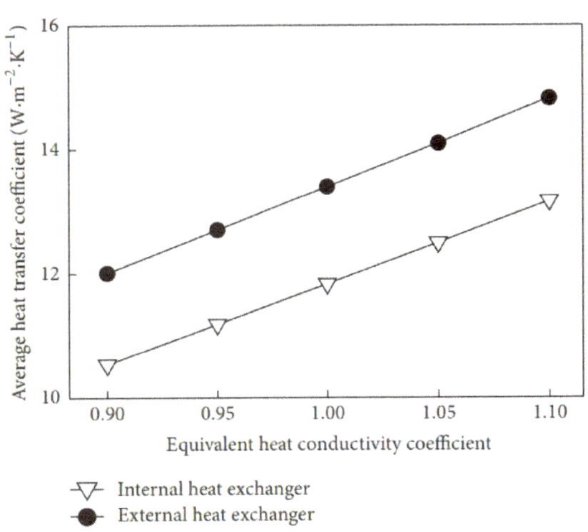

Figure 11: Variations of the average heat transfer coefficient with different equivalent heat conductivity coefficients.

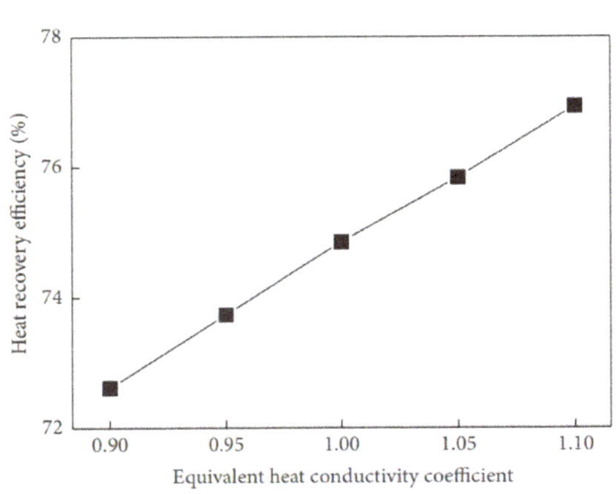

Figure 12: Variations of the heat recovery efficiency with different equivalent heat conductivity coefficients.

Effect of Equivalent Specific Heat

The normal value of specific heat is the effective specific heat when the calcined petroleum coke porosity is 0.7. The definition of equivalent specific heat is a ratio of the objective specific heat and the normal value.

Figure 13 shows the effects of equivalent specific heat on calcined petroleum coke temperature at the heat exchanger outlet. Figure 13 shows that when the equivalent specific heat increases from 0.9 to 1.1, the average temperature of calcined petroleum coke increases from 501 K to 535 K and the maximum temperature increases from 595 K to 644 K. The main reason is that when the equivalent specific heat increases, the initial total heat quantity heat increases and when the heat conductivity coefficient of calcined petroleum coke is a constant, the calcined petroleum coke temperature at the heat exchanger outlet increases.

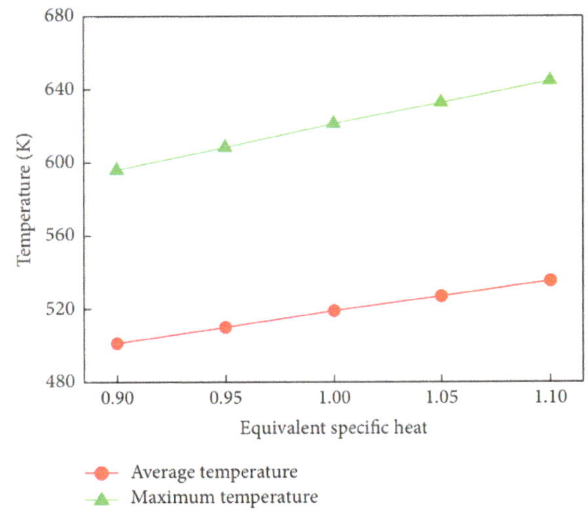

Figure 13: Variations of the temperature at heat exchanger outlet with different equivalent specific heat.

Figure 14 shows the effects of equivalent specific heat on average heat transfer coefficient. As can be seen from Figure 14, when the equivalent specific heat increases from 0.90 to 1.10, the average heat transfer coefficients of the internal and external heat exchanger decrease by 0.16 and 0.28 W/(m^2·K), respectively. In other words, the effect is very small.

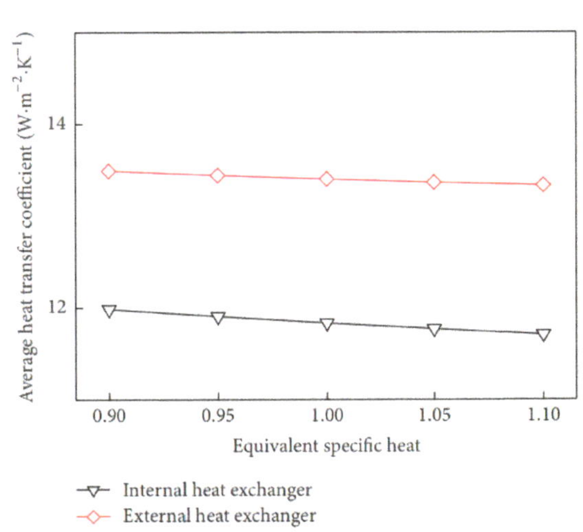

Figure 14: Variations of the average heat transfer coefficient with different equivalent specific heat.

Figure 15 shows the effects of the equivalent specific heat on heat recovery efficiency. Figure 15 shows that the heat recovery efficiency decreases by 4.1% when the equivalent specific heat increases from 0.90 to 1.10. The main reason is that when the heat transfer time is a constant, with the increase of equivalent specific heat, the initial total heat quantity heat increases, the heat transfer sufficiency decreases, and the heat recovery efficiency reduces.

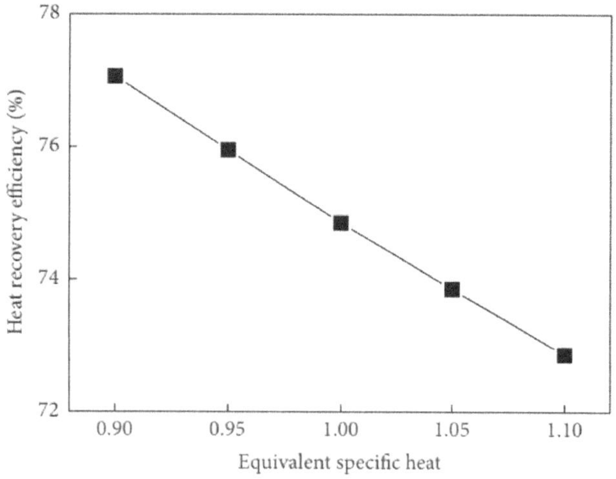

Figure 15: Variations of the heat recovery efficiency with different equivalent specific heat.

CONCLUSIONS

The main conclusions of the present study are as follows.

(1) The temperature of calcined petroleum coke is low near the wall of the heat exchanger tube, and the temperature differences are small in different planes, while, in the center of the internal and external heat exchangers, the temperature of calcined petroleum coke is high and the temperature differences of different planes are great.

(2) When the porosity increases from 0.58 to 0.79, the average temperature of calcined petroleum coke at the outlet of the heat exchanger decreases from 541 K to 489 K and the maximum temperature decreases from 650 K to 582 K. The average heat transfer coefficients of the internal and external heat exchanger decrease by 6.4 and 5.6 W/(m^2·K), respectively. The heat recovery efficiency increases by 5.2%.

(3) With the increase of the equivalent heat conductivity coefficient (from 0.9 to 1.1), the average temperature reduces by 35 K and the maximum temperature decreases by 50 K, the average heat transfer coefficients of the internal and external heat exchanger increase by 2.8 and 2.6 W/(m^2·K), respectively, and the heat recovery efficiency increases by 4.3%.

(4) If the other conditions remain the same, when the equivalent specific heat increases from 0.9 to 1.1, the average temperature of calcined petroleum coke increases from 501 K to 535 K and the maximum temperature increases from 595 K to 644 K. The average heat transfer coefficient of the internal and external heat exchanger decreases by 0.16 and 0.28 W/(m^2·K), respectively. The heat recovery efficiency decreases by 4.1%.

Conflict of Interests

The authors declare that there is no conflict of interests regarding the publication of this paper.

ACKNOWLEDGMENTS

This work was supported by Shandong Provincial Natural Science Foundation, China (ZR2013EEQ005), and Shandong Provincial Science and Technology Development Program, China (2013GGX10404).

REFERENCES

1. D. Bob, BP Statistical Review of World Energy, BP, London, UK, 2014.
2. Q. Zhao, Y. Wang, and X. Wang, "Technical advances status of China's waste heat utilization," Industrial Boiler, vol. 25, no. 5, pp. 8–15, 2009.
3. B. Zheng, Y. Liu, and Z. Wang, "Study on thermo-physical properties of calcined petroleum coke,"Carbon Techniques, vol. 32, no. 2, pp. 33–35, 2013.
4. J.-J. Cai, J.-J. Wang, and C.-X. Chen, "Recovery of residual-heat integrated steelworks," Iron & Steel, vol. 42, no. 6, pp. 1–7, 2007.
5. Y. Ding and D.-M. Shi, "High-efficiency utilization of waste heat at fully integrated steel plant," Iron & Steel, vol. 46, no. 10, pp. 88–98, 2011.
6. B. Wang, X.-C. Wang, Y.-C. Yuan, and Q.-P. Zhou, "Advances in the study of the blast furnace slag waste heat recovery technologies," Journal of Engineering for Thermal Energy and Power, vol. 29, no. 2, pp. 113–121, 2014.
7. Y. Chen, Numerical simulation study on waste heat recovery system of high temperature steel slag [M.S. thesis], Shandong University, Shandong, China, 2014.
8. F. Liu, Research on a novel waste heat recovery equipment for blast furnace slag and heat transfer numerical simulation [M.S. thesis], Qingdao Technological University, Shandong, China, 2010.
9. F. Herz, I. Mitov, E. Specht, and R. Stanev, "Experimental study of the contact heat transfer coefficient between the covered wall and solid bed in rotary drums," Chemical Engineering Science, vol. 82, pp. 312–318, 2012.
10. T. Atmakidis and E. Y. Kenig, "Numerical analysis of mass transfer in packed-bed reactors with irregular particle arrangements," Chemical Engineering Science, vol. 81, pp. 77–83, 2012.
11. R. Sundaresan and A. K. Kolar, "Axial heat transfer correlations in a circulating fluidized bed riser,"Applied Thermal Engineering, vol. 50, no. 1, pp. 985–996, 2013.
12. X.-Y. Guo, H.-S. Chai, and D.-H. Chao, "Numerical simulation of large particle fluidized bed and comparison of gas-particle heat transfer models," Journal of University of Shanghai for Science and Technology, vol. 34, no. 1, pp. 81–87, 2012.
13. W. Woodside and J. H. Messmer, "Thermal conductivity of porous media. I. Unconsolidated sands,"Journal of Applied Physics, vol. 32, no. 9, pp. 1688–1699, 1961.

CHAPTER 15

INFLUENCE OF CROSS-DIFFUSION ON SLIP FLOW AND HEAT TRANSFER OF CHEMICALLY REACTING UCM FLUID BETWEEN POROUS PARALLEL PLATES WITH HALL AND ION SLIP CURRENTS

Odelu Ojjela, Pravin Kashyap Kambhatla, N. Naresh Kumar, and Samir Kumar Das

Department of Applied Mathematics, Defence Institute of Advanced Technology (Deemed University), Pune 411025, India

ABSTRACT

The present paper deals with the Hall and ion slip currents on an incompressible unsteady free convection flow and heat transfer of an upper convected Maxwell fluid between porous parallel plates with Soret and Dufour effects by considering the velocity slip and convective boundary conditions. Assume that there are periodic injection and suction at the lower and upper plates, respectively. The temperature and concentration at the lower and upper plates change periodically with time. The flow field equations are reduced to nonlinear ordinary differential equations by using similarity transformations and a semi-analytical-numerical solution has been obtained by the differential transform method. The velocity components, temperature distribution, and concentration with respect to different fluid and geometric parameters are discussed in detail and presented in the form of graphs. It is observed that the Biot number increases the temperature and concentration of the fluid. Further, the concentration of the fluid is enhanced whereas the temperature decreases with increasing slip. The present results are compared with the existing literature and are found to be in good agreement.

INTRODUCTION

The flow through porous channels is of great importance in both engineering and biological flows. Examples of this are found in aerodynamic heating, electrostatic precipitation, soil mechanics, food preservation, polymer technology, petroleum industry, the mechanics of the cochlea in the human ear, the flow of blood in the arteries, and artificial dialysis. The flows due to the periodic injection/suction at the boundary are of immense importance because of their wide range of applications. Several researchers have studied the incompressible laminar fluid flow between two porous parallel plates due to its mathematical simplicity and the flows can be analyzed easily in theoretical and experimental studies. Berman [1] examined the incompressible viscous fluid flow in a two-dimensional uniformly porous channel and a series solution was obtained for small values of the Reynolds number. Later, White Jr. et al. [2] considered the steady incompressible laminar viscous fluid flow between porous parallel plates with uniform suction or injection and the problem was analyzed for

Citation: Odelu Ojjela, Pravin Kashyap Kambhatla, N. Naresh Kumar, and Samir Kumar Das, "Influence of Cross-Diffusion on Slip Flow and Heat Transfer of Chemically Reacting UCM Fluid between Porous Parallel Plates with Hall and Ion Slip Currents," Journal of Computational Engineering, vol. 2016, Article ID 9160956, 14 pages, 2016. doi:10.1155/2016/9160956.

a wide range of suction Reynolds number. Walker and Davies [3] analyzed the mass transfer in an incompressible laminar viscous fluid flow between permeable parallel plates and a confluent hypergeometric function was used to obtain the solution. Hamza [4] investigated an incompressible laminar viscous fluid flow between two parallel rectangular as well as circular plates and obtained an approximate analytical solution by regular perturbation technique. The unsteady micropolar fluid flow between two parallel porous plates with periodic suction and injection was inspected by Srinivasacharya et al. [5]. The steady flow of chemically reacting micropolar fluid through a permeable channel was studied by Sheikholeslami et al. [6] and an analytical solution was obtained by using the homotopy perturbation method. Chen and Zhu [7] considered the slip effects on a Couette Poiseuille flow of a Bingham fluid between porous parallel plates and the reduced flow field equations are solved analytically. Singh and Pathak [8] analyzed the Hall currents and thermal radiation effects on unsteady MHD slip flow in a vertical rotating porous channel. Bhat and Mittal [9] inspected the Hall and ion slip currents on an electrically conducting viscous fluid flow through a parallel plate channel in the presence of uniform magnetic field and observed that the temperature of the fluid is reduced by Hall currents. Srinivasacharya and Kaladhar [10] examined analytically the Hall and ion slip currents on the mixed convective flow of couple stress fluid through a vertical parallel plate channel. Singh and Gorla [11] investigated the effects of thermal diffusion and heat source on free convection flow of a viscous fluid past an infinite vertical porous plate with Hall currents. Raptis et al. [12] studied the effects of Grashof number and permeability parameter on the free convection flow of a viscous fluid through a porous medium between two parallel plates. Abdulaziz and Hashim [13] investigated the heat and mass transfer of free convective flow of micropolar fluid in a vertical porous parallel plate channel and observed that the velocity decreases for increasing of Reynolds number. Makinde and Aziz [14] considered the convective boundary conditions for a MHD mixed convection fluid flow experiencing a first-order chemical reaction over a vertical plate embedded in a porous medium and a numerical solution is obtained by using shooting technique along with a sixth-order Runge-Kutta integration.

Yao et al. [15] analyzed an incompressible laminar viscous fluid flow and heat transfer over a stretching/shrinking sheet with convective boundary conditions and have found that the convective boundary condition results in a temperature slip at the wall. RamReddy et al. [16] studied an incompressible laminar free convective micropolar fluid flow along a permeable vertical plate with the convective boundary condition and obtained a numerical solution using spectral quasilinearization method. Postelnicu [17] examined the effects of Soret and Dufour on a two-dimensional steady stagnation point flow of a free convective Darcian fluid with suction/injection and the governing partial differential equations are reduced to nonlinear ordinary differential equations using similarity transformations and then solved by the Keller box method. Hayat et al. [18] investigated the three-dimensional flow of a chemically reacting viscous fluid over exponentially stretching sheet with Soret and Dufour effects and observed that the thermal boundary layer increases for higher Dufour number while concentration boundary layer was increased for higher Soret values. Chamkha and Rashad [19] have numerically investigated the unsteady heat and mass transfer of an electrically conducting mixed convective chemically reacting viscous fluid flow over a rotating vertical cone with thermal-diffusion and diffusion-thermo effects. The MHD mixed convective heat and mass transfer of couple stress fluid through a porous channel with periodic suction and injection in the presence of cross-diffusion effects was considered by Ojjela and Naresh Kumar [20] and the reduced governing equations are solved numerically by the method of quasilinearization. The systematic framework of rate type viscoelastic fluids was developed by Oldroyd [21]. The unsteady three-dimensional flow of time dependent UCM fluid over a stretching sheet was studied by Awais et al. [22] and the series solution was obtained by HAM.

Choi et al. [23] considered the two-dimensional steady incompressible laminar suction flow of a UCM fluid in a porous channel with the combined effects of inertia and viscoelasticity and obtained both analytical and numerical solutions. Hayat et al. [24] found an analytical solution for an electrically conducting Maxwell fluid peristaltic flow in a porous space. Abbas et al. [25] studied an incompressible magnetohydrodynamic two-dimensional boundary layer flow of UCM fluid through a rectangular porous channel using the homotopy analysis method. Mukhopadhyay and Gorla [26] analyzed the two-dimensional MHD flow and mass transfer of UCM fluid over an unsteady stretching sheet with first-order constructive/destructive chemical reaction and obtained a numerical solution by shooting method. Hayat and Abbas [27] inspected the effects of Deborah's number, Schmidt number, and chemical reaction parameter on velocity and concentration fields in the flow of chemically reacting UCM fluid between porous parallel plates. The unsteady MHD flow of Maxwell fluid over an impulsively stretching sheet was analyzed by Alizadeh-Pahlavan and Sadeghy [28] and the flow field equations are solved by the homotopy analysis method. Differential transform method is first used by Zhou [29] in solving circuit problems. The theory of two-dimensional differential transform method was proposed by Chen and Ho [30]. Ayaz [31] studied the solutions of differential equations using differential transform method in a three-dimensional platform. Recently, several researchers have obtained solutions for various problems in fluid mechanics by differential transform method due to its simplicity and high accuracy. Mosayebidorcheh et al. [32] have investigated the effect of mass transfer on an incompressible laminar upper convected Maxwell fluid in a porous channel with high permeability medium and the reduced flow field equations are solved by differential transform method. Sheikholeslami et al. [33] have considered an incompressible steady nanofluid flow and heat transfer between two horizontal parallel plates with thermophoresis and Brownian motion under the influence of uniform magnetic field and the governing equations are reduced into nondimensional ordinary differential equations using similarity transformation and then solved by differential transform method. Sheikholeslami and Ganji [34] have studied the effect of Brownian motion on a nanofluid in the presence of variable magnetic field and the solution is obtained by differential transform method. Hatami and Jing [35] have analyzed two different problems, the unsteady two-dimensional squeezing nanofluid flow and heat transfer between two parallel plates and a non-Newtonian free convective nanofluid flow, and heat transfer between two vertical flat plates and obtained a solution for both by differential transform method and the results are compared with Runge-Kutta fourth-order method. Esmail and Taha [36] have found a solution to the atmospheric dispersion equation by using DTM and the results are shown to be in good agreement with field data.

In the present paper, the effects of chemical reaction and Hall and ion slip currents on an unsteady incompressible free convective slip flow of an upper convected Maxwell fluid through a porous channel with Soret and Dufour in the presence of the convective boundary condition are considered. The reduced governing nonlinear ordinary differential equations are solved by differential transform method (DTM). The effects of various fluid and geometric parameters on nondimensional velocity components, temperature distribution, and concentration are discussed in detail and shown in the form of graphs. The present results are compared with Bujurke et al. [37] for the Newtonian fluid and presented in the form of a table.

FORMULATION OF THE PROBLEM

Consider an unsteady incompressible slip flow of an electrically conducting upper convected Maxwell fluid between two parallel plates which are separated by "h." Assume $V_1e^{i\omega t}$ and $V_2e^{i\omega t}$ are periodic injection and suction at the lower and upper plates, respectively. Also the temperature at the upper plate is maintained with $T_2e^{i\omega t}$, while the lower plate is governed by convective boundary condition and the concentrations at the lower and upper plates are $C_1e^{i\omega t}$ and $C_2e^{i\omega t}$, respectively, as shown in Figure 1.

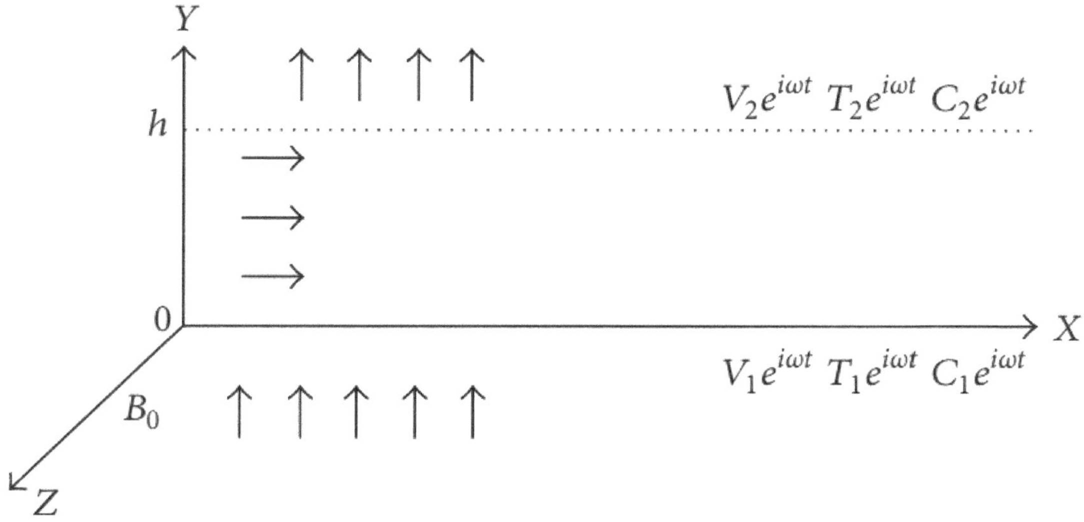

Figure 1: Schematic diagram of fluid flow between porous parallel plates.

The equations governing the heat and mass transfer of a UCM fluid flow in the presence of strong magnetic field and in the absence of body forces are [22, 28]

$$\frac{\partial u}{\partial x}+\frac{\partial v}{\partial y}=0, \tag{1}$$

$$\begin{aligned}\rho\Bigg[&\frac{\partial u}{\partial t}+u\frac{\partial u}{\partial x}+v\frac{\partial u}{\partial y}\\&+\beta\left(u^2\frac{\partial^2 u}{\partial x^2}+v^2\frac{\partial^2 u}{\partial y^2}+2uv\frac{\partial^2 u}{\partial x\partial y}\right)\Bigg]=-\frac{\partial P}{\partial x}\\&+\mu\nabla^2 u-\sigma B_0^2\left(\frac{(1+\beta_i\beta_e)u-\beta_e v}{(1+\beta_i\beta_e)^2+\beta_e^{\,2}}\right)+\rho g\beta_T\left(T\right.\\&\left.-T_1e^{i\omega t}\right)+\rho g\beta_C\left(C-C_1e^{i\omega t}\right),\end{aligned} \tag{2}$$

$$\begin{aligned}\rho\Bigg[&\frac{\partial v}{\partial t}+u\frac{\partial v}{\partial x}+v\frac{\partial v}{\partial y}\\&+\beta\left(u^2\frac{\partial^2 v}{\partial x^2}+v^2\frac{\partial^2 v}{\partial y^2}+2uv\frac{\partial^2 v}{\partial x\partial y}\right)\Bigg]=-\frac{\partial P}{\partial y}\\&+\mu\nabla^2 v-\sigma B_0^2\left(\frac{(1+\beta_i\beta_e)v+\beta_e u}{(1+\beta_i\beta_e)^2+\beta_e^{\,2}}\right),\end{aligned} \tag{3}$$

$$\rho c\left[\frac{\partial T}{\partial t}+u\frac{\partial T}{\partial x}+v\frac{\partial T}{\partial y}\right]=k\left(\frac{\partial^2 T}{\partial x^2}+\frac{\partial^2 T}{\partial y^2}\right)$$
$$+\mu\left(2\left(\frac{\partial u}{\partial x}\right)^2+2\left(\frac{\partial v}{\partial y}\right)^2+\left(\frac{\partial u}{\partial y}+\frac{\partial v}{\partial x}\right)^2\right)$$
$$+\sigma B_0^2\left(\frac{u^2+v^2}{(1+\beta_i\beta_e)^2+\beta_e{}^2}\right)+\frac{\rho D_1 k_T}{c_s}\left(\frac{\partial^2 C}{\partial x^2}\right.$$
$$\left.+\frac{\partial^2 C}{\partial y^2}\right), \tag{4}$$

$$\rho c\left[\frac{\partial C}{\partial t}+u\frac{\partial C}{\partial x}+v\frac{\partial C}{\partial y}\right]=D_1\left(\frac{\partial^2 C}{\partial x^2}+\frac{\partial^2 C}{\partial y^2}\right)$$
$$+\frac{D_1 k_T}{T_m}\left(\frac{\partial^2 T}{\partial x^2}+\frac{\partial^2 T}{\partial y^2}\right)-k_2\left(C-C_1 e^{i\omega t}\right). \tag{5}$$

The associated boundary conditions are as in (6)

$$u(x,\lambda,t)=0,$$
$$v(x,\lambda,t)=V_1 e^{i\omega t},$$
$$-k\frac{\partial T}{\partial y}=h_1\left(T-T_1 e^{i\omega t}\right),$$
$$C(x,\lambda,t)=C_1 e^{i\omega t}$$
$$\text{at } y=0, \tag{6}$$
$$u(x,\lambda,t)=-\frac{\sqrt{k_1}}{\sigma_1}\frac{\partial u}{\partial y},$$
$$v(x,\lambda,t)=V_2 e^{i\omega t},$$
$$T(x,\lambda,t)=T_2 e^{i\omega t},$$
$$C(x,\lambda,t)=C_2 e^{i\omega t}$$
$$\text{at } y=h.$$

The governing equations are reduced to nondimensional equations using the following similarity transformations [20]:

$$u(x,\lambda,t)=\left(\frac{U_0}{a}-\frac{V_2 x}{h}\right)f'(\lambda)e^{i\omega t},$$
$$v(x,\lambda,t)=V_2 f(\lambda)e^{i\omega t},$$
$$T(x,\lambda,t)$$
$$=\left(T_1+\frac{\mu v_2}{\rho hc}\left(\phi_1(\lambda)+\left(\frac{U_0}{aV_2}-\frac{x}{h}\right)^2\phi_2(\lambda)\right)\right)e^{i\omega t},$$
$$C(x,\lambda,t)$$
$$=\left(C_1+\frac{\dot{n}_A}{hv}\left(g_1(\lambda)+\left(\frac{U_0}{aV_2}-\frac{x}{h}\right)^2 g_2(\lambda)\right)\right)e^{i\omega t}, \tag{7}$$

where $f(\lambda)$, $\phi_1(\lambda)$, $\phi_2(\lambda)$, $g_1(\lambda)$, and $g_2(\lambda)$ are unknown functions to be determined.

Substituting (7) in (2), (3), (4), and (5), we have

$$f^{iv}\left(1-\mathrm{Wi}\,\mathrm{Re}\,f^2\cos\phi\right)=-\mathrm{Re}\cos\phi\left(f'f''-ff'''\right)$$
$$-2\mathrm{Wi}\,\mathrm{Re}\left(f'^2 f''+ff''^2\right)\cos 2\phi$$
$$+\mathrm{Ha}^2\left(\frac{(1+\beta_i\beta_e)}{(1+\beta_i\beta_e)^2+\beta_e{}^2}\right)f''-\frac{\mathrm{Ec}\,\mathrm{Gr}}{\xi}\left(\phi_1'\right.$$
$$\left.+\xi^2\phi_2'\right)-\frac{\mathrm{Sh}\,\mathrm{Gc}}{\xi}\left(g_1'+\xi^2 g_2'\right),$$
$$\phi_1''=-\frac{\mathrm{Re}\,\mathrm{Pr}\cos\phi}{(1-\mathrm{Du}\,\mathrm{Sc}\,\mathrm{Sr})}\left(4\left(f'\right)^2\right.$$
$$\left.+\left(\frac{\mathrm{Ha}^2}{(1+\beta_i\beta_e)^2+\beta_e{}^2}\right)f^2-f\phi_1'\right)$$
$$-\frac{\mathrm{Du}\,\mathrm{Kr}}{(1-\mathrm{Du}\,\mathrm{Sc}\,\mathrm{Sr})}g_1-\frac{\mathrm{Re}\,\mathrm{Sc}\,\mathrm{Du}\cos\phi}{(1-\mathrm{Du}\,\mathrm{Sc}\,\mathrm{Sr})}fg_1'-2\phi_2,$$
$$\phi_2''=-\frac{\mathrm{Re}\,\mathrm{Pr}\cos\phi}{(1-\mathrm{Du}\,\mathrm{Sc}\,\mathrm{Sr})}\left(\left(f''\right)^2\right.$$
$$\left.+\left(\frac{\mathrm{Ha}^2}{(1+\beta_i\beta_e)^2+\beta_e{}^2}\right)\left(f'\right)^2+2f'\phi_2-f\phi_2'\right)$$
$$-\frac{\mathrm{Du}\,\mathrm{Kr}}{(1-\mathrm{Du}\,\mathrm{Sc}\,\mathrm{Sr})}g_2-\frac{\mathrm{Re}\,\mathrm{Sc}\,\mathrm{Du}\cos\phi}{(1-\mathrm{Du}\,\mathrm{Sc}\,\mathrm{Sr})}\left(fg_2'\right.$$
$$\left.-2f'g_2\right),$$
$$g_1''=-2g_2+\frac{\mathrm{Kr}}{(1-\mathrm{Du}\,\mathrm{Sc}\,\mathrm{Sr})}g_1+\frac{\mathrm{Re}\,\mathrm{Sc}\,\mathrm{Du}\cos\phi}{(1-\mathrm{Du}\,\mathrm{Sc}\,\mathrm{Sr})}fg_1'$$
$$+\frac{\mathrm{Re}\,\mathrm{Sc}\;\mathrm{Pr}\;\mathrm{Sr}\cos\phi}{(1-\mathrm{Du}\,\mathrm{Sc}\,\mathrm{Sr})}\left(4\left(f'\right)^2\right.$$
$$\left.+\left(\frac{\mathrm{Ha}^2}{(1+\beta_i\beta_e)^2+\beta_e{}^2}\right)f^2-f\phi_1'\right),$$
$$g_2''=-\frac{\mathrm{Kr}}{(1-\mathrm{Du}\,\mathrm{Sc}\,\mathrm{Sr})}g_1+\frac{\mathrm{Re}\,\mathrm{Sc}\;\mathrm{Pr}\;\mathrm{Sr}\cos\phi}{(1-\mathrm{Du}\,\mathrm{Sc}\,\mathrm{Sr})}\left(\left(f''\right)^2\right.$$
$$\left.+\left(\frac{\mathrm{Ha}^2}{(1+\beta_i\beta_e)^2+\beta_e{}^2}\right)\left(f'\right)^2+2f'\phi_2-f\phi_2'\right)$$
$$+\frac{\mathrm{Re}\,\mathrm{Sc}\cos\phi}{(1-\mathrm{Du}\,\mathrm{Sc}\,\mathrm{Sr})}\left(fg_2'-2f'g_2\right), \tag{8}$$

where prime denotes the differentiation with respect to λ and the nondimensional boundary conditions in terms of f, ϕ_1, ϕ_2, g_1, and g_2 are

$$f(0)=1-a,$$
$$f'(0)=0,$$
$$g_1(0)=0,$$
$$g_2(0)=0,$$
$$\phi_1'(0)=-\gamma\phi_1(0),$$
$$\phi_2'(0)=-\gamma\phi_2(0),$$
$$f(1)=1,$$

$$f''(1) = -\frac{f'(1)}{\mathrm{Sl}},$$

$$\phi_1(1) = \frac{1}{\mathrm{Ec}},$$

$$\phi_2(1) = 0,$$

$$g_1(1) = \frac{1}{\mathrm{Sh}},$$

$$g_2(1) = 0. \tag{9}$$

SOLUTION OF THE PROBLEM

The nonlinear equations (8) along with the boundary conditions (9) are transformed into recurrence equations (11)–(15) and initial values (16) using the transformations in (10)

$$f \longrightarrow F_1[k],$$

$$\phi_1 \longrightarrow F_2[k],$$

$$\phi_2 \longrightarrow F_3[k],$$

$$g_1 \longrightarrow F_4[k],$$

$$g_2 \longrightarrow F_5[k],$$

$$f\phi_1 \longrightarrow \sum_{l=0}^{k} F_1[l]\,F_2[k-l],$$

$$f\phi_1\phi_2 \longrightarrow \sum_{s=0}^{k}\sum_{m=0}^{k-s} F_1[s]\,F_2[m]\,F_3[k-s-m],$$

$$F_1[k] = \frac{1}{k!}\frac{d^k f}{dx^k}, \quad f(x) = \sum_{j=0}^{k} F_1[j]\,x^j, \quad \frac{d^n f}{dx^n} \longrightarrow \frac{(k+1)}{k!}F_1[k+n],$$

$$F_2[k] = \frac{1}{k!}\frac{d^k \phi_1}{dx^k}, \quad \phi_1(x) = \sum_{j=0}^{k} F_2[j]\,x^j,$$

$$F_3[k] = \frac{1}{k!}\frac{d^k \phi_2}{dx^k}, \quad \phi_2(x) = \sum_{j=0}^{k} F_3[j]\,x^j,$$

$$F_4[k] = \frac{1}{k!}\frac{d^k g_1}{dx^k}, \quad g_1(x) = \sum_{j=0}^{k} F_4[j]\,x^j,$$

$$F_5[k] = \frac{1}{k!}\frac{d^k g_2}{dx^k}, \quad g_2(x) = \sum_{j=0}^{k} F_5[j]\,x^j, \tag{10}$$

$$F_1[k+4] = \Bigg(\mathrm{Re}\cos\phi\sum_{l=0}^{k}\big((l+3)(l+2)(l+1)F_1[l+3]\,F_1[k-l] - (l+2)(l+1)F_1[l+2]\,(k-l+1)F_1[k-l+1]\big) + \mathrm{Re\,Wi}\cos 2\phi$$

$$\cdot\sum_{s=0}^{k-1}\sum_{m=0}^{k-s}(s+4)(s+3)(s+2)(s+1)F_1[m]\,F_1[s+4]\,F_1[k-m-s] - 2\,\mathrm{Re\,Wi}\cos 2\phi$$

$$\cdot\sum_{s=0}^{k}\sum_{m=0}^{k-s}\big((m+1)(s+1)F_1[m+1]\,F_1[s+1]\,F_1[k-m-s] + (m+2)(m+1)(s+2)(s+1)F_1[m+2]\,F_1[s+2]\,F_1[k-s-m]\big)$$

$$+\frac{\mathrm{Ha}^2(1+\beta_i\beta_e)}{(1+\beta_i\beta_e)^2+\beta_e^2}(k+2)(k+1)F_1[k+2] - \frac{\mathrm{Ec\,Gr}}{\xi}(k+1)\big(F_2[k+1]+\xi^2F_3[k+1]\big) - \frac{\mathrm{Sh\,Gc}}{\xi}(k+1)\big(F_4[k+1]$$

$$+\xi^2F_5[k+1]\big)\Bigg)\Big(\big(1-F_1[0]^2\,\mathrm{Re\,Wi}\cos 2\phi\big)(k+4)(k+3)(k+2)(k+1)\Big)^{-1}, \tag{11}$$

$$\begin{aligned}F_2[k+2] = \Bigg(&-\frac{\mathrm{Re}\,\mathrm{Pr}\cos\phi}{(1-\mathrm{Du}\,\mathrm{Sc}\,\mathrm{Sr})}\sum_{l=0}^{k}(4(l+1)F_1[l+1](k-l+1)F_1[k-l+1]-(l+1)F_2[l+1]F_1[k-l])-\frac{\mathrm{Re}\,\mathrm{Pr}\cos\phi}{(1-\mathrm{Du}\,\mathrm{Sc}\,\mathrm{Sr})}\\ &\cdot\sum_{l=0}^{k}\left(\frac{\mathrm{Ha}^2}{(1+\beta_i\beta_e)^2+\beta_e^{\,2}}\right)F_1[l]F_1[k-l]-\frac{\mathrm{Du}\,\mathrm{Kr}}{(1-\mathrm{Du}\,\mathrm{Sc}\,\mathrm{Sr})}F_4[k]-\frac{\mathrm{Re}\,\mathrm{Sc}\,\mathrm{Du}\cos\phi}{(1-\mathrm{Du}\,\mathrm{Sc}\,\mathrm{Sr})}\sum_{l=0}^{k}(l+1)F_4[l+1]F_1[k-l]-2F_3[k]\Bigg)((k\\ &+2)(k+1))^{-1},\end{aligned} \tag{12}$$

$$\begin{aligned}F_3[k+2] = \Bigg(&-\frac{\mathrm{Re}\,\mathrm{Pr}\cos\phi}{(1-\mathrm{Du}\,\mathrm{Sc}\,\mathrm{Sr})}\Bigg(\sum_{l=0}^{k}((l+2)(l+1)F_1[l+2](k-l+2)(k-l+1)F_1[k-l+2])\\ &+\sum_{l=0}^{k}\left(\frac{\mathrm{Ha}^2}{(1+\beta_i\beta_e)^2+\beta_e^{\,2}}\right)(l+1)F_1[l+1](k-l+1)F_1[k-l+1]+\sum_{l=0}^{k}2(l+1)F_1[l+1]F_3[k-l]-(l+1)F_3[l+1]F_1[k-l]\Bigg)\\ &-\frac{\mathrm{Du}\,\mathrm{Kr}}{(1-\mathrm{Du}\,\mathrm{Sc}\,\mathrm{Sr})}F_5[k]-\frac{\mathrm{Re}\,\mathrm{Sc}\,\mathrm{Du}\cos\phi}{(1-\mathrm{Du}\,\mathrm{Sc}\,\mathrm{Sr})}\sum_{l=0}^{k}((l+1)F_5[l+1]F_1[k-l]-2(l+1)F_1[l+1]F_5[k-l])\Bigg)((k+2)(k+1))^{-1},\end{aligned} \tag{13}$$

$$\begin{aligned}F_4[k+2] = \Bigg(&-2F_5[k]-\frac{\mathrm{Kr}}{(1-\mathrm{Du}\,\mathrm{Sc}\,\mathrm{Sr})}F_4[k]-\frac{\mathrm{Re}\,\mathrm{Sc}\,\mathrm{Du}\cos\phi}{(1-\mathrm{Du}\,\mathrm{Sc}\,\mathrm{Sr})}\sum_{l=0}^{k}(l+1)F_4[l+1]F_1[k-l]+\frac{\mathrm{Re}\,\mathrm{Sc}\ \mathrm{Pr}\ \mathrm{Sr}\cos\phi}{(1-\mathrm{Du}\,\mathrm{Sc}\,\mathrm{Sr})}\\ &\cdot\sum_{l=0}^{k}\left(4(l+1)F_1[l+1](k-l+1)F_1[k-l+1]+\left(\frac{\mathrm{Ha}^2}{(1+\beta_i\beta_e)^2+\beta_e^{\,2}}\right)F_1[l]F_1[k-l]-(l+1)F_2[l+1]F_1[k-l]\right)\Bigg)((k+2)(k\\ &+1))^{-1},\end{aligned} \tag{14}$$

$$\begin{aligned}F_5[k+2] = \Bigg(&\frac{\mathrm{Re}\,\mathrm{Sc}\cos\phi}{(1-\mathrm{Du}\,\mathrm{Sc}\,\mathrm{Sr})}\sum_{l=0}^{k}((l+1)F_5[l+1]F_1[k-l]-2(l+1)F_1[l+1]F_5[k-l])+\frac{\mathrm{Re}\,\mathrm{Sc}\,\mathrm{Du}\cos\phi}{(1-\mathrm{Du}\,\mathrm{Sc}\,\mathrm{Sr})}\\ &\cdot\sum_{l=0}^{k}((l+2)(l+1)F_1[l+2](k-l+2)(k-l+1)F_1[k-l+2])+\frac{\mathrm{Re}\,\mathrm{Sc}\ \mathrm{Pr}\ \mathrm{Sr}\cos\phi}{(1-\mathrm{Du}\,\mathrm{Sc}\,\mathrm{Sr})}\\ &\cdot\sum_{l=0}^{k}\left(\left(\frac{\mathrm{Ha}^2}{(1+\beta_i\beta_e)^2+\beta_e^{\,2}}\right)(l+1)F_1[l+1](k-l+1)F_1[k-l+1]\right)+\frac{\mathrm{Re}\,\mathrm{Sc}\ \mathrm{Pr}\ \mathrm{Sr}\cos\phi}{(1-\mathrm{Du}\,\mathrm{Sc}\,\mathrm{Sr})}\sum_{l=0}^{k}(2(l+1)F_1[l+1]F_3[k-l]\\ &-(l+1)F_3[l+1]F_1[k-l])-\frac{\mathrm{Kr}}{(1-\mathrm{Du}\,\mathrm{Sc}\,\mathrm{Sr})}F_5[k]\Bigg)((k+2)(k+1))^{-1}.\end{aligned} \tag{15}$$

The transformed equations of the boundary conditions are

$$\begin{aligned}&F_1[0]=1-a,\\ &F_1[1]=0,\\ &F_2[0]=n_3,\\ &F_3[0]=0,\\ &F_4[0]=0,\\ &F_5[0]=0,\\ &F_1[2]=n_1,\\ &F_1[3]=n_2,\\ &F_2[1]=\frac{\gamma}{\mathrm{Ec}},\\ &F_3[1]=n_4,\\ &F_4[1]=n_5,\\ &F_5[0]=n_6.\end{aligned} \tag{16}$$

The values in (16) are used in the recurrence relations (11)– (15) and n_1, n_2, n_3, n_4, n_5, and n_6 are calculated such that they satisfy the boundary conditions at λ=1; then we obtain nondimensional velocity components, temperature distribution, and concentration for various parameters by substituting these constants into the respective Taylor series.

RESULTS AND DISCUSSION

In order to understand the flow characteristics in a better way, the numerical results for the nondimensional velocity components, temperature, and concentration with respect to various fluid and geometric parameters such as Weissenberg number Wi, slip parameter Sl, Biot number γ, Soret number Sr, Dufour number Du, Hall parameter β_e, ion slip parameter β_i, and Prandtl number Pr are calculated in the domain [0, 1] and presented in the form of graphs. A comparative study for the case of Newtonian fluid has been studied and the skin friction values using differential transform method and long series with polynomial coefficients as reported by Bujurke et al. [37] are presented in Table 1.

Table 1: Comparison of the present nondimensional skin friction values with Bujurke et al. [37] for Newtonian case

Serial number	a	Re	$f''(0)$	
			Bujurke et al. [37]	Present
1	0.04106	13.8833	0.0953	0.0954
2	0.15420	31.3577	0.3190	0.3395
3	0.00790	41.0960	0.0166	0.0176
4	0.16480	52.9262	0.3300	0.3593
5	0.01310	67.3621	0.0261	0.0298

The effect of Weissenberg number Wi on velocity components, temperature distribution, and concentration is presented in Figure 2. It is observed that as Wi increases the radial velocity and concentration also increase towards the upper plate and the velocity in X-direction increases towards the center of the plates and then decreases. However, the temperature decreases near the lower plate and then increases towards the upper plate. It is a known fact that for low Reynolds number flows ($0 < \text{Re} < 5$) the viscoelastic effect dominates the inertial effect.

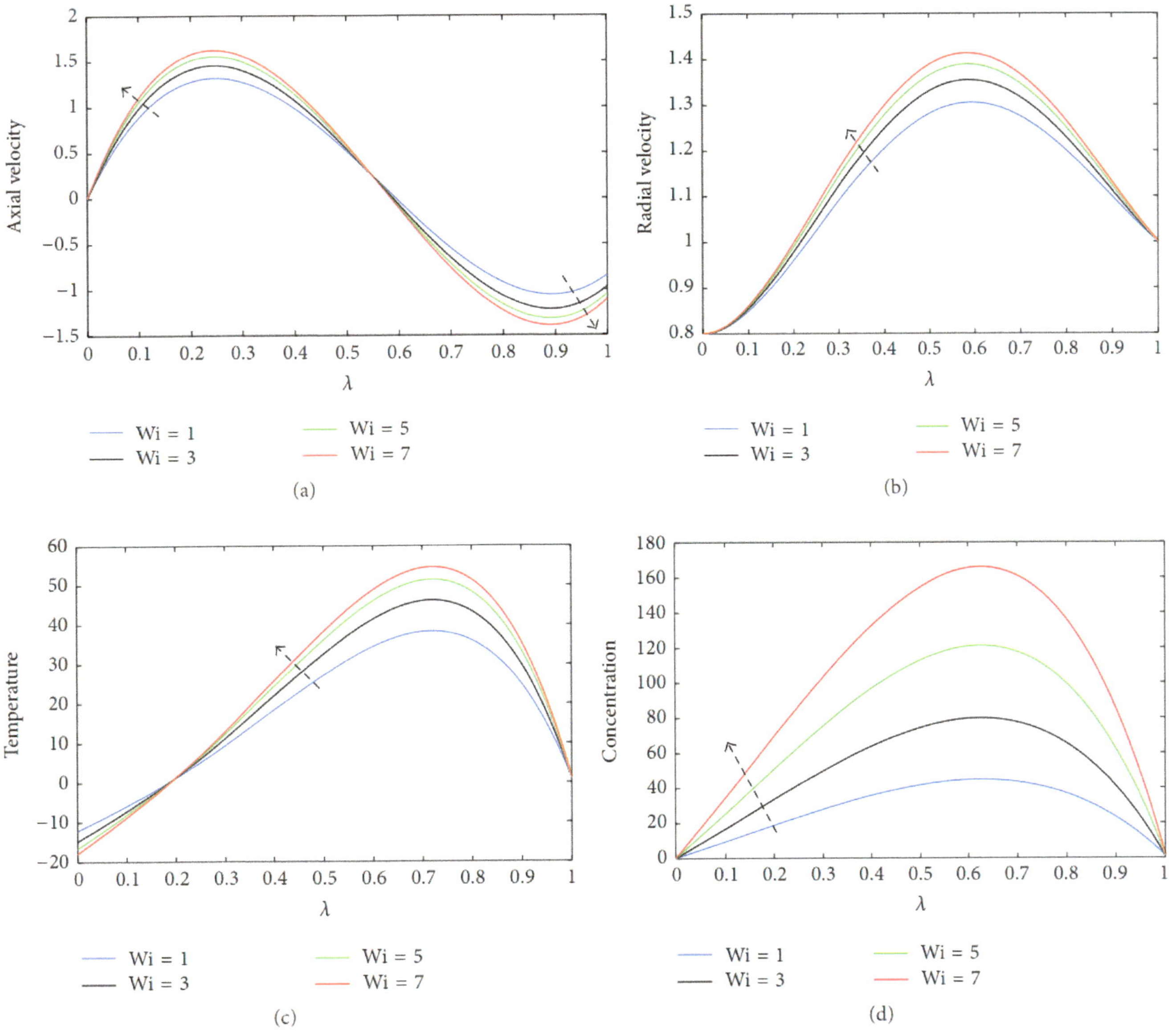

Figure 2: Effect of Wi on (a) axial velocity, (b) radial velocity, (c) temperature, and (d) concentration at $a = 0.2$; Re = 3; $\phi = 0.4$; Ha = 4; $\beta i = 8$; $\beta e = 0.2$; Gr = 10; Gc = 10; $\gamma = 0.2$; Du = 0.2; Sr = 0.2; Sc = 0.22; Kr = 0.2; Ec = 1; Sh = 1; Pr = 0.71; Sl = 0.2.

Figure 3 gives the effect of slip parameter Sl on velocity components, temperature, and concentration. It is evident that as Sl increases, the velocity components and concentration are following the similar trend of Wi whereas the temperature is decreasing. It is due to the fact that the specific permeability decreases the velocity components of the fluid.

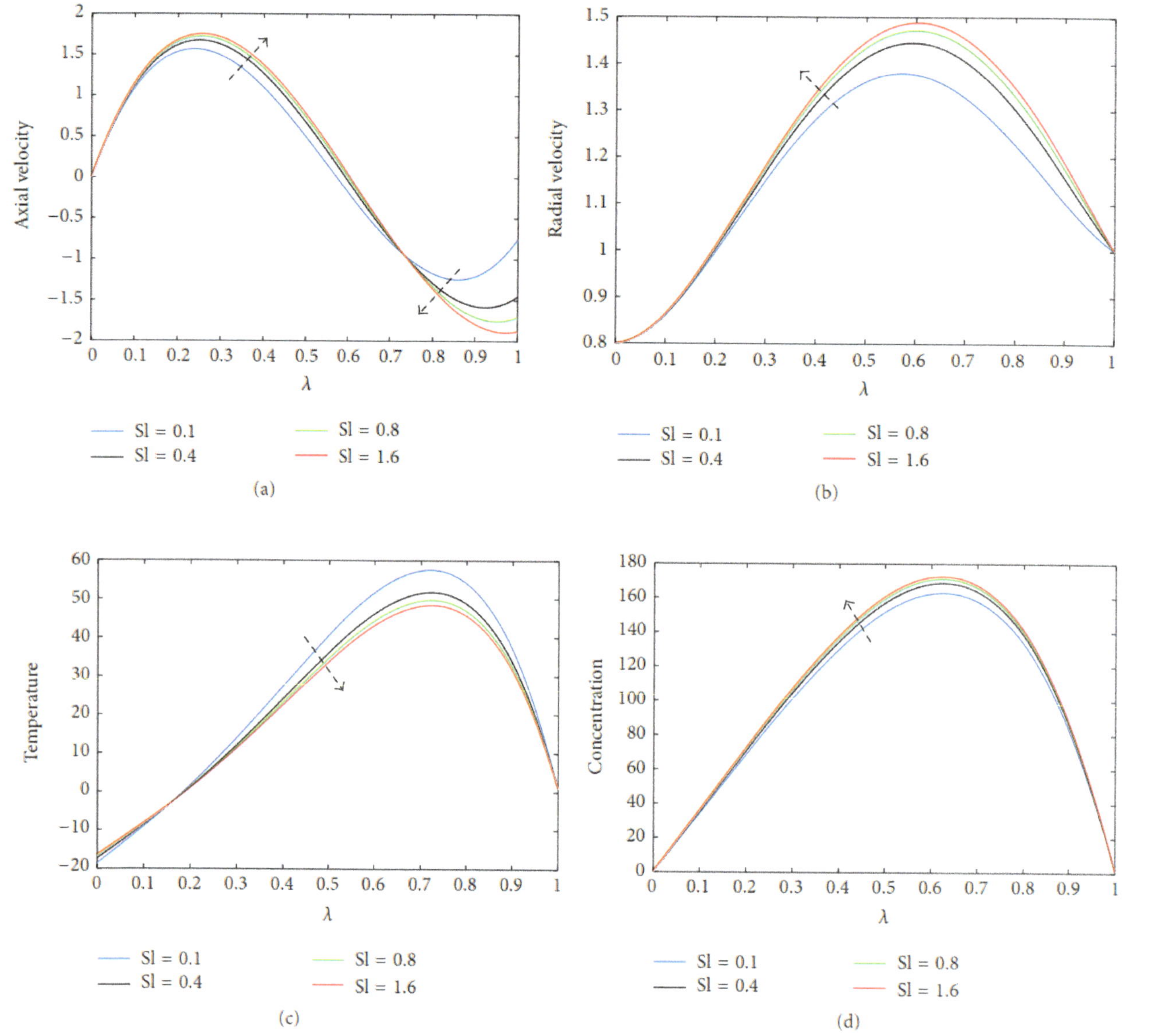

Figure 3: Effect of slip parameter on (a) axial velocity, (b) radial velocity, (c) temperature, and (d) concentration at $a = 0.2$; Re = 3; Wi = 7; $\phi = 0.4$; Ha = 4; $\beta i = 8$; $\beta e = 0.2$; Gr = 10; Gc = 10; $\gamma = 0.2$; Du = 0.2; Sr = 0.2; Sc = 0.22; Kr = 0.2; Ec = 1; Sh = 1; Pr = 0.71.

The effect of Biot number γ on velocity components, temperature, and concentration is displayed in Figure 4. It is noticed that the radial velocity, temperature, and concentration profiles are decreasing with increasing γ and the axial velocity is decreasing towards the center of the plates and then increases. Since γ is inversely proportional to the thermal conductivity of the fluid, the heat transfer between the plates is decreased.

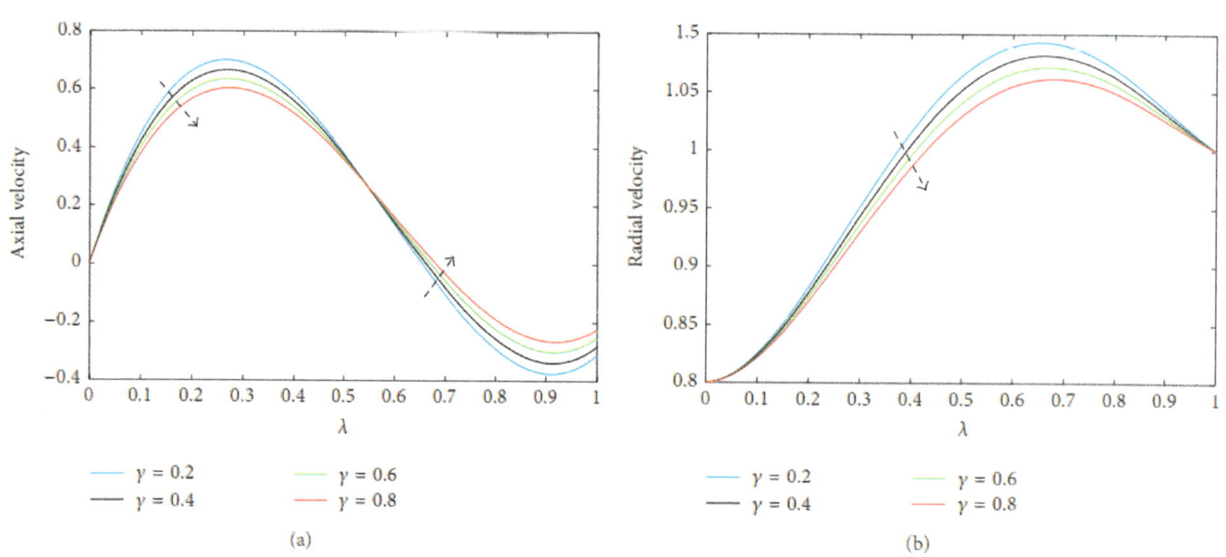

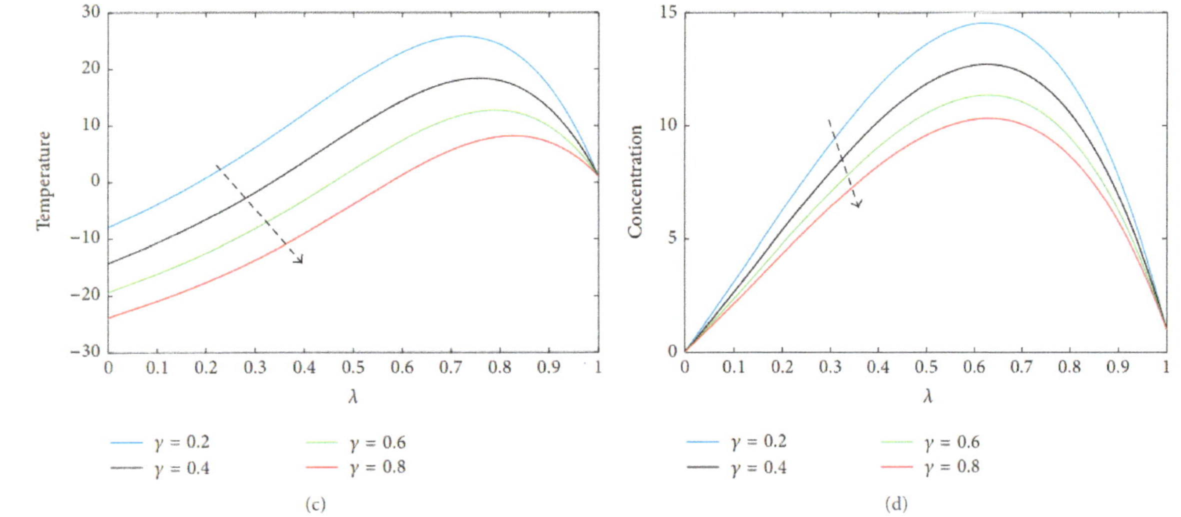

Figure 4: Effect of Biot number on (a) axial velocity, (b) radial velocity, (c) temperature, and (d) concentration at $a = 0.2$; Re = 3; Wi = 1; ϕ = 0.4; Ha = 10; $\beta i = 2$; $\beta e = 0.4$; Gr = 10; Gc = 10; Du = 0.2; Sr = 0.2; Sc = 0.22; Kr = 0.2; Ec = 1; Sh = 1; Pr = 3; Sl = 0.2.

The effect of Soret number Sr on velocity components, temperature distribution, and concentration is studied in Figure 5. As Sr increases, the axial velocity decreases up to the center of the channel and then increases towards the upper plate, whereas the radial velocity and concentration of the fluid decrease towards the upper plate. However, the temperature distribution increases initially and then decreases. This is because of the fact that the mean heat transfer between the plates decreases with the increasing of thermal diffusion.

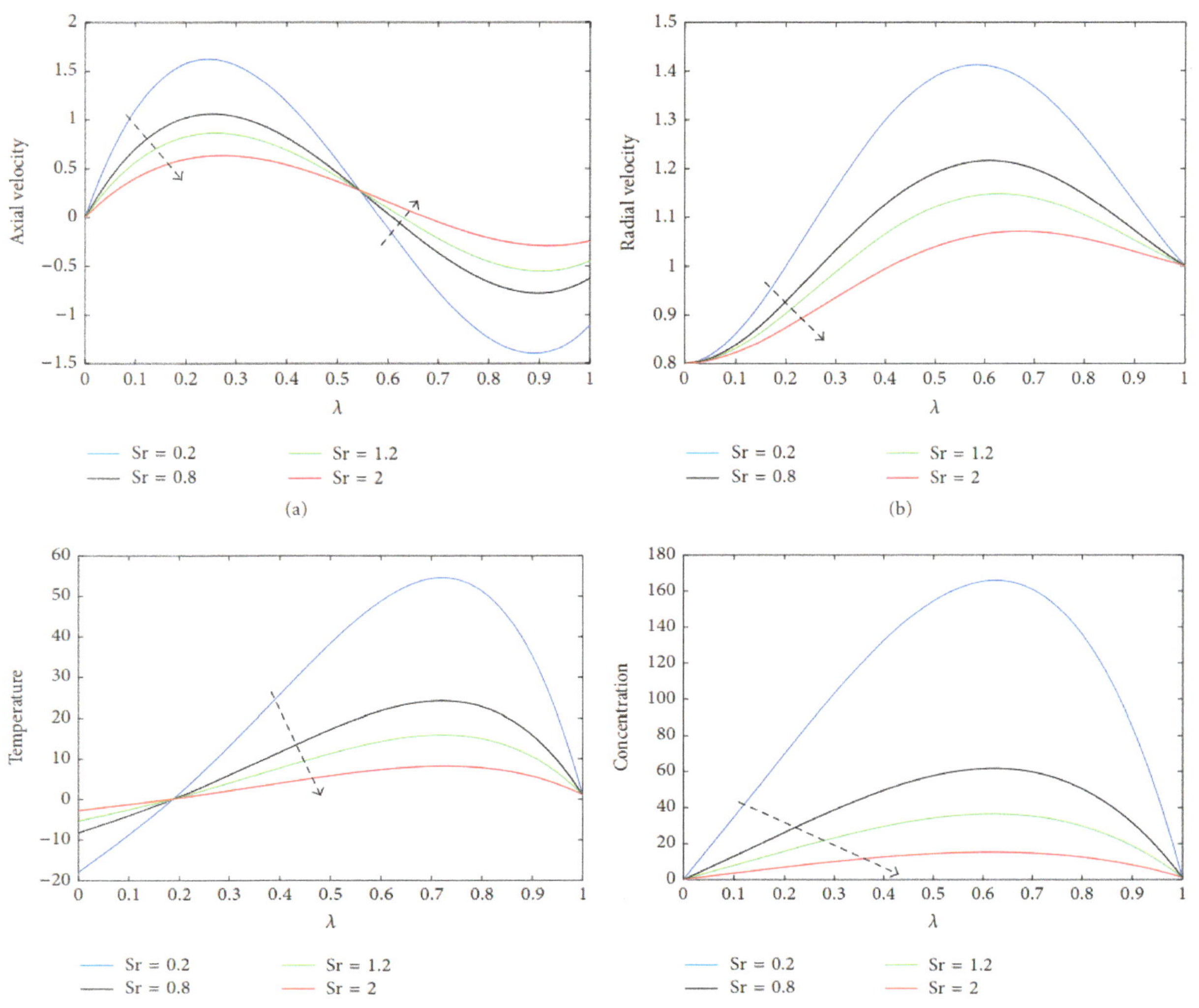

Figure 5: Effect of Sr on (a) axial velocity, (b) radial velocity, (c) temperature, and (d) concentration at $a = 0.2$; Re = 3; Wi = 7; $\phi = 0.4$; Ha = 4; $\beta i = 8$; $\beta e = 0.2$; Gr = 10; Gc = 10; $\gamma = 0.2$; Du = 0.2; Sc = 0.22; Kr = 0.2; Ec = 1; Sh = 1; Pr = 0.71; Sl = 0.2.

The effect of Dufour number Du on the velocity components, temperature, and concentration is shown in Figure 6. It is inferred that as Du increases, the radial velocity and concentration also increase whereas the temperature decreases. However the axial velocity increases towards the center of the channel and then decreases towards the upper plate. This is due to the fact that the concentration of the fluid is enhanced with diffusion in the temperature.

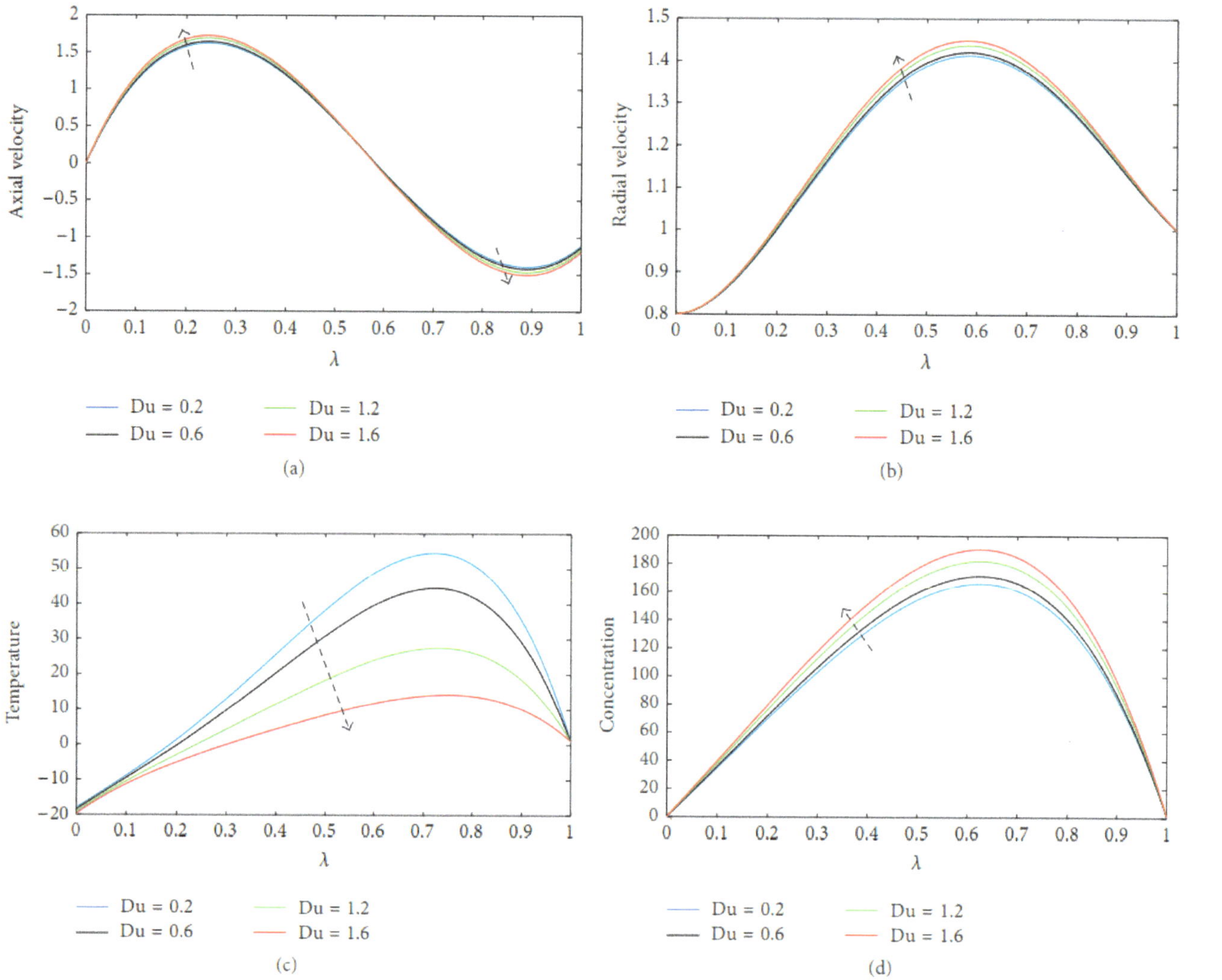

Figure 6: Effect of Du on (a) axial velocity, (b) radial velocity, (c) temperature, and (d) concentration at $a = 0.2$; Re = 3; Wi = 7; $\phi = 0.4$; Ha = 4; $\beta i = 8$; $\beta e = 0.2$; Gr = 10; Gc = 10; $\gamma = 0.2$; Sr = 0.2; Sc = 0.22; Kr = 0.2; Ec = 1; Sh = 1; Pr = 0.71; Sl = 0.2.

The effect of Hall parameter β_e on velocity components, temperature distribution, and concentration is shown in Figure 7. β_e increases the radial velocity and concentration also increases towards the upper plate, whereas the axial velocity increases towards the middle of the channel and then decreases. However, the temperature decreases near the lower plate and then increases towards the upper plate. This is because the effective conductivity decreases the damping force on the flow field.

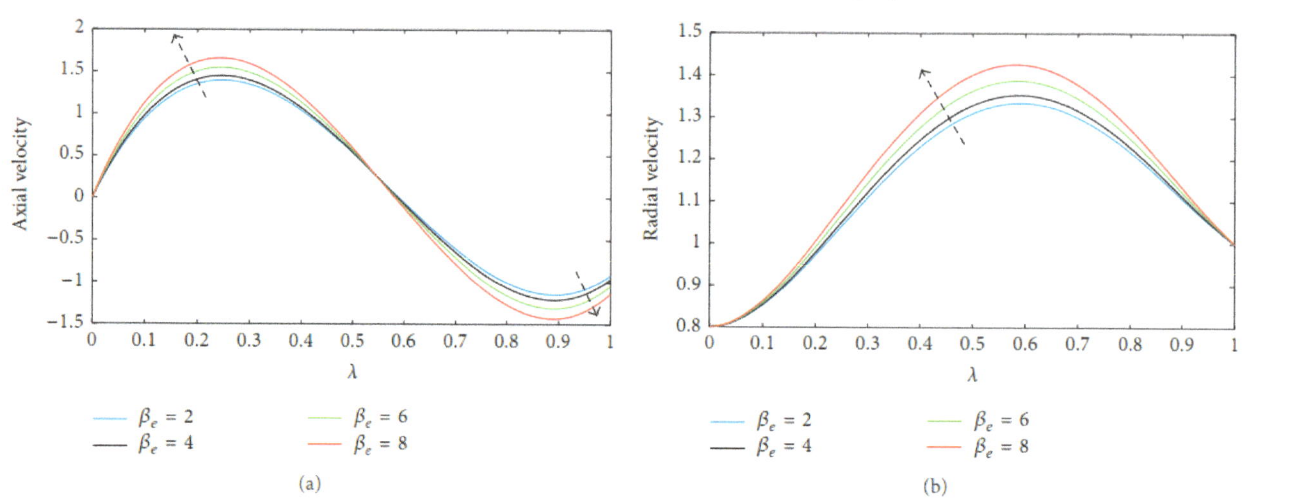

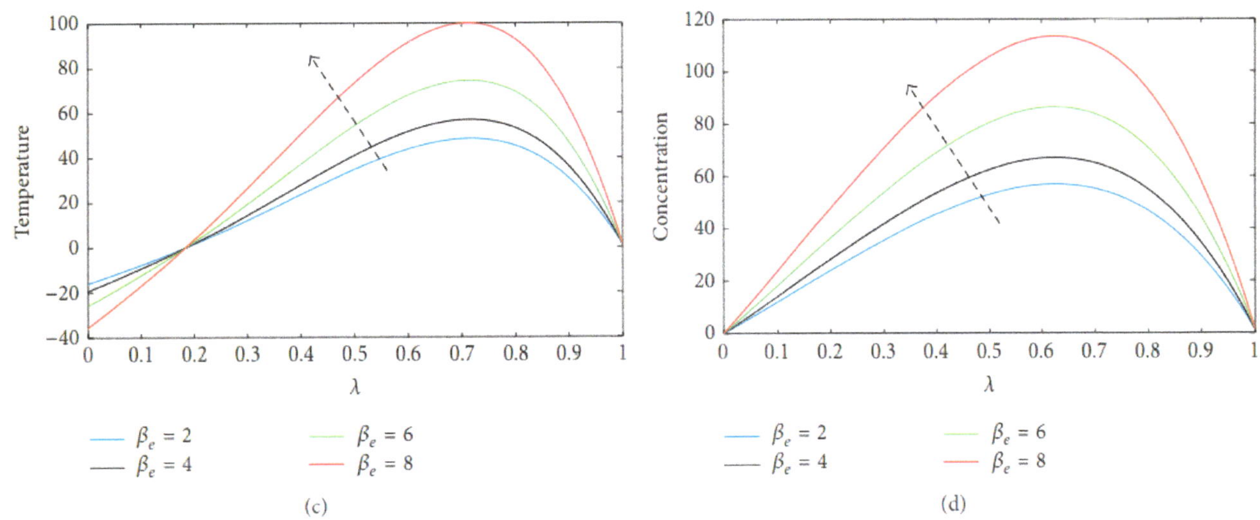

Figure 7: Effect of βe on (a) axial velocity, (b) radial velocity, (c) temperature, and (d) concentration at $a = 0.2$; Re = 3; Wi = 1; $\phi = 0.4$; Ha = 4; $\beta i = 0.2$; Gr = 10; Gc = 10; $\gamma = 0.2$; Du = 0.2; Sr = 0.2; Sc = 0.22; Kr = 0.2; Ec = 1; Sh = 1; Pr = 0.71; Sl = 0.2.

Figure 8 represents the effect of ion slip parameter β_i on velocity components, temperature, and concentration. It is observed that as β_i increases, the velocity components, temperature, and concentration are following the opposite trend of β_e. This is because the increase in the ion slip scales down the effect of Lorentz force [38].

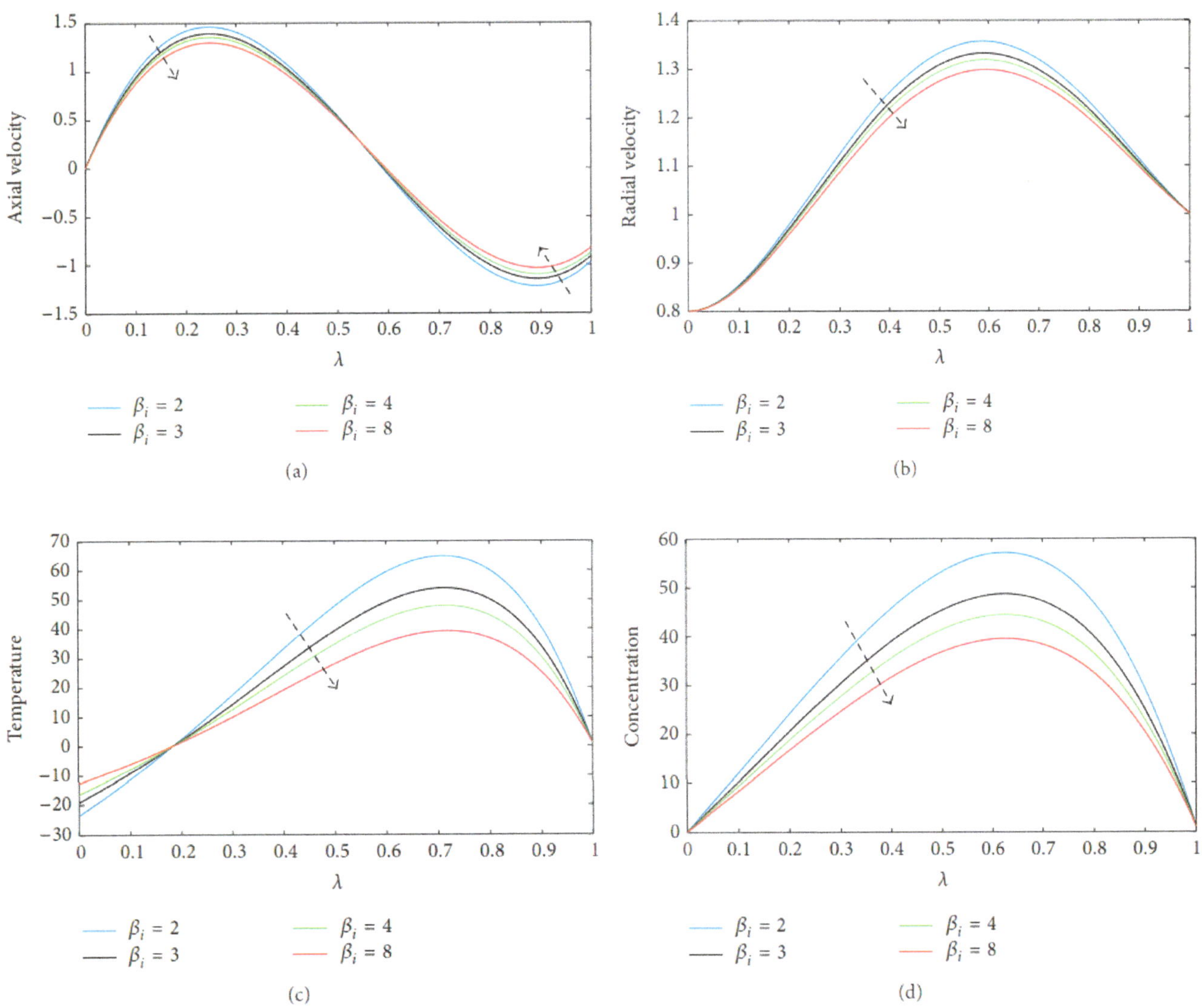

Figure 8: Effect of βi on (a) axial velocity, (b) radial velocity, (c) temperature, and (d) concentration at $a = 0.2$; Re = 3; Wi = 1; $\phi = 0.4$; Ha = 10; $\beta e = 0.4$; Gr = 10; Gc = 10; $\gamma = 0.2$; Du = 0.2; Sr = 0.2; Sc = 0.22; Kr = 0.2; Ec = 1; Sh = 1; Pr = 0.71; Sl = 0.2.

The effect of Prandtl number Pr on velocity components, temperature, and concentration is presented in Figure 9. It is analyzed that as Pr increases the axial velocity decreases towards the center of the plates and then increases, whereas the radial velocity and concentration are decreasing from the lower plate to the upper plate. However, the temperature increases near the lower plate and then decreases towards the upper plate. Physically, if Pr increases the thermal diffusivity decreases and this leads to the reduction in the heat transfer ability at the thermal boundary layer.

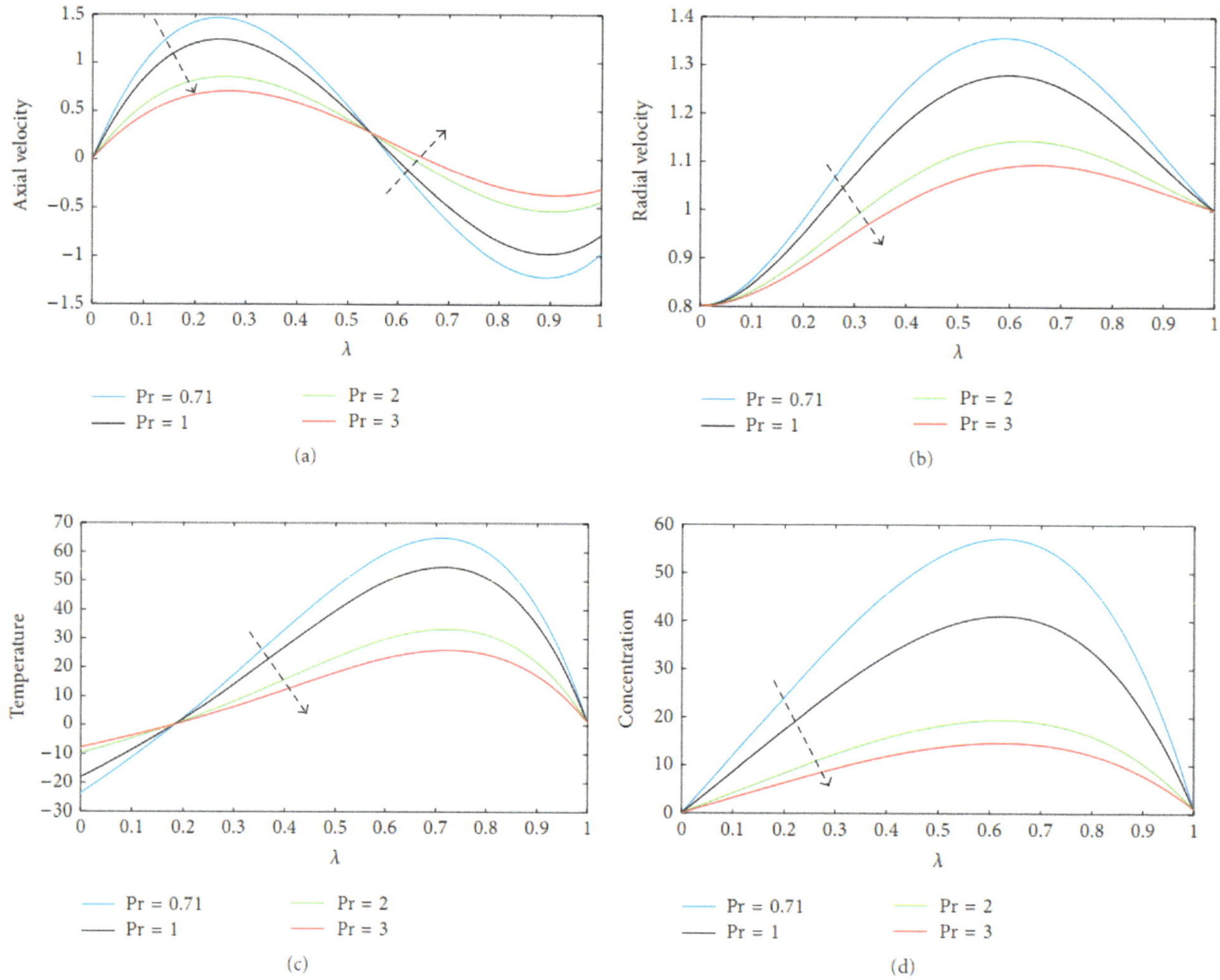

Figure 9: Effect of Pr on (a) axial velocity, (b) radial velocity, (c) temperature, and (d) concentration at $\alpha = 0.2$; Re = 3; Wi = 1; $\phi = 0.4$; Ha = 10; $\beta i = 2$; $\beta e = 0.4$; Gr = 10; Gc = 10; $\gamma = 0.2$; Du = 0.2; Sr = 0.2; Sc = 0.22; Kr = 0.2; Ec = 1; Sh = 1; Sl = 0.2.

CONCLUSIONS

The effects of thermal-diffusion and diffusion-thermo on an unsteady free convective chemically reacting upper convected Maxwell fluid flow between two porous parallel plates with Hall and ion slip currents under the influence of slip and convective boundary conditions are considered. The flow field equations are reduced to nonlinear ordinary differential equations by using similarity transformations and then solved by differential transform method. The results are obtained for the nondimensional velocity components, temperature, and concentration distributions with respect to various fluid and geometric parameters such as Wi, Sl, γ, Sr, Du, β_e, β_i, and Pr and we observe the following:

- The concentration of the fluid is enhanced with slip parameter, whereas the temperature of the fluid is decreased with increasing of Biot number.
- Soret and Dufour parameters exhibit opposite trend on concentration distribution.
- The Weissenberg number and Hall parameter show similar effects on velocity components.
- The Hall and ion slip parameters exhibit reverse trend on temperature and concentration distributions.
- The present results are compared with the previously published work [37] and are found to be in good agreement.

These results have possible applications in science and technology such as fermentation, biorheology, geophysics, polymer industry, composite processing, food processing, and petroleum industries.

ACKNOWLEDGMENTS

One of the authors (Pravin Kashyap Kambhatla) is grateful to the University Grants Commission, Government of India, for providing financial support in the form of Senior Research Fellowship (F.2-18/2012(SA-I)).

REFERENCES

1. A. S. Berman, "Laminar flow in channels with porous walls," Journal of Applied Physics, vol. 24, pp. 1232–1235, 1953.
2. F. M. White Jr., B. F. Barfield, and M. J. Goglia, "Laminar flow in a uniformly porous channel," Journal of Applied Mechanics, vol. 25, no. 4, pp. 613–617, 1958.
3. G. Walker and T. Davies, "Mass transfer in laminar flow between parallel permeable plates," AIChE Journal, vol. 20, no. 5, pp. 881–889, 1974.
4. E. A. Hamza, "Suction and injection effects on a similar flow between parallel plates," Journal of Physics D: Applied Physics, vol. 32, no. 6, pp. 656–663, 1999.
5. D. Srinivasacharya, J. V. Ramana Murthy, and D. Venugopalam, "Unsteady stokes flow of micropolar fluid between two parallel porous plates," International Journal of Engineering Science, vol. 39, no. 14, pp. 1557–1563, 2001.
6. M. Sheikholeslami, M. Hatami, and D. D. Ganji, "Micropolar fluid flow and heat transfer in a permeable channel using analytical method," Journal of Molecular Liquids, vol. 194, pp. 30–36, 2014.
7. Y.-L. Chen and K.-Q. Zhu, "Couette–Poiseuille flow of Bingham fluids between two porous parallel plates with slip conditions," Journal of Non-Newtonian Fluid Mechanics, vol. 153, no. 1, pp. 1–11, 2008.
8. K. D. Singh and R. Pathak, "Effect of slip conditions and Hall current on an oscillatory convective MHD flow in a rotating vertical porous channel with thermal radiation," International Journal of Applied Mathematics and Mechanics, vol. 9, pp. 60–77, 2013.
9. A. N. Bhat and M. L. Mittal, "Heat transfer in a MHD channel with uniform wall heat flux—effects of Hall and ion slip currents," International Journal of Heat and Mass Transfer, vol. 23, no. 7, pp. 919–926, 1980.
10. D. Srinivasacharya and K. Kaladhar, "Mixed convection flow of couple stress fluid between parallel vertical plates with Hall and Ion-slip effects," Communications in Nonlinear Science and Numerical Simulation, vol. 17, no. 6, pp. 2447–2462, 2012.
11. A. K. Singh and R. S. R. Gorla, "Free convection heat and mass transfer with Hall current, Joule heating and thermal diffusion," Heat and Mass Transfer, vol. 45, no. 11, pp. 1341–1349, 2009.·
12. A. Raptis, C. Massalas, and G. Tzivanidis, "Hydromagnetic free convection flow through a porous medium between two parallel plates," Physics Letters A, vol. 90, no. 6, pp. 288–289, 1982.
13. O. Abdulaziz and I. Hashim, "Fully developed free convection heat and mass transfer of a micropolar fluid between porous vertical plates," Numerical Heat Transfer; Part A: Applications, vol. 55, no. 3, pp. 270–288, 2009.
14. O. D. Makinde and A. Aziz, "MHD mixed convection from a vertical plate embedded in a porous medium with a convective boundary condition," International Journal of Thermal Sciences, vol. 49, no. 9, pp. 1813–1820, 2010.
15. S. Yao, T. Fang, and Y. Zhong, "Heat transfer of a generalized stretching/shrinking wall problem with convective boundary conditions," Communications in Nonlinear Science and Numerical Simulation, vol. 16, no. 2, pp. 752–760, 2011.
16. C. RamReddy, T. Pradeepa, and D. Srinivasacharya, "Similarity solution for free convection flow of a micropolar fluid under convective boundary condition via Lie scaling group transformations," Advances in High Energy Physics, vol. 2015, Article ID 650813, 16 pages, 2015.
17. A. Postelnicu, "Heat and mass transfer by natural convection at a stagnation point in a porous medium considering Soret and Dufour effects," Heat and Mass Transfer, vol. 46, no. 8-9, pp. 831–840, 2010.
18. T. Hayat, T. Muhammad, S. A. Shehzad, and A. Alsaedi, "Soret and Dufour effects in three-dimensional flow over an exponentially stretching surface with porous medium, chemical reaction and heat source/sink," International Journal of Numerical Methods for Heat and Fluid Flow, vol. 25, no. 4, pp. 762–781, 2015.
19. A. J. Chamkha and A. M. Rashad, "Unsteady heat and mass transfer by MHD mixed convection flow from a rotating vertical cone with chemical reaction and Soret and Dufour effects," The Canadian Journal of Chemical Engineering, vol. 92, no. 4, pp. 758–767, 2014. ·
20. O. Ojjela and N. Naresh Kumar, "Unsteady MHD mixed convective flow of chemically reacting and radiating couple stress fluid in a porous medium between parallel plates with Soret and Dufour effects,"Arabian Journal for Science and Engineering, vol. 41, no. 5, pp. 1941–1953, 2016.
21. J. G. Oldroyd, "On the formulation of rheological equations of state," Proceedings of the Royal Society of London, Series A: Mathematical, Physical and Engineering Sciences, vol. 200, pp. 523–591, 1950.
22. M. Awais, T. Hayat, A. Alsaedi, and S. Asghar, "Time-dependent three-dimensional boundary layer flow of a Maxwell fluid," Computers and Fluids, vol. 91, pp. 21–27, 2014.
23. J. J. Choi, Z. Rusak, and J. A. Tichy, "Maxwell fluid suction flow in a channel," Journal of Non-Newtonian Fluid Mechanics, vol. 85, no. 2-3, pp. 165–187, 1999.
24. T. Hayat, N. Ali, and S. Asghar, "Hall effects on peristaltic flow of a Maxwell fluid in a porous medium,"Physics Letters A, vol. 363, no. 5-6, pp. 397–403, 2007.
25. Z. Abbas, M. Sajid, and T. Hayat, "MHD boundary-layer flow of an upper-convected Maxwell fluid in a porous channel," Theoretical and Computational Fluid Dynamics, vol. 20, no. 4, pp. 229–238, 2006.
26. S. Mukhopadhyay and R. S. Gorla, "Unsteady MHD boundary layer flow of an upper convected Maxwell fluid past a stretching sheet with first order constructive/destructive chemical reaction," Journal of Naval Architecture and Marine Engineering, vol. 9, no. 2, pp. 123–133, 2012.
27. T. Hayat and Z. Abbas, "Channel flow of a Maxwell fluid with chemical reaction," Zeitschrift für ange-wandte

Mathematik und Physik, vol. 59, no. 1, pp. 124–144, 2008.

28. A. Alizadeh-Pahlavan and K. Sadeghy, "On the use of homotopy analysis method for solving unsteady MHD flow of Maxwellian fluids above impulsively stretching sheets," Communications in Nonlinear Science and Numerical Simulation, vol. 14, no. 4, pp. 1355–1365, 2009.

29. J. K. Zhou, Differential Transformation and Its Applications for Electrical Circuits, Huazhong University Press, Wuhan, China, 1986.

30. C. K. Chen and S. H. Ho, "Solving partial differential equations by two-dimensional differential transform method," Applied Mathematics and Computation, vol. 106, no. 2-3, pp. 171–179, 1999.

31. F. Ayaz, "Solutions of the system of differential equations by differential transform method," Applied Mathematics and Computation, vol. 147, no. 2, pp. 547–567, 2004.

32. S. Mosayebidorcheh, M. Vatani, D. D. Ganji, and T. Mosayebidorcheh, "Investigation of the viscoelastic flow and species diffusion in a porous channel with high permeability," Alexandria Engineering Journal, vol. 53, no. 4, pp. 779–785, 2014.

33. M. Sheikholeslami, M. M. Rashidi, D. M. Al Saad, F. Firouzi, H. B. Rokni, and G. Domairry, "Steady nanofluid flow between parallel plates considering thermophoresis and Brownian effects," Journal of King Saud University—Science, 2015.

34. M. Sheikholeslami and D. D. Ganji, "Nanofluid flow and heat transfer between parallel plates considering Brownian motion using DTM," Computer Methods in Applied Mechanics and Engineering, vol. 283, pp. 651–663, 2015.

35. M. Hatami and D. Jing, "Differential transformation method for newtonian and non-newtonian nanofluids flow analysis: compared to numerical solution," Alexandria Engineering Journal, 2016.

36. S. H. Esmail and M. M. Taha, "Application of the differential transform method to the advection-diffusion equation in three-dimensions," Progress in Physics, vol. 12, no. 3, pp. 205–210, 2016.

37. N. M. Bujurke, V. S. Madalli, and B. G. Mulimani, "Long series analysis of laminar flow through parallel and uniformly porous walls of different permeability," Computer Methods in Applied Mechanics and Engineering, vol. 160, no. 1-2, pp. 39–56, 1998.

38. G. W. Sutton and A. Sherman, Engineering Magnetohydrodynamics, McGraw-Hill, New York, NY, USA, 1965.

CHAPTER 16

COMPUTATIONALLY EFFICIENT ASSESSMENTS OF THE EFFECTS OF RADIATIVE TRANSFER, TURBULENCE RADIATION INTERACTIONS, AND FINITE RATE CHEMISTRY IN THE MACH 20 REENTRY F FLIGHT VEHICLE

Gautham Krishnamoorthy and Lauren Elizabeth Clarke

Department of Chemical Engineering, University of North Dakota, Harrington Hall Room 323, 241 Centennial Drive, Grand Forks, ND 58202-7101, USA

ABSTRACT

Effects of finite rate chemistry, radiative heat transfer, and turbulence radiation interactions (TRI) are assessed in a fully coupled manner in simulations of the Mach 20 Reentry F flight vehicle. Add-on functions were employed to compute a Planck mean absorption coefficient and the temperature self-correlation term (for TRI effects) in the optically thin shock layer. Transition onset was induced by specifying a wall roughness height at the experimentally observed transition location. The chemistry was modeled employing eight elementary reactions and an equilibrium approach allowing species to relax towards their chemical equilibrium values over the process characteristic time scale. The wall heat fluxes in the turbulent region, density, and velocity profiles compared reasonably well against measurements as well as similar calculations reported previously. The density predictions were more sensitive to the choice of modeling options than the velocities. The radiative source term magnitude agreed closely with its measurements deduced from shock tube experiments. The TRI model predicted a 60% enhancement in emission due to temperature fluctuations in the turbulent boundary layer. While the variations in density and velocity predictions among the models diminished along the length of the body, the O and NO prediction variations extended well into the turbulent boundary layer.

INTRODUCTION

Planetary entry vehicles are subjected to significant aerothermal heating as they dissipate their kinetic energy into the atmosphere of their destination planet. The very high velocity of an earth bound reentry vehicle causes viscous dissipation and the formation of a shock wave resulting in very high air-temperatures surrounding the vehicle. Intense thermochemical processes occur in the region between the body surface and the shock. The molecules in the surrounding air dissociate and might be far from their equilibrium state [1, 2]. These reactions may be further catalyzed by the products resulting from the decomposition and ablation of the vehicle's thermal protection system (TPS). For earth reentry, the concentrations of N and O and their recombination to produce NO molecules surrounding the vehicle in particular have been determined to be important towards determining the shock and

Citation: Gautham Krishnamoorthy and Lauren Elizabeth Clarke, "Computationally Efficient Assessments of the Effects of Radiative Transfer, Turbulence Radiation Interactions, and Finite Rate Chemistry in the Mach 20 Reentry F Flight Vehicle," Journal of Computational Engineering, vol. 2016, Article ID 8401249, 14 pages, 2016. doi:10.1155/2016/8401249.

boundary layer characteristics and for providing more reliable TPS sizing requirements [3, 4]. Further, at certain reentry conditions radiative transfer can contribute both to the surface heat fluxes and to the cooling of the shock and boundary layers surrounding the vehicle depending on the spectral emission and absorption properties of the surrounding molecules [5, 6]. Computational Fluid Dynamics (CFD) through a coupled modeling of nonequilibrium, reacting flows with radiative heat transfer can yield valuable insights into the flow characteristics within these hypersonic flight environments [7, 8]. However, due to the fact that radiative heat transfer calculations are inherently expensive as a result of its strong spatial, directional, and wave-length dependencies [9, 10], the radiation field is often computed in a 1D domain. Further, the coupling between the radiation and the thermal field is accomplished in a loose coupling/decoupled manner [11]. These approaches are justified by facts that, at high reentry velocities, the radiative heat flux to the stagnation point is significant. The shock stand-off distance is small compared with the nose radius of vehicles, and a spatial change of physical properties is much more severe in the direction normal to the shock than in the tangential direction. Since it is well known from simulations of combustion phenomena that a strong coupling between the fluid dynamics and radiative heat transfer is necessary to accurately predict the concentrations of minor species such as O, N, and NO [12, 13], this may hold true in hypersonic flows as well. Similarly, combustion simulations have also examined and highlighted the importance of turbulence-chemistry and turbulence radiation interaction (TRI) models on these minor specie concentrations [14, 15]. Therefore, the primary goals of this paper are as follows:

To demonstrate a simple methodology for providing initial assessments of radiative transfer and TRI effects in a computationally efficient manner in hypersonic flow simulations for an optically thin shock layer.

- To provide an initial assessment of fully taking into account radiative transfer and TRI effects on the density, velocity, and O and NO predictions surrounding a hypersonic vehicle.
- To assess the differences between employing finite rate chemistry and equilibrium based thermochemical models on the density, velocity, and O and NO predictions surrounding the vehicle [16, 17].

The hypersonic flow scenario specifically chosen for the purpose is the Reentry F flight vehicle which is a cone with a 5-degree half angle with a small spherical nose tip (Figure 1) [18]. This has been a valuable dataset for evaluating the predictions of several CFD codes. The criteria that make this a good test case for this investigation are as follows:

- The gas temperature was found to be more than 6000 K at the nose tip with dissociation of oxygen and nitrogen occurring. Further dissociation of oxygen in the boundary layer was observed as a result of viscous dissipation resulting in temperatures around 3000 K [19]. The high temperatures make it a good test case to investigate the sensitivity of the predictions to equilibrium chemistry, finite rate chemistry, and radiation models.
- At an altitude of 24.4 km (which is the conditions investigated in this study), the laminar to turbulence transition occurred at about 62.5% the total wall length translating to 2.5 m along the length of the simulated vehicle. This implies a significant interplay between turbulence, chemistry, and radiative transfer spanning the remaining 1.5 m of the simulated cone providing sufficient length of the domain to investigate the effects of TRI on the flow field.
- Local thermodynamic equilibrium (LTE) can be assumed at this altitude. When shock passes through air, the energy goes initially into the translational energy of the molecules and is later redistributed into the rotational and vibrational energy levels. At 24.4 km this energy redistribution rapidly results from collisions but can be much slower at higher altitudes. However, LTE is a good approximation at altitudes between 15 km and 45 km [20]. Therefore, this alleviates the need to solve separate transport equations for the rotational and vibrational temperatures and subsequently use a geometric average of those temperatures to estimate the reaction rates, radiative transfer.
- In general, the pathlines (the path followed by a fluid element) after the strongly dissociating shock layer will determine the chemical composition at any given location [2]. Therefore, at large distances downstream of the bow shock the chemistry is assumed to relax towards equilibrium. Considering the large length of the vehicle (4 m), the differences between the finite rate and equilibrium predictions can be quantified and the validity of this assumption can be assessed at a number of locations along the body including the turbulent boundary layer.
- The availability of numerical results encompassing this scenario from previous investigators is needed to assess the validity of our simulation techniques.

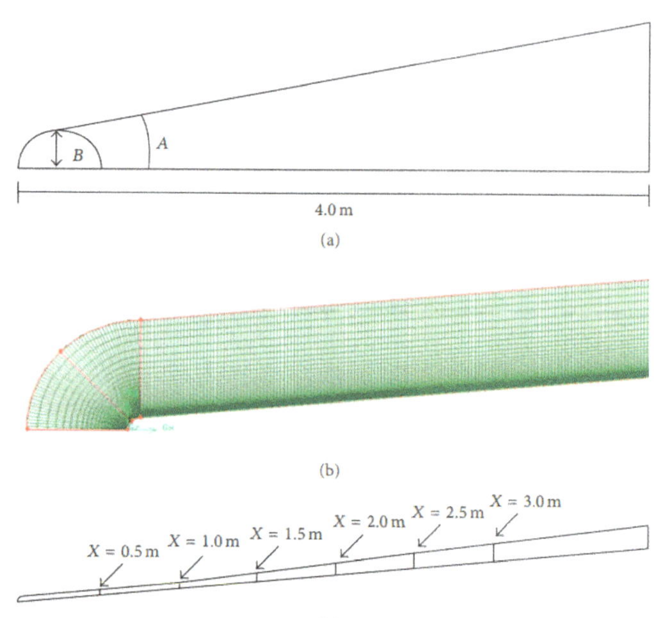

Figure 1: (a) Geometry of the reentry F vehicle. A 5° half-angle cone ("A") with a spherical nose tip of radius 0.00254 m (represented as "B") and a total length of 4.0 m; (b) the mesh associated with the vehicle geometry employed in this study; (c) the entire computational domain along with the lines where the results were monitored.

Previous simulations of this reentry condition have been carried out by several researchers [21–26] with a focus mainly on predicting the convective heat transfer rates on the surface. Since the contribution of surface radiative fluxes at these reentry conditions was expected to be negligible, these previous investigations assumed equilibrium chemistry and neglected radiative transfer. Consequently, finite rate chemistry, radiative transfer, and TRI effects on the temperature and species concentration have not been rigorously assessed. For instance, Wood et al. [21] provided three equilibrium air numerical solutions employing thin-layer Navier Stokes equation methods and viscous shock layer solution methods employing equilibrium air chemistry and neglecting the effects of radiative transfer. The viscous shock layer method was also determined to work well in the study by Thompson et al. [22]. Hamilton et al. [23] employed a variable entropy boundary layer code in conjunction with equilibrium chemistry. Recent studies have emphasized the fact that the prediction of transitional heating from this experiment is beyond the current state of the art [27]. Therefore, as a result of judicious a posteriori modeling of the transitional region (i.e., beginning and end of the transition region), the numerical predictions of Sutton et al. [24] were in excellent agreement with the transitional heating measurements. Further, by carrying out several detailed predictions of the Reentry F flight data, Wurster et al. [25] and more recently Barnhardt and Candler [26] have emphasized the sensitivities of nose bluntness and the angle of attack considered in the simulations on the heating rate predictions.

METHODS

The simulations in this study were carried out employing the density based solver framework of the commercial code ANSYS FLUENT (version 12) [30]. Further, user-defined functions (UDFs) were developed to compute a temperature and pressure dependent absorption coefficient and model the TRI. In order to assess the influence of different models on the boundary and shock layer predictions, four scenarios were investigated:

- Finite rate chemistry without radiative transfer.
- Fully coupled finite rate chemistry calculations with radiative transfer.
- Fully coupled finite rate chemistry and radiative transfer calculations accounting for TRI.
- Chemical equilibrium calculations without radiative transfer/TRI where the species were allowed to relax to their chemical equilibrium state depending on the characteristic (convective/diffusive + chemical) time scale associated with the process.

Geometry and Meshing

In order to get an initial assessment of the impact of different models, 2D axisymmetric flow conditions were assumed as in previous studies [19] by neglecting the experimentally reported angle of attack of 0.14 degrees. Further, the ablation of the nose tip was neglected. The nose tip radius (B, in Figure 1) was found to increase from 0.00254 m to 0.00343 m during the experiments. However, neglecting this has a minimal impact on the monitor locations along the body. The entire domain was meshed employing 280,000 cells employing a boundary layer meshing strategy. Comparing this against the 83,200 cells employed in a previously published simulation of these conditions [19] indicates the adequacy of this mesh resolution to obtain grid independent results.

Radiative Transfer and TRI Modeling

Since the shock layer surrounding the vehicle was determined to be optically thin ($kL \ll 1$), where k is the absorption coefficient (in m^{-1}) and L the path-length (in m), an appropriately determined Planck mean absorption coefficient was computed based on the local pressure and temperature [6] and implemented as a user-defined function (UDF). This is similar to the approach employed in the combustion modeling community where Planck mean absorption coefficients are routinely employed to couple radiation with the flow field in optically thin flames particularly when radiation is not the dominant energy transfer mechanism [31]. We also like to point out that the goal here was to provide an initial assessment on radiative transfer on the temperature and species concentration and not on a highly accurate radiative transfer calculation. A highly accurate spectral 1D calculation in itself would be 1 to 4 orders of magnitude longer than the time to compute a fluid solve [7].

To fully account for the effects of TRI within the context of time-averaged flow simulations (such as Reynolds Averaged Navier Stokes Equation Models) requires time-averaging of the radiative transport equation (RTE) written in terms of the radiative intensity (I). In the absence of scattering, this may be written as

$$\nabla \cdot \left(\overline{I}(\vec{r},\hat{s})\,\hat{s}\right) = -\overline{kI}(\vec{r},\hat{s}) + \overline{kI_b}(\vec{r},\hat{s}). \tag{1}$$

In (1) "$\vec{r}$" and "$\hat{s}$" are the position and directional vectors, respectively, and Ib is the blackbody emissive power. Equation (1) results in several "closure terms" that require some form of modeling to fully account for TRI, mainly the temperature self-correlation (TSC) and absorption-temperature correlation (second term on the right hand side of (1)) and the absorption-intensity correlation (first term on the right hand side of (1)) [32]. However, for an optically thin shock layer, the thin eddy approximation may be invoked, which neglects the absorption-intensity correlation. With the additional assumption of computing the absorption coefficients based on mean species and temperature values we get

$$\nabla \cdot \left(\overline{I}(\vec{r},\hat{s})\,\hat{s}\right) = -k\overline{I}(\vec{r},\hat{s}) + k\overline{I_b}(\vec{r},\hat{s}). \tag{2}$$

The validity of these approximations has also been shown in recent studies of hypersonic turbulent boundary layers where the effects of absorption TRI due to radiation emitted in the shock layer were minimal [33]. Further, calculations in this study indeed showed that magnitude of emission was more than three orders of magnitude larger than the absorption term. Therefore, only the TSC term (the time-averaged $\overline{I_b}$ term in (2)) was employed to account for TRI effects. The time-averaged blackbody emissive power may be expanded in Taylor's series about the mean temperature as [32]

$$I_b(T) = \sum_{n=0}^{\infty} \frac{T'^n}{n!} \frac{d^n}{dT^n} I_b(T)\Big|_{\overline{T}}, \tag{3}$$

where T' is the fluctuation in the temperature about the mean. Time-averaging equation (3) results in an expression for the time-averaged emissive power as

$$\overline{I_b}(T) = \sum_{n=0}^{\infty} \frac{\overline{T'^n}}{n!} \frac{d^n}{dT^n} I_b(T)\Big|_{\overline{T}}. \tag{4}$$

Expanding (4) results in the following expression:

$$\frac{\overline{T^4}}{\overline{T}^4} = \left\{1 + 6\frac{\overline{T'^2}}{\overline{T}^2} + 4\frac{\overline{T'^3}}{\overline{T}^3} + \frac{\overline{T'^4}}{\overline{T}^4}\right\}. \tag{5}$$

If the probability density functions (pdfs) of the temperature fluctuations can be assumed to be Gaussian, then odd-order terms can be assumed to be small or zero as suggested by data from the literature [34]. Further, experimental evidence points to the fact that [35]

$$\frac{\overline{T^4}}{\overline{T}^4} \gg \frac{\overline{T'^4}}{\overline{T}^4}. \tag{6}$$

Therefore, the TSC term in this study was modeled as

$$\frac{\overline{T^4}}{\overline{T}^4} = \left\{1 + 6\frac{\overline{T'^2}}{\overline{T}^2}\right\}. \tag{7}$$

The turbulence was modeled in this study using the realizable k-ε turbulence model. A transport equation for the temperature variance (T_{var}) was solved [30] employing the computed values of turbulent viscosity (μ_t), turbulent kinetic energy (k), and the turbulence dissipation rates (ε) as

$$\frac{\partial(\rho T_{\text{var}})}{\partial t} + \nabla \cdot [\rho \vec{v} T_{\text{var}}] = \nabla\left[\frac{\mu_t}{\sigma_t}\nabla T_{\text{var}}\right] + C_g \mu_t (\nabla T)^2 - C_d \rho \frac{\varepsilon}{k}(T_{\text{var}}). \tag{8}$$

Constants σ_t, C_g, and C_d in (4) take the values 0.85, 2.86, and 2.0, respectively [30].

Chemistry Modeling

To account for finite rate chemistry effects, five species were considered: N_2, O_2, NO, N, and O. The temperatures encountered in this study were below the ionization temperatures (>9000 K) of the species. In this study, the kinetic rate constants corresponding to the two temperature model of Park [1] were evaluated employing the translational temperature. In the Park model, an average of the translational and vibrational temperatures needs to be employed to evaluate the kinetic rates. However, since LTE can be assumed at the altitudes considered in this study, the need to solve a separate transport equation for the vibrational temperature is alleviated [20]. The rate constants employed in this study are listed in Table 1. The species transport properties were estimated employing the kinetic theory of gases employing Lennard-Jones parameters [30]. The specific heat of each gas species was computed from polynomial curve-fits encompassing the temperature range 300 K to 15000 K listed in Sockalingam and Tabiei [28, 29].

Table 1: Reactions and rate constants employed in this study [28]

Reaction	Preexponential factor		Activation energy (J/kmol)		Temperature exponent	
	M = O_2, N_2, NO	M = N, O	M = O_2, N_2, NO	M = N, O	M = O_2, N_2, NO	M = N, O
$N_2 + M \rightarrow 2N + M$	7.0E + 18	3.0E + 19	9.4114E + 08	9.4114E + 08	−1.6	−1.6
$O_2 + M \rightarrow 2O + M$	2.0E + 18	1.0E + 19	4.9468E + 08	4.9468E + 08	−1.5	−1.5
$NO + M \rightarrow N + O + M$	5.0E + 12	1.1E + 14	6.2771E + 08	6.2771E + 08	0.0	0.0
$N_2 + O \rightarrow NO + N$	6.4E + 14		3.1926E + 08		−1.0	
$NO + O \rightarrow O_2 + N$	8.4E + 09		1.6171E + 08		0.0	

The general conservation equation for the mass fraction of specie "i" (Y_i) can be written as

$$\frac{\partial}{\partial t}(\rho Y_i) + \nabla \cdot (\rho \vec{v} Y_i) = -\nabla \cdot \vec{J}_i + R_i, \tag{9}$$

where "ρ" denotes the density and "$\vec{J}$" and "R" represent the diffusive fluxes and reaction source term of the species "i," respectively. Based on an established solution obtained from solution to (9) employing the finite rate chemistry model, the deviation of the species compositions and temperature from their thermochemical equilibrium state was determined by allowing the compositions to relax to their chemical equilibrium values [30]. This was undertaken by allowing the reaction source term in (9) to be modeled as

$$R_i = \rho \frac{Y_i^{\text{eq}} - Y_i}{\tau_{\text{char}}}, \tag{10}$$

where Y_i is the mass fraction of species i and $Y^{eq}\ i$ is the equilibrium mass fraction of species i (determined as a function of the local species, temperature, and pressure) and a characteristic time scale (τ_{char}). τ_{char} in turn is modeled as the sum of a convective/diffusion time scale in the computational cell and a chemical time scale determined from the chemical kinetic mechanism. Equation (10) basically implies that the species reaction towards their chemical equilibrium state over the τchar period associated with the process.

Turbulence Modeling and Initiating the Onset of Transition

Predicting the onset of transition in this hypersonic flow regime has proven to be really challenging. However, fairly accurate predictions of the surface heat flux profiles have been obtained by judicious specification of the location for the onset of transition [19, 24, 27] with typical prediction accuracies ranging from 20 to 25% [27].

The turbulence was modeled in this study employing the realizable k-ε turbulence model and the wall temperature of the vehicle was assumed to be constant at 500 K [19]. Since there is a factor of 3 variation among current empirical correlations for the extent and onset of transition (as a function of the edge Mach number) [36], the onset of transition at a distance of 2.5 m along the wall was artificially induced in this study by specifying a roughness height of 5×10^{-4} m along the wall from 2.5 m to 4 m. Wall roughness is one of the driving mechanisms by which disturbances can be induced in the boundary layer to initiate transition [27]. The predicted wall heat flux profiles in the turbulent boundary layer region are shown in Figure 2. It is worth noting that although the prediction accuracies at the peak heating locations are within the range reported in previous studies [19], the goal here is not a rigorous assessment of turbulence or transition prediction methods but on assessing the impacts of chemistry and radiative transfer models. We can also see in Figure 2 that as far as the surface heat fluxes are concerned, results from the different chemistry models (relaxation to chemical equilibrium versus finite rate chemistry) as well as the inclusion of radiative transfer/TRI all yield identical predictions. A complete summary of the modeling options employed in this study is listed in Table 2.

Table 2: Summary of physical models employed in the simulations

Physical models	Primary modeling option
Density	Ideal gas
Specific heat	Piecewise polynomial [29]
Thermal conductivity and mass diffusivity	Kinetic theory
Viscosity	Sutherland law
Wall boundary condition	No slip (fluid flow), constant temperature of 500 K (energy equation)
Far field boundary conditions	Gauge pressure: 2761 pascal Mach number: 19.97
Turbulence	Realizable k-ε
Turbulence initiation	Roughness height of 5×10^{-4} m specified in the wall region 2.5 m to 4 m
Turbulence boundary conditions	Intensity: 0.01% Viscosity ratio: $2.29E - 09$
Radiative transport equation solver	Finite volume radiation model (angular resolution, theta × phi: 3 × 3)
Planck mean absorption coefficient (Kp) [6]*	$3.7516 \times 10^{-6} \cdot (P)^{1.31} \cdot \exp(5.18 \times 10^{-4}T - 7.13 \times 10^{-9}T^2)$
Gas phase kinetics	Table 1
Relaxation to chemical equilibrium (kinetic mechanism for chemical time scale determination)	Table 1
Turbulence-radiation interactions*	Equation (7)

*These models were implemented as user-defined functions (UDFs) in ANSYS FLUENT [30].

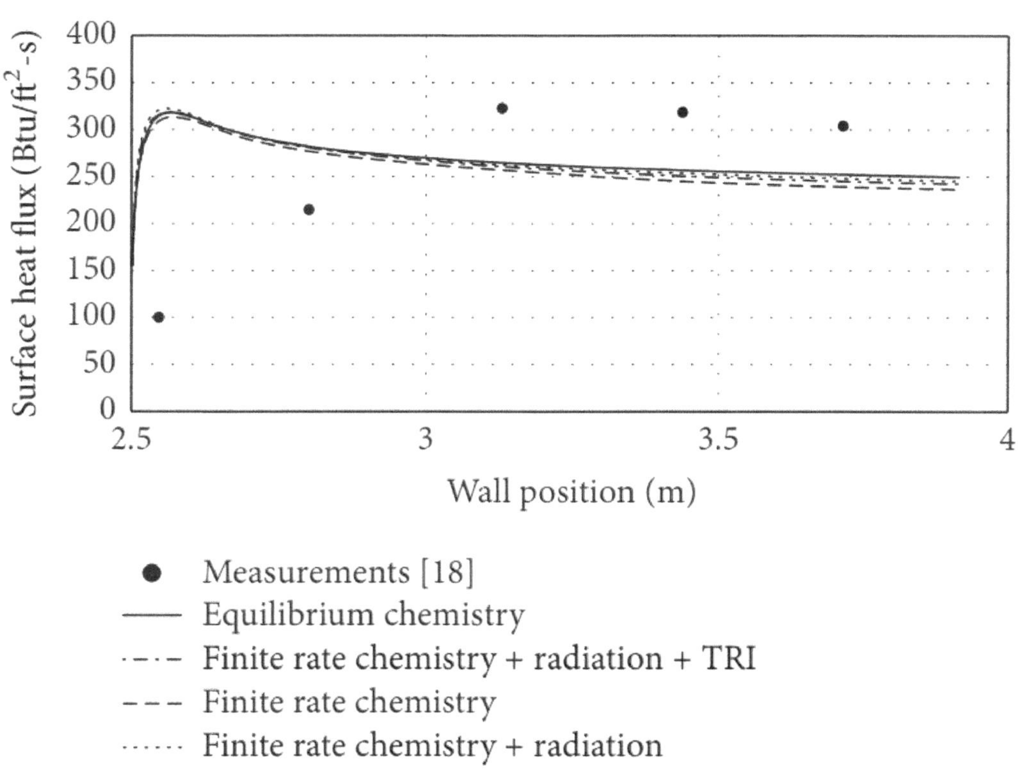

Figure 2: Wall heat flux profiles in the turbulent boundary layer predicted by the realizable k-ε turbulent model (with a specified roughness height).

RESULTS AND DISCUSSION

The results reported in this study correspond to the flow conditions at an altitude of 24.4 km that has been frequently investigated by other studies. The free-stream conditions corresponding to this altitude are adapted from [19] and are summarized in Table 3. Since the goal of this paper is to assess the impact of different chemistry and radiative transfer modeling options on the temperature and specie fields in the laminar and turbulent boundary and shock layers, our initial goal was to establish a shock and boundary layer field across the body that replicated the experimental scenario reasonably well. To ascertain this, we employed experimental measurements as well as simulations of this scenario carried out by previous investigators.

Table 3: Free stream conditions (adapted from [19])

Mach number (M_∞)	19.97
Angle of attack	0 degrees
Density (ρ_∞)	0.043523 kg/m^3
Temperature (T_∞)	221.034 K
Pressure (P_∞)	2761.41 N/m^2
Turbulent intensity	0.01%
Turbulent viscosity ratio	2.29*E* – 09

Shock Layer Profiles: Comparison against Previously Reported Solutions [21]

Since there are no previously reported solutions of temperatures and species concentrations across the boundary and shock layers for this geometry, an initial assessment of the validity of our modeling methodology was made by comparing our velocity and density predictions against those reported in the literature. The only published profiles of these variables at this altitude (24.4 km) in the open literature to the author's best knowledge is the equilibrium air solutions provided by (Langley Aerothermodynamic Upwind Relaxation Algorithm) LAURA [21]. However, a few differences in the modeling methodologies employed by Wood et al. [21] and this study are to be noted:

- The calculations employing LAURA [21] were performed with a spherically blunted nose of 3.43 mm, whereas our calculations were performed with a blunted nose of 2.54 mm radius (as shown in Figure 1).
- LAURA solutions in [21] take into consideration the angle of attack of 0.14 degrees, whereas this was neglected in this study. The impact of including this small angle of attack on the numerical predictions has been discussed by Barnhardt and Candler [26].
- Wood et al. [21] assumed a wall temperature of 467.94 K as opposed to 500 K that was considered in this study.

It is also worth noting that the calculations employing LAURA were reported only up to a distance of 2.0 m along the plate where the flow conditions were laminar [21].

The relative density (ρ/ρ_∞) profiles predicted by the different modeling options across the shock layer in the laminar region are shown in Figure 3 at different axial locations from the nose cone. First, we note clear differences between the modeling scenarios on the relative density profiles. However, these differences reduce downstream of the bow shock along the vehicle body. Second, there are some differences between our equilibrium calculations and the equilibrium solutions reported in [21]. These may be attributed to the differences in the modeling methodology employed by Wood et al. [21] that have been listed above along with the differences in the thermochemical database employed in the calculations. Wood et al. [21] note that 7–10 percent discrepancies in the density profiles can arise due to differences in the thermochemical database alone. However, Figure 3 shows that the agreement between our equilibrium predictions and those employing LAURA improves at larger distances (2.0 m) as the influence of the differences in the nose bluntness goes away.

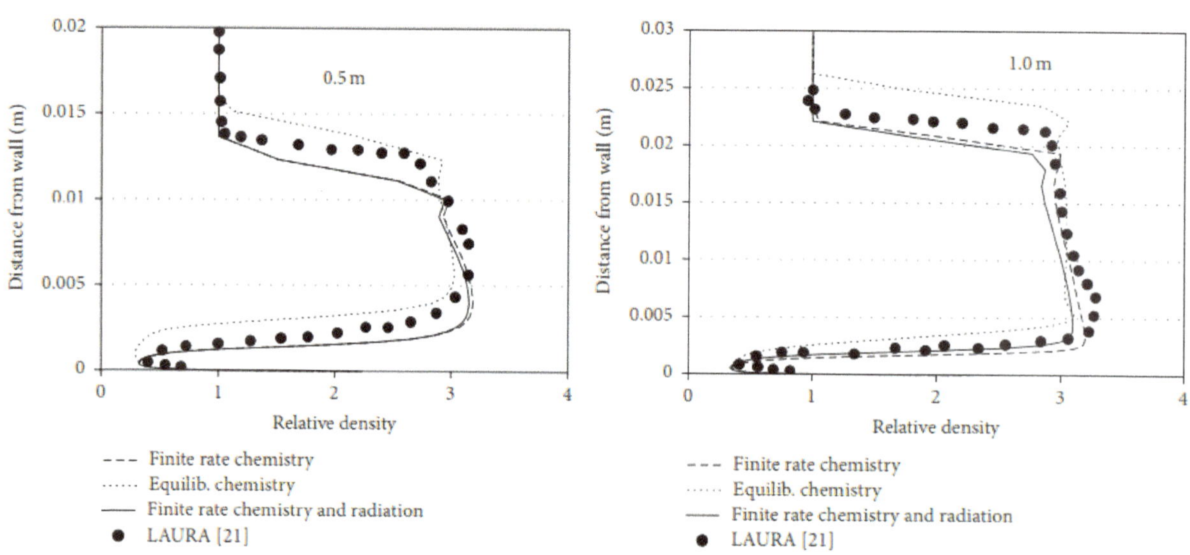

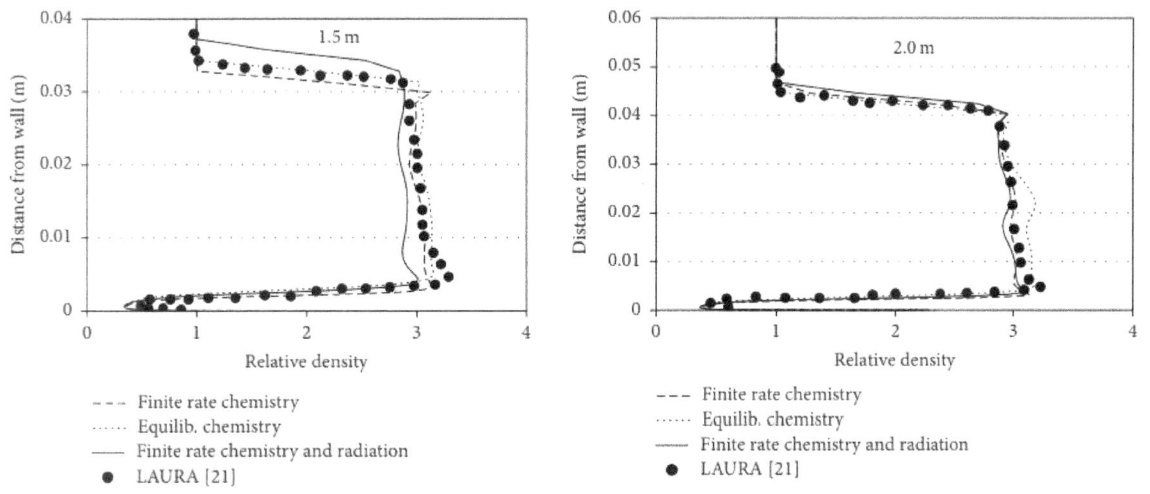

Figure 3: Relative density (ρ/ρ_∞) profiles in the shock layer at different axial locations from the nose cone.

The relative tangential velocity ($/C_\infty$) profiles (where C_∞ is 297.3 m/s) across the shock layer at different axial locations from the nose cone are shown in Figure 4. First, the differences in the modeling scenarios appear to have had a greater impact on the density profiles (cf. Figure 3) than the velocity profiles. A similar observation was reported in Wood et al. [21], where differences in the thermochemical models employed to compute the chemical equilibrium influenced the density profiles more strongly than the velocities. Again, the agreement between our predictions and the predictions reported in [21] is only seen to improve at larger distances away from the nose as the influence of nose bluntness goes away.

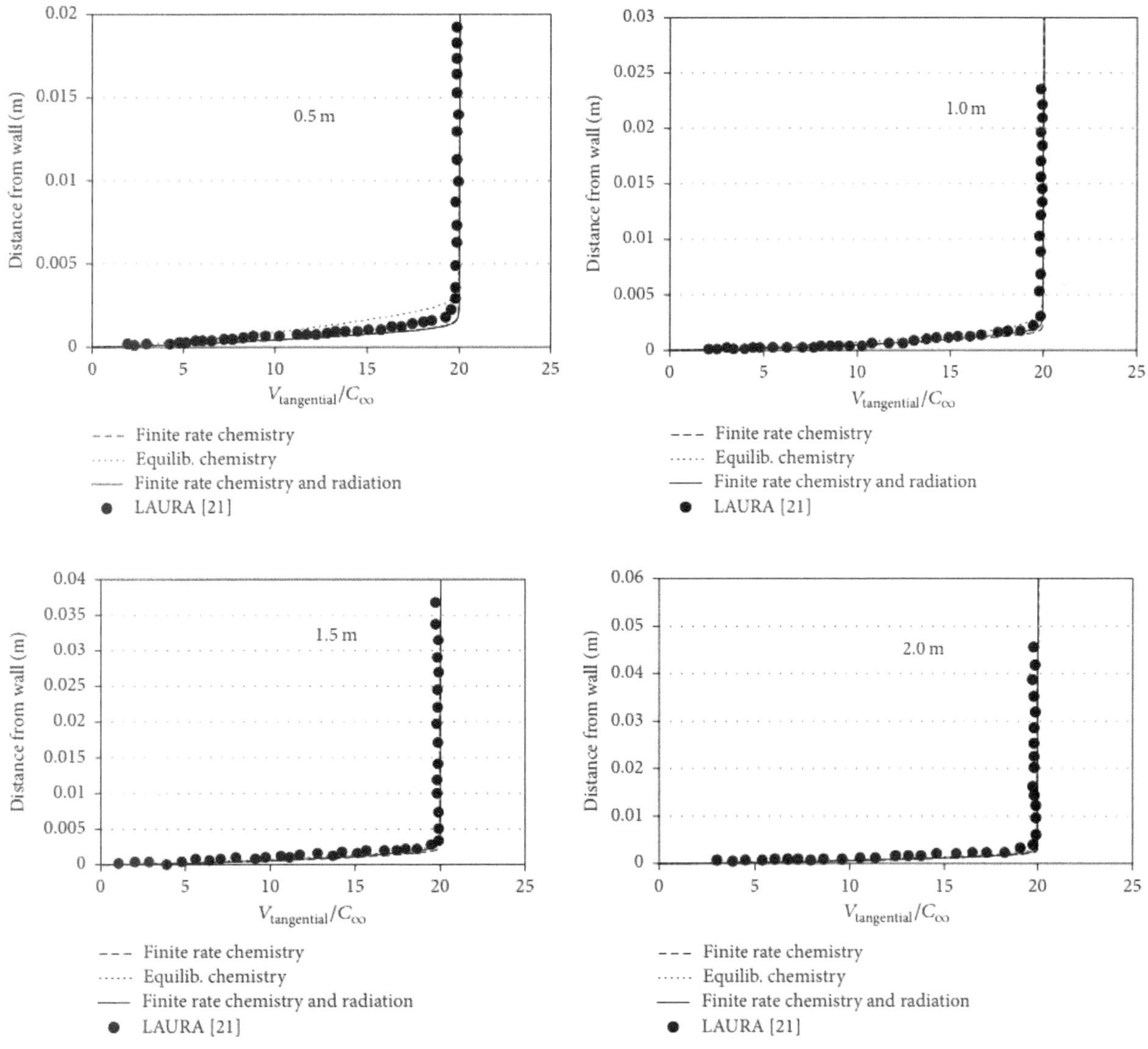

Figure 4: Relative velocity ($V_{tangential}/C_\infty$) profiles in the shock layer at different axial locations from the nose cone.

Boundary Layer Profiles: Comparison against Previously Reported Solutions [21]

Next, the impact of the different modeling scenarios in the boundary layer is investigated to examine the prediction differences more closely. Figures 5 and 6 compare the relative density and velocity profiles across the laminar boundary layer at different locations along the body. Similar to the observations in the shock layer, the differences between the modeling scenarios as well as the solutions from LAURA [21] are more apparently close to the nose and diminish at larger distances along the vehicle body. However, the inclusion of radiative transfer in the calculations appears to have a greater impact with an increase in axial distances downstream of the bow shock. This indicates that the magnitude of radiative emission as well as its region of influence grows with axial distance along the vehicle body. These are assessed next.

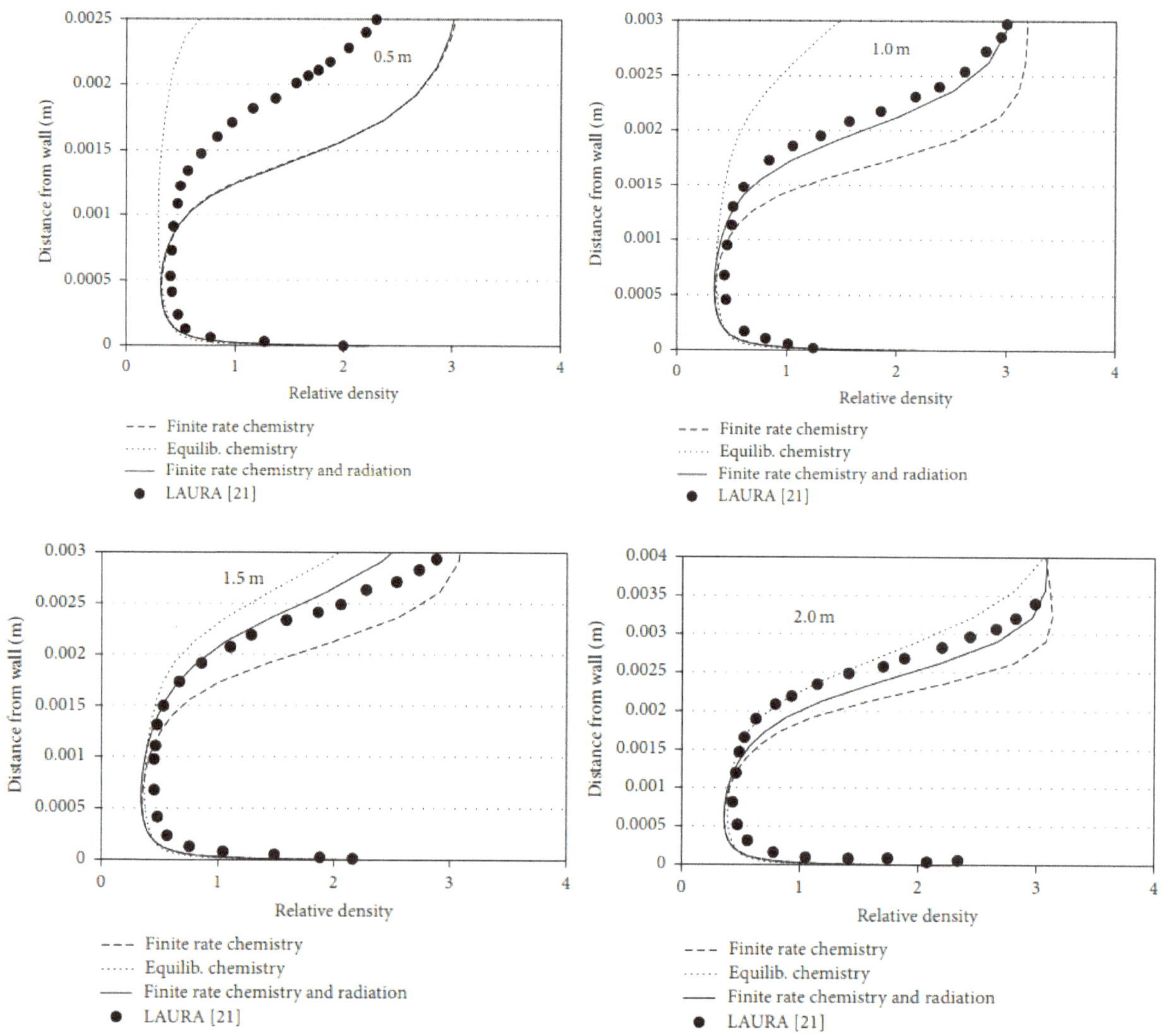

Figure 5: Relative density (ρ/ρ_∞) profiles in the boundary layer at different axial locations from the nose cone.

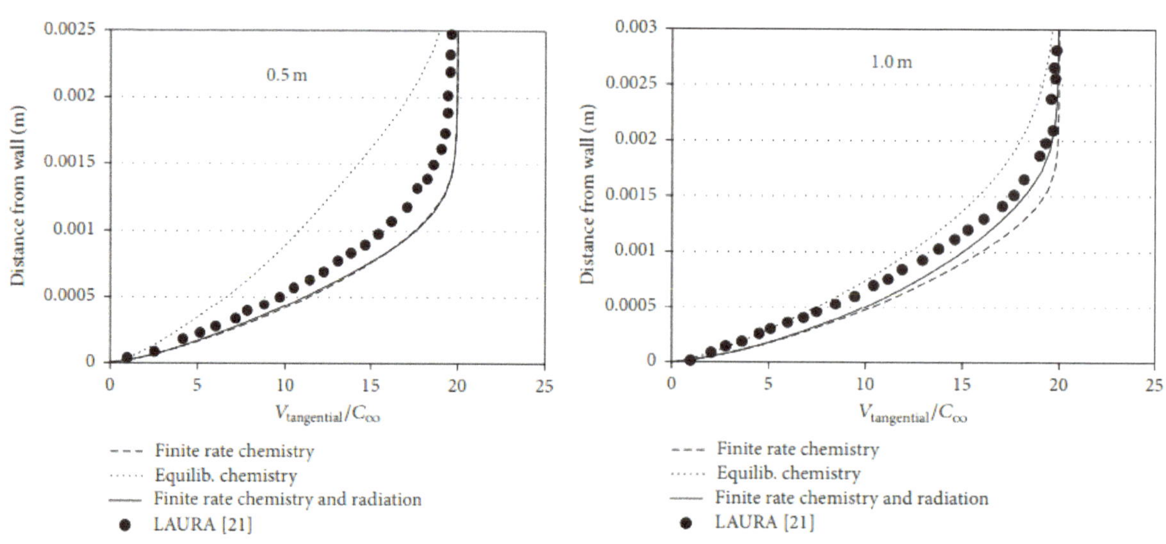

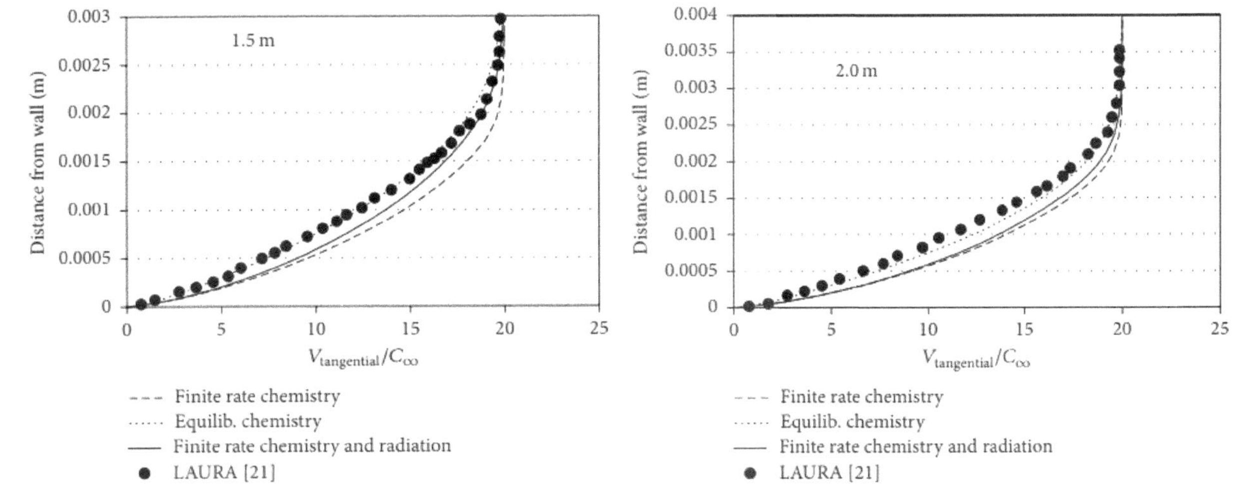

Figure 6: Relative velocity ($V_{\text{tangential}}/C_{\infty}$) profiles in the boundary layer at different axial locations from the nose cone.

Radiative Transfer Calculations and the Effect of TRI

The variables of interest in most radiative transfer analysis are the distributions of radiative heat flux vectors $(q(\vec{r}))$ and the radiative source terms $(-\nabla \cdot q(\vec{r}))$. Distributions of the radiative heat flux vectors are necessary to perform accurate energy balances on any solid/liquid interfaces within the solution domain and at the boundaries. The radiative source term describes the conservation of radiative energy within a control volume and goes into the energy balance equation associated with the fluid flow. Therefore, these variables couple radiation with the other physical processes that occur in a multiphysics simulation. The radiative source term represents the radiative energy balance between absorption and emission within a control volume approximated using the same Planck mean absorption coefficient for the absorption and emission term as

$$-\nabla \cdot q(\vec{r}) = k\left(G(\vec{r}) - 4\pi I_b(\vec{r})\right), \tag{11}$$

where I_b represents blackbody emission and G is the incident radiation (representing absorption) computed by integrating the radiative intensity (I) across all directions as

$$G(\vec{r}) = \int_{4\pi} I(\vec{r},\vec{s})\, d\Omega. \tag{12}$$

Previous studies have highlighted the importance of accounting for the self-absorption on the shock layer even when the wall heat transfer is dominated by convection [37–39]. First, radiative emission and self-absorption can decrease the magnitude of the convective fluxes [38] and influence the shock stand-off distance when coupled strongly with ablating flows [39]. In this study the finite volume radiation model (which is a conservative variant of the discrete ordinates method) was employed to estimate $(-\nabla \cdot q)$.

Plots of the radiative source term are shown in Figure 7 at two locations each along the laminar and turbulent boundary layers.The magnitude of $(-\nabla \cdot q)$ at different temperatures and densities has been estimated from shock tube measurements and reported previously by Kivel [40]. At the densities and temperatures encountered in this study, $-\nabla \cdot q$ was measured to range between 4,000 and 10,000 W/m^3 which compares well with the values shown in Figure 7. We also note that the magnitude of $-\nabla \cdot q$ as well as its "zone of influence" increases along the length of the vehicle. If Γ represents the ratio of radiative heat fluxes to the convective heat fluxes, Γ was found to be 0.001 in the laminar region and 0.003 in the turbulent region.

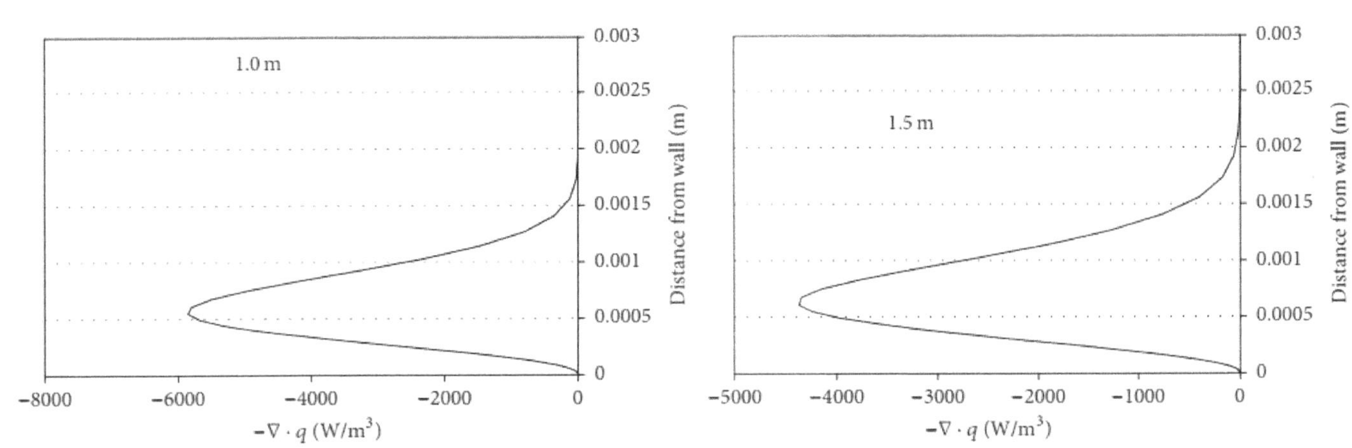

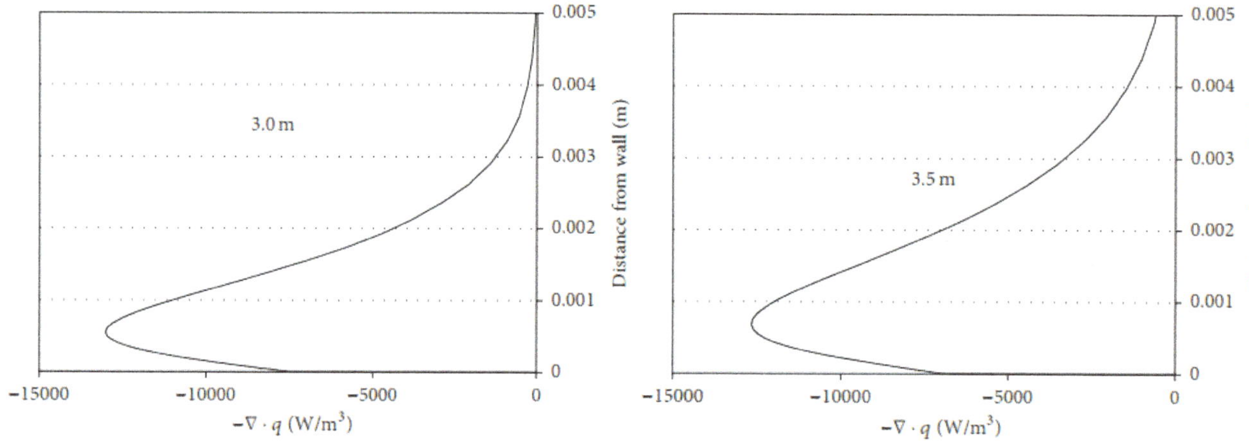

Figure 7: Radiative source ($-\nabla \cdot q$) term profiles in the boundary layer (finite rate chemistry and radiation) at different axial locations from the nose cone.

Figure 8 investigates the effects of TRI at two axial locations along the turbulent boundary layer. The emission enhancement factor is the right hand side of (7) and quantifies the increase in emission due to the turbulent fluctuations in temperature. Up to a 60% increase in emission is observed at some locations along the turbulent boundary layer. However, Figure 8 shows that the magnitude of the radiative source term and the impact of TRI are not large enough to have any bearing on the turbulent boundary layer.

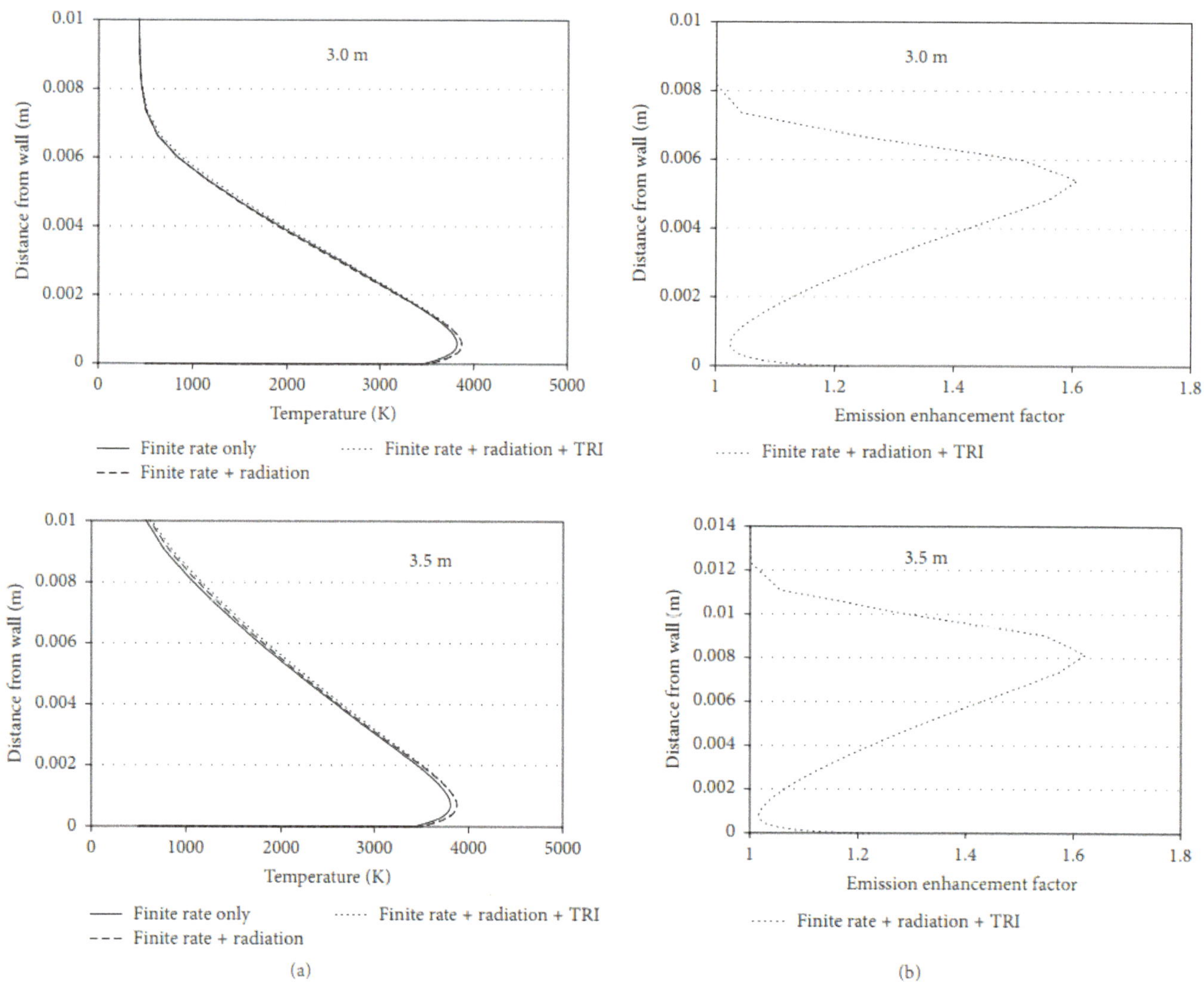

Figure 8: Temperature profiles (a) and the enhancement in emission (b) at two axial locations from the nose cone across the turbulent boundary layer.

Species Profiles across the Boundary Layer

The concentrations of the species along the boundary layer can significantly impact the absorption characteristics and shock stand-off distances. Further, accurate characterization of the ablation phenomena requires accurate determination of the species concentrations. It is generally perceived that at large distances from the bow shock, the chemistry within a fluid parcel tends towards their equilibrium concentrations. Temperatures over 4000 K result in significant dissociation of the O molecule. Figure 9 shows the O mole fraction profiles in the boundary layer. At a V_∞ of 6 km/s which is close to these reentry conditions, the equilibrium concentration of O downstream of the shock has been reported to be about 0.3 [20] which compares well with the values shown in Figure 9. It is interesting to note the significant differences in the O mole fraction predictions between the two chemistry models with the finite rate chemistry predicting superequilibrium concentrations. The differences between the chemistry models however seem to improve with increase in distance downstream of the bow shock. The NO mole fraction predictions from the two chemistry models are shown in Figure 10. Contrary to the O concentrations, the NO mole fractions do not attain their equilibrium values even well into the turbulent boundary layer.

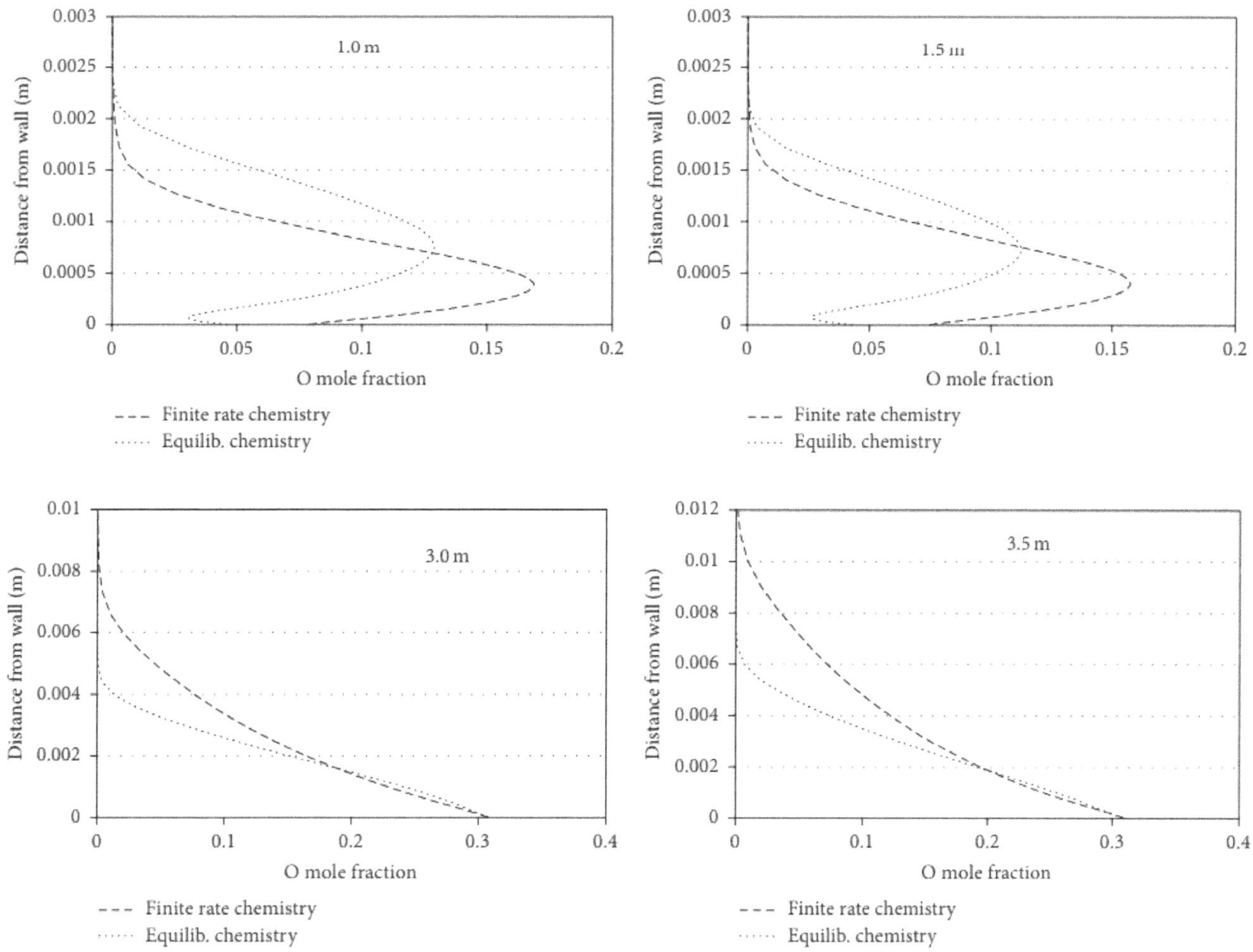

Figure 9: O mole fraction profiles in the boundary layer at different axial locations from the nose cone.

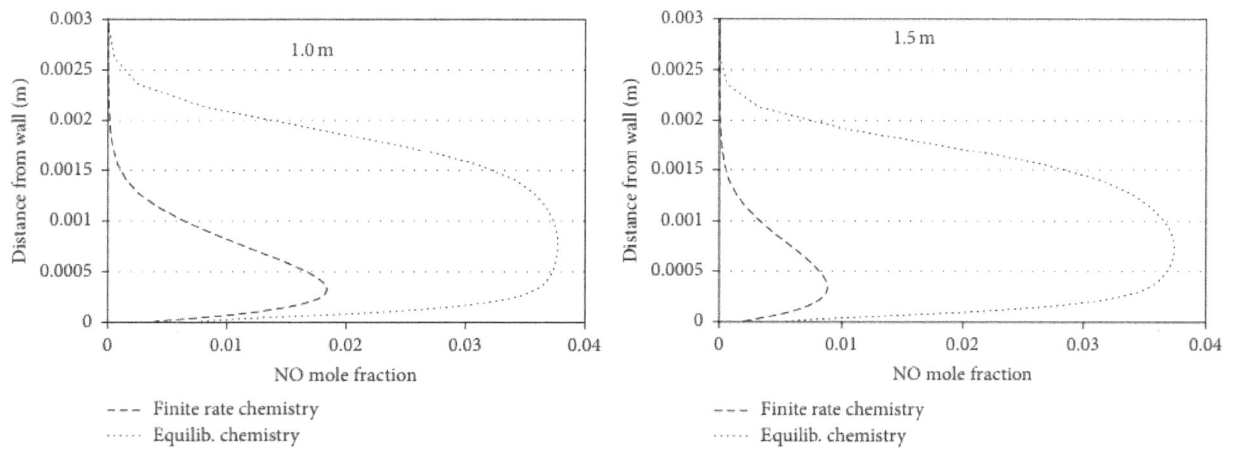

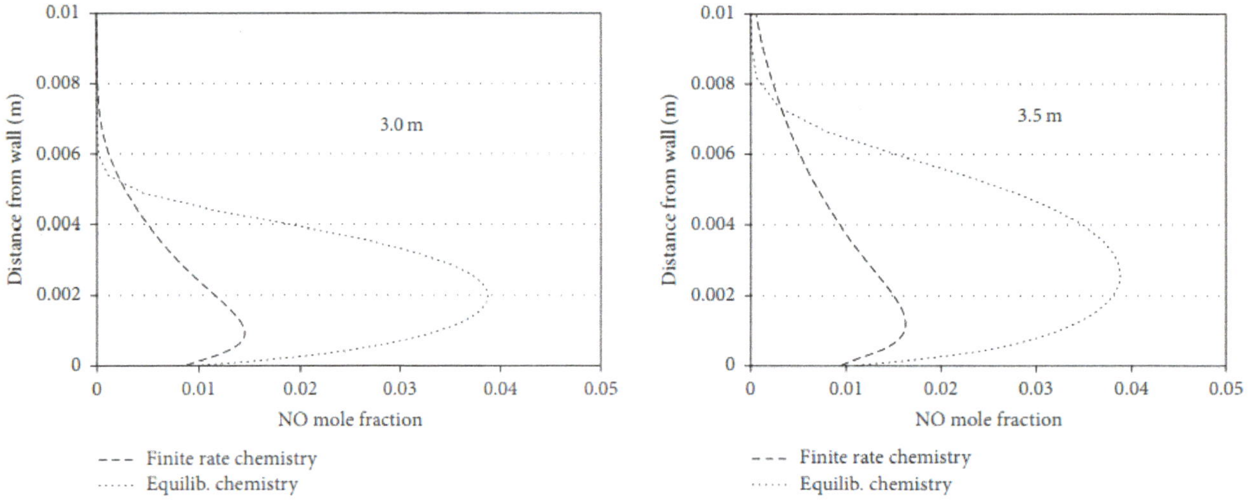

Figure 10: NO mole fraction profiles in the boundary layer at different axial locations from the nose cone.

CONCLUSIONS

A methodology for performing fully coupled, multidimensional calculations of finite rate chemistry, radiative heat transfer, and turbulence radiation interactions (TRI) in hypersonic reentry scenarios in a computationally efficient manner was demonstrated. By considering the Mach 20 Reentry F vehicle as a case study, the impact of different modeling options on density, velocity, temperature, and O and NO predictions in the shock and boundary layers was assessed. Radiative transfer calculations were performed using the finite volume radiation model employing add-on functions to compute a Planck mean absorption coefficient and the temperature self-correlation term (for TRI effects) in the optically thin shock layer. The onset of transition was induced by specifying a wall roughness height beyond the experimentally observed transition location (2.5 m). The chemistry was modeled employing 8 elementary reactions and an equilibrium approach where the species were allowed to relax towards their chemical equilibrium values over the characteristic time scale associated with the process. From the results of this study, the following conclusions may be drawn:

- The wall heat fluxes in the turbulent region compared reasonably well against measurements and were not significantly impacted by the choice of modeling options investigated in this study.
- The density and velocity profiles across the boundary and shock layers compared well against similar calculations reported in previous studies [21]. Density predictions were more sensitive to the choice of modeling options than the velocities.
- The magnitude of the radiative source term employing the Planck mean absorption coefficient compared well against its measurements deduced from shock tube experiments [40]. The emission contribution to the radiative source term was more than three orders of magnitude greater than the absorption term, further justifying the usage of Planck mean absorption coefficient and modeling only the temperature self-correlation term to account for TRI.
- The TRI model predicted up to a 60% enhancement in emission due to temperature fluctuations in the turbulent boundary layer Further, the value of Γ (the ratio of radiative heat fluxes to the convective heat fluxes) increased from 0.001 in the laminar region to 0.003 in the turbulent region. However, including the effects of radiation or TRI did not have any bearing on the temperature predictions in the boundary layer for this reentry scenario.
- While the variations in density and velocities between the different chemistry modeling options diminished along the length of the body, the O and NO prediction variations extended well into the turbulent boundary layer.

ACKNOWLEDGMENTS

This work was partially funded by a North Dakota Space Grant Fellowship to Lauren Elizabeth Clarke. The authors thank a former undergraduate student, Mr. Anthony Marshik, for having carried out some preliminary work on this project.

REFERENCES

1. C. Park, Nonequilibrium Hypersonic Aerothermodynamics, John Wiley & Sons, New York, NY, USA, 1990.
2. J. D. Anderson, Hypersonic and High Temperature Gas Dynamics, AIAA, 2000.
3. R. A. Copeland, J. B. Pallix, and D. A. Stewart, "Surface-catalyzed production of NO from recombination of N and O atoms," Journal of Thermophysics and Heat Transfer, vol. 12, no. 4, pp. 496–499, 1998.
4. D. A. Pejaković, J. Marschall, L. Duan, and M. P. Martin, "Nitric oxide production from surface recombination of oxygen and nitrogen atoms," Journal of Thermophysics and Heat Transfer, vol. 22, no. 2, pp. 178–186, 2008.
5. E. E. Whiting and C. Park, "Radiative heating at the stagnation point of the AFE vehicle," NASA Technical Memorandum 102829, NASA, 1990.
6. S. M. Scala and D. H. Sampson, Heat Transfer in Hypersonic Flow with Radiation and Chemical

Reactions, General Electric Co., Missile and Space Division, Philadelphia, Pa, USA, 1963.

7. P. A. Gnoffo, "A perspective on computational aerothermodynamics at NASA," in Proceedings of the 16th Australasian Fluid Mechanics Conference, Gold Coast, Australia, December 2007.

8. C. Park, "Hypersonic aerothermodynamics: past, present and future," International Journal of Aeronautical and Space Sciences, vol. 14, no. 1, pp. 1–10, 2013.

9. M. Perrin, G. Colonna, G. D'Ammando et al., "Radiation models and radiation transfer in hypersonics,"The Open Plasma Physics Journal, vol. 7, no. 1, pp. 114–126, 2014.

10. C. Park, Non-Equilibrium Air Radiation NEQAIR Program: User›s Manual, Ames Research Center, NASA, 1985.

11. M. J. Wright, D. Bose, and J. Olejniczak, "Impact of flowfield-radiation coupling on aeroheating for Titan aerocapture," Journal of Thermophysics and Heat Transfer, vol. 19, no. 1, pp. 17–27, 2005.

12. P. J. Coelho, O. J. Teerling, and D. Roekaerts, "Spectral radiative effects and turbulence/radiation interaction in a non-luminous turbulent jet diffusion flame," Combustion and Flame, vol. 133, no. 1-2, pp. 75–91, 2003.

13. L. Wang, D. C. Haworth, S. R. Turns, and M. F. Modest, "Interactions among soot, thermal radiation, and NOx emissions in oxygen-enriched turbulent nonpremixed flames: a computational fluid dynamics modeling study," Combustion and Flame, vol. 141, no. 1-2, pp. 170–179, 2005.

14. S. M. Correa, "Turbulence-chemistry interactions in the intermediate regime of premixed combustion,"Combustion and Flame, vol. 93, no. 1-2, pp. 41–60, 1993.

15. P. J. Coelho, "Numerical simulation of the interaction between turbulence and radiation in reactive flows," Progress in Energy and Combustion Science, vol. 33, no. 4, pp. 311–383, 2007.

16. D. Bose and G. V. Candler, "Simulation of hypersonic flows using a detailed nitric oxide formation model," Physics of Fluids, vol. 9, no. 4, pp. 1171–1181, 1997.

17. S. Kumar, Numerical Simulation of Chemically Reactive Hypersonic Flows, Shaker, Aachen, Germany, 2005.

18. R. L. Wright and E. V. Zoby, "Flight boundary layer transition measurements on a slender cone at Mach 20," in Proceedings of the 10th American Institute of Aeronautics and Astronautics, Fluid and Plasmadynamics Conference, Albuquerque, NM, USA, June 1977.

19. C. J. Roy and F. G. Blottner, "Assessment of one- and two-equation turbulence models for hypersonic transitional flows," Journal of Spacecraft and Rockets, vol. 38, no. 5, pp. 699–710, 2001.

20. E. Bauer, "Physics of high-temperature air. Part 1: basics," Final Report, Institute for Defense Analyses, Alexandria, Va, USA, 1990.

21. W. A. Wood, C. J. Riley, and C. F. Cheatwoodz, "Reentry-F flowfield solutions at 80,000 Ft.," NASA Technical Memorandum 112856, NASA, 1997.

22. R. A. Thompson, E. V. Zoby, K. E. Wurster, and P. A. Gnoffo, "Aerothermodynamic study of slender conical vehicles," Journal of Thermophysics and Heat Transfer, vol. 3, no. 4, pp. 361–367, 1989.

23. H. H. Hamilton II, D. R. Millman, and R. B. Greendyke, "Finite-difference solution for laminar or turbulent boundary layer flow over axisymmetric bodies with ideal gas, CF4, or equilibrium air chemistry," Technical Paper 3271, NASA, 1992.

24. K. Sutton, E. V. Zoby, and H. H. Hamilton, "Overview of CFD methods and comparison with flight aerothermal data," in Proceedings of the Symposium on Validation of Computational Fluid Dynamics, CP-437, vol. 2, p. 16, AGARD, Lisbon, Portugal, May 1988.

25. K. E. Wurster, E. V. Zoby, and R. A. Thompson, "Flowfield and vehicle parameter influence on results of engineering aerothermal methods," Journal of Spacecraft and Rockets, vol. 28, no. 1, pp. 16–22, 1991.

26. M. Barnhardt and G. V. Candler, "Detached-eddy simulation of the reentry-F flight experiment," Journal of Spacecraft and Rockets, vol. 49, no. 4, pp. 691–699, 2012.

27. S. P. Schneider, "Hypersonic laminar–turbulent transition on circular cones and scramjet forebodies,"Progress in Aerospace Sciences, vol. 40, no. 1-2, pp. 1–50, 2004.

28. S. Sockalingam and A. Tabiei, "Fluid/thermal/chemical non-equilibrium simulation of hypersonic reentry vehicles," The International Journal of Multiphysics, vol. 3, no. 3, pp. 293–306, 2009.

29. S. Sockalingam, Coupling of fluid thermal simulation for nonablating hypersonic reentry vehicles using commercial codes FLUENT and LS-DYNA [M.S. thesis], University of Cincinnati, 2008.

30. ANSYS, ANSYS Fluent 12.0 Theory Guide, ANSYS, Canonsburg, Pa, USA, 2009.

31. http://www.sandia.gov/TNF/radiation.html.

32. S. P. Burns, "Turbulence radiation interaction modeling in hydrocarbon pool fire simulations," Tech. Rep. SAND99-3190, Sandia National Labs, Albuquerque, NM, USA, 1999.

33. A. Feldick, L. Duan, M. Modest, P. Martín, and D. Levin, "Influence of interactions between turbulence and radiation on transmissivities in hypersonic turbulent boundary layers," in Proceedings of the 48th AIAA Aerospace Sciences Meeting, AIAA 2010-1185, Orlando, Fla, USA, January 2010.

34. F. N. Frenkiel and P. S. Klebanoff, "Probability distributions and correlations in a turbulent boundary-layer," Physics of Fluids, vol. 16, no. 6, pp. 725–737, 1973.

35. G. Cox, "Some measurements of fire turbulence," Fire and Materials, vol. 1, no. 3, pp. 116–122, 1976.

36. S. P. Schneider, "Flight data for boundary-layer transition at hypersonic and supersonic speeds," Journal of Spacecraft and Rockets, vol. 36, no. 1, pp. 8–20, 1999.

37. P. D. Thomas, "Transparency assumption in hypersonic radiative gas dynamics," AIAA Journal, vol. 3, no. 8, pp. 1401–1407, 1965.

38. H. Hoshizaki and K. H. Wilson, "Viscous, radiating shock layer about a blunt body," AIAA Journal, vol. 3, no. 9, pp. 1614–1622, 1965.

39. C. D. Engel, R. C. Farmer, and R. W. Pike, "Ablation and radiation coupled viscous hypersonic shock layers," AIAA Journal, vol. 11, no. 8, pp. 1174–1181, 1973.

40. B. Kivel, "Radiation from hot air and its effect on stagnation-point heating," Journal of the Aerospace Sciences, vol. 28, no. 2, pp. 96–102, 1961.

CHAPTER 17

DEVELOPMENT BOILING TO SPRINKLED TUBE BUNDLE

Petr Kracík and Jiří Pospíšil

Institute of Power Engineering, Brno University of Technology, Faculty of Mechanical Engineering, Technická 2896/2, 616 69 Brno, The Czech Republic

ABSTRACT

This paper presents results of a studied heat transfer coefficient at the surface of a sprinkled tube bundle where boiling occurs. Research in the area of sprinkled exchangers can be divided into two major parts. The first part is research on heat transfer and determination of the heat transfer coefficient at sprinkled tube bundles for various liquids, whether boiling or not. The second part is testing of sprinkle modes for various tube diameters, tube pitches and tube materials and determination of individual modes' interface. All results published so far for water as the falling film liquid apply to one to three tubes for which the mentioned relations studied are determined in rigid laboratory conditions defined strictly in advance. The sprinkled tubes were not viewed from the operational perspective where there are more tubes and various modes may occur in different parts with various heat transfer values. The article focuses on these processes. The tube is located in a low-pressure chamber where vacuum is generated using an exhauster via ejector. The tube consists of smooth copper tubes of 12 mm diameter placed horizontally one above another.

MATHEMATICAL MODEL OF HEAT TRANSFER OF BOILING LIQUID

Lorenz and Yung created a mathematical model published at **[1]** in 1979 for prediction of a heat transfer coefficient at the surface of a single sprinkled horizontal tube using a simplified geometry. The results of for instance **[2, 3, 4]** were compared with this model. The mathematical calculation considers a single vertical half of a horizontal tube that has been unrolled.

Citation: Petr Kracík and Jiří Pospíšil, "Development boiling to sprinkled tube bundle," http://dx.doi.org/10.1051/epjconf/201611402060.

The mathematical model further takes into account that the temperature of the liquid falling on a heated tube equals the temperature of a saturated liquid. On these grounds two basic regions are formed at the tube during consequent heating. The first one is a transition convective region where boiling develops and the second one represents a fully developed boiling where evaporation occurs. The total heat transfer coefficient at the surface of a sprinkled horizontal tube equals a sum of partial heat transfer coefficients [at nucleate boiling effect (α_{ob}), at thermal developing region (α_{od}) and at fully developed region (α_{oc})] which are defined below and corrected by length for which their mean value has been set against the total studied length (of the unrolled area)

$$\bar{\alpha}_o = \alpha_{ob} + \alpha_{od} \cdot \frac{L_d}{L} + \alpha_{oc} \cdot \left(1 - \frac{L_d}{L}\right) \tag{1}$$

The heat transfer coefficient at nucleate boiling effect (also referred to as boiling nucleus) is based on a sufficient heat transfer at which the liquid starts to boil. According to [1] higher heat transfers with boiling temperature are achieved in thin liquid films than in steam environment and a conservative approach is the application of a relation defined by Rohsenow [5] who uses in his equation besides other things latent heat [enthalpy of vaporization l_v (J·Kg^{-1})]

$$\alpha_{ob} = \frac{\mu \cdot l_v}{C_{sf}^3 \cdot \sqrt{\frac{g_0 \cdot \sigma}{g \cdot \rho}}} \cdot \left(\frac{c_p}{l_v \cdot Pr}\right)^3 \cdot \Delta T^2 \tag{2}$$

and parameter C_{sf} is a function of the surface tension of the liquid and of the surface which the liquid flows down from. Parameter C_{sf} for water and various surfaces is for example in [6, tab. 9.1].

Cooper (1984) [7] created relatively simple relation for experimental data of heat transfer at nucleate boiling effect that is only a function of reduced pressure, molar mass and roughness of the surface of a sprinkled tube bundle. Cooper formed several identical functions. One of them applies to common logarithm

$$\alpha_{ob} = \left\{90 \cdot \frac{\Delta T_{ws}^{0.67}}{\sqrt{M}} \cdot \frac{p_r^m}{[-\log_{10}(p_r)]^{0.55}}\right\}^{3.0303} \tag{3}$$

where p_r (–) is a reduced pressure, i.e. pressure in the tube bundle's environment divided by critical pressure of the liquid that flows around the tube bundle. T_{ws} (K) stands for the overheating temperature, i.e. the temperature of the tube wall is deducted by the temperature of saturation of the liquid flowing around it at a given environment pressure. The capital "M (kg·kmol^{-1})" represents molar mass and the exponent m (–) is dependent on the roughness of the tube surface

$$m = 0.12 - 0.2 \cdot \log_{10}\left(\frac{Ra}{0.4}\right) \tag{4}$$

Ra (mm) stands for an average (mean) arithmetical deviation of the studied profile of the given surface calculated from the roughness profile. More in [8]

The thermal developing region where the heat is used for liquid overheating and where boiling does not occur yet, but evaporation does, is limited by the length L_d (m) [1]

$$L_d = \frac{c_p \cdot \sqrt[3]{\Gamma^4}}{4 \cdot \pi \cdot \rho \cdot \lambda} \cdot \sqrt{\frac{3 \cdot \mu}{g \cdot \rho^2}} \tag{5}$$

this equation is derived for laminar to slightly turbulent flow and the mean value of a heat transfer coefficient at the tubes' surface in this region is [1]

$$\alpha_{od} = \frac{3}{8} \cdot c_p \cdot \frac{\Gamma}{L_d} \tag{6}$$

which is based on energy balance.

For a fully developed region at a vertical smooth tube the following two equations are recommended by [1] for a calculation of a heat transfer coefficient at laminar (and also pseudo laminar) and at turbulent flow according to Chun and Seban [9]

$$\alpha_{oc,lam} = 0.821 \cdot \sqrt[3]{\frac{\lambda^3 \cdot g}{\nu^2}} \cdot \sqrt[0.22]{\frac{\mu}{4 \cdot \Gamma}} \tag{7}$$

$$\alpha_{oc,tur} = 0.003\,8 \cdot \sqrt[3]{\frac{\lambda^3 \cdot g}{\nu^2}} \cdot \sqrt[0.4]{\frac{4 \cdot \Gamma}{\mu}} \cdot \sqrt[0.65]{\frac{\nu}{a}} \tag{8}$$

The laminar mode considers the influence of ripple and waves as well that have an effect on the increase of heat transfer and the decrease of film's thickness. Both equations are a function of the Reynolds and Prandtl number and they should be applied for a condition at a constant heat transfer or at constant thermal conditions at the tube wall. Correlation between the equations (7) and (8) creates a transition boundary defined as a functional dependence of the Reynolds number on the Prandtl number.

$$Re_{f,tr} > \frac{5\,800}{Pr^{1.06}} \tag{9}$$

This equation does not define a real boundary between pseudo laminar and turbulent mode, but the transition between the equations (7) and (8) only.

MATHEMATICAL MODEL FOR TWO-PHASE FLOW

Chien and Cheng (2006) [3] state that the model of Chun and Seban (1971) [9] does not comply well with practically measured results for various coolants, i.e. shows a significant error. For these reasons they suggest applying this model, which draws on the fact that the calculation of heat transfer coefficient at a tube surface should also consider occurrence of bubbles, to a horizontal tube. The model is based on the superposition principle, i.e. the final heat transfer coefficient consists of multiple elements which add up to create its resultant. This model has been developed by Chen (1966) [10] and it is given by the basic equation

$$\alpha_o = S \cdot \alpha_{ob} + \alpha_{cv} = S \cdot \alpha_{ob} + E \cdot h_l \tag{10}$$

which is identical also for flow inside a channel, where S (–) represents the suppression factor of heat transfer coefficient at nucleate boiling effect (a_{ob}) The factor is determined by the equation (3). A modified form of the mentioned equation which is defined by Cooper [7] can be used which uses natural logarithm and the overheating temperature is replaced by the heat transfer density.

$$\alpha_{ob} = 93.5 \cdot \frac{\dot{q}^{0.67}}{\sqrt{M}} \cdot \frac{p_r^{\,0.12-0.434\,3\cdot\ln\left(\frac{Ra}{0.4}\right)}}{[-0.434\,3 \cdot \ln(p_r)]^{0.55}} \tag{11}$$

For the equation (11) Chien and Cheng [3] state the deviation ±10% at the largest height of protrusions Ra=0.1 mm.

The quantity a_{cv} in the equation (10) is a convective heat transfer coefficient that equals the product of a two-phase convection enhancement factor E (–) and the heat transfer coefficient for a single-phase liquid flow. Chien and Cheng (2006) [3] suggest to use the following functions to calculate the suppression factor and the twophase convection enhancement factor

$$S = \frac{1}{1 + 2.56 \cdot 10^{-6} \cdot \left(Re_f \cdot E^{1.25}\right)^{1.17}} \quad (12)$$

$$\left.\begin{aligned} E &= 1\,[-] && for \quad \frac{1}{X_{tt}} < 1\,[-] \\ E &= 2.35 \cdot \left(0.213 + \frac{1}{X_{tt}}\right)^{0.736} && for \quad \frac{1}{X_{tt}} \geq 1\,[-] \end{aligned}\right\} \quad (13)$$

where X_tt [–] is the Martinelli parameter

$$X_{tt} = \left(\frac{1-x}{x}\right)^{0.875} \cdot \left(\frac{\mu_f}{\mu_g}\right)^{0.125} \cdot \left(\frac{\rho_g}{\rho_f}\right)^{0.5} \quad (14)$$

MEASUREMENT APPARATUS

For the purposes of examination of the heat transfer at sprinkled tube bundles a test apparatus has been constructed; see the diagram at Fig. 1. A tube bundle at which a heat transfer from a heated water flowing inside the tubes into a falling film liquid is studied is placed in a vessel where low pressure is created by an exhauster through an ejector.

The test apparatus chamber is a cylindrical vessel of the length 1.2 m with three apertures in which the tube bundle of the examined length 940.0 mm is inserted. The tube bundle is installed in two fitting metal sheets which define the sprinkled area. The bundle consists of eight copper tubes of the diameter 12.0 mm situated horizontally one above another, with a distribution tube above them with apertures of the diameter from 1.0 mm to 9.2 mm. The bundle can be operated using only the first four or six tubes too.

Two closed loops are connected to the chamber. A heating one and a sprinkling one. The heating liquid flowing inside the tubes is intended for overpressure up to 1.0 MPa. The second loop contains flowing falling film liquid. There is a pump, a regulation valve, a flow meter and plate heat exchangers attached to both loops.

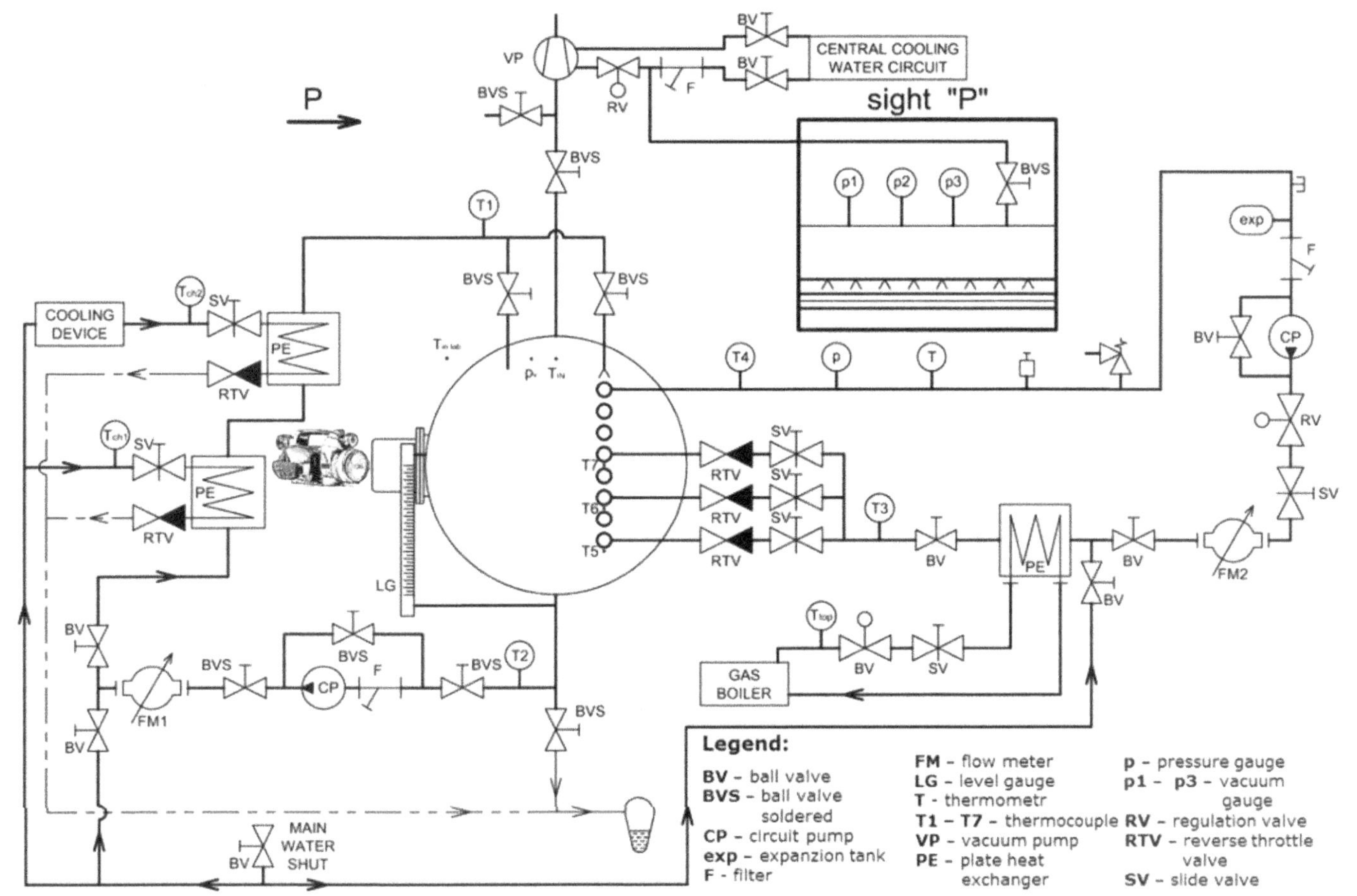

Figure 1. Measurement Apparatus Diagram.

The plate heat exchanger at the heating loop is connected to a gas boiler which supplies heat to the heating liquid. The sprinkling loop uses two plate heat exchangers. In the first one the falling film liquid is cooled by a cold drinkable water from water mains and the falling film liquid is cooled in the second exchanger by a drinkable water cooled in a cooler which regulates the temperature up to 1.0 °C. In order to enable visual control the heating loop also includes a manometer and a thermometer. The thermal status in individual loops is measured by wrapped unearthed T-type thermocouples on the agents' input and output from the vessel. All thermocouples have been calibrated in the CL1000 Series calibration furnace which maintains a given temperature with the accuracy ±0.15 °C. None of the thermocouples has exceeded the error ±0.5 °C within the studied range from 28 °C to 75 °C. That is why the total error at temperature measurement is set uniformly for all thermocouples along the whole studied range ±0.65 °C.

There are three vacuum gauges measuring the low pressure. The first vacuum gauge is designed for visual control and it is a mercury meter, the second one is a digital vacuum gauge Baumer TED6 and enables measuring within the whole desired low-pressure range, but it is less accurate with lower pressure values. To allow a precise measuring of the low spectrum a third digital vacuum gauge of the range 2.0 kPa – 0 Pa is used. The accuracy of a vacuum gauge the results of which have been used for the assessment is 0.5% from the measured range, i.e. ±0.5 kPa.

Electromagnetic flow meters Flomag 3000 attached to both loops measure the flow rate. The flow meters' range is 0.007 8 ÷ 0.942 4 $l \cdot s^{-1}$, where the accuracy is 0.5% from the measured range, i.e. ± 0.004 67 $l \cdot s^{-1}$. All examined quantities are either directly (thermocouples) or via transducers scanned by measuring cards DAQ 56.

METHODOLOGY OF DATA ASSESSMENT

The assessment of the measured data is based on the thermal balance between the operation liquid circulating inside the tubes and a sprinkling loop according to the law of conservation of energy. Heat transfer is realized by convection, conduction and radiation. In lower temperatures the heat transferred by radiation is negligible, therefore it is excluded from further calculations. The calculation of the studied heat transfer coefficient is based on the Newton's heat transfer law and Fourier's heat conduction law that have been used to form the following relation

$$\alpha_o = \frac{1}{2 \cdot \pi \cdot r_o \cdot \left[\frac{1}{k_S} - \frac{1}{2 \cdot \pi \cdot \alpha_i \cdot r_i} - \frac{\ln\left(\frac{r_o}{r_i}\right)}{2 \cdot \pi \cdot \lambda_S} \right]} \quad (15)$$

where a_o ($W \cdot m^{-2} \cdot K^{-1}$) is the heat transfer coefficient at the sprinkled tubes' surface, i ($W \cdot m^{-2} \cdot K^{-1}$) is the heat transfer coefficient at the inner side of a tube set for a fully developed turbulent flow according to [11, page 59 and 138], r_o; r_i (m) are outer and inner tube radii, l_S ($W \cdot m^{-1} \cdot K^{-1}$) is thermal conductivity and k_S ($W \cdot m^{-1} \cdot K^{-1}$) is heat admittance based on the above mentioned laws governing heat transfer which is calculated from heat balance of the heating side of the loop, that is why the following must be valid:

$$\dot{Q}_{34} = k_S \cdot L \cdot \Delta T_{ln} = \dot{M}_{34} \cdot c_p \left(\frac{t_3 + t_4}{2} \right) \cdot (t_3 - t_4) \quad (16)$$

where M_{34} (kg·s-1) is the mass flow of heating water, c_p ($J \cdot kg^{-1} \cdot K^{-1}$) is the specific heat capacity of water at constant pressure related to the mean temperature inside the loop, L (m) is the total length of the bundle and T_{ln} (K) is a logarithmic temperature gradient where a counter-current exchanger was considered.

The above mentioned calculation is valid only for heat transfer from water flowing inside the tubes into a falling film liquid which does not boil at the surface or where insignificant boiling may occasionally occur at the last tube. In case boiling occurs at a major part of the tube bundle, i.e. the tube bundle is sprinkled by a saturated water or higher at a corresponding pressure in a chamber p_V (Pa), two-phase flow should be considered. The amount of developed steam can be calculated according to the law of conservation of mass. I.e. the sum of the mass flow rate of steam M_v ($kg \cdot s^{-1}$) and the mass flow rate of the water flowing down after the exchanger M_f ($kg \cdot s^{-1}$) must equal the mass flow rate flowing from a distribution tube M1 ($kg \cdot s^{-1}$). However, there are two unknown variables in this equation. That is why it is necessary to apply the law of conservation of energy as well. I.e. the sum of the energy necessary to develop steam Q_v (W) and the energy transferred to a falling film water Q_f (W) must equal the energy supplied by the heating liquid Q_{34} (W). Then the following is valid for the calculation of the amount of developed steam:

$$\dot{M}_V = \frac{\dot{Q}_{34} - \dot{M}_1 \cdot [i_2(t_2; x = 0) - i_1(t_1; x = 0)]}{i_{IN}(t_{IN}; p_v) - i_2(t_2; x = 0)} \quad (17)$$

under the condition that the temperature of the falling film water t1 (°C) in a distribution tube is max. at the saturation boundary or slightly below it. In the mentioned equation, i_{IN} ($J \cdot kg^{-1}$) stands for the enthalpy of a developed steam cushion at the temperature tin (°C), and i_2 ($J \cdot kg^{-1}$) stands for the enthalpy of a heated falling film water at the temperature t_2 (°C), after the last tube the test tube bundle.

The studied heat transfer coefficient can be determined on the basis of the Newton›s law of heat transfer where the wall temperature is set as a mean temperature of the heating water and the temperature of the liquid flowing around it as a mean value of the temperature of a falling film water flowing out of the distribution tube and the temperature of a steam cushion.

$$\left. \begin{aligned} \dot{Q}_V &= \alpha_{o,v} \cdot S \cdot (T_w - T_\infty) \\ \Rightarrow \alpha_{o,v} &= \frac{\dot{M}_V \cdot [i_{IN}(t_{IN}; p_v) - i_1(t_1; x = 0)]}{S \cdot \left(\frac{t_3 + t_4}{2} - \frac{t_{in} + t_1}{2} \right)} \end{aligned} \right\} \quad (18)$$

Where S (m^2) stands for a heat-transfer region, i.e. the surface of the tube bundle.

EXPERIMENTS RESULTS

Experiments published in this paper have focused on the influence of a heat-exchanging surface size on boiling development at a sprinkled tube bundle at two thermal levels. A tube bundle has been assembled consisting of eight tubes positioned horizontally one above another where only the first four, six or all eight tubes were heated. The input heating liquid temperature (t3) was maintained at the level of approx. 40°C or 50°C with one degree centigrade deviation and at the flow rate (V1) of 7.2 liters per minute with a deviation up to 0.1 liter per minute.

The falling film liquid flow rate was adjusted and maximum low pressure was created in the vessel by an exhauster for each experiment. The difference in pressures at individual thermal levels was given by the fact that with more vapor generation of a higher thermal level the exhauster's absorption capacity decreased. The heat source itself was regulated by the temperature of the falling film water at the distribution tube's output. Summary of basic data gained by measurement is provided in Table 1. Apart from the quantities mentioned above the table also contains falling film liquid mass flow rate related to the sprinkled zone length (Γ_1), Reynolds number of the liquid phase (Re_f), Prandtl number of the liquid phase (Pr_f), pressure in a chamber (p_v) during the whole experiment, specific heat transfer derived from the heating liquid for boiling (q“) that was determined on the basis of the law of conservation of energy and mass, and a total heat transfer derived from the heating liquid (q).

Due to the fact that the waveform of a falling film liquid heating on the bundle at tested flow rates is approximately linear, it was possible to apply linear approximation to find the borderline of saturated liquid at the bundle. Figure 2 shows four sets of points in relation to the Prandtl number of the liquid phase and on the falling film liquid temperature at the distribution tube’s output, deducted by the temperature of saturation corresponding to the current pressure in a chamber. Points “0%” mark a state when the falling film water temperature flowing out of the exchanger reaches a saturation borderline. Points “50%” mean that the saturation state occurs between the 4th and 5th tube in case of an 8-tube bundle etc. The results for the Prandtl number up to approx. 4.7 come from experiments where the heating water at the bundle’s input was approx. 50°C. Higher values of the Prandtl number correspond to the heating water of approx. 40°C.

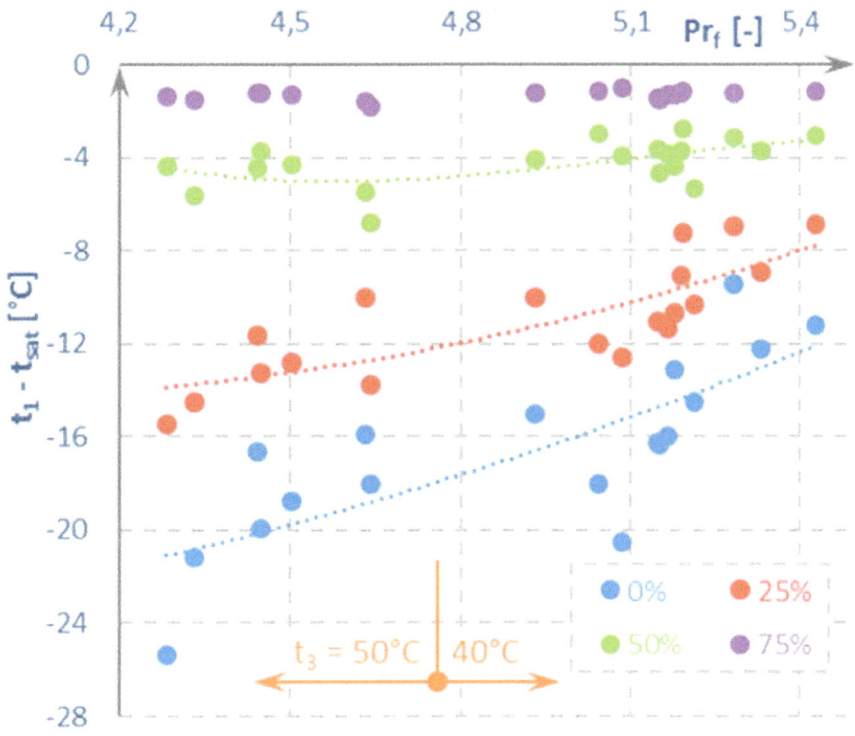

Figure 2. Boiling Development at Sprinkled Tube Bundle.

Table 1. Basic information about the experiments

	t_3 (°C)	V_1 (l/min)	Γ_1 (kg/(s.m))	Re_f (-)	Pr_f (-)	p_v (kPa)	q" (kW/m^2)	q (kW/m^2)
4 tubes	40.4 ± 0.3	4.04 ± 0.03	0.0713 ± 0.0005	361.8	5.3	4.4 ± 0.1	3.8 ± 0.9	12.1 ± 2.5
	40.4 ± 0.3	6.03 ± 0.01	0.1063 ± 0.0003	544.5	5.3	4.5 ± 0.2	3.4 ± 1.0	12.4 ± 3.5
	40.3 ± 0.4	4.03 ± 0.00	0.0710 ± 0.0001	369.0	5.2	4.7 ± 0.3	4.4 ± 0.9	10.4 ± 2.0
	40.4 ± 0.2	6.03 ± 0.01	0.1063 ± 0.0002	531.0	5.4	4.2 ± 0.2	4.8 ± 1.2	14.5 ± 3.8
	50.3 ± 0.4	4.03 ± 0.01	0.0708 ± 0.0003	405.5	4.6	6.2 ± 0.4	8.0 ± 1.6	18.3 ± 3.9
	50.1 ± 0.3	6.03 ± 0.02	0.1062 ± 0.0005	610.5	4.6	6.2 ± 0.5	7.3 ± 1.4	19.8 ± 4.5
6 tubes	40.6 ± 0.2	4.03 ± 0.02	0.0710 ± 0.0003	367.3	5.2	4.7 ± 0.1	3.4 ± 0.6	7.6 ± 1.1
	40.5 ± 0.2	6.02 ± 0.02	0.1062 ± 0.0003	552.1	5.2	4.7 ± 0.2	3.5 ± 1.1	9.4 ± 2.7
	40.3 ± 0.3	4.06 ± 0.01	0.0715 ± 0.0001	374.4	5.1	4.8 ± 0.2	2.7 ± 0.5	6.8 ± 1.9
	40.5 ± 0.3	5.02 ± 0.01	0.0885 ± 0.0003	462.0	5.2	4.8 ± 0.2	3.2 ± 0.8	8.3 ± 2.5
	50.2 ± 0.4	4.03 ± 0.03	0.0710 ± 0.0006	416.7	4.5	6.7 ± 0.5	6.8 ± 1.6	12.6 ± 2.8
	50.1 ± 0.5	5.01 ± 0.02	0.0881 ± 0.0003	524.2	4.5	6.9 ± 0.3	6.5 ± 1.4	13.5 ± 2.7
	50.4 ± 0.2	6.05 ± 0.01	0.1064 ± 0.0002	633.9	4.4	7.0 ± 0.6	7.3 ± 1.6	15.2 ± 3.4
8 tubes	40.5 ± 0.2	4.02 ± 0.01	0.0708 ± 0.0001	370.2	5.2	4.8 ± 0.2	2.5 ± 0.7	6.6 ± 1.7
	39.9 ± 0.7	4.02 ± 0.04	0.0708 ± 0.0007	375.2	5.1	5.0 ± 0.2	1.9 ± 1.0	5.5 ± 2.1
	40.2 ± 0.5	5.04 ± 0.01	0.0887 ± 0.0001	473.8	5.0	5.1 ± 0.2	1.8 ± 0.8	5.3 ± 1.9
	40.4 ± 0.2	6.03 ± 0.02	0.1063 ± 0.0003	554.4	5.2	4.7 ± 0.2	2.1 ± 0.8	6.8 ± 2.4
	40.4 ± 0.5	8.08 ± 0.00	0.1423 ± 0.0001	776.5	4.9	5.4 ± 0.2	1.4 ± 0.6	5.4 ± 1.8
	50.2 ± 0.3	4.04 ± 0.04	0.0710 ± 0.0007	436.4	4.3	7.6 ± 0.3	4.4 ± 0.8	9.7 ± 1.6
	50.3 ± 0.3	6.02 ± 0.02	0.1058 ± 0.0003	645.9	4.3	7.4 ± 0.5	4.2 ± 0.8	10.7 ± 2.3

Figure 2 clearly shows that at the temperature difference of -3 and more, the boiling occurs at more than 60% of the tube bundle's height (considered from the last tube upwards). Simultaneously, the heat transfer coefficient was being stabilized at this value. That is why an average value and a standard deviation have been calculated starting at this borderline up for the purpose of further comparison. Figure 3 shows heat transfer coefficients of the vapor phase that have been determined according to the equation (18) in relation to the Reynolds number. Red tones represent an 8-tube bundle, green tones represent a 6-tube bundle and blue tones represent a 4-tube bundle. A lighter tone displays t_3 = 50°C and a darker tone displays t_3 = 40°C. This legend is applied to the following diagrams too, with the exception of the last one.

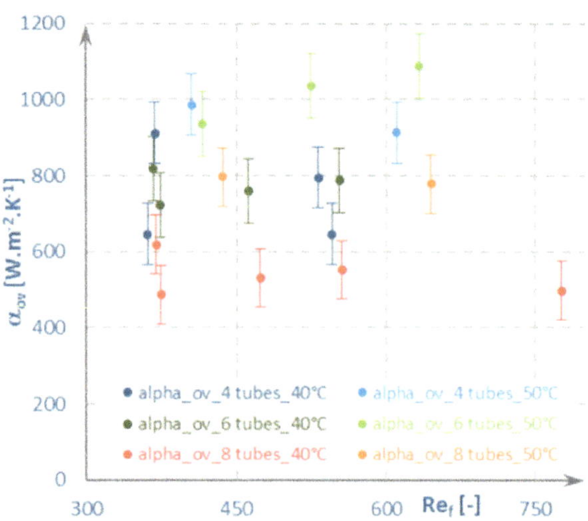

Figure 3. Heat transfer coefficient of vapor phase.

A deviation is highlighted at individual points that is uniform at the given number of tubes. It was determined according to the average value of individual deviations increased by a standard deviation of this average value. Deviations ranged between 7 and 25%.

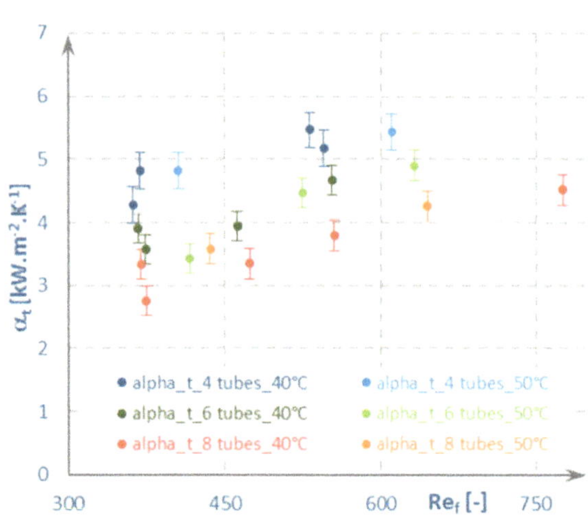

Figure 4. Total heat transfer coefficient.

The liquid phase coefficient given by the equation (15) as well as the vapor phase coefficient is derived from total energies at the input and output of the bundle. For this reason the coefficients are valid for the whole bundle and they represent its mean value. On the basis of the superposition principle it is therefore possible to total up both coefficients, thus creating a total mean heat transfer coefficient at the tube's surface which is displayed in Figure 4 also in relation to the Reynolds number of the liquid phase. Figure 5 shows the values of the generated vapor amount in relation to the Reynolds number.

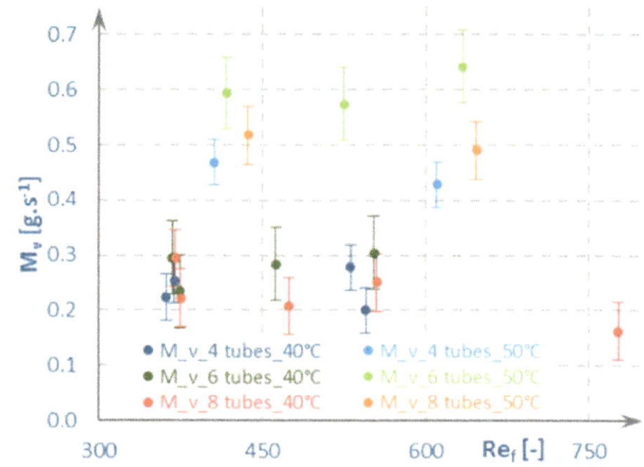

Figure 5. Generated vapor amount.

The last Figure 6 displays all values of the total mean coefficient in single color (green) and compares them with two mathematical models given in the theoretical part. Equations of these models were completed by the state values gained by our measurement. Blue values have been calculated according to the Lorenz and Yung model, i.e. by the basic equation (1), and red values have been calculated according to the Chien and Cheng model (2006), i.e. by the basic equation (10).

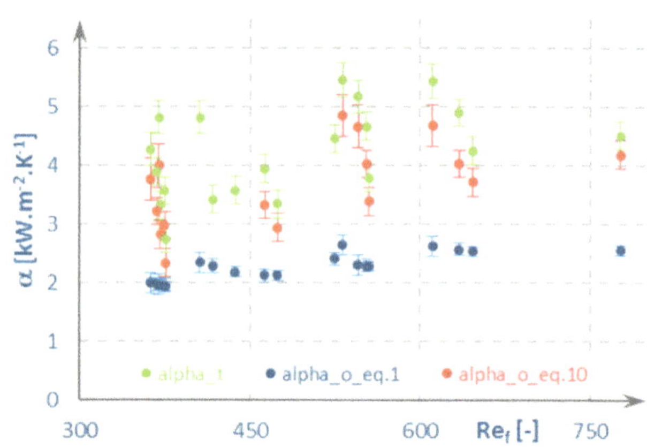

Figure 6. Results comparison with other authors.

CONCLUSION

The objective of this paper was to describe the influence of heat-exchanging surface size on boiling development at a sprinkled tube bundle at two thermal levels of a heating liquid that flows inside the tubes. Boiling development, i.e. its consequent testing on a bundle, was realized using the temperature of a falling film water that has been sprinkled on a tube bundle. It has been proved that the lowest values of the heat transfer coefficient of the vapor phase occurred at the 8-tube bundle and fluctuated around 550 W/(m^2 .K). On the contrary, the highest values (almost double) have been reached at the 6-tube bundle. The vapor phase coefficient as well as the generated vapor amount showed to be functionally independent on the value of the total falling film liquid flow rate.

While the convenient number of tubes cannot be clearly determined at the thermal level of 40°C because the vapor amount fluctuated around the mean value of 0.25 g/s, the higher thermal level provides more clear-cut results. The highest values have been reached with six tubes where the vapor amount fluctuated around approx. 0.6 g/s. 8 tubes follow where the amount was approx. 0.5 g/s and the mean value at four tubes oscillated around 0.45 g/s. Higher amounts of generated vapor at six tubes can be justified by a better distribution of the temperature field at the bundle which is visible at the boiling development. Whereas the gradient increases between six and eight tubes, it falls sharply due to further reduction of the heat-exchanging surface (best visible at 50 and 75%). The total heat transfer coefficient at the tube surface consists mostly of the liquid phase coefficient. That is why it is also functionally dependent on the value of the falling film liquid flow rate. The highest values have been reached at a 4-tube bundle. Values of a 6-tube bundle were approx. 20% lower and the difference at an 8-tube bundle was up to approx. 40%.

Coefficients suggested by us have been compared with two other models. State values gained by our measurement have been put into these models. The first model by Chien and Cheng did not correspond very well, i.e. the calculated values were lower in the range of approx. 30 to 56%. Contrary to this, the second model showed a relatively good concordance. Values were lower only in the range of approx. 7 to 17%.

ACKNOWLEDGMENT

This work is an output of research and scientific activities of NETME CENTRE PLUS (LO1202) by financial means from the Ministry of Education, Youth and Sports under the „National Sustainability Programme I“.

REFERENCES

1. J. J. Lorenz and D. Yung. Journal of Heat Transfer, vol. 101, issue 1, pp. 178-180, (1979).
2. W. H. Parken, L. S. Fletcher, V. Sernas and J. C. Han. Journal of Heat Transfer, vol. 112, issue 3, pp. 744-750, (1990).
3. L.-H. Chien and Ch.-H. Cheng, HVAC, vol. 12, issue 1, pp. 69-87, (2006).
4. J. J. Lorenz and D. Yung, Proceedings, Fifth Ocean Thermal Energy Conversion Conference, Miami Beach, Florida, pp. 46-69, (1978).
5. W. M. Rohsenow, Journal of Heat Transfer, vol. 94, issue 2, pp. 255-256, (1972.
6. J. R. Thome, Wolverine Tube, Inc. <http://www.wlv.com/products/databook/db3/DataBookIII.pdf>, (2004-2010)
7. M. G. Cooper, 1st UK National Conference on Heat Transfer, vol. 86, pp. 785-793, (1984).
8. ČSN EN ISO 4287, Praha: ČSNI, 04/1999.
9. K. R. Chun and R. A. Seban, Journal of Heat Transfer, vol. 93, issue 4, pp. 391-396, (1971).
10. J. C. Chen, Industrial & Engineering Chemistry Process Design and Development, vol. 5, issue 3, pp. 322-329, (1966).
11. M. Jícha, M. Issue 1. 160 p. ISBN 80-214-2029-4, (2001).

CHAPTER 18

DESIGN AND CONSTRUCTION OF A SPIRAL HEAT EXCHANGER

Jamshid Khorshidi, Salman Heidari

Department of Engineering, Hormozgan University, Bandar Abbas, Iran

ABSTRACT

In this article, the performance and applications of a Spiral Plate Heat Exchanger are demonstrated. Also, governing equation of heat transfer phenomena in such heat exchangers is discussed. Regarding the governing equations, a LAB-sized model of this type of heat exchanger was designed and constructed. Galvanized Iron sheets were used as the heat transfer surfaces. Two Galvanized Iron sheets were rolled together around a central core and, as a result, two separated channels were made. Also, a predesign simulation of the heat exchanger was done using the Fluent software to predict the performance of the heat exchanger. First the geometry was made using Gambit software environment then the model was analyzed through Fluent. Because of less fouling, easier cleaning and high heat transfer coefficient, Spiral Heat Exchanger is a good alternative to the other types of heat exchangers, especially when it's going to handle high fouling flows or highly viscous fluids. Low fouling rate of the heat exchanger, reduces the need of cleaning and therefore the out of service will be decreased. In the constructed heat exchanger, Nusselt number increases as the mass flow rate increases. Average Nusselt number is about 100 that is very good.

INTRODUCTION

Heat exchangers are approximately the most used parts in chemical processes and can be seen in almost all industrial units. These are the tools providing the possibilities for heat transfer between two or more fluids. Heat transfer can be done in arrangements of liquid-liquid, liquid-gas and gas-gas. Heat exchangers are used in order to cool a hot fluid or to warm a cold fluid or both together. Heat exchangers are used in a wide variety of applications such as in power plants, refineries, petrochemical industries, process industries, food and drug industries, heating and air conditioning. Spiral heat exchangers have got especial usage in handling high viscous fluids, high solid content flows and two-phase solid-liquid flows.

Cleaning after fouling is one of the most expensive issues related to heat exchangers that lead to waste the time and money. As an example, the costs of fouling in the USA industries reach up to 5 billion dollars a year.

Among different types of heat exchangers that are used for cooling and heating the fluids in industries, Spiral Plate Heat

Citation: Khorshidi, J. and Heidari, S. (2016) Design and Construction of a Spiral Heat Exchanger. Advances in Chemical Engineering and Science, 6, 201-208. doi: 10.4236/aces.2016.62021.

Exchanger (SPHE) has an especial place. As shown in Figure 1, a SPHE consists of two sheets that were rolled around a central rod and therefore two separated concentric channels are made.

The ends of the channels are sealed through welding. There are two possibilities to seal the sides of the heat exchanger: using bolts and gaskets to fasten the covering sheets to the heat exchanger or welding the covering sheets to the heat exchanger. Both have their own advantages and disadvantages. For the case of bolts and gaskets one can open the heat exchanger easier and do the maintenance with more comfort. But the sealing is more complicated and working parameters are restricted to those of the gasket. On one hand, welding will extend working parameter possibilities. On the other hand, it makes the maintenance more difficult. Anyway, we decided to use welding approach.

Each of the concurrent and countercurrent trends of flows is available. Almost always the hot fluid enters the heat exchanger in the center and quits at the periphery. Cold fluid can enter either from the center or from the periphery, and therefore, this choice will lead to concurrent or countercurrent flows, respectively. It is obvious that countercurrent arrangement has better heat transfer performance.

Figure 1 shows the most common arrangement type for flows. Distance between heat transfer surfaces can be chosen from 5 to 25 mm to optimize the velocity and heat transfer behavior. Commonly, heat capacity and heat transfer surface area of this type of heat exchanger are up to 900 Kw and 500 m^2, respectively for one unit. For higher capacities that are especially required for cellulosic and coke industries, parallel and series arrangements can be used. SPHE can work under pressure up to 15 bar and temperature up to 450 degrees Celsius.

One passaged path of the flow, spiral shape of the path, and existence of studs among the channels will lead to turbulent flows even at velocities lower than anticipated. This and spiral nature of the flow cause permanent impact of the fluid particles to the heat transfer surfaces, and so, prevent the fouling. As fouling increases, the local velocity of fluid increases and this higher speed will remove the fouling. It must be mentioned here that for the heat exchangers that have more than one passage to flow through, such as for the case of shell and tube heat exchanger, when fouling occurs in one passage, flow will go through another passage and the speed in first passage decreases and the fouling will grow even more.

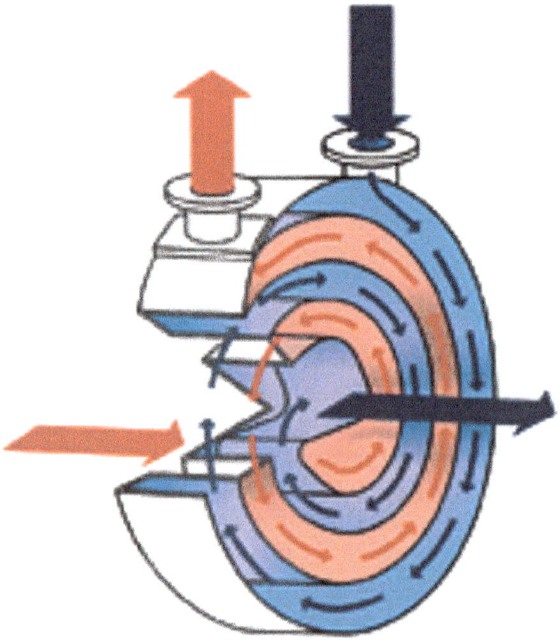

Figure 1. Spiral plate heat exchanger.

LITERATURE REVIEW

In the past years some efforts have been done to use Spiral Heat Exchangers and to simulate the flow in them. Oil extraction system pipes were studied by G. E. Kondhalkar and V. N. Kapatkat. They found out that spiral tube could be used to reduce the costs [1] . Thermal performance and pressure drop of the helical-coil heat exchangers were considered by P. Naphon. This work showed that mass flow rate of both the hot and cold fluids flows had considerable effects on the heat exchanger performance [2] . P. M. Deshpande and S. Dawande studied hydrodynamics of horizontal spiral coil tube. Their effort led to the result that the transverse vibrations of the conical coiled tubes have noticeable effect on the entrance tube vibration frequency [3] . M. S. Tandale and S. M. Joshi designed a spiral heat exchanger to recover the waste heat from producer gas. They realized that several forces (viscous, buoyancy, and centrifugal) influenced the flow in spiral tubes [4] . Mathematical analysis of the transverse vibration of conical spiral tube was done by Y. Ke et al. Experiments showed that the difference between the heat transfer coefficients that were calculated by experimental procedure and by analytical approach was about 12% [5] . M. P. Nueza et al. studied the design approach for the spiral heat exchangers. They made it clear that the heat transfer rate of the spiral heat exchanger is more than that of the other types of heat exchangers [6] . Design of spiral tube heat exchangers was studied by P. E. Minton. This work led to some relations usable for designing heat exchangers [7] . Y. Mori and W. Nakayama worked on forced convective heat transfer in curved pipes for high dean numbers. Flow and temperature fields were studied theoretically and experimentally. They thought that the flow was divided into two parts: a main central flow and a thin boundary layer near the walls [8] . C. E. Kalb and J. D. Seader discussed heat and mass transfer phenomena for viscous flow in curved circular pipes. They did numerical analysis for uniform wall heat flux, while uniform peripheral temperature was supposed, for the Dean number range of 1 - 1200 and Prandtl number range of 0.005 - 1500 and the curvature ratios of 10 - 100 for developed velocity and temperature fields. They found out that the curvature ratio parameter had little effect on the Nusselt number no matter what the Prandtl number was [9] .

EXCHANGER DESIGN

The design of heat exchangers is done in two steps. First, thermal and fluid dynamical design and then mechanical design. Thermal and fluid dynamical design is the main and substantial step to heat exchanger design. This calculates the required heat transfer surface and pressure drop for the heat exchanger. The goal of mechanical design is to have a heat exchanger able to tolerate working conditions. Actually, this step is going to determine physical properties and shape of the heat exchanger.

Thermal Design

During heat exchanger design often overall heat transfer coefficient is used that connects the rate of heat transfer to the temperature difference of the system. Actually, all the resistances to the heat transfer including that of cold fluid film, wall thickness, hot fluid film, and fouling resistances in both sides are compacted in one coefficient. Finally, the relation for heat transfer is as relation (1).

$$Q = U * A * F * \Delta T_{LMTD} \tag{1}$$

where: A is the heat transfer surface area, F is a correction coefficient, DT_{LMTD} is the mean logarithmic temperature difference, and U is the overall heat transfer coefficient. U is defined as below:

$$U = \frac{1}{\frac{1}{h_h} + \frac{t_w}{k_w} + \frac{1}{h_c} + r_i + r_o} \tag{2}$$

where: h_h and h_c are heat transfer coefficients of the hot and cold fluids, kw is the conductivity of wall (heat transfer surface), r_i and r_o are the fouling resistances of both sides, and t_w is the wall thickness.

Some experimental relations are available to calculate h_h and h_c for heat exchanger [1] :

For radial laminar flow:

$$h = 1.86 * c * G * [Re]^{-2/3} \left[\frac{L}{D_e}\right]^{-1/3} \left[\frac{z_f}{z_b}\right]^{-0.14} \tag{3}$$

and for radial turbulent flow:

$$h = \left(1 + 3.54 * \frac{D_e}{D_h}\right) * 0.023 * c * G * [Re]^{-0.2} * [Pr]^{-2/3} \tag{4}$$

where: D_e is the equivalent diameter (ft), D_h is the spiral diameter (ft), L is the length of each heat transfer surface (ft), z is viscosity (cp), c refers to specific heat capacity $\left(\frac{\text{Btu}}{\text{lb}^\circ\text{F}}\right)$, G is mass flow rate $\left(\frac{\text{lb}}{\text{hr} \cdot \text{ft}^2}\right)$, h is heat transfer coefficient $\left(\frac{\text{lb}}{\text{lb} \cdot \text{hr} \cdot \text{ft}^2}\right)$, Re is the replacement for Reynolds Number, and Pr denotes Prandtl Number.

Re is calculated by: $Re = \frac{G \cdot D_e}{z}$.

If H is the heat exchanger height, s is the distance between surfaces in each channel, and $\dot{m}$ is the mass flow rate, we have:

$$G = \frac{\dot{m}}{H \cdot s} \tag{5}$$

$$D_e = \frac{2H \cdot s}{H + s} \cong 2s. \tag{6}$$

As H >> s then the equivalent diameter is twice the distance between channels.

Mechanical Design

Spiral plate heat exchangers are built by rolling two parallel long sheets around a central bar to make a spiral shape. The free final edges of channels then will weld together to seal the end of channels. The distance between metal sheets is kept using studs that are welded to the sheets. The length of studs can vary from 5 to 25 mm. hence, with respect to the mass flow rate, different distances between the sheets can be chosen during the design period. In each channel, hot or cold fluid path, secondary flows are developed that lead to better mixing and therefore heat transfer rate is increased and fouling is decreased. These heat exchangers are compact but their complicated construction procedure cause higher primary construction costs. Heat transfer surface area of these heat exchangers is from 0.5 to 500 m^2. Maximum working pressure and temperature often restricted to 15 bar and 450 degrees Celsius, respectively. Applying new technologies may increase the maximum temperature up to 850 degrees Celsius. Using available equations and with respect to experimental requirements, a small LAB- sized model of the heat exchanger was made. Galvanized-Iron sheets were used. Properties of Galvanized-Iron are given in Table 1.

This type of sheet was used regarding its availability, acceptable price and good thermodynamic properties. The distances between two sheets were set to be 15 mm for one channel and 25 mm for another one. Figure 2 shows a view of the channels. The ends are welded. Also, the sides are closed by welding some sheets.

Figure 3 shows the completed heat exchanger.

Air pressure tests were done to check if there are some leakages. This is done by filling the heat exchanger up with air of the pressure of 2 bar gauge. After 12 hours, no change in the pressure was noticed. So we decided that the heat exchanger has no leakage.

Table 1. Properties of galvanized-iron

ρ	$7700 - 8100 \, \frac{\text{kg}}{\text{m}^3}$
k	$11.2 - 48.3 \, \frac{\text{W}}{\text{m} \cdot \text{k}}$
α	$(9 - 27) \times 10^{-6} \, \frac{1}{\text{k}}$

Figure 2. Hot and cold channels.

Figure 3. Completed heat exchanger.

SIMULATION

To check the usability of heat exchanger, a geometric model of it was made using Gambit software and then it was analyzed by means of Fluent software. Table 2 shows the flow parameters were used in Fluent.

To check the network independency, 3 networks including 16,000, 32,000, and 48,000 nodes were used. Considering the results, the network with 32,000 nodes opted to be used. In network making, we stressed the network near the highly changing areas such as entrances and walls. Figure 4 shows a view of the model.

Fluid is incompressible. The following boundary conditions were used: No slip condition on walls, temperature of the environment considered as 306 k. Heat transfer coefficient of air was set to be $h = 5 \, \frac{\text{W}}{\text{m}^2 \cdot \text{k}}$. For internal walls we had: $-k \frac{\partial T}{\partial r} = h \Delta T$. For cold and hot outlet, the pressure was assumed to be atmospheric. Table 3 reviews some of the Fluent results.

Temperature distribution at surfaces of the distances of 0.1, 0.2, 0.3, and 0.4 m from the entrance of the heat exchanger are shown in Figure 5. In analysis the worst case that is concurrent flow is considered.

As it is obvious from Figure 5, hot fluid is cooled as it flows through the heat exchanger, and cold fluid is warmed before leaving the heat exchanger. One can see from the figure that as the hot fluid passes the heat exchanger for each 0.1 m it advances its temperature drops about 1 degree Celsius and this amount is added to the temperature of the cold fluid.

To show a better view of the influence of the velocity on the Nusselt number, the Nusselt number versus velocity is shown in Figure 6.

This chart shows that for each velocity, Nusselt number of cold fluid is more than that of hot fluid. Also, for both fluids, Nusselt numbers increases as the velocity increases. This effect is the result of increasing the heat transfer coefficient by velocity. In this picture all the properties of two fluids are the same but the temperature. So, if the temperature of the cold fluid increases, and approaches that of hot fluid, its Nusselt number decreases.

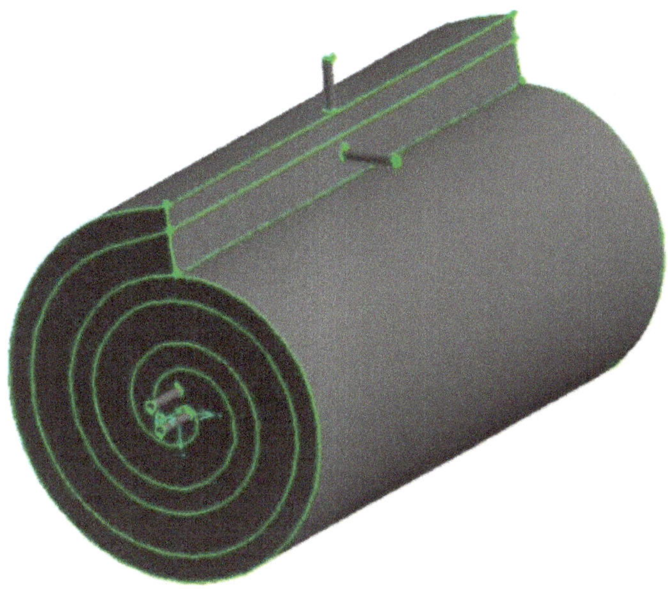

Figure 4. Heat exchanger model.

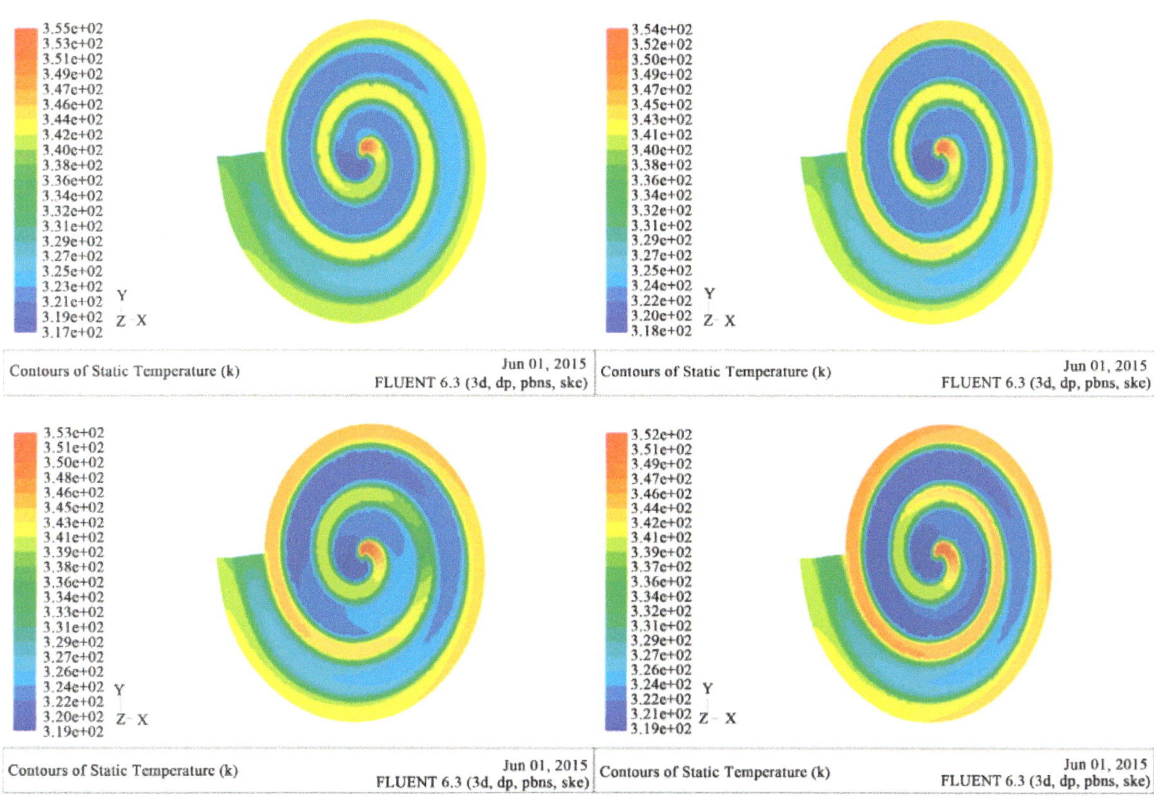

Figure 5. Several surfaces temperatures.

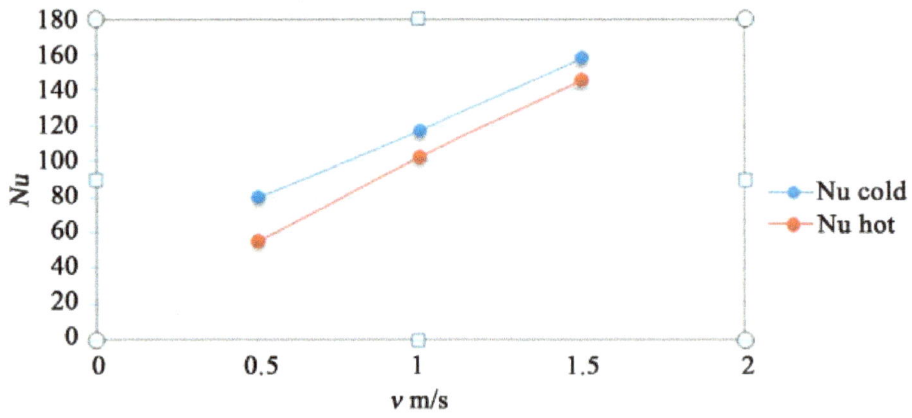

Figure 6. Nuselt vs velocity.

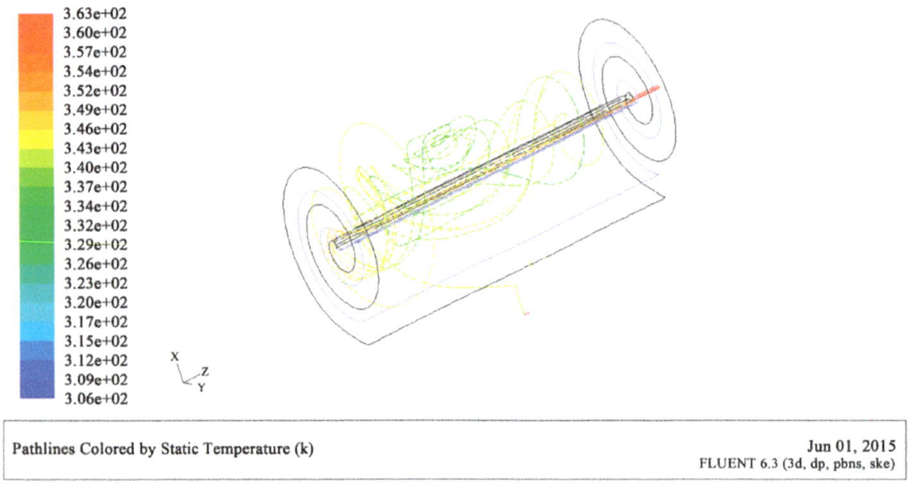

Figure 7. Particle movement path.

Table 2. Flow parameters for fluent

Row	$v_{in,\,hot}\left(\frac{m}{s}\right)$	$v_{in,\,cold}\left(\frac{m}{s}\right)$	$T_{in,\,hot}\,(k)$	$T_{in,\,cold}\,(k)$
1	0.5	0.5	353	301
2	1	1	363	306
3	1.5	1.5	373	311

Table 3. Review of fluent simulation outlet

Row	$T_{in,\,hot}\,(k)$	$T_{in,\,hot}\,(k)$	$T_{out,\,hot}\,(k)$	$T_{out,\,cold}\,(k)$	Nu_{hot}	Nu_{cold}	$v\left(\frac{m}{s}\right)$
1	353	301	331.46	322	130.28	143.12	1
2	353	306	333.52	325.47	102.13	116.18	1
3	353	311	335.59	328.40	80.97	94.22	1
4	363	301	337.23	326.76	128.22	148.39	1
5	363	306	339.37	334.50	102.25	116.42	1
6	363	311	342.18	331.86	89.11	96.04	1
7	373	306	345.10	333.89	103.09	118.06	1
8	363	306	334.48	334.51	55.33	80.47	0.5
9	363	306	342.12	326.87	146.50	158.98	1.5

Furthermore, mixing of the flow and the path of particles movement in heat exchanger were studied. Figure 7 shows that the particles path is highly random and mixing is excellent. In the other hand, this leads to turbulent flow at speed much lower than that of straight path. Also, particles move in radial, axial and swirly paths.

CONCLUSION

In this work, a LAB-sized Spiral Plate Heat Exchanger was built. Galvanized-Iron was used because of its good thermodynamic properties, availability and acceptable price. No leakage was determined using air pressure test. This article studies the distribution of temperature and Nusselt number through Gambit and Fluent software. Heat transfer from hot fluid to cold fluid was done along the heat exchanger. The constructed heat exchanger will show good heat transfer behavior in both channels as can be interpreted from software outlets. Heat transfer is going to be very well along the heat exchanger. Fluid flow will become completely turbulent. Especial shape of the heat exchanger and secondary flows lead to random movement of the fluid particles in all directions so that axial, radial, and swirly movements exist together that mix the fluid very well and force the flow to be turbulent. Nusselt number magnitude is higher in the entrances because of the more temperature difference but it decreases as fluids go forward through the heat exchanger toward the exit points and as the temperature difference decreases. This is unanimous to the physics of heat transfer. Since the simulation of heat exchanger performance is possible, one can study the optimization of the heat exchanger. Velocity, Nusselt and temperature distribution predictions agree with the anticipated conditions. Regarding all the materials mentioned above, using this type of heat exchanger is economical and due to its all other advantages its usage is recommended.

REFERENCES

1. Kondahkar, G.E. and Kapatkat, V.N. (2012) Performance Analysis of Spiral Tube Heat Exchanger Used in Oil Extraction System. International Journal of Modern Engineering Research, 2, 930-936.
2. Naphon, P. (2007) Thermal Performance and Pressure Drop of the Helical-Coil Heat Exchangers with and without Helically Crimped Fins. International Communications in Heat and Mass Transfer, 34, 321-330. http://dx.doi.org/10.1016/j.icheatmasstransfer.2006.11.009
3. Deshpande, P.M. and Dwande, S. (2012) Study of Hydrodynamics of Horizontal Spiral Coil Tube. International Journal of Advanced Engineering Research and Studies, 1, 112-114.
4. Tandale, M.S. and Joshi, S.M. (2008) Design of Heat Exchanger for Waste Heat Recovery from Producer Gas. Proceedings of the International Conference on Heat and Mass Transfer, Acapulco, January 2008, 1790-2769.
5. Ke, Y., Qi, G.P., Cay, S.Y. and Bo, B.W. (2010) Mathematical Analysis of Transverse Vibration of Conical Spiral Tube Bundle with External Fluid Flow. Journal of Hydrodynamics, 22, 816-822. http://dx.doi.org/10.1016/S1001-6058(09)60121-9
6. Nueza, M.P., Polley, G.T., Davalos, L.C. and Rodriguez, G.M. (2007) Design Approach for Spiral Heat Exchanges. Institution of Chemical Engineering, 85, 322-327.
7. Minton, P.E. (1971) Designing Spiral Tube Heat Exchangers. Process Heat Exchange, 136-144.
8. Mori, Y. and Nakayama, W. (1965) Study on Forced Convective Heat Transfer in Curved Pipes (1st Report, Laminar Region). International Journal of Heat and Mass Transfer, 8, 67-82. http://dx.doi.org/10.1016/0017-9310(65)90098-0
9. Kalb, C.E. and Seader, J.D. (1972) Heat and Mass Transfer Phenomena for Viscous Flow in Curved Circular Tubes. International Journal of Heat and Mass Transfer, 15, 801-817. http://dx.doi.org/10.1016/0017-9310(72)90122-6

CHAPTER 19

LATTICE BOLTZMANN SIMULATION OF AIRFLOW AND MIXED CONVECTION IN A GENERAL WARD OF HOSPITAL

Taasnim Ahmed Himika, Md. Farhad Hasan, and Md. Mamun Molla

Department of Mathematics & Physics, North South University, Dhaka 1229, Bangladesh

ABSTRACT

In the present investigation the airflow and heat transfer for mixed convection have been simulated for a model general ward of hospital with six beds and partitions using the Lattice Boltzmann Method (LBM). Three different Reynolds numbers 100, 250, and 350 have been considered. Bounce-back condition has been applied at the wall. Results have been represented in three different case studies and the changes have been discussed in terms of streamlines and isotherms. Code validation has also been included before going through the simulation process and it shows good agreement with previously published papers when the comparison is made on average Nusselt number. Results show that the pattern of indoor airflow is varied in each and every case study due to the effect of mixed convection flow and placement of partition. In addition, the changes in average rate of heat transfer indicate that patients closer to inlet get the most air and feel better and if any patient does not need much air, he or she should be kept near the outlet to avoid temperature related complications.

INTRODUCTION

The study of mixed convection fluid flows has become an attention-grabber due to considerable practical interest in many engineering applications, such as nuclear reactors [1], crystal growth, solar collectors [2], and reservoirs and lakes [3]. Apart from these, mixed convection appears to be a concerning issue while designing modern ventilation system in high-rise buildings and health care centers. The main purpose of room air-ventilation system is to provide comfortable and good indoor air quality with minimum energy consumption so that the residents inside do not feel stuffy or uncomfortable. So, to design efficient ventilation system, necessary designing or simulation tools are required in order to accurately visualize the airflow pattern and know the appropriate positions of inlets and outlets in any complex geometry.

The combined effect of natural and forced convection in fluid flows has been analyzed by many researchers over decades. The fact behind their curiosity lies in the buoyancy effect that can be the dominant factor in determining heat transfer. Since

Citation: Taasnim Ahmed Himika, Md. Farhad Hasan, and Md. Mamun Molla, "Lattice Boltzmann Simulation of Airflow and Mixed Convection in a General Ward of Hospital," Journal of Computational Engineering, vol. 2016, Article ID 5405939, 15 pages, 2016. doi:10.1155/2016/5405939.

experimental scale modeling is very expensive, researchers continued their study with the aid of traditional Computational Fluid Dynamics (CFD) model simulation technique. But the main problem with this traditional method involves solving the Navier-Stokes (NS) equations that needs more CPU time and memory. Researchers found that the solution lies in the Lattice Boltzmann Model (LBM), an alternative Computational Fluid Dynamics (CFD) model that does the same job by solving discrete Boltzmann equation which is a faster process than solving the NS equations [4]. LBM has originated from Lattice Gas Automata (LGA) [5] which is used to simulate fluid dynamics by tracing the movement of single-particle distribution function considering the space, time, and particle velocities to be discrete.

The reasons behind LBM's popularity are as follows: because it has simple mathematical modeling of any complex system, the boundaries are treated in simple manner, it is a suitable model for parallel computation, and, apparently, it saves 30 percent of the total CPU time which would have been occupied by nonlinear Riemann problem in CFD [6]. LBM is also considered to be very easy and efficient method over finite volume, finite-difference, and finite element techniques. Many experiments and researches were performed to explore the effectiveness of LBM. Peng et al. [7] simplified thermal LBM to study incompressible thermal flows. Lee et al. [8] researched the incompressible lattice Bhatnagar-Gross-Krook (BGK) model to simulate axisymmetric geometry. Sidik [9] calculated the temperature field and velocity field utilizing two different distribution functions for each field and developed an incompressible 2D and 3D thermohydrodynamics for the LBM. The main aim of Mondal and Li's [10] investigation was to observe the effect that volumetric radiation has on the natural convection in a square cavity containing an absorbing, emitting, and scattering medium. Szucki and Suchy [11] studied a method concerning local viscosity changes in single-relaxation time (SRT) LBM. Moreover, LBM was successful in various fluid dynamical problems including flows in porous media [12], magnetohydrodynamics [13], immiscible fluids [14], and turbulence [15]. LBM can also be applied to investigate indoor airflow. In 2002 Crouse et al. [16] were the first and probably the only ones to implement LBM in indoor convective airflow analysis.

An investigation was carried out by Zhang et al. [17] who studied the airflow inside an airliner ward performing experimental mathematics and numerical simulations. The ward was assumed to be half occupied with 28 seats and twin-aisle. The authors used the Reynolds averaged Navier-Stokes equations based on the RNG $k-\epsilon$ model to solve the air velocity, air temperature, and gas contaminant concentration and applied a Lagrangian technique to model the particle transport. The numerical results showed good agreement quantitatively with their experimental data. However, they confessed that it is very challenging and difficult to achieve a complete and accurate validation for the complex flow in a ward. Later, Zhang and Lin [18] used LBM to analyze indoor airflows. They worked on low Reynolds number indoor airflow fields in a model room building a partition and their computational results of LBM were found to be in agreement with the experimental data in terms of airflow velocities. They applied LBM in a relatively more complex environment, that is, a model ward with 10 beds. Although their works were solely on fluid flow, they did not consider any temperature equation.

Liu et al. [19], on the other hand, utilized CFD to examine a critical issue of hospital Operating Room (OR). They numerically simulated a horizontal airflow to control airborne particles inside a hospital OR. The investigation was performed on different horizontal airflow patterns and air quality inside an OR model of dimension 300 cm × 296 cm × 240 cm. They realized the urgency of a proper ventilation system inside a hospital OR and, so, carried on some onsite tests and CFD simulations to estimate the effectiveness of latitudinal unidirectional airflow in order to bring down infectious airborne particles. The suggested ventilation supplying ultraclean air from a horizontal plane avoided obstacles, like medical lights from upstream flow, and also lateral flow perpendicular to the airflow of thermal plume around surgeons and patient can avoid opposing effect between the two airflows in downward airflow system. However, the application of airflow in horizontal direction had not been properly examined in their paper.

All the above examples verified the capability and effectiveness of LBM over traditional CFD, but the experiments were done either considering a natural fluid flow or forced fluid flow, whereas in reality the presence of both flows can be seen simultaneously. In purely natural convection flow, the only driving force is the density gradient, and for forced convection the force is inflow condition or pressure-gradient. Since there is temperature difference between all objects, living beings, and the environment, in the presence of a gravitational field temperature gradients tend to induce density gradients that ultimately drives the fluid flow. Watzinger and Johnson [20] took the first initiative to investigate heat transfer in a turbulent mixed convective water flow, both for opposing and supporting flow conditions. Later, Prasad and Koseff [21] investigated the combined natural and forced convection heat transfer in a deep lid-driven cavity flow. Hongtau Xu et al. [22] numerically studied double diffusive mixed convection around a heated cylinder in an enclosure and Al-Sanea et al. [23] analyzed the effect of input Reynolds number and room aspect ratio on turbulent flow and ceiling transfer coefficient for mixed convection using CFD method. Zhao et al. [24] used traditional CFD method to investigate three-dimensional (3D) nonlinear multiple steady fluid flows in a slot-ventilated enclosure. Khanafer and Chamkha [25] applied finite volume method to experiment with a fluid-saturated porous medium on mixed convection flow in a lid-driven enclosure. On the other hand, Kefayati et al. [26] applied LBM to simulate magnetohydrodynamic mixed convection in a lid-driven square cavity with linearly heated wall. As the present investigation is based on mixed convection inside a hospital ward having partition in the middle, the work of Lee and Awbi [27] is worth mentioning which is on the effect of room air quality with mixing ventilation due to internal partitioning, but their observation was only on statistical analysis using traditional CFD codes. They stated that to get better room air quality the partition positioned towards the exhaust zone having a larger height and gap underneath was beneficial.

The purpose of the present work is to focus on indoor airflow simulation by LBM in a hospital ward with six beds and partitions. The airflow inside the ward is considered to be mixed convection flow and different cases have been considered as well. The room temperature has been kept lower than patients' temperature as each patient on bed has been considered to release heat continuously in the proposed model. Before going through further simulation, code validation has been presented. This paper is organized in the following way: in Section2, physical geometry of the problem has been discussed, followed by the mathematical formulation in Section3. In Section 5, the results have been discussed in detail and then conclusion has been added to Section 7.

PHYSICAL GEOMETRY

The present investigation has been applied for the mixed convection in a general ward of hospital, as shown in Figure 1.

In Figure 1(a), it is just a ward without any bed or partition; in Figure 1(b), a general ward with six model beds is shown and in Figure 1(c) that ward with six beds along with a partition is depicted. Here, h is the height and $l = 2h$ is the horizontal length of each block and the height and width of the partition are $a = 4h$ and $b = h/2$, respectively. The height of the ward is $H = 5h$ and $L = 8H$ is the length of the ward. The length of the inlet and outlet is $(h + b)/2$.

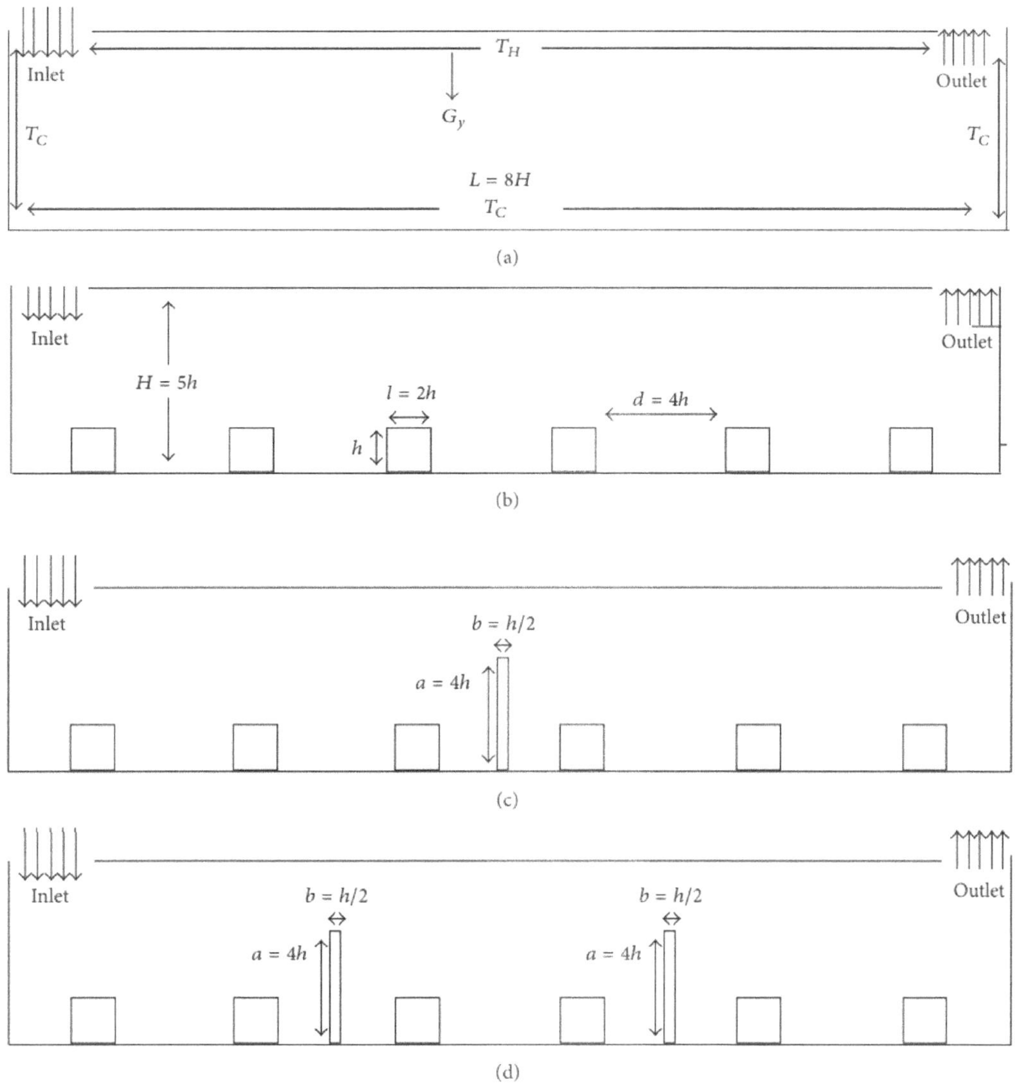

Figure 1: Schematic diagram of the geometry of hospital ward showing (a) principles applied building the model, (b) Case I without partition, (c) Case II with one partition, and (d) Case III with two partitions.

MATHEMATICAL FORMULATION

The LBM was originated from Ludwig Boltzmann's kinetic theory of gases. The basic theory of this equation is that gases or fluids can be imagined as consisting of a large number of tiny particles moving with random motion. The exchange of momentum and energy is found through particle streaming and collision between particles. So, this process can be demonstrated by the Boltzmann transport equation [28, 29]:

$$\frac{\partial f}{\partial t} + \bar{u} \cdot \overline{\nabla} f = \Omega\left(f, f^{eq}\right), \tag{1}$$

where f^{eq} is the equilibrium distribution function, respectively, $\bar{u}$ is the particle velocity, and Ω is the collision operator. Here, f represents $(\bar{x}, ei, t)$ which is the probability to locate a particle at position x with velocity vector e_i at time t.

Lattice Boltzmann Method

The LBM simplifies Boltzmann's original idea of gas dynamics by reducing the number of particles and circumscribing them to the nodes of a lattice. For a two-dimensional model, a particle is preferred to stream in a possible of 9 directions, including the one staying at rest. For this type of modeling, the most popular one is *D2Q9* model which has been considered in this work. This is the standard model for two-dimensional structure for both flow and temperature conditions. The velocities for these conditions are known as macroscopic velocities.

The discretized Lattice Boltzmann (LB) equations for both flow and temperature field can be written as follows [29]:

$$f_i(\overline{x}+\delta\overline{e}_i, t+\delta t) - f_i(\overline{x},t) = -\frac{1}{\tau_\nu}\left[f_i(\overline{x},t) - f_i^{(eq)}(\overline{x},t)\right] + \delta t \cdot F_i,$$

$$g_i(\overline{x}+\delta\overline{e}_i, t+\delta t) - g_i(\overline{x},t) = -\frac{1}{\tau_\alpha}\left[g_i(\overline{x},t) - g_i^{(eq)}(\overline{x},t)\right]. \quad (2)$$

Here, $\tau_v = 3] + 0.5$ and $\tau_\alpha = 3\alpha + 0.5$ are the single-relaxation times that control the rate of approach to equilibrium, $f_i(\overline{x}, t)$ and $g(\overline{x}, t)$ are the density distribution functions for velocity and temperature, respectively, along the direction $\overline{e}_i$ at $(\overline{x}, t)$,] is the kinematic viscosity, and α is the thermal diffusivity. The particle speed is e_i and ρ is the density.

To calculate the equilibrium distribution functions for the velocity, $f_i^{eq}(\overline{x}, t)$, and for the temperature, $g_i^{eq}(\overline{x}, t)$, the general expressions have been considered [29–31]:

$$f_i^{eq} = w_i\rho\left[1 + \frac{3(\overline{e}_i\cdot\overline{u})}{c^2} + \frac{9(\overline{e}_i\cdot\overline{u})^2}{2c^4} - \frac{3(\overline{u}\cdot\overline{u})}{2c^2}\right],$$

$$g_i^{eq} = w_iT\left[1 + \frac{3(\overline{e}_i\cdot\overline{u})}{c^2} + \frac{9(\overline{e}_i\cdot\overline{u})^2}{2c^4} - \frac{3(\overline{u}\cdot\overline{u})}{2c^2}\right], \quad (3)$$

where $\overline{u} = (u, V)$ is the fluid velocity and $c = \Delta x/\Delta t = 1$ is the lattice speed. The weighting factors w_i for *D2Q9* are given as [29]

$$w_i = \begin{cases} \frac{4}{9}, & i = 0 \\ \frac{1}{9}, & i = 1,2,3,4 \\ \frac{1}{36}, & i = 5,6,7,8. \end{cases} \quad (4)$$

The macroscopic fluid density ρ and velocity $\overline{u}$ are obtained from the moments of the distribution function as follows [28]:

$$\rho(\overline{x},t) = \sum_{i=0}^{8} f_i(\overline{x},t),$$

$$\overline{u}(\overline{x},t) = \frac{1}{\rho}\sum_{i=0}^{8}\overline{e}_i f_i(\overline{x},t) \quad (5)$$

and the temperature T is obtained as [32]

$$T = \sum_{i=0}^{8} g_i(\overline{x},t). \quad (6)$$

Boundary Conditions

Boundary conditions (BCs) are very important part for both the stability and the accuracy of any kind of numerical solution. For LB modeling, the discrete distribution functions on the boundary need to be taken care of to reflect the macroscopic BCs of the fluid. In this work, at the wall, the bounce-back condition has been applied.

Bounce-back BCs are usually used to implement no-slip conditions on the boundary. By the bounce-back condition, it is meant that when a particle of a certain fluid for discrete distribution function reaches a boundary node, the particle will scatter back to the fluid along with its incoming direction. Bounce-back condition can be of various types but on-grid bounce-back condition has been used in this paper.

Boundary Condition for Velocity

Four possible directions in two-dimensional structure—east, west, north, and south—have been considered in the proposed model and no-slip or bounce-back boundary condition is used at the walls, which has been shown in Figure 1.

At east (right), west (left), and south (bottom) walls the bounce-back condition has been applied and the resulting relations are given below in order:

$$f_3 = f_1,$$
$$f_7 = f_5,$$
$$f_6 = f_8,$$
$$f_1 = f_3,$$
$$f_5 = f_7,$$
$$f_8 = f_6,$$
$$f_2 = f_4,$$
$$f_5 = f_7,$$
$$f_6 = f_8. \quad (7)$$

On the other hand, at the north (top) wall, three different boundary conditions have been applied. At the inlet, Zou and He boundary conditions have been applied as shown in the following [33]:

$$\rho_N = \frac{(f_0 + f_1 + f_3 + 2(f_2 + f_6 + f_5))}{(1+U)},$$

$$f_4 = f_2 - \frac{2}{3}\rho_N U,$$

$$f_8 = f_6 - \frac{1}{6}\rho_N U,$$

$$f_7 = f_5 - \frac{1}{6}\rho_N U, \tag{8}$$

where U is the bulk velocity based on the Reynolds number $\mathrm{Re} \stackrel{\mathrm{def}}{=} UH/\nu$. At the outlet the following boundary conditions are used:

$$f_{4,N} = f_{4,N-1},$$

$$f_{8,N} = f_{8,N-1},$$

$$f_{7,N} = f_{7,N-1} \tag{9}$$

and bounce-back condition is used for the rest of the top walls. At the top surface and at each side of a block and partition, the following bounce-back conditions have been used:

Top surface:

$$f_{2,x_i} = f_{4,x_i},$$

$$f_{5,x_i} = f_{7,x_i},$$

$$f_{6,x_i} = f_{8,x_i}. \tag{10}$$

Left side:

$$f_{1,x_i} = f_{3,x_i},$$

$$f_{5,x_i} = f_{7,x_i},$$

$$f_{8,x_i} = f_{6,x_i}. \tag{11}$$

Right side:

$$f_{3,x_i} = f_{1,x_i},$$

$$f_{6,x_i} = f_{8,x_i},$$

$$f_{7,x_i} = f_{5,x_i}. \tag{12}$$

Boundary Condition for Temperature

Cold and hot temperature conditions are applied at the walls (see Figure 1).

At east (right) wall the cold temperature condition is $g_3 = -g_1$, $g_7 = -g_5$, and $g_6 = -g_8$.

At west (left) wall the cold temperature condition is $g_1 = -g_3$, $g_5 = -g_7$, and $g_8 = -g_6$.

At south (bottom) wall the cold temperature condition is $g_2 = -g_4$, $g_5 = -g_7$, and $g_6 = -g_8$.

At the north (top) wall in the outlet a zero gradient boundary condition is applied as follows:

$$g_{4,N} = g_{4,N-1},$$

$$g_{8,N} = g_{6,N-1},$$

$$g_{7,N} = g_{5,N-1}. \tag{13}$$

And the rest of the top wall is kept cold by using the conditions $g_4 = -g_2$, $g_8 = -g_6$, and $g_7 = -g_5$.

At the top surface of each block the hot temperature condition is used as follows:

$$g_{2,y_2} = T_w(w_2 + w_4) - g_{4,y_2},$$

$$g_{5,y_2} = T_w(w_5 + w_7) - g_{7,y_2},$$

$$g_{6,y_2} = T_w(w_6 + w_8) - g_{8,y_2}, \tag{14}$$

where y_2 indicates the top surface of the blocks and T_w is the wall temperature.

Nondimensional Number and Mixed Convection Parameter

The important nondimensional number Grashof (Gr) and Prandtl (Pr) numbers are defined as [32]

$$\mathrm{Gr} = \frac{\beta G_y H^3 (T_H - T_C)}{\nu^2},$$

$$\mathrm{Pr} = \frac{\nu}{\alpha}. \tag{15}$$

In the mixed convection the Richardson number, $\mathrm{Ri} \stackrel{\mathrm{def}}{=} \mathrm{Gr}/\mathrm{Re}^2$, controls the flow phenomena. In this simulation the force term is defined as follows [34]:

$$F = 3w_i \mathrm{Ri}(T - T_m)e_y, \tag{16}$$

where $T_m = (T_H + T_C)/2$ is the mean temperature. In this paper the fluid velocity u is nondimensionalized by the constant inlet velocity U and the nondimensional temperature is $\theta = (T - T_c)/(T_w - T_c)$.

In the present work, the Mach number (Ma) for Re = 100 is 0.05, for Re = 250 is 0.02, and for Re = 350 is 0.014. All these values are less than 0.3 which satisfy the condition of incompressible fluid flow [32]. In buoyant flow, Mach number and thermal diffusivity (α) can be defined by the following equations after fixing the values of viscosity, Prandtl number, and Reynolds number:

$$\mathrm{Ma} = \frac{\nu\sqrt{\mathrm{Gr}}}{Nc},$$

$$\alpha = \frac{\nu}{\mathrm{Pr}}. \tag{17}$$

CONVERGENCE CRITERIA

Standard LBM, being explicit time-marching in nature, requires a long time to attain steady state convergence. However, convergence of solution is one of the highlights of the procedure of recovering the Navier-Stokes equations from LBM. In this single-relaxation-time LBM, the iterative procedure is terminated when the velocity and temperature field satisfy the following convergence criteria:

$$\frac{\sum\left|\phi^{(n+1)}-\phi^{(n)}\right|}{\sum\left|\phi^{(n+1)}\right|}<10^{-9}, \tag{18}$$

where ϕ is the velocity $\bar{u}$ or temperature T and n is the iteration index and the sum is over the whole domain.

RESULTS AND DISCUSSIONS

This section has been divided into four subsections. At first, code validation has been performed for different values of Re and has been compared in tabular format. In the consequent three subsections, three case studies have been considered inside the hospital ward. In all three cases, one inlet, one outlet, and six beds have been considered. Simulations have been carried out by placing one partition and later two partitions in the room while observing the indoor airflow pattern and the field of temperature in the three situations. This section is ended with the discussion and comparison of the average rate of change of heat transfers achieved for each case study. In these cases, Re = 100, 250, and 350, Pr = 0.71, and Ri = 1 have been kept for all simulations except the code validation case. Different values of τ] generated for different Re are as follows: for Re = 100, it is 0.575, for Re = 250, it is 0.530, and for Re = 350, it is 0.521. Likewise, different values of $\tau\alpha$ are as follows: for Re = 350, it is 0.53, for Re = 250, it is 0.542, and for Re = 100, it is 0.605.

Code Validation

For code validation process, one test case is the mixed convection in a lid-driven square cavity flow for Re = 100, 400, and 100 and Gr = 100 has been considered. The present results in terms of the average Nusselt number have been compared quantitatively with the available results published in different articles that are demonstrated in Table 1. The comparison reveals a very good agreement. The results were compared with the works of Iwatsu et al. [35], Tiwari and Das [38], Kefayati et al. [26], Khanafer et al. [36], Abdelkhalek [37], Khanafer and Chamkha [25], and Waheed [39].

Table 1: Comparison of average Nusselt numbers of present work with other investigations for Gr = 100 and Pr = 0.71

Re	Ri	Present	[35]	[25]	[36]	[37]	[38]	[39]	[26]
100	0.01	1.917965	1.94	2.01	2.02	1.985	2.10	2.03116	2.09
400	0.00062	3.789101	3.84	3.91	4.01	3.8785	3.85	4.0246	4.08082
1000	0.0001	6.339137	6.33	6.33	6.42	6.345	6.33	6.48423	6.54687

Case Study I: Six Beds without Partition in the Ward

In the first case study, six beds without partition have been considered. Each bed is of the same dimension. However, beds are represented as blocks in Figure 1. For the whole computational domain a lattice size of (400 × 50) ~ ($x \times y$) has been taken in all simulations. Each block of (20×10) ~ (x×y) lattice size is used in the ward. The beds are uniformly distributed in one row. Inlet and outlet have been placed vertically on the ceiling on the left and right corners, respectively. The width of both the inlet and the outlet has been kept from 0 to 15 lattices and from 385 to 400 lattices, respectively.

Figures 2(a)–2(c) show the streamlines appended on u/U velocity contour of the airflow that enters the ward through the inlet and flows out through the outlet for Re = 100, 250, and 350, respectively. In case of Re = 100, small recirculations are seen between the blocks and also near the walls in the east and west sides of the ward due to bounce-back condition. A comparatively large recirculation is formed in immediate right to the inlet that is near the top wall. For increasing Re the length of the recirculation zones increases. As shown in Figure 2, the highest value of u/U is 1.08 and the lowest value is −0.14 for Re = 100. In case of Re = 250 and 350, the maximum values increase to 1.26 and 1.30, respectively. As Re increases, the inertia force of air increases which makes it easier to flow around in the ward and, hence, more heat will be transferred out through the outlet.

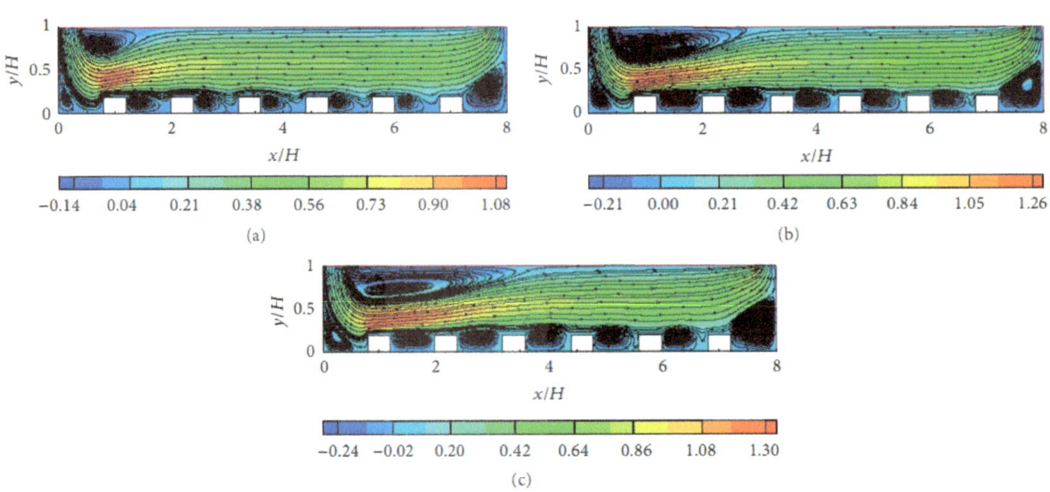

Figure 2: Streamline patterns of Case I for (a) Re = 100, (b) Re = 250, and (c) Re = 350.

Figure 3 represents the isotherms of the room. As can be seen from the simulation result, only the top surface of each block is kept heated while the gaps between the blocks are cold. The heated portion of the blocks represents patients on beds who are releasing heat continuously.

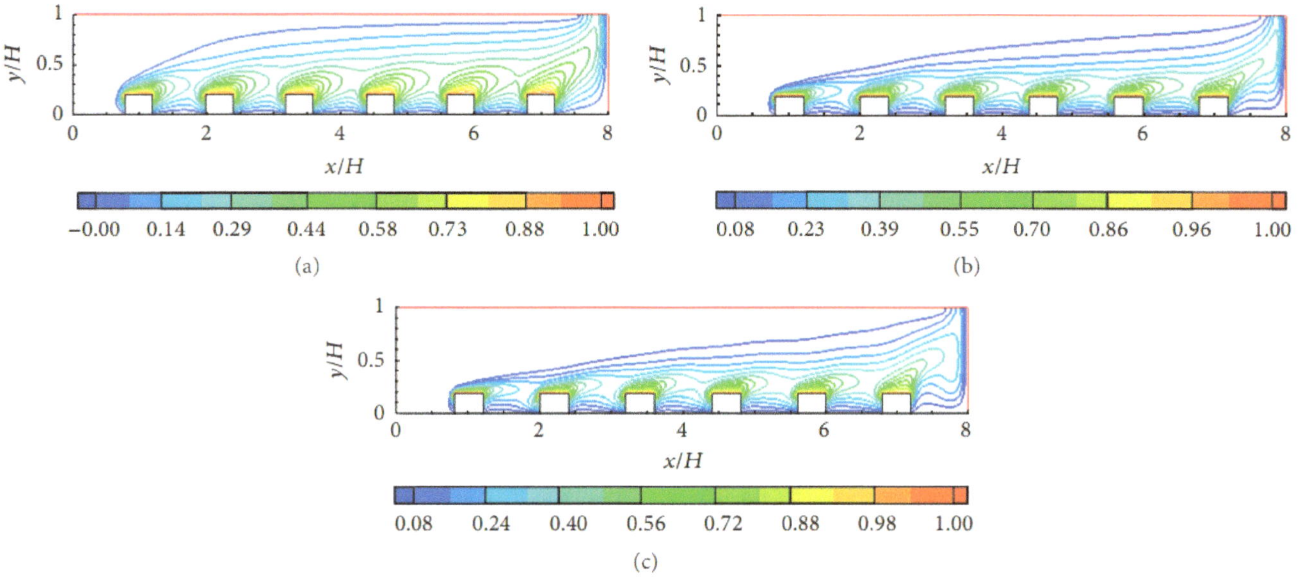

Figure 3: Isotherms of Case I for (a) Re = 100, (b) Re = 250, and (c) Re = 350.

Different velocity patterns have been shown in Figures 4(a)–4(n) demonstrating the u/U velocity profiles at different positions of x/H for the three different Re. This figure outlines the speed at which the air is flowing inside the ward. Due to the recirculation near the inlet, the velocity profiles are also seen to increase resulting in a bent-shaped plot in Figures 4(b) and 4(c). After $x/H = 3.5$ onwards the graphs are almost the same as there the flow characteristics are almost similar, except for Figure 4(n) which is slightly bent at the top forming because at this point the air is preparing to go through the outlet.

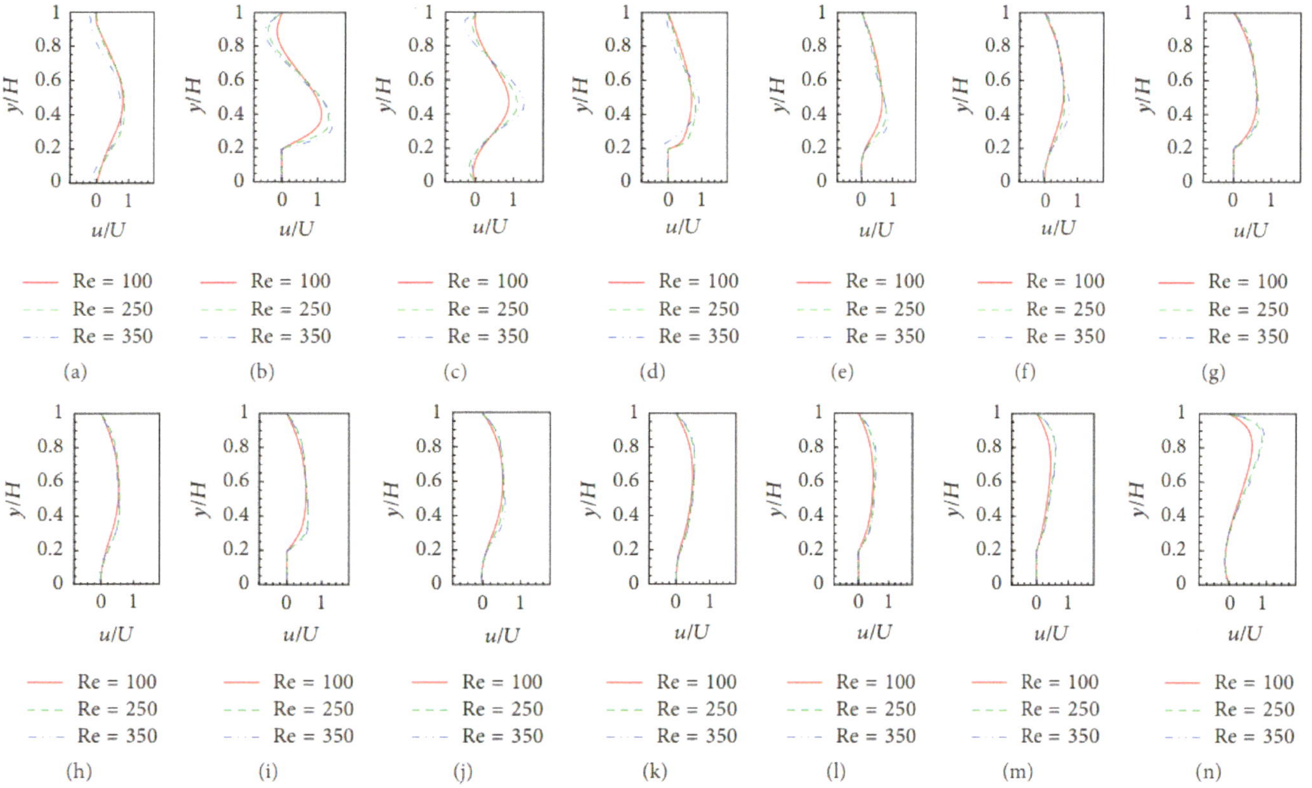

Figure 4: u/U velocity profiles in Case I for the different Re at different positions of horizontal locations (a) $x/H = 0.5$, (b) $x/H = 1$, (c) $x/H = 1.5$, (d) $x/H = 2$, (e) $x/H = 3$, (f) $x/H = 3.5$, (g) $x/H = 4$, (h) $x/H = 4.5$, (i) $x/H = 5$, (j) $x/H = 5.5$, (k) $x/H = 6$, (l) $x/H = 6.5$, (m) $x/H = 7$, and (n) $x/H = 7.5$.

Case Study II: Six Beds with One Partition

In this investigation, the previous model has been updated by adding a partition in the middle of the general ward while all other schematic measurements have been kept the same, as shown in Figure 1.The lattice size of partition is(5×40) ~ (x×y) and it is placed just after the first three blocks (beds). Simulations have been carried out for the same three Reynolds numbers. After placing a partition just like in Figure 5, recirculation is seen in Figure 5 and it shows much better result than the previous simulations. The airflow pattern is clearer this time which shows that air moves inward smoothly, passes the partition naturally without overlapping it, and nicely flows out through the outlet creating its usual recirculation near the inlet, between the blocks and near the outlet. Unlike earlier, the recirculations now are larger and more frequent, particularly after crossing the partition. For Re = 100, a large air circulation can be noticed between blocks 4 and 5 and in case of Re = 250, an even larger recirculation has appeared between blocks 5 and 6 along with a newly emerged one just beside the outlet. As for Re = 350, with the increase in convective air current more chaotic flow dynamics can be noticed before the partition appears. The airflow pattern in this region indicates that, in case of mixed convection for Re = 350, the air stays in transitional state which shows both the laminar characteristics of natural convection and the turbulent nature of forced convection, thereby representing forced convection. Moreover, after placing the partition, the peak values of the contour have risen to 2.15 from 1.08, 2.33 from 1.26, and 2.38 from 1.30 for Re = 100, Re = 250, and Re = 350, respectively.

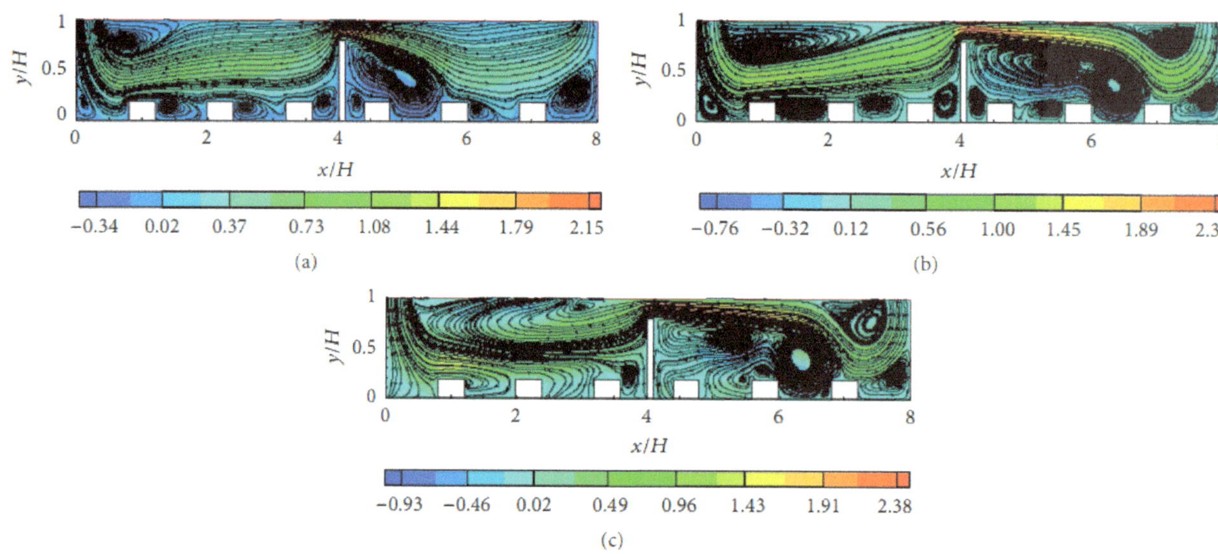

Figure 5: Streamline patterns of Case II for (a) Re = 100, (b) Re = 250, and (c) Re = 350.

Figure 6 presents the isotherms for this situation. Like in previous case, here also the top surfaces of the blocks are kept heated to represent patients on beds and the spaces between the blocks along with the partition are kept cold. The presence of the partition does affect the airflow characteristics as well as heat transfer phenomenon, which has been discussed in detail in the heat transfer part. Pr remains the same.

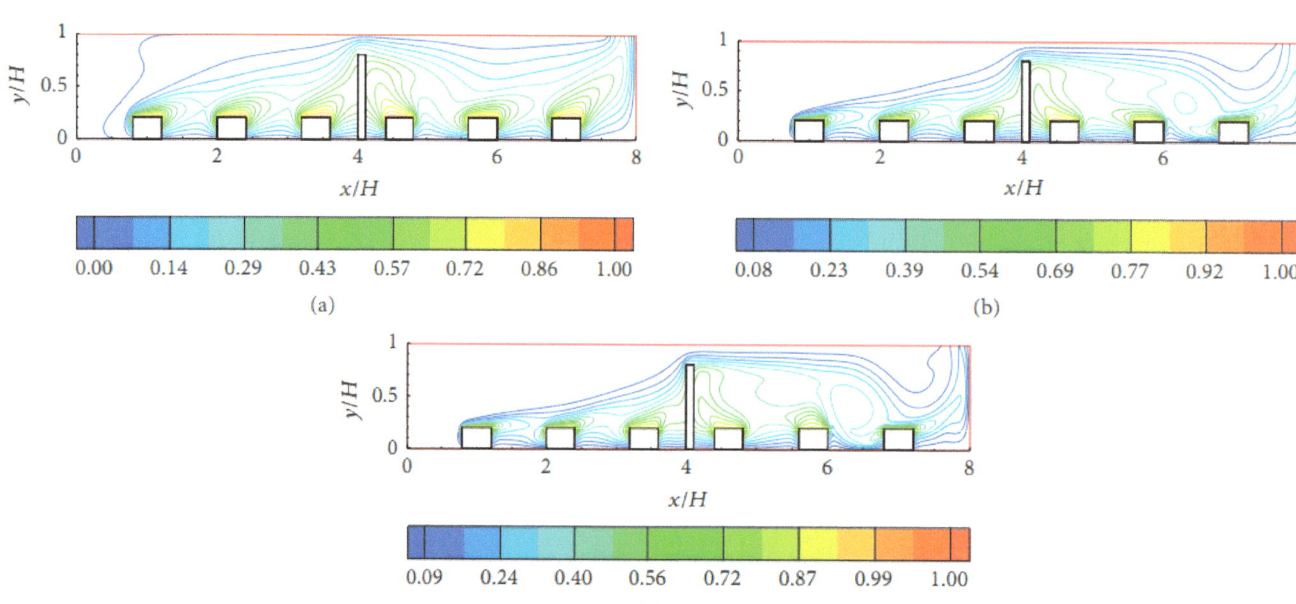

Figure 6: Isotherms of Case II for (a) Re = 100, (b) Re = 250, and (c) Re = 350.

In Figure 7, the U velocity profiles at different locations of x are discussed for the same three Re. This time the results show more curvy lines throughout the whole experiment due to the obstacle caused because of the partition. One of the obvious differences worth mentioning is in Figure 7(g) when the air faces the partition. At this point, a long vertical line with a narrow parabolic curve at the top is shown representing the height of the partition whereas the curved part illustrates that the air is crossing the partition. Even in Figure 7(h), the graph still maintains its parabolic shape and continues this pattern throughout till the end. Another noticeable change can be seen in Figures 7(j), 7(k), and 7(l), at x = 5.5, x=6, and x = 6.5, respectively, where the bents are more irregular than the previous case. This is because at these locations large recirculations have formed due to the partition, as shown in Figure 5.

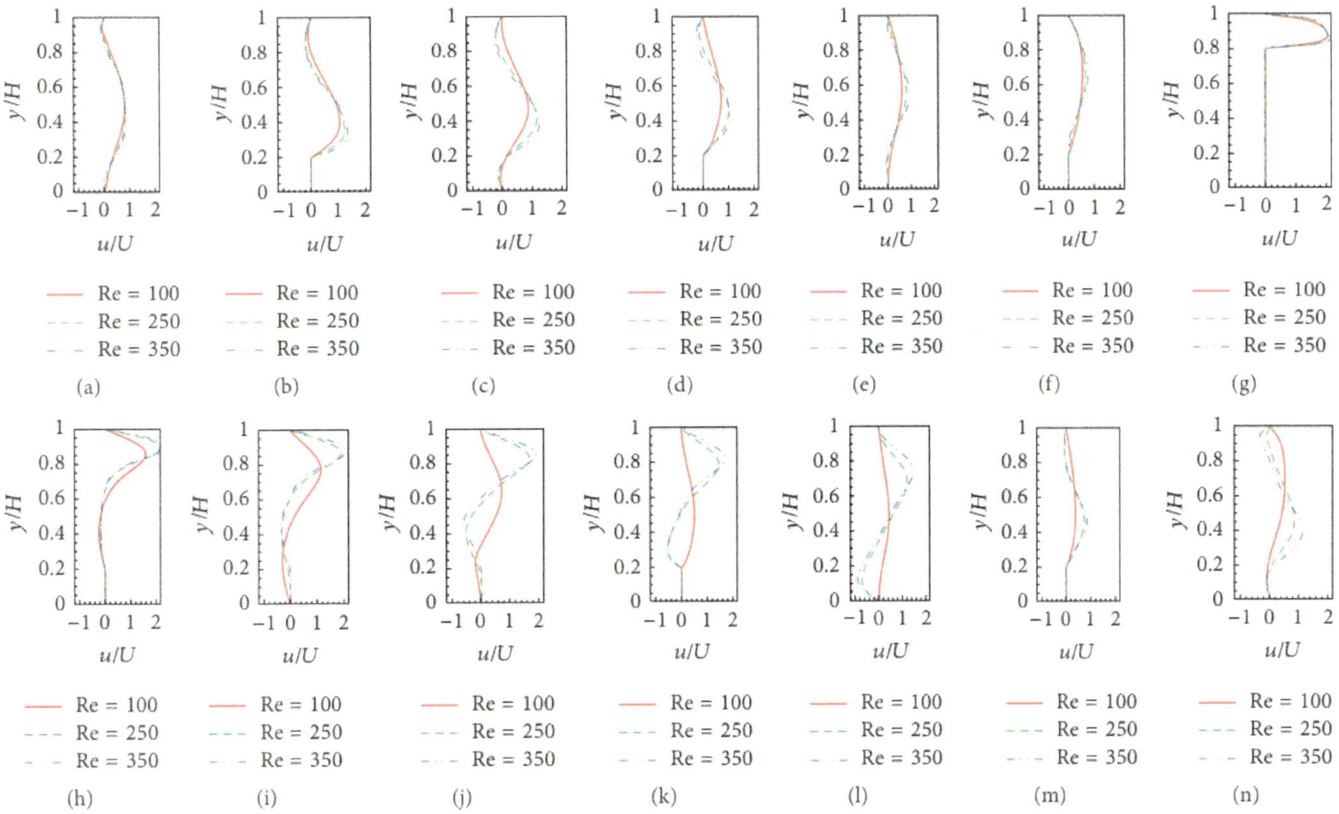

Figure 7: u/U velocity profiles in Case II for the different Re at different positions of horizontal locations (a) x/H = 0.5, (b) x/H = 1, (c) x/H = 1.5, (d) x/H = 2, (e) x/H = 3, (f) x/H = 3.5, (g) x/H = 4, (h) x/H = 4.5, (i) x/H = 5, (j) x/H = 5.5, (k) x/H = 6, (l) x/H = 6.5, (m) x/H = 7, and (n) x/H = 7.5.

According to the U and V centerline velocity distribution plots in Figures 8(a) and 8(b), respectively, there is a constriction in the middle of graph. This represents the partition that actually works like a wall where the velocity plots cease, just like the vertical walls in the extreme left and right sides. One possible advantage of this partition is that the airflow can be controlled if the air enters with a high speed through the inlet. Also it is noticeable in Figures 8(a) and 8(b) that, unlike the previous case, the partition helps to maintain the air velocity at a constant level until it exits out through the outlet. The V velocity still drops to negative range from zero at the beginning due to downward fall of air through the inlet and ascends to zero again where it maintains its constant velocity. The graphs are more chaotic after crossing the partition due to the recirculations shown in the contour map. The temperature plot in Figure 8(c) starts just like that of the previous situation, but this time instead of becoming constant after a certain level, the graph climbs up the partition, crosses it, and then drops again, eventually decreasing to zero at x=8. The rise of airflow against the partition also results in the increase in temperature at this point, therefore, the steep in the figure.

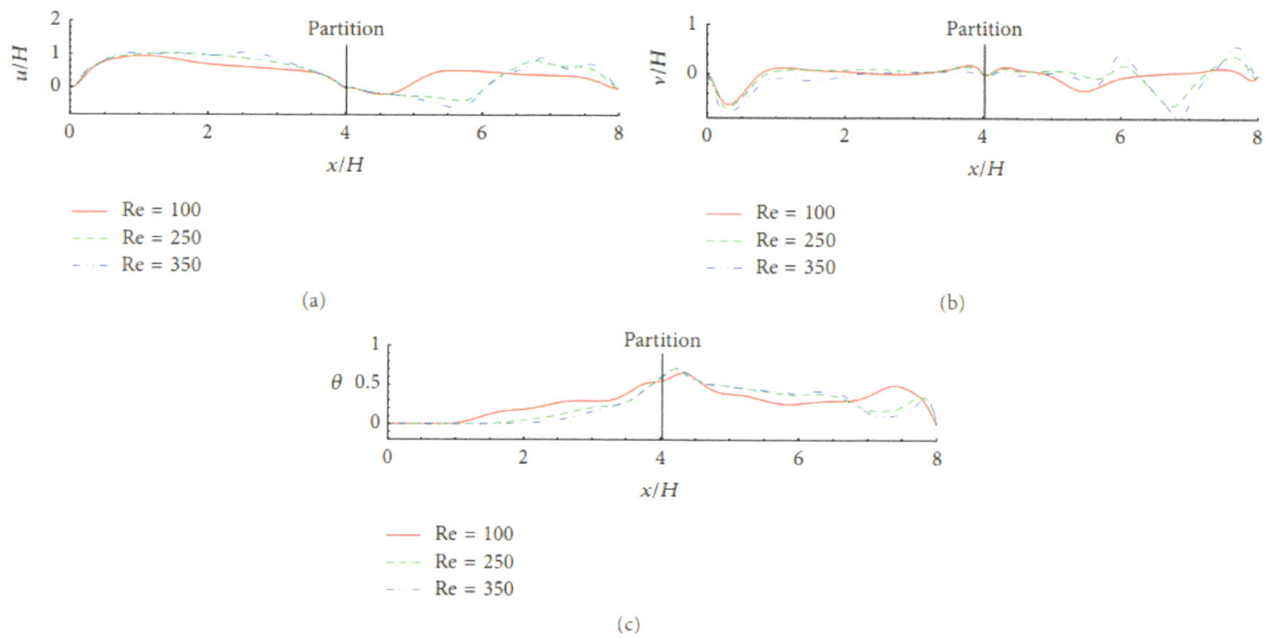

Figure 8: In Case II for Re = 100, 250 and Re = 350 the velocity distributions are (a) u/H at mid-y and (b) V/H at mid-y, and temperature distribution is (c) θ at mid-y.

Case Study III: One Inlet and One Outlet with Two Partitions

This case study has been investigated by further modifying the proposed model adding another partition in the ward. It was done to keep patients from being exposed into more air. Each partition is placed every after two blocks (beds). Both the partitions are of the same dimension, that is, $(5 \times 40) \sim (x \times y)$ lattice size, and all other measurements are kept the same as previous situations. Simulations and analysis of this case have also been carried out for the same three Re values. Streamlines pattern of Figure 9 obtained after adding another partition shows that the size and the number of recirculations increase more than the last observation. Unlike Case II, now two large recirculations have formed for the corresponding experiments with each Re. Even the maximum and minimum legend values have become wider than the previous situation; for example, for Re = 100 the 2.15 value has increased to 2.37. For Re = 250 and Re = 350 the uppermost values have risen to 2.47 and 2.52 from 2.33 and 2.38, respectively. Due to bounce-back condition, the air is now forming more recirculations against the walls and between the blocks. This augmentation of values and recirculations due to buoyancy effect is visible for every increment of Re values and addition of obstacles such as partition.

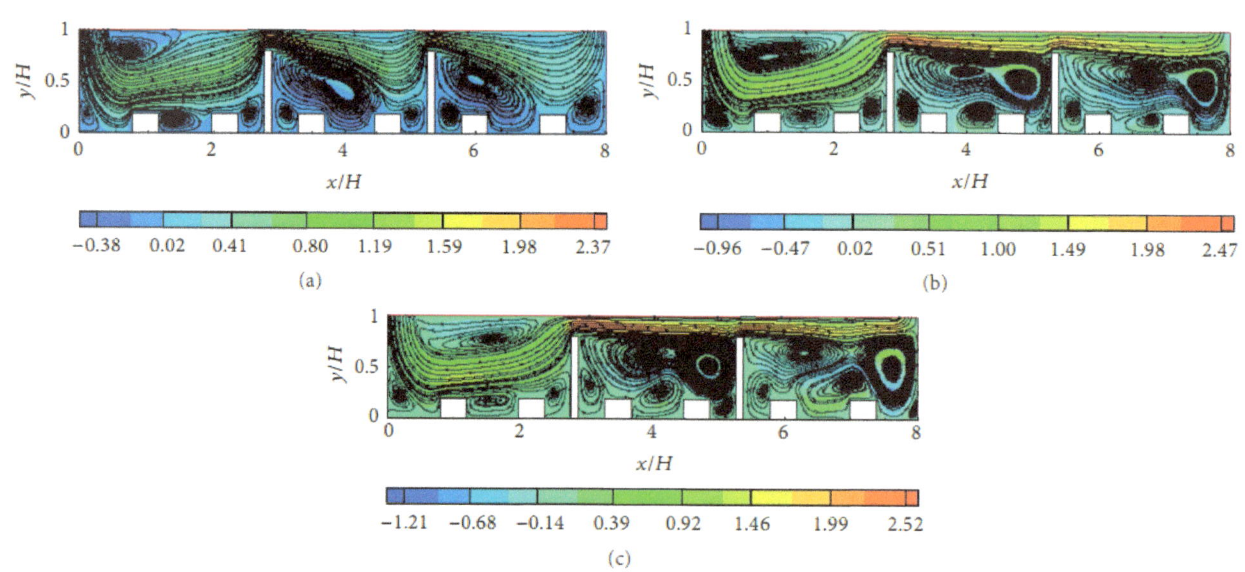

Figure 9: Streamline patterns of Case III for (a) Re = 100, (b) Re = 250, and (c) Re = 350.

The generated results of isotherms in Figure 10 give a visual idea how the heat is transferred from the blocks and passes through the outlet. Like the other cases, only the surfaces of the blocks are heated to be considered as patients who are continuously releasing heat. Everything remains the same like previous cases.

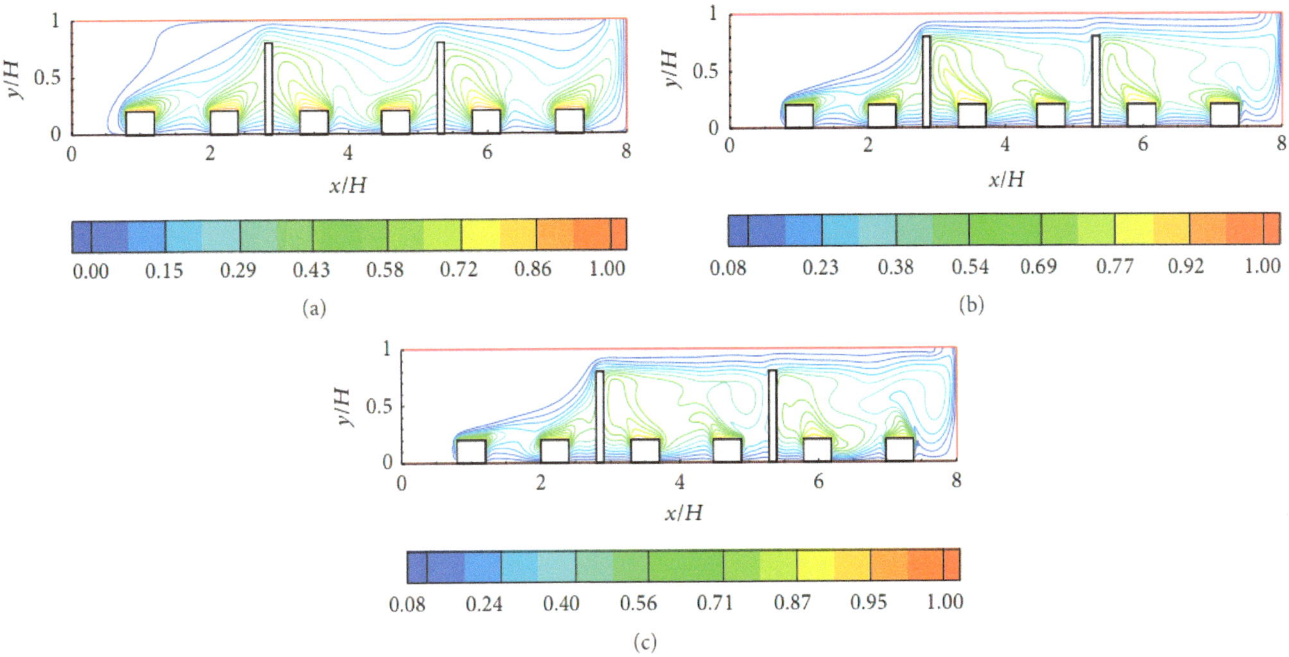

Figure 10: Isotherms of Case III for (a) Re = 100, (b) Re = 250, and (c) Re = 350.

Figure 11 provides the U velocity profiles for different positions of x. According to this the graph characteristic for a partition can be seen in Figure 11(e) at x=3 and Figure 11(j) at x = 5.5, indicating the presence of partitions at these two locations. Apart from this, at x=5 and 7.5 in Figures 11(i) and 11(n), respectively, the parabolic shapes are more curvy than the profiles at other positions caused due to recirculations in vector plots of Re = 250 and Re = 350. Figures 12(a) and 12(b) illustrate U and V velocity distribution curves in mid-y and Figure 12(c) shows the temperature profile at mid-y. All of these three plots have one thing in common that is also different than the previous case. In these plots there are two constrictions, instead of one, where the velocities or the temperature curve ceases. This is the location where the two partitions are present. But after crossing these obstacles the respective graphs continue until x=8. The temperature plot in Figure 12(c) has now two steep regions due to the partitions.

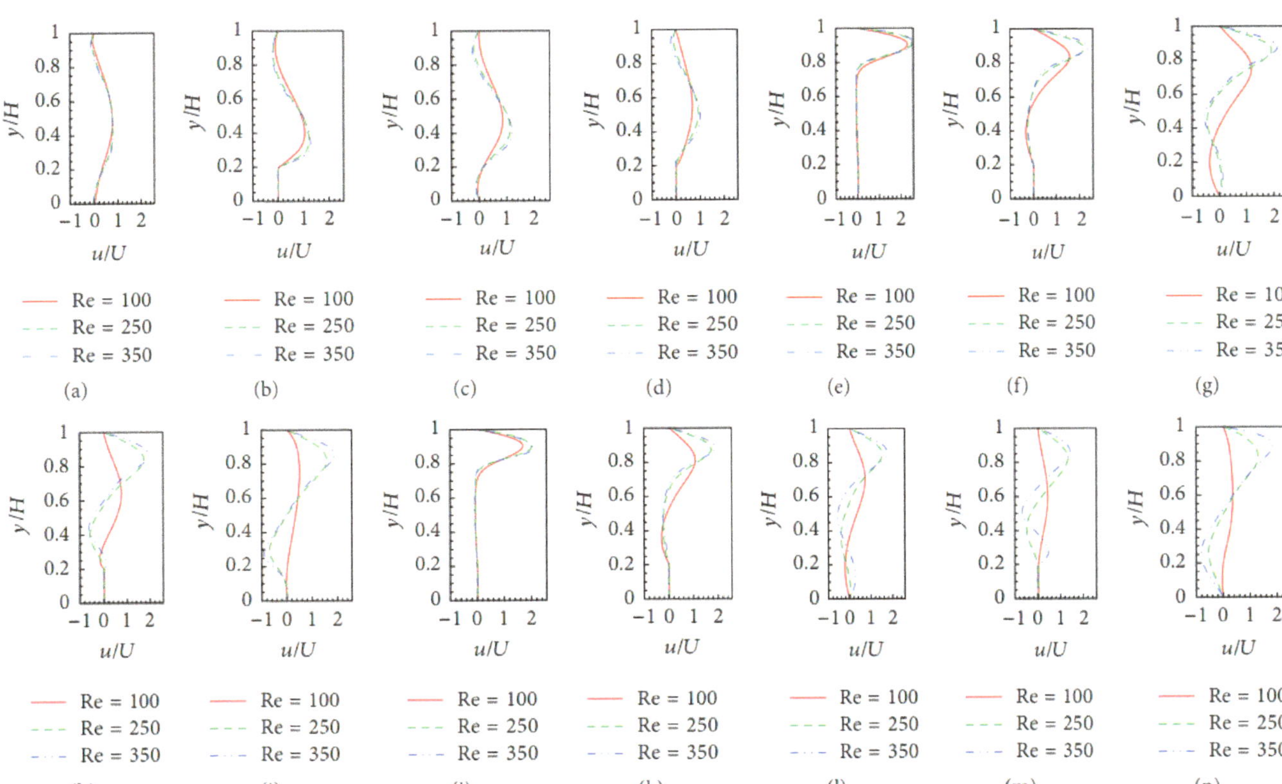

Figure 11: u/U velocity profiles in Case III for the different Re at different positions of horizontal locations (a) x/H = 0.5, (b) x/H = 1, (c) x/H = 1.5, (d) x/H = 2, (e) x/H = 3, (f) x/H = 3.5, (g) x/H = 4, (h) x/H = 4.5, (i) x/H = 5, (j) x/H = 5.5, (k) x/H = 6, (l) x/H = 6.5, (m) x/H = 7, and (n) x/H = 7.5.

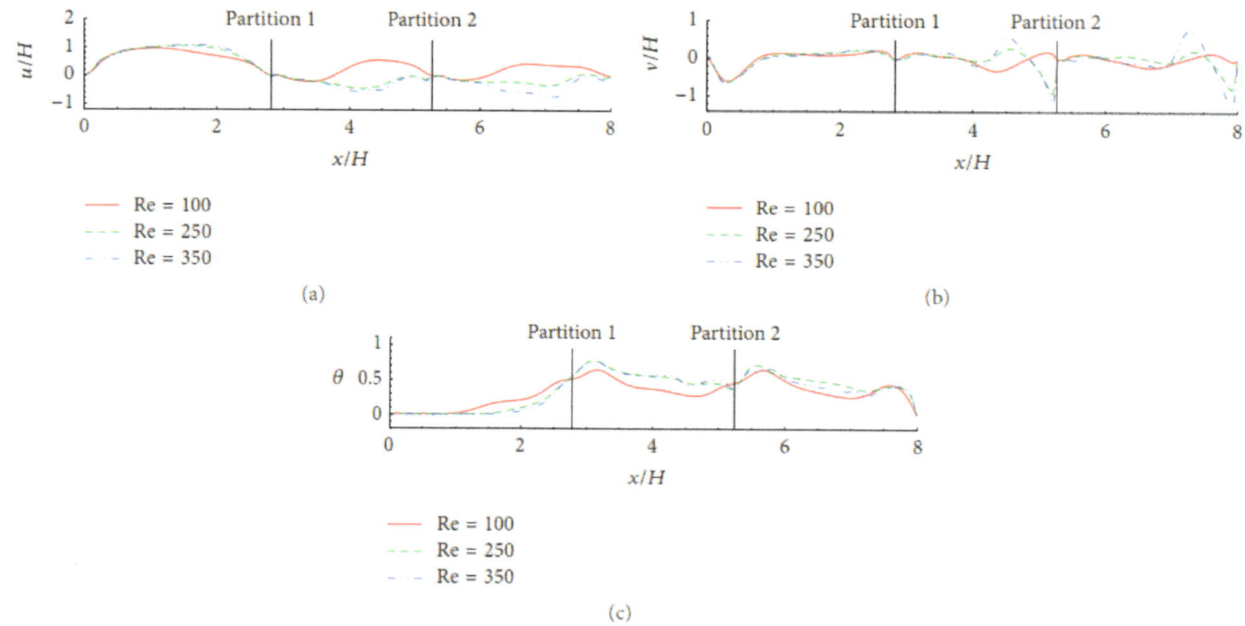

Figure 12: In Case III for Re = 100, 250 and Re = 350 the velocity distributions are (a) u/H at mid-y and (b) V/H at mid-y, and temperature distribution is (c) θ at mid-y.

Calculation of Rate of Heat Transfer

In heat transfer at a boundary (surface) within a fluid, the Nusselt number (Nu) is the ratio of convective to conductive heat transfer across (normal to) the boundary. Here, convection includes both advection and diffusion procedures. The convection and conduction heat flows are parallel to each other and to the surface normal of the boundary surface and are all perpendicular to the mean fluid flow in the simple case. This is denoted as [32]

$$\mathrm{Nu}_L = \frac{h_t L}{k}, \tag{19}$$

where h_t is the convective heat transfer coefficient of the flow, L is the characteristic length, and k is the thermal conductivity of the fluid.

However, this section discusses the average rate of change of heat transfer obtained for different Re. This rate is expressed as average Nusselt number, Nu, for each case study in Tables 2, 3, and 4. The tables also show generated local Nusselt number, $\overline{Nu}$, that follows the expression below:

$$\mathrm{Nu}\,(x) = -\left.\frac{\partial\theta}{\partial y}\right|_{y=y_2}, \tag{20}$$

where y_2 is the upper surface of each block.

Table 2: Summary of average Nusselt numbers $\overline{Nu}$ for six blocks individually and in total for three Reynolds numbers at Pr = 0.71 in case of without partition

Re	$\overline{\text{Nu}}$ (individually)						$\overline{\text{Nu}}$ (total)
	Block 1	Block 2	Block 3	Block 4	Block 5	Block 6	
100	3.0164	2.0366	1.7990	1.6667	1.5695	1.4486	11.5368
250	4.6238	2.9266	2.2996	2.1784	2.0573	1.8028	15.8885
350	5.4359	2.0344	2.4841	2.2300	2.1352	1.7329	17.5970

Table 3: Summary of average Nusselt numbers $\overline{Nu}$ for six blocks individually and in total for three Reynolds numbers at Pr = 0.71 in case of with partition

Re	$\overline{\text{Nu}}$ (individually)						$\overline{\text{Nu}}$ (total)
	Block 1	Block 2	Block 3	Block 4	Block 5	Block 6	
100	3.2979	2.0344	1.6115	1.5341	2.0111	1.6124	12.1014
250	5.0273	2.9488	1.7955	1.3628	2.1598	3.8673	17.1615
350	5.8293	3.6676	1.8670	1.4032	2.7535	5.3320	20.8527

Table 4: Summary of average Nusselt numbers $\overline{Nu}$ for six blocks individually and in total for three Reynolds numbers at Pr = 0.71 in case of two partitions

Re	$\overline{\text{Nu}}$ (individually)						$\overline{\text{Nu}}$ (total)
	Block 1	Block 2	Block 3	Block 4	Block 5	Block 6	
100	3.2791	1.8265	1.6086	1.8914	1.5810	1.7583	11.9449
250	4.9303	2.2066	1.1352	2.5811	1.3338	2.4044	14.5915
350	5.7620	2.4463	1.1924	2.9056	1.2632	2.7495	16.3191

Average, $\overline{Nu}$, is calculated by integrating the above relation over the entire range of interest and thus the resulting equation becomes

$$\overline{\text{Nu}} = \frac{1}{x_2 - x_1} \int_{x_1}^{x_2} \text{Nu}\,(x)\, dx. \tag{21}$$

Here, $(x_2 - x_1)$ is the length of each block.

For the first investigation, both the individual and average Nu obtained in Table 2 show a rising trend for each increasing Re. Since block 1 is the nearest to the inlet, it always has the highest rate of heat transfer while block 6 has the lowest value as it is located farthest from the inlet.

After placing the partition in the middle, the rate of heat transfer shown in Table 3 changes dramatically. Like Table 2, the patient at block 1 is releasing most heat for all three Re and it keeps reducing until block 5. At block 5, the values of both individual $\overline{Nu}$ and total $\overline{Nu}$ increase for Re = 250 and Re = 350, but for Re = 100, this type of airflow pattern is seen at block 4. The large recirculations were seen earlier in Figure 5. For Re = 100, recirculation is more over block 4 and that is the reason why, at block 4, individual $\overline{Nu}$ and total $\overline{Nu}$ have greater values than those of Re = 250 and Re = 350. $\overline{Nu}$ is 1.5341 at block 4 and, for the same block, these values are 1.3628 and 1.4032, respectively, for Re = 250 and Re = 350. On the other hand, if $\overline{Nu}$ at block 5 are observed, Re = 250 and Re = 350 lead the race. As per Table 3, $\overline{Nu}$ are 2.0111, 2.1598, and 2.7535 for three Re. The spike in the values of Re = 250 and Re = 350 was due to chaotic flows over block 5 in Figure 5. If changes in total $\overline{Nu}$ are observed, the values increase rapidly for an increase in Re.

For Case Study III, the shape of recirculation gets bigger due to increased Re. Above block 1, the recirculation zone comes into view and, due to that reason, $\overline{Nu}$ has the biggest edge in this situation. But as the air has tendency to pass the first partition, it moves slightly upward and that is why patient at block 2 is getting less airflow. The heat transfer rate also falls. After passing the first partition, the airflow gets its space and leads to recirculation near block 3 for Re = 100. Meanwhile, for Re = 250 and Re = 350, patients on block 4 get most air. More air means more heat transfer and that is why $\overline{Nu}$ are 2.5811 and 2.9056 in terms of block 4 for Re = 250, 350. However, the change in the values of $\overline{Nu}$ repeats in terms of block 5. For Re = 100, patient at block 5 gets most air and that is also due to vorticity, which is seen in Figure 9. For two other Re, these recirculations appear over block 6, which also lead to an increase in $\overline{Nu}$ as per Table 4.

Comparing the results after analyzing Tables 2, 3, and 4, few statements can be made. Heat transfer rate is at its highest level when one partition is present. For example, when there is not any partition, total $\overline{Nu}$ is 15.8885 for Re = 250 and it rises to 17.1615 when one partition is present. However, placing two partitions create one extra obstacle for airflow and $\overline{Nu}$ decreases to 14.5915. Same variations can be observed for Re = 100 and Re = 350 as well.

Comparing the results after analyzing Tables 2, 3, and 4, few statements can be made. Heat transfer rate is at its highest level when one partition is present. For example, when there is not any partition, total $\overline{Nu}$ is 15.8885 for Re = 250 and it rises to 17.1615 when one partition is present. However, placing two partitions create one extra obstacle for airflow and $\overline{Nu}$ decreases to 14.5915. Same variations can be observed for Re = 100 and Re = 350 as well.

Consequently, if the ward has one partition in the middle (which is pretty common in hospital nowadays), both blocks (beds) 1 and 2 will be just fine for more air. If disease is related to air, these two beds in particular should be avoided. For moderate airflow, any bed or block after the first two should be considered.

Last but not least, if three different zones are available in the ward due to two partitions, any bed between 3 and 6 should be okay for normal airflow. Blocks (beds) 1 and 2 should be kept for patients who are suffering from acute temperature related disease. By any chance, if none of those beds are available, block (bed) 4 will be a pretty good replacement.

LOCAL REYNOLDS NUMBER

Local Reynolds numbers (Re) for all obstacles have been computed at the mid of the gap between block and top wall for Case II (with one partition) at Re = 100. They are 61.54 for block 1, 35.34 for block 2, 25.54 for block 3, 36.93 for partition, 16.32 for block 4, 24.54 for block 5, and 32.19 for block 6. The above results demonstrate that as the air enters the domain and flows far away from the inlet, the inlet air velocity also reduces and, consequently, the local Re*l* values also decrease. Block 1 has the highest local Re*l* value since the inlet velocity is maximum at this position. After gradual decline of the numerals, block 4 has negative local Re*l* due to the negative velocity of the recirculation zone. The values then increase as the air flows up towards the outlet but cannot attain the maximum value due to the distance from the inlet.

CONCLUSION

In this paper, the study of airflow on mixed convection flow inside a hospital ward has been analyzed using the Lattice Boltzmann Method. The changes in the patterns of airflow have been observed in three different case studies. The code validation has been carried out for three different Re and agreement has been shown as well. In the case studies, LBM was able to capture the convective structure as the air flowed over the room and even when the partition was present. It was found that the heat transfer process and flow characteristics depend on different Re. Moreover, the presence of the partition in each case was found to have significance on the pattern of airflow in both streamlines and isotherms on a particular Ri and Gr. Some conclusions are summarized as follows:

- A proper code validation with previous numerical investigations implies that Lattice Boltzmann Method is indeed an appropriate method for studying mixed convective airflow.
- Generally, the increase in Reynolds number resulted in augmentation of heat transfer. The pattern of airflow also changed.
- As the Reynolds number increases, the Mach number decreases. The Mach number cannot be more than 0.3 to satisfy the condition of flow in incompressible fluid.
- Since local is directly proportional to the velocity of inlet flow, therefore, as the air gradually flows away from the inlet, the flow velocity reduces for which local also decreases.
- Placing partition between the blocks or beds had effects on the airflow structure and for different cases, the results were different.
- The most effective rate of heat transfer was found when one partition was present, which were due to recirculations in the contours.
- Influence of two partitions on heat transfer was less for all three Reynolds numbers, which indicates that air did not get enough space to flow freely and that led to a fall in heat transfer rate.
- Patients who are in need of most air should be kept near the inlet and patients who do not need to be exposed directly to air should be moved near the outlet.

REFERENCES

1. C. K. Cha and Y. Jaluria, "Recirculating mixed convection flow for energy extraction," International Journal of Heat and Mass Transfer, vol. 27, no. 10, pp. 1801–1812, 1984.
2. F. J. K. Ideriah, "Prediction of turbulent cavity flow driven by buoyancy and shear," Journal of Mechanical Engineering Science, vol. 22, no. 6, pp. 287–295, 1980.
3. J. Imberger and P. F. Hamblin, "Dynamics of lakes, reservoirs, and cooling ponds," Annual Review of Fluid Mechanics, vol. 14, pp. 153–187, 1982.
4. H. Huang, M. C. Sukop, and X. Lu, Multiphase Lattice Boltzmann Methods: Theory and Application, John Wiley & Sons, New York, NY, USA, 2015.
5. B. Hasslacher, Y. Pomeau, and U. Frisch, "Lattice-gas automata for the Navier-Stokes equation," Physical Review Letters, vol. 56, no. 14, pp. 1505–1508, 1986.
6. G. Vahala, P. Pavlo, L. Vahala, and N. S. Martys, "Thermal lattice-boltzmann models (TLBM) for compressible flows," International Journal of Modern Physics C, vol. 9, no. 8, pp. 1247–1261, 1998.
7. Y. Peng, C. Shu, and Y. T. Chew, "Simplified thermal lattice Boltzmann model for incompressible thermal flows," Physical Review E, vol. 68, no. 2, Article ID 026701, 2003.
8. T. S. Lee, H. Huang, and C. Shu, "An axisymmetric incompressible lattice BGK model for simulation of the pulsatile flow in a circular pipe," International Journal for Numerical Methods in Fluids, vol. 49, no. 1, pp. 99–116, 2005.
9. N. A. C. Sidik, "The development of thermal lattice Boltzmann models in incompressible limit,"Malaysian Journal of Fundamental and Applied Sciences, vol. 3, no. 2, pp. 193–202, 2008.
10. B. Mondal and X. Li, "Effect of volumetric radiation on natural convection in a square cavity using lattice Boltzmann method with non-uniform lattices," International Journal of Heat and Mass Transfer, vol. 53, no. 21-22, pp. 4935–4948, 2010.
11. M. Szucki and J. S. Suchy, "A method for taking into account local viscosity changes in single relaxation time the lattice Boltzmann model," Metallurgy and Foundry Engineering, vol. 38, pp. 33–42, 2012.
12. S. Chen and G. D. Doolen, "Lattice Boltzmann method for fluid flows," Annual Review of Fluid Mechanics, vol. 30, no. 1, pp. 329–364, 1998.
13. G. Breyiannis and D. Valougeorgis, "Lattice kinetic simulations in three-dimensional magnetohydrodynamics," Physical Review E, vol. 69, no. 6, Article ID 065702, 2004.
14. I. Halliday, L. A. Hammond, C. M. Care, K. Good, and A. Stevens, "Lattice Boltzmann equation hydrodynamics," Physical Review E, vol. 64, no. 1 I, Article ID 011208, 2001.
15. C. S. N. Azwadi and T. Tanahashi, "Development of 2-D and 3-D double-population thermal lattice boltzmann models," Matematika, vol. 24, no. 1, pp. 53–66, 2008.
16. B. Crouse, M. Krafczyk, S. Kühner, E. Rank, and C. Van Treeck, "Indoor air flow analysis based on lattice Boltzmann methods," Energy and Buildings, vol. 34, no. 9, pp. 941–949, 2002.
17. Z. Zhang, X. Chen, S. Mazumdar, T. Zhang, and Q. Chen, "Experimental and numerical investigation of airflow and contaminant transport in an airliner cabin mockup," Building and Environment, vol. 44, no. 1, pp. 85–94, 2009.
18. S. J. Zhang and C. X. Lin, "Application of lattice Boltzmann method in indoor airflow simulation," HVAC & R Research, vol. 16, no. 6, pp. 825–841, 2010.
19. J. Liu, H. Wang, and W. Wen, "Numerical simulation on a horizontal airflow for airborne particles control in

hospital operating room," Building and Environment, vol. 44, no. 11, pp. 2284–2289, 2009.

20. A. Watzinger and D. G. Johnson, "Wärmeübertragung von wasser an rohrwand bei senkrechter strömung im übergangsgebiet zwischen laminarer und turbulenter strömung," Forschung auf dem Gebiet des Ingenieurwesens A, vol. 10, no. 4, pp. 182–196, 1939.
21. A. K. Prasad and J. R. Koseff, "Combined forced and natural convection heat transfer in a deep lid-driven cavity flow," International Journal of Heat and Fluid Flow, vol. 17, no. 5, pp. 460–467, 1996.
22. H. Xu, R. Xiao, F. Karimi, M. Yang, and Y. Zhang, "Numerical study of double diffusive mixed convection around a heated cylinder in an enclosure," International Journal of Thermal Sciences, vol. 78, pp. 169–181, 2014.
23. S. A. Al-Sanea, M. F. Zedan, and M. B. Al-Harbi, "Effect of supply Reynolds number and room aspect ratio on flow and ceiling heat-transfer coefficient for mixing ventilation," International Journal of Thermal Sciences, vol. 54, pp. 176–187, 2012.
24. F.-Y. Zhao, D. Liu, and G.-F. Tang, "Multiple steady fluid flows in a slot-ventilated enclosure,"International Journal of Heat and Fluid Flow, vol. 29, no. 5, pp. 1295–1308, 2008.
25. K. M. Khanafer and A. J. Chamkha, "Mixed convection flow in a lid-driven enclosure filled with a fluid-saturated porous medium," International Journal of Heat and Mass Transfer, vol. 42, no. 13, pp. 2465–2481, 1999.
26. G. H. R. Kefayati, S. F. Hosseinizadeh, M. Gorji, and H. Sajjadi, "Lattice Boltzmann simulation of natural convection in an open enclosure subjugated to water/copper nanofluid," International Journal of Thermal Sciences, vol. 52, no. 1, pp. 91–101, 2012.
27. H. Lee and H. B. Awbi, "Effect of internal partitioning on room air quality with mixing ventilation-statistical analysis," Renewable Energy, vol. 29, no. 10, pp. 1721–1732, 2004.
28. Y. B. Bao and J. Meskas, Lattice Boltzmann Method for Fluid Simulations, Courant Institute of Mathematical Sciences, New York, NY, USA, 2011.
29. H. N. Dixit and V. Babu, "Simulation of high Rayleigh number natural convection in a square cavity using the lattice Boltzmann method," International Journal of Heat and Mass Transfer, vol. 49, no. 3-4, pp. 727–739, 2006.
30. R. Huang, H. Wu, and P. Cheng, "A new lattice Boltzmann model for solid-liquid phase change,"International Journal of Heat and Mass Transfer, vol. 59, no. 1, pp. 295–301, 2013.
31. Y. Xuan and Z. Yao, "Lattice boltzmann model for nanofluids," Heat and Mass Transfer, vol. 41, no. 3, pp. 199–205, 2005.
32. G. R. Kefayati, M. Gorji-Bandpy, H. Sajjadi, and D. D. Ganji, "Lattice Boltzmann simulation of MHD mixed convection in a lid-driven square cavity with linearly heated wall," Scientia Iranica, vol. 19, no. 4, pp. 1053–1065, 2012.
33. Q. Zou and X. He, "On pressure and velocity boundary conditions for the lattice Boltzmann BGK model," Physics of Fluids, vol. 9, no. 6, pp. 1591–1598, 1997.
34. A. A. Mohamad and A. Kuzmin, "A critical evaluation of force term in lattice Boltzmann method, natural convection problem," International Journal of Heat and Mass Transfer, vol. 53, no. 5-6, pp. 990–996, 2010.
35. R. Iwatsu, J. M. Hyun, and K. Kuwahara, "Mixed convection in a driven cavity with a stable vertical temperature gradient," International Journal of Heat and Mass Transfer, vol. 36, no. 6, pp. 1601–1608, 1993.
36. K. M. Khanafer, A. M. Al-Amiri, and I. Pop, "Numerical simulation of unsteady mixed convection in a driven cavity using an externally excited sliding lid," European Journal of Mechanics B: Fluids, vol. 26, no. 5, pp. 669–687, 2007.
37. M. M. Abdelkhalek, "Mixed convection in a square cavity by a perturbation technique," Computational Materials Science, vol. 42, no. 2, pp. 212–219, 2008.
38. R. K. Tiwari and M. K. Das, "Heat transfer augmentation in a two-sided lid-driven differentially heated square cavity utilizing nanofluids," International Journal of Heat and Mass Transfer, vol. 50, no. 9-10, pp. 2002–2018, 2007.
39. M. A. Waheed, "Mixed convective heat transfer in rectangular enclosures driven by a continuously moving horizontal plate," International Journal of Heat and Mass Transfer, vol. 52, no. 21-22, pp. 5055–5063, 2009.

CHAPTER 20 STUDY ON ENGINEERING DESIGN AND COST PERFORMANCE OF BOREHOLE HEAT EXCHANGERS

Lei Yu and Jiemin Cheng

College of Population, Resources and Environment, Shandong Normal University, Jinan 250014, China

ABSTRACT

The ground source heat pump (GSHP) is an environmentally friendly and energy-efficient technology, and the investigations on the simulation models of borehole heat exchanger (BHE) have made progress. However, the researches on engineering design and economic performance based on the simulation models of BHE are a few. This paper firstly describes the interior and exterior heat transfer according to the existing achievements, and the corresponding characteristics of models are explained. Afterwards, the detailed design procedures of BHE size are provided and the cost performance is highlighted. Accordingly, an actual engineering project which employs GSHP system as the air conditioning system is researched; the necessary expenses on BHE, heat pump units, and indoor end equipment are all taken into account. The initial investment, operating cost, and other factors that exert influences on the cost performance are demonstrated. In addition, comparisons between different systems are conducted to show the advantages of GSHP. By means of thermal transfer analysis and the design of engineering project, the practicability, energy saving, and environmental protection of GSHP are brought into play, which means the promotion of this technology is the general trend of development of air condition system.

INTRODUCTION

The ground source heat pump (GSHP) system is the most advanced green air-conditioning technology in the world, because GSHP is favorable to achieve environmental protection and energy-saving. The underground low-grade energy is utilized with the help of heat pump unit and therefore the heating, cooling, and domestic hot water of buildings are all achieved [1]. More specifically, the thermal transfer is conducted from the circulating liquid of the heat exchange tube to the underground medium, which means the renewable energy existing in underground space can be made use of. Underground mediums locating at the depth of more than ten meters keep nearly constant temperature all the year round because they are free of the thermal disturbance that comes from the outside environment. The depth of underground mediums can achieve one hundred meters while GSHP is employed. Underground temperatures are lower and higher than outdoor air's temperatures in summer and in winter, respectively [2, 3]. Therefore, the technical barriers such as

Citation: Lei Yu and Jiemin Cheng, "Study on Engineering Design and Cost Performance of Borehole Heat Exchangers,"Mathematical Problems in Engineering, vol. 2015, Article ID 870380, 11 pages, 2015. doi:10.1155/2015/870380.

frosting that appears at air source heat pump can be overcome and then the system efficiency is improved significantly. For example, the 4 kW heat quantity is acquired if only 1 kW electric energy is consumed; accordingly, the energy efficiency ratio is larger than 4. In addition, GSHP that employs borehole heat exchanger (BHE) as the underground heat transfer elements does not extract groundwater and thus the system is a new energy-saving technology keeping sustainable development. The whole system usually consists of several parts including BHEs, the heat pump unit, and end equipment, and sometimes domestic hot water is contained either.

Also, the advantages of GSHP should be followed with interest and they are summarized as several aspects. Firstly, high-efficiency and energy-saving are undeniable because running efficiency of GSHP is 30~50% larger than that of traditional air-conditioning system. Secondly, there is no boiler and the corresponding room and therefore the problems of both air pollution and greenhouse gases emission are prevented. Thirdly, an equipment can serve several purposes. Next, the objectives of beautifying building and household metering are both realized when the GSHP is adopted in buildings. Last but not least, the characteristics of the safety and reliability should be emphasized as the life spans of BHE and heat pump unit, respectively, attain 50 and 20 years. Figure 1 shows the schematic diagram of BHEs that are responsible for the heat or cooling load of the building [4].

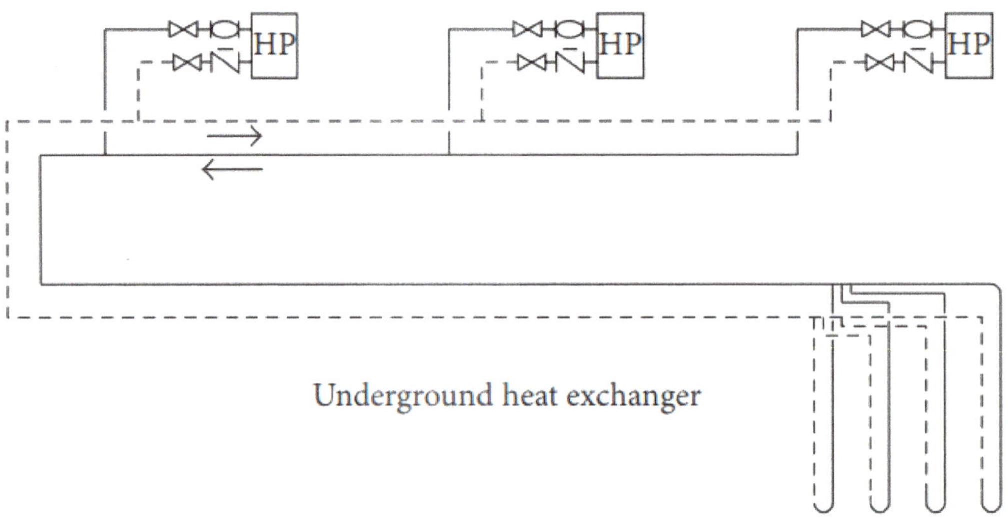

Figure 1: The schematic diagram of underground heat exchanger employed for cooling or heating.

It can be observed that the underground heat transfer components of one engineering project are composed of a number of BHEs. No matter whether to study the theoretical models or to analyze the reasonable design of GSHP system, it is related to the BHEs.

The simulation models describing the heat transfer inside and outside borehole have gained more and more attention, and they have been put to use in theoretical calculation or engineering design. Apart from that, the design procedures of obtaining the BHE's reasonable size are becoming increasingly applicable. It is a pity that the analyses on engineering projects based on both thermal transmission models and design procedures are a few, and this is a noticeable research gap. The novelty and originality of the paper are to investigate design procedure and cost performance of engineering projects according to the heat transfer simulation models. Afterwards, the detailed design plane for an actual project is conducted to certify the necessity of applying GSHP system; the comparison between GSHP and other systems is made to show the significant economic performance of GSHP system.

SIMULATION MODELS OF HEAT TRANSFER

Heat Transfer Inside Borehole GHE

Borehole is drilled and then the U-tubes are installed into it [5]. The vertical BHE is produced if the backfilling materials are poured, and these materials have the effects of sealing and improving heat transfer performance. The heat transfer domain from U-tube to underground medium is divided into two parts, that is, the areas inside and outside borehole [6]. The steady-state mode can be put into use to deal with the heat exchange inside GHE because the geometric size and thermal capacity are relatively small compared with those outside borehole. The equivalent diameter is employed to simplify the model and it is usually applied to explore the heat transfer between U-tube and the borehole wall [7]; the principle is that all branch pipes are simplified as a pipe with larger diameter; therefore, the two-dimensional conduction can be regarded as a one-dimension radial heat exchange.

The heat conductivities of backfilling material, pipe material, and underground medium are, respectively, k_b, k_p, and k_s. The radius of borehole, external radius of pipe, and inner radius of pipe are, respectively, r_b, r_0, and r_i. The layout drawing of U-tube inside borehole and the corresponding equivalent diameter are both shown in Figure 2.

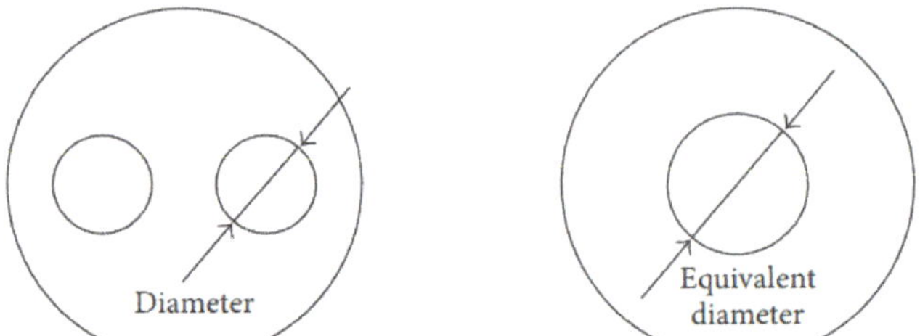

Figure 2: The diameter and the corresponding equivalent diameter of U-tube.

The equivalent diameter of U-tubes is approximately $\sqrt{n}d_0$ in case the number of branch tubes is n; thus, the thermal-convection resistance from the circulating fluid to the inner wall of tube is expressed in (1) while the convection coefficient is h:

$$R_f = \frac{1}{2\pi r_i h}. \tag{1}$$

Then, the thermal resistance between inner wall and external wall of tube is concluded in

$$R_t = \frac{1}{2\pi k_p} \ln \frac{\sqrt{n}r_0}{\sqrt{n}r_0 - (r_0 - r_i)}. \tag{2}$$

Afterwards, the heat is further conveyed to the borehole wall when the backfilling material acts as the thermal conductor, and the corresponding resistance is shown in

$$R_b = \frac{1}{2\pi k_b} \ln \frac{r_b}{\sqrt{n}r_0}. \tag{3}$$

The total resistance inside borehole should be summarized according to the analyses above, and the expression of energy equation is given in (4) while the temperatures of circulating liquid and the borehole wall are, respectively, t_f and t_b. There is no doubt that the parameters such as driving force of heat transfer, thermal resistance and heat flux are shown clearly. Three components constitute the total resistance inside BHE, and the original source of heat energy is the circulating liquid. q_l is the heat transfer rate of BHE, that is, the heat transfer quantity per meter BHE:

$$t_f - t_b = \left(R_f + R_t + R_b\right) q_l. \tag{4}$$

Heat Transfer outside Borehole GHE

Line Heat Source Model

Obviously, the heat continues to be transmitted to far locations of underground medium. The heat transfer process outside BHE plays a significant role in carrying energy, because the borehole's diameter is so small that the heat capacity inside BHE is little and the thermal resistance outside borehole is much larger than that inside borehole. The investigation on heat transfer outside borehole is becoming increasingly important and therefore the reasonable mathematical simulation models should be taken into account. The borehole's depth is far larger than its diameter and thus BHE can be regarded as a line heat source [8], which means the radius of BHE is ignored when the heat transfer outside borehole is studied. The diagram of line heat source is described in Figure 3.

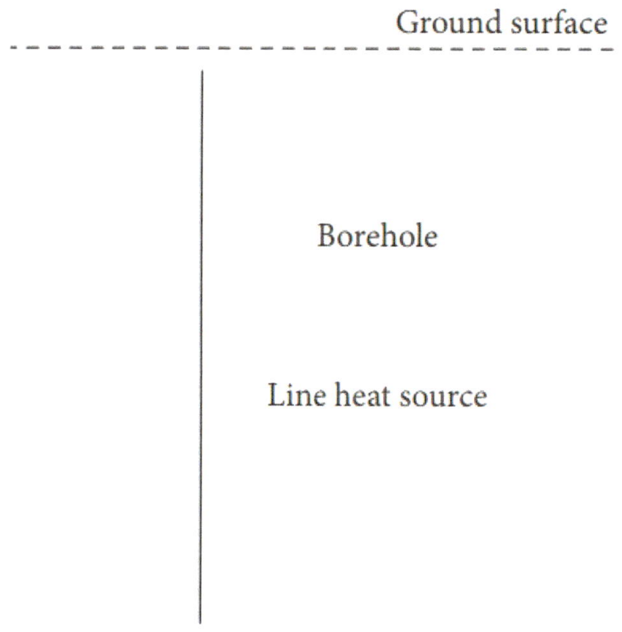

Figure 3: The line heat source adopted for BHE.

Accordingly, a most widely used one-dimensional model for this purpose is Kelvin's line source model, in which the borehole is replaced by a line heat source with its radial dimension neglected. However, the actual depth of any BHE is finite and then the finite line heat source is proposed; the Kelvin's model lays a firm foundation for the research of the finite line heat source [9]. In order to investigate the conduction around the borehole, Green's function theory is employed and the corresponding expression is given in (5) showing the temperature response of any point (x, y, z) except heat source in the underground medium at time τ, and the point heat source locating at (x', y', z') emits heat at time τ' [10]:

$$G\left(x, y, z, \tau; x', y', z', \tau'\right) = \frac{1}{8\left[\pi a\left(\tau - \tau'\right)\right]^{3/2}} \cdot \exp\left\{-\frac{\left(x - x'\right)^2 + \left(y - y'\right)^2 + \left(z - z'\right)^2}{4a\left(\tau - \tau'\right)}\right\}. \tag{5}$$

After that, the temperature response induced by the finite line heat source is obtained in (6), where erfc is the complementary error function and $t0$ is the initial temperature of the underground medium [11]:

$$t - t_0 = \frac{q_l}{4\pi k} \cdot \int_0^H \left\{ \frac{\operatorname{erfc}\left[\sqrt{(x - x')^2 + (y - y')^2 + (z - z')^2}/2\sqrt{a\tau}\right]}{\sqrt{(x - x')^2 + (y - y')^2 + (z - z')^2}} - \frac{\operatorname{erfc}\left[\sqrt{(x - x')^2 + (y - y')^2 + (z - z')^2}/2\sqrt{a\tau}\right]}{\sqrt{(x - x')^2 + (y - y')^2 + (z + z')^2}} \right\} dz'. \tag{6}$$

To facilitate the expression, the nondimensional parameters are employed and the corresponding forms are $\Theta = (t - t_0)/q_l$, $X = x/r_b$, $Y = y/r_b$, $Z = z/r_b$, $X' = x'/r_b$, $Y' = y'/r_b$, $Z' = z'/r_b$, $H = h/r_b$, $Fo = a\tau/r_b^2$. Equation (6) is translated into

$$\Theta = \frac{1}{4\pi}\int_0^H \left\{ \frac{\operatorname{erfc}\left[\sqrt{(X-X')^2+(Y-Y')^2+(Z-Z')^2}/2\sqrt{Fo}\right]}{\sqrt{(X-X')^2+(Y-Y')^2+(Z-Z')^2}} - \frac{\operatorname{erfc}\left[\sqrt{(X-X')^2+(Y-Y')^2+(Z+Z')^2}/2\sqrt{Fo}\right]}{\sqrt{(X-X')^2+(Y-Y')^2+(Z+Z')^2}} \right\} dZ'. \quad (7)$$

The temperature responses with the time are shown in Figure 4 according to the calculation results of (7).

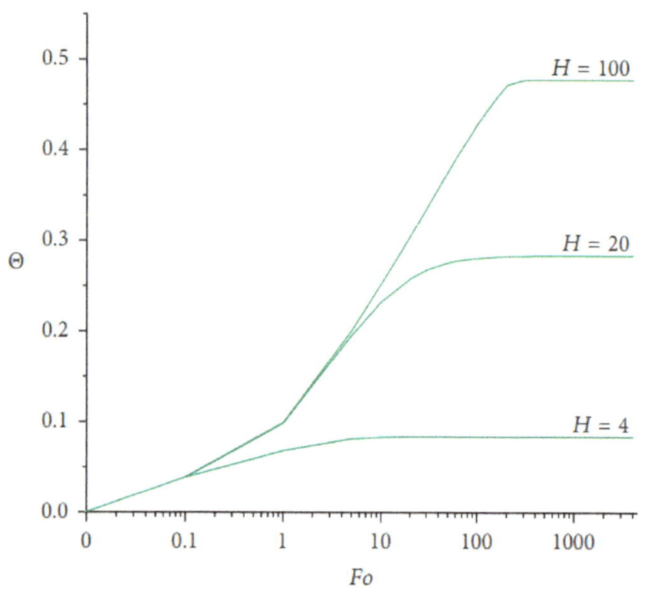

Figure 4: The temperature responses of finite line heat source with the time.

It is found that all the temperature responses will be stable at last. Furthermore, the larger the length, the fiercer the response.

Cylindrical Heat Source Model

The heat capacity inside borehole can be ignored as the diameter is extremely small compared to its depth, but the existence of inner heat transfer is undisputed. Accordingly, the line heat source model is not an ideal choice for simulating the heat transfer of BHE. The solid cylindrical heat sources have been proposed in recent years; in this model, it is supposed that the cylinder is no longer a cavity but is filled with the medium identical to that out of the cylinder [12–14]. The whole infinite domain is composed of a homogeneous medium and therefore the heat capacity of borehole is taken into consideration. The diagram of cylindrical heat source is shown in Figure 5.

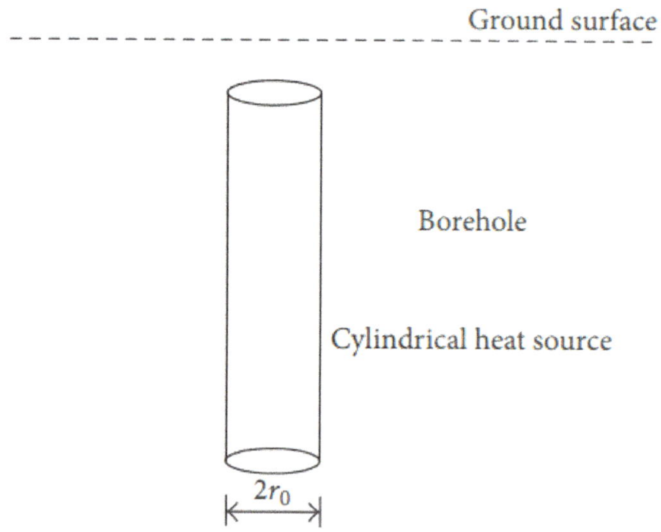

Figure 5: The cylindrical heat source adopted for BHE.

The Green function is still the basic method to obtain the temperature response caused by the heat source, and the temperature response of any point in the underground medium except heat source is shown in

$$t - t_0 = \frac{q_l}{8\pi k}\int_0^{\tau}\frac{1}{(\tau-\tau')}I_0\left[\frac{\sqrt{x^2+y^2}r_b}{2a(\tau-\tau')}\right] \cdot \exp\left[-\frac{\sqrt{x^2+y^2}+r_b^2}{4a(\tau-\tau')}\right] \cdot \left\{\operatorname{erfc}\left[\frac{z-H}{2\sqrt{a(\tau-\tau')}}\right] + \operatorname{erfc}\left[\frac{z+H}{2\sqrt{a(\tau-\tau')}}\right] - 2\operatorname{erfc}\left[\frac{z}{2\sqrt{a(\tau-\tau')}}\right]\right\} d\tau', \quad (8)$$

where I_0 is the zero-order modified Bessel function and the borehole is supposed to be arranged at z-axis. Again, dimensionless method is used to simplify the expression and the corresponding formula is given in

$$\Theta = \frac{1}{8\pi}\int_0^{Fo}\frac{dFo'}{(Fo-Fo')}I_0\left[\frac{\sqrt{X^2+Y^2}}{2(Fo-Fo')}\right] \cdot \exp\left[-\frac{X^2+Y^2+1}{4(Fo-Fo')}\right] \cdot \left\{\operatorname{erfc}\left[\frac{Z-H}{2\sqrt{Fo-Fo'}}\right] + \operatorname{erfc}\left[\frac{Z+H}{2\sqrt{Fo-Fo'}}\right] - 2\operatorname{erfc}\left[\frac{Z}{2\sqrt{Fo-Fo'}}\right]\right\}. \quad (9)$$

The temperature responses with the time are shown in Figure 6, and the regular patterns and trends are similar with those of line heat source; that is, all responses must attain the steady states and the response value increase with the borehole length if the time is constant.

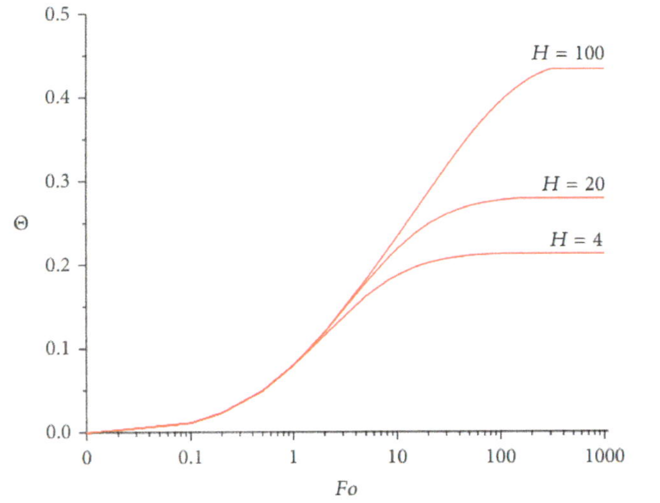

Figure 6: The temperature responses of cylindrical heat source with the time.

As far as the differences between line and cylindrical sources are concerned, the temperature responses changing with time are described in Figure 7 when the radius and depth of borehole are constant.

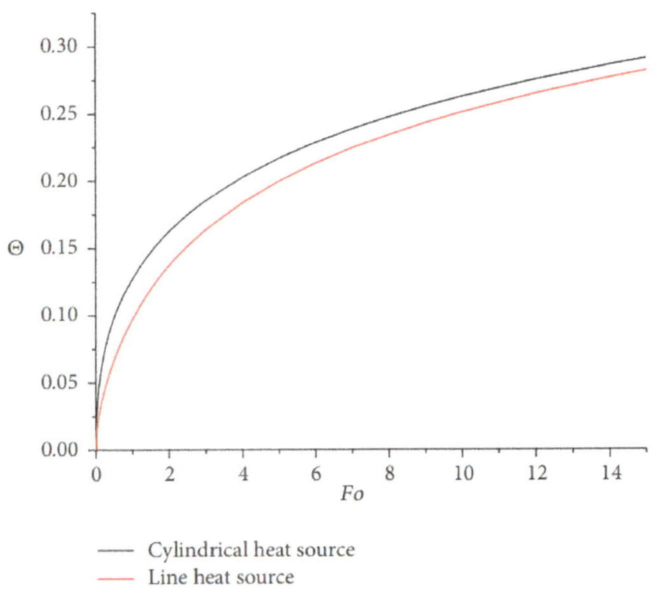

Figure 7: The comparison of temperature responses between line and cylindrical heat sources.

It is noticed that the temperature response of cylindrical heat source is always larger than that of line heat source. The dimensionless temperature response is $\Theta = k(t - t_0)/q_l$, and the heat transfer rate of line heat source becomes larger while the actual temperature rise is stable. Therefore, the calculational length of BHEs by means of line heat source is shorter than that depending on cylindrical heat source when the design of BHEs is conducted.

The Validation of Heat Transfer Experiments

Line and cylindrical heat source models are usually adopted to simulate the heat transfer of BHE, and their rationalities can be validated by means of experiments. A hot water tank with electronic heater is employed and the power of tank is constant in the process of experiment. Figure 8 shows the diagram of experiment and the BHE with U-tube is used. The circulating fluid enters the tank and is heated by electronic heater and then flows through the U-tube of borehole to exchange heat with the underground medium. The water pump drives the circulating liquid to flow circularly. Thermal sensors are set at the surface of borehole to record the temperature response with the time. The initial temperature and thermophysical properties of underground medium are known before conducting the experiment. In addition, the inlet and outlet temperatures of circulating liquid are recorded to calculate the heat transfer rate q_l. There are three sensors, one is set at the depth's middle location of the outer surface of borehole, the other two are employed to record the inlet and outlet temperatures of circulating liquid, and the accuracy of thermal sensors is ±0.1%. The diameter of BHE is 130 mm and the flow rate of fluid is around 0.3liters/s.

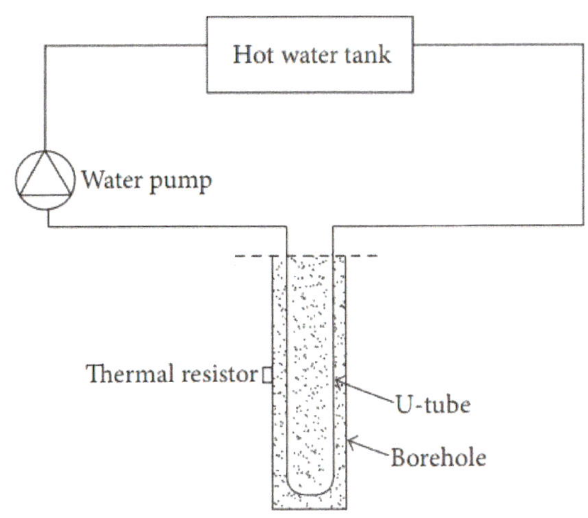

Figure 8: Diagram of experiment of temperature response.

$\Theta = k(t - t_0)/q_l$, and the dimensionless temperature response can be obtained based on the actual temperature recorded by the thermal sensor with the time. Then, the theoretical dimensional temperature responses of line and cylindrical heat sources can be obtained according to (7) and (9), respectively. Accordingly, the comparisons between experimental data and simulation results are shown in Figure 9.

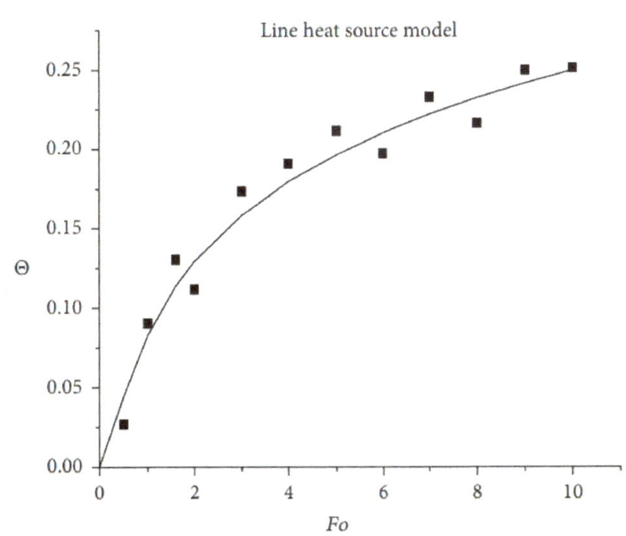

(a)

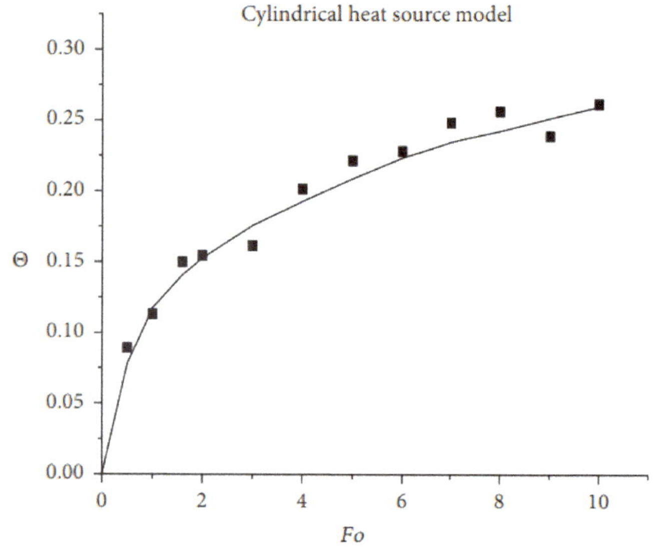

(b)

Figure 9: The comparisons between theoretical results and experimental data.

The linear graph stands for the calculation results by means of simulation models and the scattergram means the experimental data. It is found that the deviation between theoretical result and experimental value is smaller if the cylindrical heat source model is employed; this is because the inner heat capacity of BHE is considered. As a result, cylindrical heat source model is more reasonable for the study of the engineering design or cost performance.

THE ENGINEERING DESIGN FOR BOREHOLE GHES

Section 2 presents the necessary theoretical knowledge of simulating heat transfer of BHE, and these contents illustrate the fundamental principles of heat transfer of BHE. There is no doubt that engineering design can put this technology into practice, and the significant design procedures and the corresponding design parameters should be emphasized to optimize the whole system. There are a number of considerations that need to be studied and they are listed in the following.

The Test of Thermal Property of Underground Medium

The thermal physical properties are very important because they can be used to calculate the size of BHE, and the field measurement combined with parameters estimation is employed to obtain the actual thermal physical parameters [15]. The on-site measuring devices connected with U-tube are shown in Figure 10.

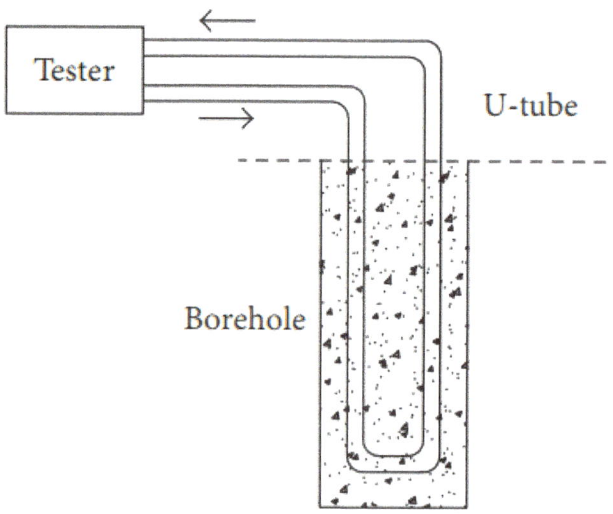

Figure 10: The equipment recording the parameters of circulating liquid.

The object function reporting the sum of variance is established and the expression is given in

$$f = \sum_{i=1}^{n} \left(T_{\mathrm{cal},i} - T_{\exp,i}\right), \tag{10}$$

where $T_{\mathrm{cal},i}$ and $T_{\exp,i}$ are, respectively, the circulating liquid's average temperatures of mathematical model and experimental recorded data [16]. Because the parameters of thermal physical parameters of underground medium are contained in the simulation model, the value of thermal physical parameters can be obtained by means of reverse calculation method while the value of f achieves the minimum.

The Cooling and Heat Load Assumed by UHEs

The heat of the building is abstracted and then is discharged into underground in summer, thereby the maximal heat release appears at the time of the maximal cooling load of buildings. The expression of the heat release is the maximal cooling load × (1 + 1/COP), where COP is the coefficient of performance of the heat pump unit.

In winter, the ultimate value of heat absorption occurs while the heat load of buildings reaches the maximum, and the corresponding formula is the maximal heating load × (1 − 1/COP), where COP means the heating performance coefficient. Comparing the heat release with the heat absorption, the final total length of BHE is determined according to the large one.

The Calculation of Pressure Drop

The purpose of calculating pressure drop is to choose the water pump so that the liquid can achieve circulation flow. As far as the BHEs are concerned, to obtain the linear head loss and partial resistance is the basis of determining the circulating pump. W, A, and d are, respectively, the rate of flow, area of cross section, and inner diameter of tube; therefore, the velocity V is obtained according to

$$V = \frac{W}{3600 \times A} = \frac{4W}{3600 \times \pi \times d^2}. \tag{11}$$

Because the turbulence is prerequisite while the liquid flows during the running process, the corresponding Reynolds number, that is, Re, should be larger than 2300 to ensure this state. The expression of Re is shown in

$$Re = \frac{\rho V d}{\mu}, \tag{12}$$

where ρ and μ are, respectively, the density and coefficient of liquid's dynamic viscosity and the calculation of Re is significant before studying the head loss. Next, (13) offers the linear heat loss and it involves several parameters:

$$P_L = 0.158 \times \rho^{0.75} \times \mu^{0.25} \times d^{1.25} \times V^{1.75} \times L, \tag{13}$$

where L denotes the length of tube of liquid.

Partial loss usually occurs when fluid flows through the elbows, equipment, and so on, and (14) shows the detailed expression

$$P_p = 0.158 \times \rho^{0.75} \times \mu^{0.25} \times d^{1.25} \times V^{1.75} \times E, \tag{14}$$

where E means the equivalent length of the elbows, equipment, and so on.

Afterwards, the total pressure drop is equal to the sum of linear head loss and partial loss [17, 18].

The Joining of BHEs

It is unquestionable that there are a number of BHEs for engineering project, which means BHEs are distributed row after row or column after column; the connection of BHEs is a technology that importance should be attached to. The subcatchment devices are commonly employed to connect a group of BHEs and there are many groups. Then, all distributors are linked to the room of heat pump. The liquids of a group of BHEs are connected by collecting breeching and then enters the room of heat pump. Later, the liquid returns and then is distributed by the separator to flow through tubes of boreholes in parallel. The material of U-tube is high-density polyethylene and this material has favorable conductivity, corrosion resistance, and long service life. The diagram describing the connection of a group of BHEs is shown in Figure 11.

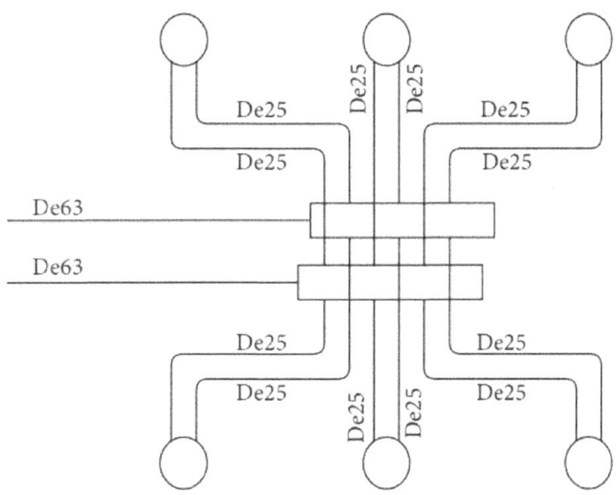

Figure 11: The connection of a group of BHEs.

As a rule, the external diameter of every U-tube is usually 25 mm or 32 mm, and the HDPE tubes which have the external diameter of 50 mm or 63 mm are often used as the main tube of a group of BHEs.

The Backfilling Material inside Borehole

Backfilling material lies in the area between U-tube and borehole wall, and it enhances the heat transfer and stops the surface water from groundwater permeation. The reasonable backfilling material guarantees the heat transfer performance of BHEs. The heat resistance inside borehole is inevitably increased in case the backfilling material has undesirable conduction performance; therefore, the total lengths of BHEs are lengthened and both investment and operating costs are added. Researchers have paid more and more attention to the improvement of backfilling material with the development of GSHP technology.

The main characteristics of backfilling material consist of conduction coefficient, uniformity, stability, and antihot pressing [19–21]. Cement, sand, and bentonite are usually employed to compose the material and the corresponding conduction performance depends on the ratios of these elements. The overseas research indicated that the heat conductivity can arrive at 2.0 W/(m·K), and in recent years the civil experts or scholars found that the value can be further improved and even achieved 2.1 W/(m·K) [22]. The influence that backfilling material exerts on the length of BHEs is obvious even if the value of heat conductivity is increased a little.

THE ANALYSIS ON THE ACTUAL PROJECTS OF GSHP SYSTEM

Although Sections 2 and 3, respectively, present the theoretical simulation models and engineering design of BHEs, the practicability, economical efficiency, and energy-saving of GSHP technology should be validated by actual engineering projects.

The Project Information

This building is a library and it belongs to public construction, the floor and air-conditioning areas are, respectively, 26800 m^2 and 21300 m^2, and the height is nearly 30 m. Table 1 lists the information of the areas and air-conditioning loads.

Table 1: The areas and corresponding loads

Building nature	Building area (m^2)	Air-conditioning area (m^2)	Cooling load (kW)	Heating load (kW)
Library	26800	21300	2378	2308

The First Design Plan

Both cooling and heat loads are assumed by BHEs so that the unity of cold and heat sources is achieved. There are two heat pump units that are responsible for cooling and heating and the corresponding parameters are given in Table 2.

Table 2: Parameters of the heat pump unit of GSHP system

Operating mode	Cooling/heating capacity (kW)	Temperature of U-tube circulating liquid (°C)	Temperature of cooling water (°C)	Pressure drop of evaporator (kPa)	Pressure drop of condenser (kPa)	Electricity consumption (kW)
Cooling	1130	12/7	15/23	98	13.7	193
Heating	1473	50/45	15/10	98	13.7	336

The total length of BHEs is 36640 m according to the analysis and calculation; there are 458 boreholes with the diameter of 140 mm and every one has the depth of 80 m, and the row and column intervals are both 4 m.

The Second Design Plan

The direct-fired lithium-bromide refrigerating machine (DFLBRM) is employed for cooling and heating. The cooling tower is set in the open air to act as the heat discharge equipment for the cooling condition, and the gas is fired to become the driving heat source so that the heating objective is achieved. The characteristics of this plan can be summarized as several points: firstly, it can save electricity but not save energy; secondly, the cold and heat sources are united; thirdly, the requirements of the automatic control and safety are very high.

According to the cooling and heat loads, two DFLBRMs are selected and the corresponding parameters are listed in Table 3.

Table 3: Parameters of the direct-fired lithium-bromide refrigerating machine

Cooling capacity (kW)	Heating capacity (kW)	Cold water			Cooling water			The maximum amount of fuel		Distribution capacity	Solution capacity	Running weight
		Rate of flow (m^3/h)	Pressure drop (MPa)	Diameter of joint pipe (mm)	Rate of flow (m^3/h)	Pressure drop (MPa)	Diameter of joint pipe (mm)	Cooling 10^4 kcal	Heating 10^4 kcal	kW	ton	kg
1454	1121	250	0.07	200	304	0.11	250	94.6	104	9.1	4.1	20.4

Comparisons of Costs between the Planes

Tables 4 and 5, respectively, describe the investments of the first and the second design planes.

Table 4: Initial cost of GSHP system

Equipment	Number	Unit price (¥10^4)	Sum price (¥10^4)
Heat pump units	2	119.365	238.73
Cold water pump	2	1.55	3.1
Cooling water pump	2	1.55	3.1
Water separator (outdoor circulation)	1	0.98	0.98
Water collector (outdoor circulation)	1	0.98	0.98
Barrier diaphragm-type pressurized water tank	1	0.69	0.69
Installation of UHEs			492.85
Total			740.43

Table 5: Initial cost of the direct-fired lithium-bromide refrigerating machine system

Equipment	Number	Unit price (¥10^4)	Sum price (¥10^4)
DFLBRM	2	135	270
Cooling tower	2	11.05	22.1
Cold water pump	2	1.55	3.1
Cooling water pump	2	1.55	3.1
Gas	29779.9	0.0025	74.4498
Gas pipe laying work			100
Total			472.75

In this project, the thermal conductivity of backfilling material is 1.5 W/(m·K) and the corresponding installation cost of BHEs is ¥492.85 × 10^4. The thermal conductivity can exert influence on the cost, and the corresponding cost decreases with the increase of thermal conductivity, the detailed information is shown in Figure 12.

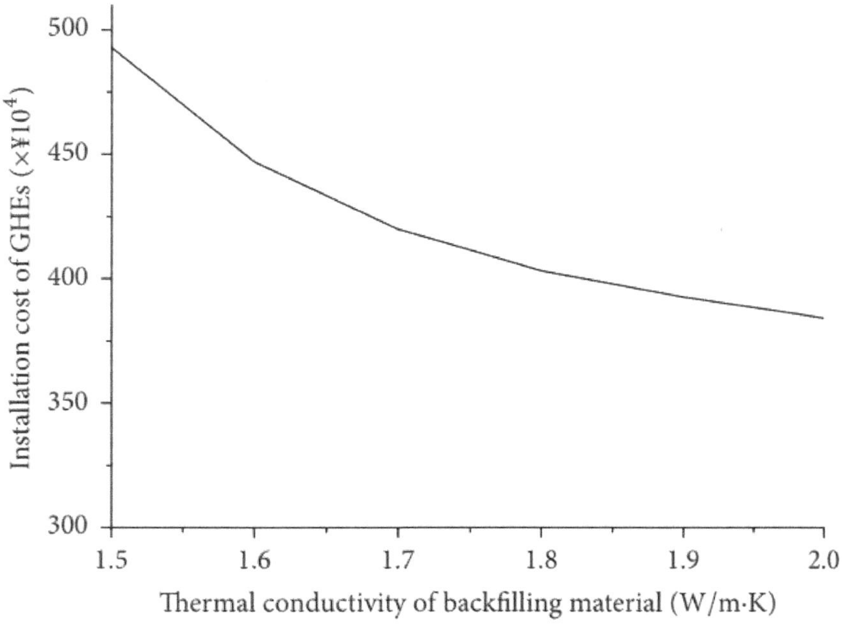

Figure 12: The installation cost of GHEs changes with the thermal conductivity of backfilling material.

It should be admitted that the initial cost of GSHP is obviously higher than that of DFLBRMs, but the running cost of these two different modes can embody the advantage of GSHP technology. And the initial cost of GSHP can be further reduced while the thermal conductivity of backfilling material is increased, and the installation cost of BHEs is obtained according to the minimal value of conductivity. The detailed information of operating costs is shown in Tables 6 and 7.

Table 6: Operating cost of GSHP system

Equipment	Number	Electricity consumption	Running time	Energy consumption	Unit price	Operating cost per day	Operating cost per month	Operating months	Operating cost per year
		kW	h	kWh	¥/kWh	¥10^4	¥10^4		¥10^4
Heat pump units	Cooling 2	193	12	4632	0.6	0.27792	8.3376	3	25.0128
	Heating 2	336	12	8064	0.6	0.48384	14.5152	4	58.0608
Cold water pump	2	37	12	888	0.6	0.05328	1.5984	7	11.1888
Cooling water pump	2	37	12	888	0.6	0.05328	1.5984	7	11.1888
				Total					105.4512

Table 7: Operating cost of DFLBRMs system

	Number	Electricity (gas) consumption	Running time	Energy consumption	Unit price	Operating cost per day	Operating cost per month	Operating months	Operating cost per year
		kW (m^3/h)	h	kW	¥	¥10^4	¥10^4		¥10^4
DFLBRMs	Cooling 2	9.1	12	218.4	0.6	0.013104	0.39312	3	1.17936
	Heating 2	9.1	12	218.4	0.6	0.013104	0.39312	4	1.57248
Cooling tower	2	25	12	600	0.6	0.036	1.08	3	3.24
Cold water pump	2	37	12	888	0.6	0.05328	1.5984	3	4.7952
Cooling water pump	2	37	12	888	0.6	0.05328	1.5984	7	11.1888
Gas cooling		267	12		2.2	0.70488	21.1464	3	63.4392
Heating		288	12		2.2	0.76032	22.8096	4	91.2384
				Total					176.6534

The significant data of tables make clear that the initial cost of GSHP system is expensive; however, the favorable benefits in terms of energy-saving and environmental protection are certainly achieved [23]. In addition, the most attention-getting point is that the running cost is obviously lower and the payback period is less than 4 years compared with the DFLBRMs. What is more, the installation cost of BHEs can be reduced while the thermal conductivity of backfilling material increases, and this can further improve the economic performance of the whole GSHP system.

CONCLUSIONS

The vertical BHEs are typical underground components that are in charge of heat release and heat abstraction; the paper demonstrates the fundamental principles of heat transfer inside and outside borehole. The research on theoretical knowledge lays a firm foundation for the engineering design of GSHP projects. The construction of underground installation is complicated and therefore a series of technological process are explained in detail, including steps and parameters. The discussions of theoretical heat transfer and actual engineering knowledge enable researchers to understand how to make full use of GSHP. In addition, an actual engineering project that employs GSHP system is analyzed; as far as GSHP and DFLBRMs are concerned, not only the investments but also the operating costs are compared with each other. By means of comparisons, it should be highlighted that the initial expense of GSHP is higher but the running cost is lower, which means the added part of initial cost can be saved by way of saving running cost. The initial cost of GSHP can be reduced further while the thermal conductivity of backfilling material is increased, and this can let the cost performance of GSHP be improved further. According to the detailed illustration, GSHP will make greater progress and gain more and more attention in the future.

ACKNOWLEDGMENT

The authors would like to thank the National Natural Science Fund Committee, China, for financial support (41171251).

REFERENCES

1. A. Bagdanavicius and N. Jenkins, "Power requirements of ground source heat pumps in a residential area," Applied Energy, vol. 102, pp. 591–600, 2013.
2. D. Nairen, Ground-Coupled Heat Pump Technology, Higher Education Press, Beijing, China, 1st edition, 2006.

3. Y. Lu, Y. Liu, X. Li, and Y. Kang, "A new method of drilling long boreholes in low permeability coal by improving its permeability," International Journal of Coal Geology, vol. 84, no. 2, pp. 94–102, 2010.
4. A. M. Omer, "Ground-source heat pumps systems and applications," Renewable and Sustainable Energy Reviews, vol. 12, no. 2, pp. 344–371, 2008.
5. B. Sanner, C. Karytsas, D. Mendrinos, and L. Rybach, "Current status of ground source heat pumps and underground thermal energy storage in Europe," Geothermics, vol. 32, no. 4, pp. 579–588, 2003.
6. N. R. Diao, H. Y. Zeng, and Z. H. Fang, "Improvement in modeling of heat transfer in vertical ground heat exchangers," HVAC&R Research, vol. 10, no. 4, pp. 459–470, 2004.
7. F. Jun, Diaonairen, and F. Zhaohong, "Analysis of heat disturbance between the pipes of the U-tube geothermal heat exchangers," Journal of Shandong University of Architecture and Engineering, vol. 19, no. 1, pp. 1–4, 2004.
8. T. V. Bandos, Á. Montero, E. Fernández et al., "Finite line-source model for borehole heat exchangers: effect of vertical temperature variations," Geothermics, vol. 38, no. 2, pp. 263–270, 2009.
9. H. Yang, P. Cui, and Z. Fang, "Vertical-borehole ground-coupled heat pumps: a review of models and systems," Applied Energy, vol. 87, no. 1, pp. 16–27, 2010. ·
10. H. S. Carslaw and J. C. Jeager, Conduction of Heat in Solids, Oxford Press, Oxford, UK, 2nd edition, 1959.
11. H. Y. Zeng, N. R. Diao, and Z. H. Fang, "A finite line-source model for boreholes in geothermal heat exchangers," Heat Transfer—Asian Research, vol. 31, no. 7, pp. 558–567, 2002.
12. L. Junhong, Z. wenke, and F. Zhaohong, "Solid cylindrical heat source model for pile ground heat exchangers with spiral coils," Journal of Shandong Jianzhu University, vol. 25, no. 2, pp. 95–100, 2010.
13. Y. Man, H. Yang, N. Diao, J. Liu, and Z. Fang, "A new model and analytical solutions for borehole and pile ground heat exchangers," International Journal of Heat and Mass Transfer, vol. 53, no. 13-14, pp. 2593–2601, 2010.
14. P. Cui, X. Li, Y. Man, and Z. Fang, "Heat transfer analysis of pile geothermal heat exchangers with spiral coils," Applied Energy, vol. 88, no. 11, pp. 4113–4119, 2011. ·
15. Y. Liang, Z. Jian, Z. Haidong, et al., "Preliminary analysis of the methods for ground thermal properties calculations based on line heat source model combined with parameter estimation," Refrigeration Air Conditioning & Electric Power Machinery, vol. 6, pp. 7–11, 2011.
16. M. Yu and Z. Fang, "A method for the on-site testing of average thermo-physical parameters of underground rock soil," Journal of Engineering for Thermal Energy and Power, vol. 17, no. 5, pp. 489–492, 2002.
17. Caneta Research, Technical Guide for Engineering of Ground Source Heat Pump, China Building Industry Press, Beijing, China, 2001.
18. Ministry of Construction of the People›s Republic of China, "Technical code for ground source heat pump system," Tech. Rep. GB50366-2005, China Building Industry Press, Beijing, China, 2005.
19. E. H. Spilker, Ground-Coupled Heat Pump Loop Design Using Thermal Conductivity Testing and the Effect of Different Backfill Materials on Vertical Bore Length, American Society of Heating, Refrigerating and Air-Conditioning Engineers, Atlanta, Ga, USA, 1998.
20. X. Li, Z. Chen, and J. Zhao, "Simulation and experiment on the thermal performance of U-vertical ground coupled heat exchanger," Applied Thermal Engineering, vol. 26, no. 14-15, pp. 1564–1571, 2006.
21. H. Wang, C. Qi, H. Du, and J. Gu, "Improved method and case study of thermal response test for borehole heat exchangers of ground source heat pump system," Renewable Energy, vol. 35, no. 3, pp. 727–733, 2010.
22. P. E. Mikkelsen, "Cement-bentonite grout backfill for borehole instruments," Geotechnical News, vol. 20, no. 4, pp. 38–42, 2002.
23. S. J. Self, B. V. Reddy, and M. A. Rosen, "Geothermal heat pump systems: status review and comparison with other heating options," Applied Energy, vol. 101, pp. 341–348, 2013.

CHAPTER 21

NUMERICAL STUDY OF MIXED CONVECTION OF THE NANOFLUIDS IN TWO-SIDED LID-DRIVEN SQUARE CAVITY WITH A PAIR OF TRIANGULAR HEATING CYLINDERS

Zoubair Boulahia, Abderrahim Wakif, and Rachid Sehaqui

Faculty of Sciences Aïn Chock, Laboratory of Mechanics, Hassan II University, BP 5366, Maarif, 20100 Casablanca, Morocco

ABSTRACT

A numerical study is carried out concerning mixed convection of the nanofluid in two-sided lid-driven square cavity with a pair of triangular heat sources. The upper and bottom moving walls are thermally insulated while the left and right walls are cooled at constant temperature. Two-dimensional Navier-Stokes and energy equations are solved using the finite volume discretization method with SIMPLE algorithm. The method used is validated against previous works. Two cases were considered depending on the direction of moving walls. Effects of various design parameters such as Richardson number ($0.1 \leq Ri \leq 100$), nanoparticle volume fraction ($0 \leq \varphi \leq 0.05$), and size ($25\ nm \leq d_p \leq 145\ nm$) and type (Cu, Al_2O_3, TiO_2) of nanoparticles on the heat transfer rate are investigated. The results of this investigation illustrate that, by reducing the diameter of the nanoparticles and Ri, the heat transfer rate increases. Moreover, it is found that by changing horizontal direction of the moving walls the heat transfer rate variation is negligible

INTRODUCTION

Mixed convection problem with lid-driven flows in enclosures occurs in many engineering applications such as heating and cooling nuclear systems of reactors, lubrication technologies, cooling of electronic devices, and heat exchangers. There are many studies concerning the heat transfer in rectangular or square cavities driven by buoyancy and shear which have been studied extensively in the literature. Work of Oztop et al. [1] can be mentioned as example of such studies, in which they numerically studied mixed convection in square cavities with two moving walls. Their results suggest that when the vertical walls move upwards in the same direction, the heat transfer decreases significantly compared to when the vertical walls move in opposite directions. Work of Khanafer and Aithal [2] was concentrated on the effect of mixed convection flow and heat transfer characteristics in a lid-driven cavity with a circular body inside. The results showed that the average Nusselt number increases with an increase in the radius of the cylinder for various Richardson numbers and the optimal heat transfer results are obtained when placing the cylinder near the bottom wall. Islam et al. [3] investigated a numerical study on a lid-driven cavity with a heated square blockage.

Citation: Zoubair Boulahia, Abderrahim Wakif, and Rachid Sehaqui, "Numerical Study of Mixed Convection of the Nanofluids in Two-Sided Lid-Driven Square Cavity with a Pair of Triangular Heating Cylinders," Journal of Engineering, vol. 2016, Article ID 8962091, 8 pages, 2016. doi:10.1155/2016/8962091.

They found that size, location, and Richardson number of the heater eccentricities affect the average Nusselt number of heater. Kalteh et al. [4] considered laminar mixed convection of nanofluid in a lid-driven square cavity with a triangular heat source. Their simulations indicate that increasing the volume fraction, the nanoparticles diameter, and Reynolds number leads to an increase in average Nusselt number.

El Abdallaoui et al. [5] performed a numerical simulation of natural convection between a decentered triangular heating cylinder and centered triangular heating cylinder [6] in a square outer cylinder filled with a pure fluid or a nanofluid using the lattice Boltzmann method. The results indicate that the horizontal displacement from the centered position to decentered position leads to a considerable increase of heat transfer at weak Rayleigh and the vertical displacement has most important effect on heat transfer at high values of Rayleigh. Oztop et al. [7] investigated mixed convection in a lid-driven cavity with a cylindrical blockage inside. The vertical left sidewall was moving up or down. The computation was carried out for wide ranges of Richardson number, inner cylinder diameter, and center and location of the inner cylinder. They concluded that the most effective parameter on flow field and temperature distribution was the direction of the moving lid.

In the present study, the main aim is to examine the mixed convection heat transfer of nanofluid in a square enclosure with a pair of triangular heat sources. The first case under investigation is characterized by the numerical models used in our study. The computational procedure elaborated in this study is validated against the numerical results of other investigations. We studied the effects of various design parameters such as Richardson number ($0.1 \leq Ri \leq 100$), nanoparticle volume fraction ($0 \leq \varphi \leq 0.05$), and size (25 nm $\leq d_p \leq$ 145 nm) and type (Cu, Al_2O_3, TiO_2) of nanoparticles on the heat transfer. The new models of the thermal conductivity and effective viscosity investigated by Corcione [8] are used to estimate thermophysical proprieties of the nanofluid. Our numerical results are presented in the form of plots of isotherms, streamlines, and average Nusselt numbers to show the influence of nanofluid and design parameters.

PROBLEM STATEMENT

The studied configurations and coordinate system of the considered enclosure in the present study are shown in Figure 1. It is a two-sided lid-driven square cavity with a pair of triangular heating cylinders, filled with nanofluids. The triangular heat source is maintained at a temperature T_h = 310K. The vertical walls are cooled at constant temperatures T_c = 290K. Two different cases were considered as shown in Figure 1. In Case 1, the upper wall (adiabatic) is moving to the right while bottom wall (adiabatic) is moving to the left. In Case 2, the direction of the two moving walls is the same. It is assumed that the nanofluid is Newtonian, incompressible, and laminar and the base fluid and the nanoparticles are in a thermal equilibrium state. The thermophysical properties of the nanofluid used in this study are evaluated at the average fluid temperature $(T_c + T_h)/2$ as listed in Table 1.

Table 1: Thermophysical properties of water and nanoparticles at T = 300K [14]

	C_p (J/KgK)	ρ (Kg/m^3)	K (W/mK)	$\beta \times 10^5$ (K^{-1})	$\mu \times 10^6$ (kgm^{-1}s^{-1})	d_p (nm)
Copper (Cu)	385	8933	401	1.67	—	25, 85, 145
Al_2O_3	765	3970	36	0.85	—	25
TiO_2	710	4157	8.4	0.9	—	25
Water (H_2O)	4179	997.1	0.613	27.6	855	0.385

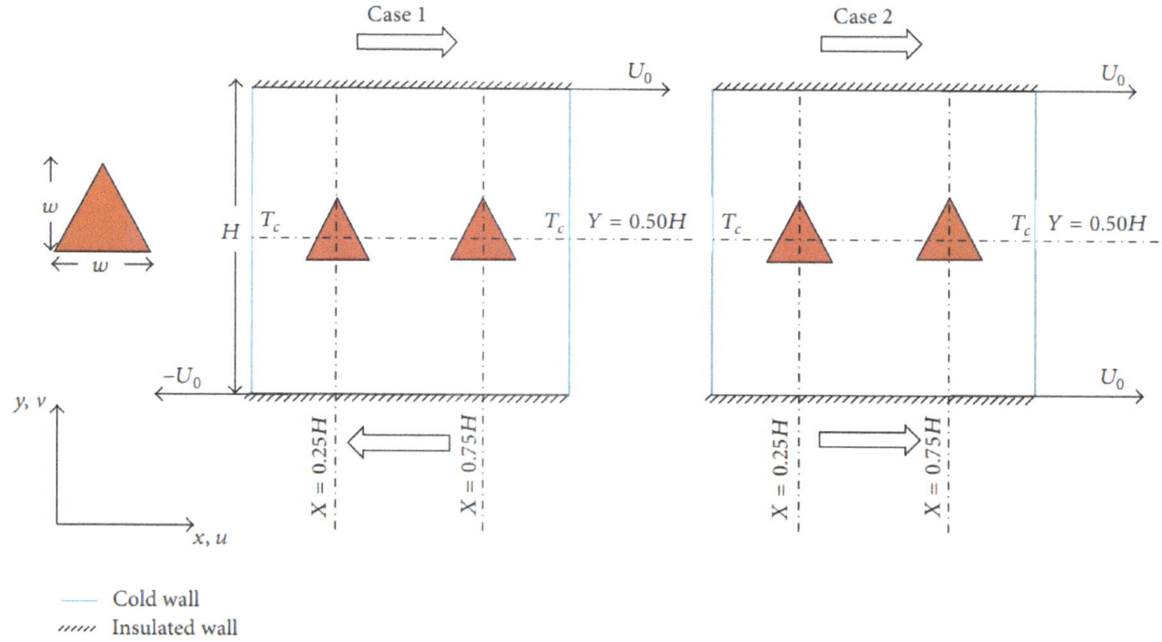

Figure 1: Schematic of the cavity with a pair of triangular heat sources and boundary conditions.

MATHEMATICAL FORMULATION

The governing equations including the two-dimensional transient equations of the continuity, momentum, and energy for an incompressible flow are expressed in the following format:

$$\frac{\partial u}{\partial x} + \frac{\partial v}{\partial y} = 0,$$

$$\frac{\partial u}{\partial t} + u\frac{\partial u}{\partial x} + v\frac{\partial u}{\partial y} = -\frac{1}{\rho_{\text{nf}}}\frac{\partial p}{\partial x} + \frac{\mu_{\text{nf}}}{\rho_{\text{nf}}}\left(\frac{\partial^2 u}{\partial x^2} + \frac{\partial^2 u}{\partial y^2}\right),$$

$$\frac{\partial v}{\partial t} + u\frac{\partial v}{\partial x} + v\frac{\partial v}{\partial y} = -\frac{1}{\rho_{\text{nf}}}\frac{\partial p}{\partial y} + \frac{\mu_{\text{nf}}}{\rho_{\text{nf}}}\left(\frac{\partial^2 v}{\partial x^2} + \frac{\partial^2 v}{\partial y^2}\right) + \frac{(\rho\beta)_{\text{nf}}}{\rho_{\text{nf}}}(T - T_c),$$

$$\frac{\partial T}{\partial t} + u\frac{\partial T}{\partial x} + v\frac{\partial T}{\partial y} = \alpha_{\text{nf}}\left(\frac{\partial^2 T}{\partial x^2} + \frac{\partial^2 T}{\partial y^2}\right), \quad (1)$$

where the nanofluid effective density, heat capacity, thermal expansion coefficient, and thermal diffusivity are calculated from the following equations [9, 10]:

$$\rho_{\text{nf}} = (1-\varphi)\rho_f + \varphi\rho_s,$$

$$(\rho C_p)_{\text{nf}} = (1-\varphi)(\rho C_p)_f + \varphi(\rho C_p)_s,$$

$$(\rho\beta)_{\text{nf}} = (1-\varphi)(\rho\beta)_f + \varphi(\rho\beta)_s,$$

$$\alpha_{\text{nf}} = \frac{k_{\text{nf}}}{(\rho C_p)_{\text{nf}}}. \quad (2)$$

Corcione model [8–10] for dynamic viscosity and thermal conductivity of the nanofluid is given by

$$\mu_{\text{nf}} = \frac{\mu_f}{\left(1 - 34.87\left(d_p/d_f\right)^{-0.3}\varphi^{1.03}\right)},$$

$$\frac{k_{\text{nf}}}{k_f} = 1 + 4.4\text{Re}_B^{0.4}\text{Pr}^{0.66}\left(\frac{T}{T_{fr}}\right)^{10}\left(\frac{k_p}{k_f}\right)^{0.03}\varphi^{0.66},$$

$$\text{Re}_B = \frac{\rho_f u_B d_p}{\mu_f},$$

$$u_B = \frac{2k_b T}{\pi\mu_f d_p^2}. \quad (3)$$

All terms are defined in Nomenclature.

The boundary conditions for mixed convection are written as

$$u = -U_0 \text{ or } U_0,\quad v = 0,\quad \frac{\partial T}{\partial y} = 0, \quad \text{on bottom wall of the cavity},$$

$$u = U_0,\quad v = 0,\quad \frac{\partial T}{\partial y} = 0, \quad \text{on upper wall of the cavity},$$

$$v = 0,\quad v = 0,\quad T = T_c, \quad \text{on right wall of the cavity},$$

$$u = 0,\quad v = 0,\quad T = T_c, \quad \text{on left wall of the cavity}. \quad (4)$$

The following dimensionless variables for mixed convection are defined based on properties of pure fluid:

$$\tau = \frac{t}{H/U_{\text{ref}}},\quad X = \frac{x}{H},\quad Y = \frac{y}{H},\quad U = \frac{u}{U_{\text{ref}}},\quad V = \frac{v}{U_{\text{ref}}},\quad P = \frac{p}{\rho_{\text{nf}}U_{\text{ref}}^2},\quad \theta = \frac{T - T_c}{T_h - T_c}, \quad (5)$$

where U_{ref} is considered to be U_0 for mixed convection. Dimensionless numbers for the system are defined as

$$\mathrm{Re} = \frac{u_{\mathrm{ref}} H}{\nu_{\mathrm{nf}}},$$

$$\mathrm{Ri} = \frac{\mathrm{Gr}}{\mathrm{Re}^2},$$

$$\mathrm{Gr} = \frac{g\beta_f (T_h - T_c) H^3}{\nu_f^2},$$

$$\mathrm{Ra} = \mathrm{Gr} \cdot \mathrm{Pr} = \frac{g\beta_f (T_h - T_c) H^3}{\alpha_f \nu_f},$$

$$\mathrm{Pr} = \frac{\nu_f}{\alpha_f}. \quad (6)$$

The governing equations (1) are written in the following dimensionless form:

$$\frac{\partial U}{\partial X} + \frac{\partial V}{\partial Y} = 0,$$

$$\frac{\partial U}{\partial \tau} + U\frac{\partial U}{\partial X} + V\frac{\partial U}{\partial Y} = -\frac{\partial P}{\partial X} + \frac{1}{\mathrm{Re}} \frac{\rho_f}{\rho_{\mathrm{nf}}} \frac{\mu_{\mathrm{nf}}}{\mu_f} \left(\frac{\partial^2 U}{\partial X^2} + \frac{\partial^2 U}{\partial Y^2} \right),$$

$$\frac{\partial V}{\partial \tau} + U\frac{\partial V}{\partial X} + V\frac{\partial V}{\partial Y} = -\frac{\partial p}{\partial Y} + \frac{1}{\mathrm{Re}} \frac{\rho_f}{\rho_{\mathrm{nf}}} \frac{\mu_{\mathrm{nf}}}{\mu_f} \left(\frac{\partial^2 V}{\partial X^2} + \frac{\partial^2 V}{\partial Y^2} \right) + \mathrm{Ri}\frac{(\rho\beta)_{\mathrm{nf}}}{\rho_{\mathrm{nf}}\beta_f}\theta,$$

$$\frac{\partial \theta}{\partial \tau} + U\frac{\partial \theta}{\partial X} + V\frac{\partial \theta}{\partial Y} = \frac{1}{\mathrm{Re} \cdot \mathrm{Pr}} \frac{\alpha_{\mathrm{nf}}}{\alpha_f} \left(\frac{\partial^2 \theta}{\partial X^2} + \frac{\partial^2 \theta}{\partial Y^2} \right). \quad (7)$$

Dimensionless form of the boundary conditions can be written as

$U = 1$ or -1,

$V = 0$,

$\frac{\partial \theta}{\partial y} = 0$,

on bottom wall of the cavity,

$U = 1$,

$V = 0$,

$\frac{\partial \theta}{\partial y} = 0$,

on upper wall of the cavity,

$V = 0$,

$V = 0$,

$\theta = 0$,

on right wall of the cavity,

$U = 0$,

$V = 0$,

$\theta = 0$,

on left wall of the cavity. (8)

The total mean Nusselt number of all cavity's wall is defined as

$$\overline{\mathrm{Nu}}_{\mathrm{tot}} = \frac{1}{H} \int_0^H \frac{k_{\mathrm{nf}}(\varphi)}{k} \left\{ \left| \frac{\partial \theta}{\partial X} \right|_{\mathrm{left}} + \left| \frac{\partial \theta}{\partial X} \right|_{\mathrm{right}} \right\} dY. \quad (9)$$

NUMERICAL DETAILS

The discretization procedure of the governing equations (7) and boundary conditions described by (8) is based on a finite volume formulation, given by Patankar [11] on a staggered grid. SIMPLE (Semi-Implicit Method for Pressure Linked Equations) is used to solve the coupled pressure velocity equation while Hybrid Differencing Scheme (HDS) of Spalding [12] is used for the convective terms. Line by line application of TDMA (Tridiagonal Matrix Algorithm) method [12] is applied on equation systems until sum of the residuals became less than 10^{-6}.The developed algorithm was implemented in FORTRAN program.

Grid Independence Study

In order to determine a proper grid for the numerical simulation, a square cavity filled with Cu-water nanofluid (φ = 0.05) having a pair of triangular heat sources with size w = 0.2 is analyzed in two extreme Richardson numbers (Ra = 0.1 and 100). The mean Nusselt number obtained using different grid numbers for particular cases is presented in Table 2. As can be observed from the table, a uniform 103 × 103 grid is sufficiently fine for the numerical calculation.

Table 2: Effect of the grid size on $\overline{\mathrm{Nu}}_{\mathrm{tot}}$ for the cavity filled with the Cu-water nanofluid (φ = 0.05) having a pair of triangular heat sources with size $W = w/H = 0.2$ for case 1 (see Figure 1).

Ri	63 × 63	83 × 83	103 × 103	123 × 123
0.1	19.781	20.295	20.351	20.357
100	8.921	9.377	9.438	9.439

Validations

The present numerical scheme was validated against various numerical results available in the literature, two different heat convection problems are chosen. The first case is the numerical results of Iwatsu et al. [13] and Oztop et al. [7] for a top heated moving lid and bottom cooled square cavity filled with air (Pr = 0.71). A 100 × 100 mesh was used and the computations were done for three different Richardson numbers. Table 3 demonstrates an excellent comparison of the average Nusselt number between the present results and the numerical results found in the literature [7–13] with a maximum discrepancy of about 1.6%. The second case is a mixed convection flow and heat transfer characteristics in a lid-driven cavity with a square heater inside. Figure 2 illustrates a comparison of the isotherms and streamlines between the present results and the results reported by Islam et al. [3] at Reynolds number (Re = 100) and at Richardson numbers (Ri = 0.1).

Table 3: Comparison of $\overline{Nu}$ at the hot lid between the present results and those reported in the literature

Ri	Average Nusselt number at the hot lid		
	Iwatsu et al. [13]	Oztop et al. [7]	Present
1.0	1.34	1.30	**1.36**
0.0625	3.62	3.63	**3.68**
0.01	6.29	6.34	**6.29**

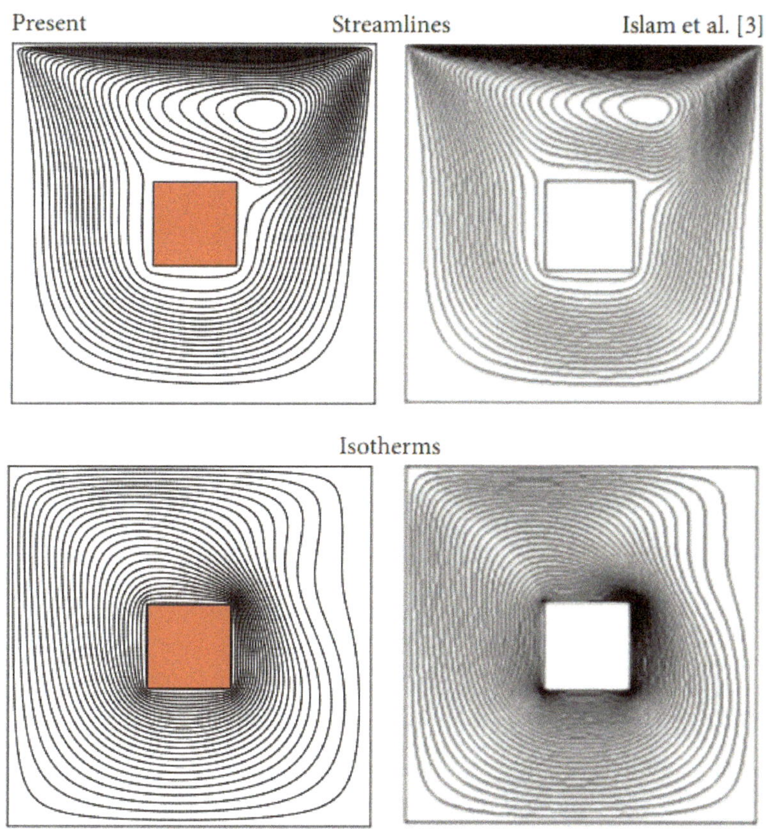

Figure 2: Comparison of the streamlines and isotherms between the present results and that of Islam et al. [3] ($h/H = 0.25$, Re = 100, and Ri = 0.1).

RESULTS AND DISCUSSION

In this section, the numerical results for mixed convection heat transfer of the nanofluid in a square cavity with horizontal moving lids are discussed. Results pertinent to the effects of the moving wall direction of motion, volume fraction ($0 \leq \varphi \leq 0.05$), diameter (25 nm $\leq d_p \leq$ 145 nm), and type (Cu, Al_2O_3, and TiO_2) of the nanoparticles on the fluid flow and heat transfer rate are presented for the square cavity with a pair of triangular heat sources placed horizontally (see Figure 1, Cases 1 and 2). Size of triangular heat sources is fixed at $w = 0.2H$.

Figures 3(a) and 3(b) show the effects of the moving wall direction of motion on the streamlines and isotherms for various values of Ri ranging from 0.1 to 100 while Grashof number is fixed at Gr = 104 . The cavity is filled with Cuwater (φ = 5%), and for comparisons the streamlines and the isotherms for pure fluid and nanofluid are shown by dashed line and solid line, respectively. As shown in Figure 3, some differences are observed in streamlines and isotherms of pure fluid and nanofluid. We can explain this by the higher viscosity of nanofluid compared to that of the pure fluid which increases the diffusion of momentum in the nanofluid. Figure 3(a) shows that, at Ri = 0.1 for Case 1, the flow field is characterized by a clockwise vortex created in the upper half of the enclosure, and another created in the lower half. These two vortexes are mostly generated by horizontal moving lids. At this Ri, we can see in Figure 3(b) that the concentration of the isotherms, close to the pair of triangular heat sources, increases, which indicates that the heat transfer is through forced convection. By increasing Ri, the effects of natural convection and buoyancy force become stronger leading to formation of a counterclockwise rotating eddy close to the left wall of the cavity. The effects of moving wall decrease by increasing Ri. The uniformly distributed isotherms at high Ri show that the main heat transfer mechanism is through the natural convection.

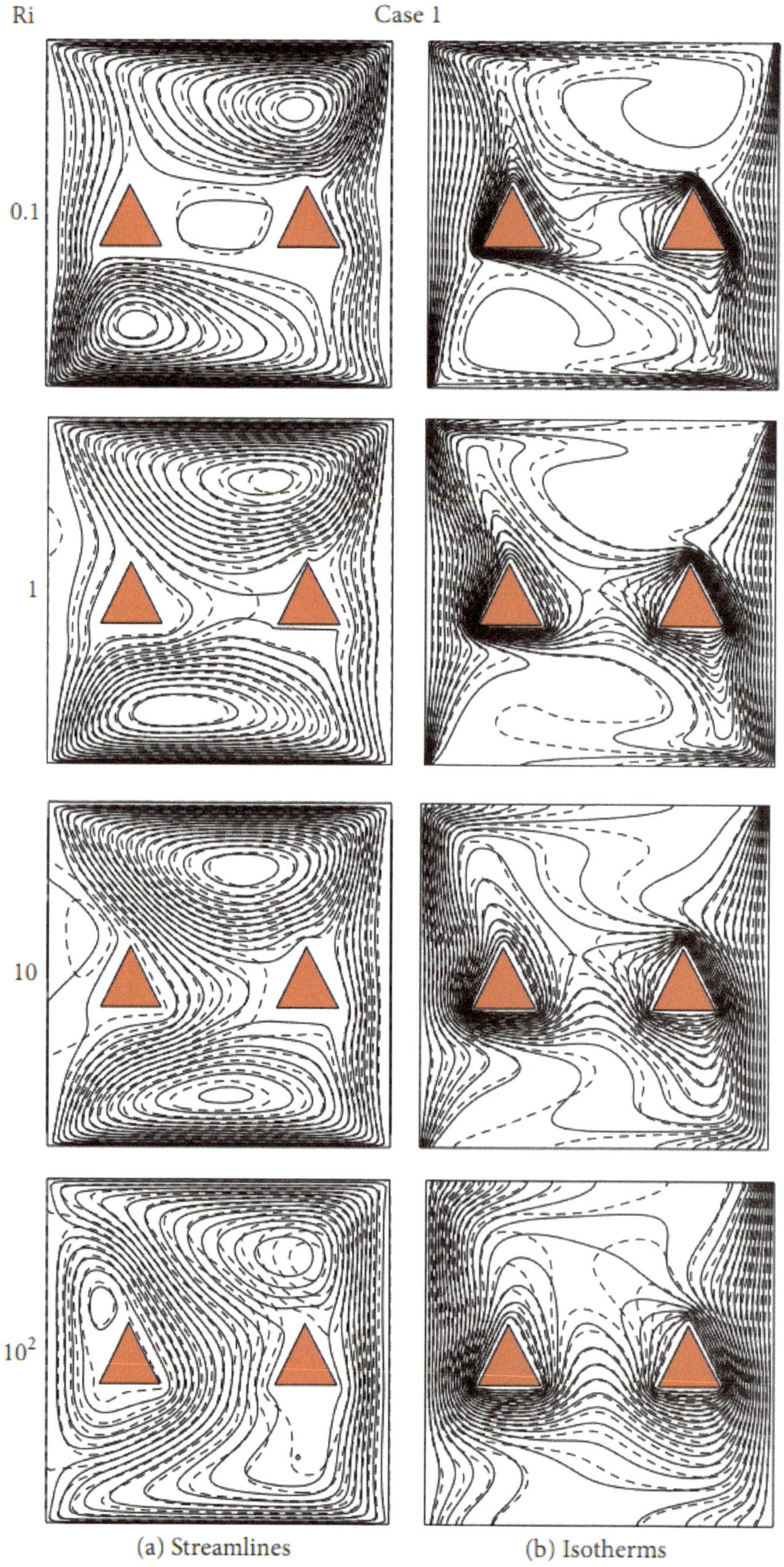

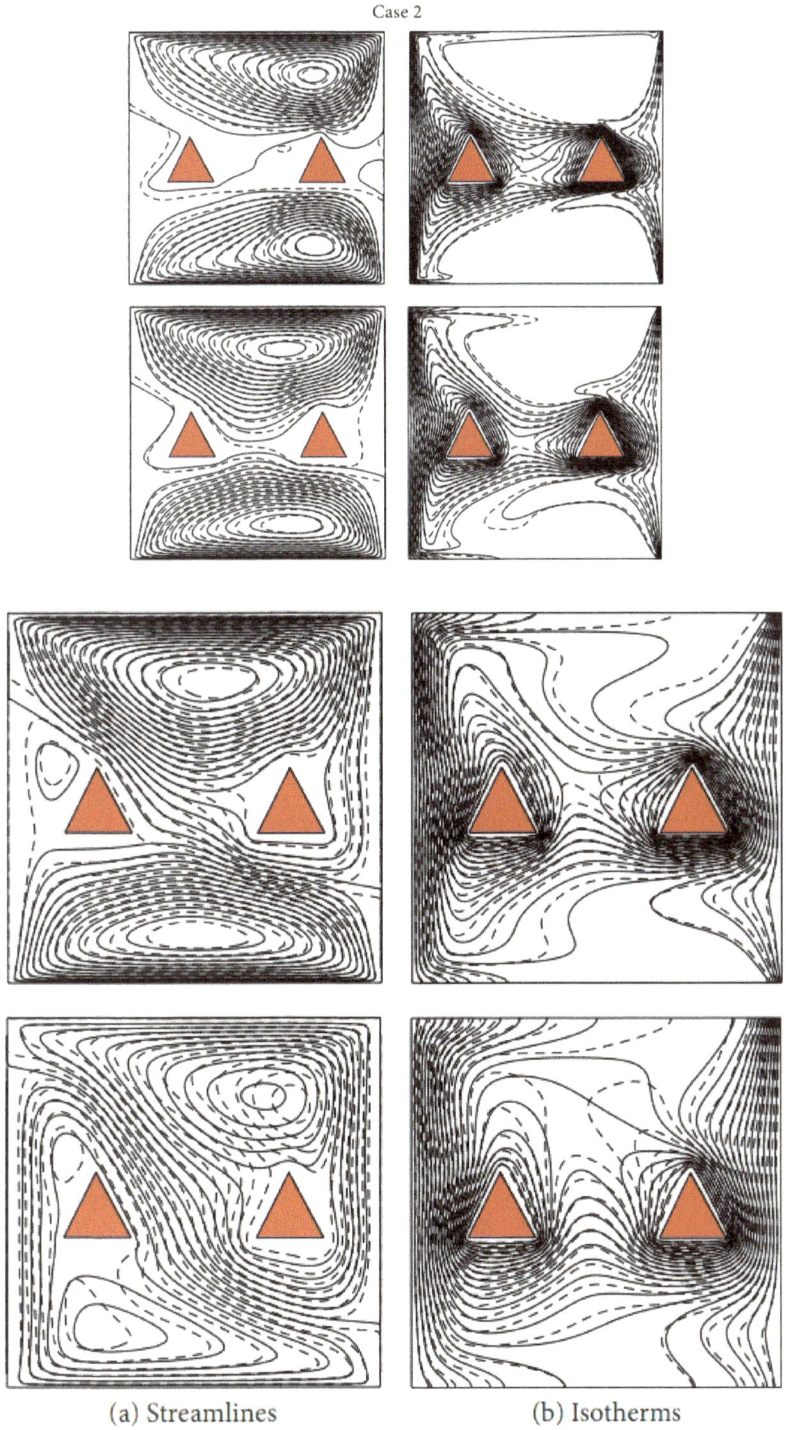

Figure 3: (a) Streamlines and (b) isotherms inside the cavity with a pair of triangular heat sources ($W = w/H = 0.2$) filled with Cu-water pure fluid (dashed line) and Cu-water nanofluid (solid line) with φ=5% and at different Ris for Cases 1 and 2. Gr = 10^4.

Figure 3(a) shows that, at low Ri for Case 2, the flow field is characterized also by a clockwise vortex created in the upper half of the enclosure, and a counterclockwise vortex created in the lower half. These two vortices move to the right side due to an increase in lids speed. Figure 3(b) shows that the concentration of the isotherms, close to the pair of triangular heat sources and left wall of the cavity, increases. The isotherms move to the left side and the diffusion of heat to the right side decreases with increasing the speed of the horizontal lids (decreasing Ri). This observation shows the higher forced convection effects in the right side of the cavity. When Ri increases to 100, the effects of moving wall decrease due to the enhancement in the buoyancy force. A secondary eddy appears on the left indicates that the natural convection becomes more dominant there.

Figure 4 shows values of $\overline{Nu}_{tot}$ for the illustrated cases in Figure 3. According to Figure 4(b), it is clear that the heat transfer rate increases with decreasing the nanoparticle diameter and the highest values of the heat transfer rate occur at 25 nm diameter. It is also noticed that by increasing volume fraction of nanoparticles and decreasing Ri the heat transfer rate increases. At Ri > 10 the effects of natural convection are dominant so there is an optimum volume fraction of nanoparticles which maximize the heat transfer which is about 3% in the most cases.

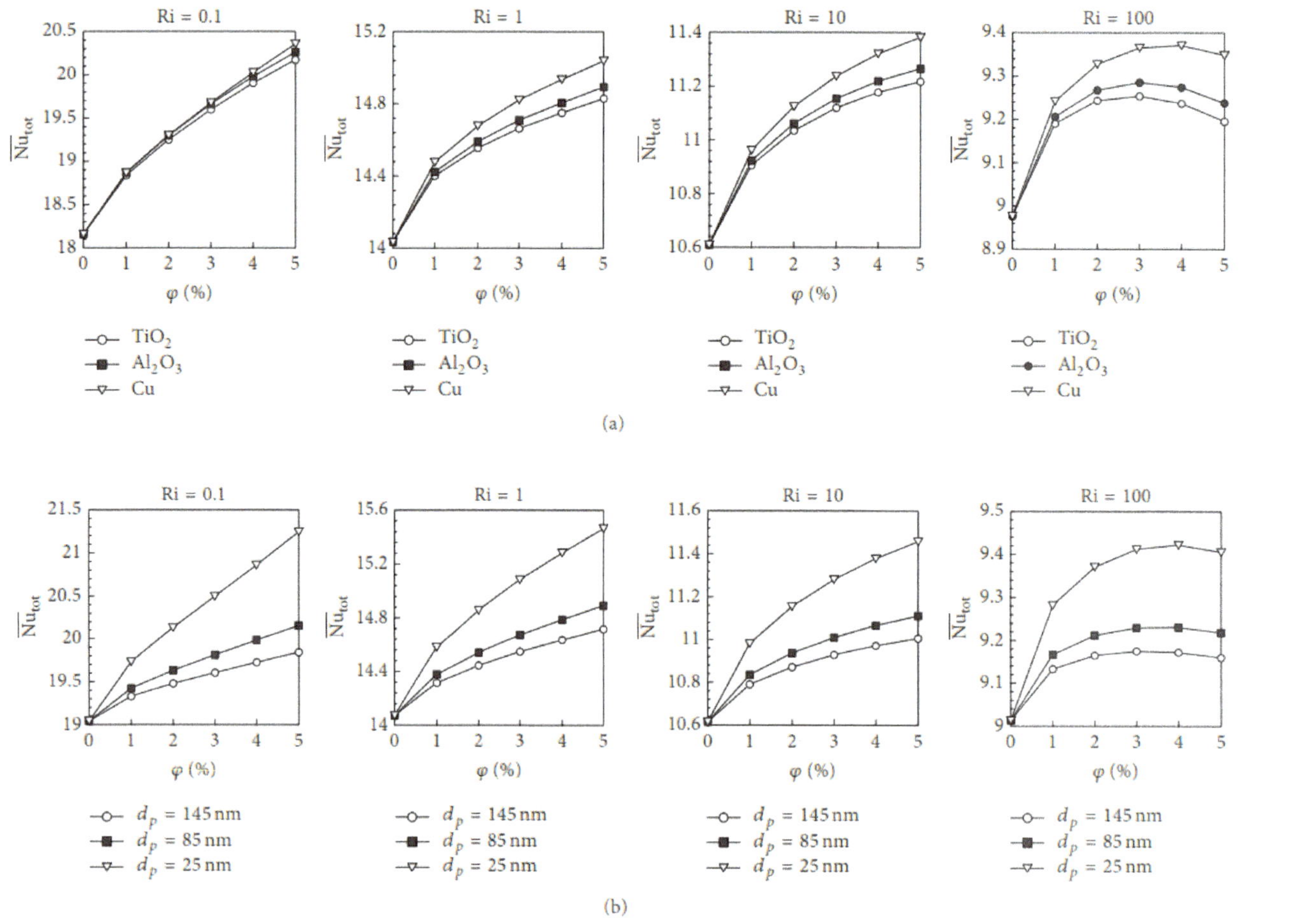

Figure 4: Variations $\overline{Nu}_{tot}$ with respect to the volume fraction of the nanoparticles at different Richardson numbers, diameters, and type of nanoparticles for (a) Case 1 and (b) Case 2. Gr = 10^4.

Figure 4(a) shows the effects of using different types of nanofluids on $\overline{Nu}_{tot}$. In all Richardson numbers nonsignificant variation on the heat transfer rate is observed when different kinds of nanofluids are used. It can be seen that among all nanoparticles those nanoparticles with higher thermal conductivity (such as Cu) produce slightly higher rate of heat transfer compared to nanoparticles with lower thermal conductivity (such as TiO_2). It is clear from Figure 4 that, at all Ri, if the direction of the two moving walls is same, the heat transfer rate increases. From Cases 2 to 1, the achieved reduction of $\overline{Nu}_{tot}$ is approximately 4.67%, 0.28%, 0.09%, and 0.44%, at Ri = 0.1, 1, 10, and 100, respectively. Consequently, the heat transfer rate variation between these two cases is negligible.

CONCLUSION

This study investigates mixed convection of the nanofluid in two-sided lid-driven square cavity with a pair of triangular heating cylinders. The effects of various design parameters such as Richardson number ($0.1 \leq Ri \leq 100$), nanoparticle volume fraction ($0 \leq \varphi \leq 0.05$), and size ($25\ nm \leq d_p \leq 145\ nm$) and type (Cu, Al_2O_3, TiO_2) of nanoparticles on the heat transfer rate are investigated. According to the presented results, the following conclusions are drawn:

- By increasing volume fraction of nanoparticles (φ) and increasing Richardson number (Ri), the heat transfer rate increases.
- The heat transfer rate increases with decreasing the nanoparticle diameter and the highest values of the heat transfer rate occur at 25 nm diameter.
- In all Richardson numbers nonsignificant variation on the heat transfer rate is observed when different kinds of nanofluids are used. It can be seen that among all nanoparticles those nanoparticles with higher thermal conductivity (such as Cu) produce slightly higher rate of heat transfer compared to nanoparticles with lower thermal conductivity (such as TiO_2).
- At Ri > 10 the effects of natural convection are dominant so there is an optimum volume fraction of nanoparticles which maximize the heat transfer which is about 3% in the most cases.
- The effect of changing direction of the two horizontal moving walls on the heat transfer rate $\overline{Nu}_{tot}$ is negligible.

Nomenclature

C_p: Specific heat, $Jkg^{-1}K^{-1}$

d_p: Diameter of the nanoparticle, m

d_f: Diameter of the base fluid molecule, m

g: Gravitational acceleration, ms^{-2}

Gr: Grashof number ($= g\beta\Delta TH3/v^2$)

H: Enclosure height, m

k: Thermal conductivity, Wm−1K−1

kb: Boltzmann's constant = 1.38066 × 10^{-23}

$\overline{Nu}_{tot}$: Heat transfer of cavity's wall

p: Dimensional pressure, Nm−2

P: Dimensionless pressure

Pr: Prandtl number (=]/αf)

Ra: Rayleigh number (= $g\beta(T_h - T_c)H^3/\alpha_f v_f$)

Re_B: Brownian-motion Reynolds number

Re: Reynolds number (= U_0H/])

Ri: Richardson number (= Gr/Re^2)

T: Temperature, K

t: Time, s

τ: Dimensionless time ($t/H/U_{ref}$)

u, V: Velocity components, ms−1

u_B: Brownian velocity of the nanoparticle, ms^{-1}

U, V: Dimensionless velocity components

w: Dimensional width of the triangular heater

W: Dimensionless width of the triangular heater

x, y: Cartesian coordinates, m

X, Y: Dimensionless Cartesian coordinates $(x,)/H$.

Greek Symbols

α: Thermal diffusivity, m^2s^{-1}

β: Thermal expansion coefficient, K^{-1}

θ: Dimensionless temperature

μ: Dynamic viscosity, $kgm^{-1}s^{-1}$]: Kinematic viscosity, m^2s^{-1}

ρ: Density, kgm^{-3}

φ: Volume fraction of the nanoparticles

ψ: Stream function $(= -\int_{Y_0}^{Y} U\,\partial Y + \psi(X, Y_0))$.

REFERENCES

1. H. F. Oztop, I. Dagtekin, and A. Bahloul, "Comparison of position of a heated thin plate located in a cavity for natural convection," International Communications in Heat and Mass Transfer, vol. 31, no. 1, pp. 121–132, 2004. · ·
2. K. Khanafer and S. M. Aithal, "Laminar mixed convection flow and heat transfer characteristics in a lid driven cavity with a circular cylinder," International Journal of Heat and Mass Transfer, vol. 66, pp. 200–209, 2013. · ·
3. A. W. Islam, M. A. R. Sharif, and E. S. Carlson, "Mixed convection in a lid driven square cavity with an isothermally heated square blockage inside," International Journal of Heat and Mass Transfer, vol. 55, no. 19-20, pp. 5244–5255, 2012. · ·
4. M. Kalteh, K. Javaherdeh, and T. Azarbarzin, "Numerical solution of nanofluid mixed convection heat transfer in a lid-driven square cavity with a triangular heat source," Powder Technology, vol. 253, pp. 780–788, 2014. · ·
5. M. El Abdallaoui, M. Hasnaoui, and A. Amahmid, "Numerical simulation of natural convection between a decentered triangular heating cylinder and a square outer cylinder filled with a pure fluid or a nanofluid using the lattice Boltzmann method," Powder Technology, vol. 27, pp. 193–205, 2015.
6. M. El Abdallaoui, M. Hasnaoui, and A. Amahmid, "Lattice-boltzmann modeling of natural convection between a square outer cylinder and an inner isosceles triangular heating body," Numerical Heat Transfer, Part A: Applications, vol. 66, no. 9, pp. 1076–1096, 2014. · ·
7. H. F. Oztop, Z. Zhao, and B. Yu, "Fluid flow due to combined convection in lid-driven enclosure having a circular body," International Journal of Heat and Fluid Flow, vol. 30, no. 5, pp. 886–901, 2009. · ·
8. M. Corcione, "Empirical correlating equations for predicting the effective thermal conductivity and dynamic viscosity of nanofluids," Energy Conversion and Management, vol. 52, no. 1, pp. 789–793, 2011. · ·
9. M. Corcione, "Heat transfer features of buoyancy-driven nanofluids inside rectangular enclosures differentially heated at the sidewalls," International Journal of Thermal Sciences, vol. 49, no. 9, pp. 1536–1546, 2010. · ·
10. Z. Haddad, H. F. Oztop, E. Abu-Nada, and A. Mataoui, "A review on natural convective heat transfer of nanofluids," Renewable and Sustainable Energy Reviews, vol. 16, no. 7, pp. 5363–5378, 2012. · ·
11. S. V. Patankar, Numerical Heat Transfer and Fluid Flow, McGraw-Hill, Washington, DC, USA, 1980.
12. D. B. Spalding, "A novel finite difference formulation for differential expressions involving both first and second derivatives," International Journal for Numerical Methods in Engineering, vol. 4, no. 4, pp. 551–559, 1972. · ·
13. R. Iwatsu, J. M. Hyun, and K. Kuwahara, "Mixed convection in a driven cavity with a stable vertical temperature gradient," International Journal of Heat and Mass Transfer, vol. 36, no. 6, pp. 1601–1608, 1993.
14. F. P. Incropera and D. P. DeWitt, Introduction to Heat Transfer, John Wiley & Sons, New York, NY, USA, 2002.

CHAPTER 22

MAGNETO BINARY NANOFLUID CONVECTION IN POROUS MEDIUM

Jyoti Sharma,[1] Urvashi Gupta,[2] and R. K. Wanchoo[2]

[1]Energy Research Centre, Panjab University, Chandigarh 160014, India

[2]Dr. S. S. Bhatnagar University Institute of Chemical Engineering and Technology, Panjab University, Chandigarh 160014, India

ABSTRACT

The effect of an externally impressed magnetic field on the stability of a binary nanofluid layer in porous medium is considered in this work. The conservation equations related to the system are solved using normal mode technique and Galerkin method to analyze the problem. The complex expressions are approximated to get useful results. Mode of heat transfer is stationary for top heavy distribution of nanoparticles in the fluid layer and top heavy nanofluids are very less stable than regular fluids. Oscillatory motions are possible for bottom heavy distribution of nanoparticles and they are not much influenced by properties of different nanoparticles. A comparative analysis of the instability of water based nanofluids with metallic (Cu, Ag) and semiconducting (TiO_2, SiO_2) nanoparticles under the influence of magnetic field is examined. Semiconducting nanofluids are found to be more stable than metallic nanofluids. Porosity destabilizes the layer while solute difference (at the boundaries of the layer) stabilizes it. Magnetic field stabilizes the fluid layer system significantly.

INTRODUCTION

The onset of instability in a horizontal fluid layer, which is heated uniformly from below, has its origin in the experiments of Bénard [1]. He observed that at the onset of instability, the fluid layer resolves into a number of equal cells and they align themselves to form a regular hexagonal pattern. Motivated by the experimental observations of Bénard, the theoretical foundations for a correct interpretation of the problem were laid by Lord Rayleigh [2]. He showed that the convection appears when a certain dimensionless parameter known as Rayleigh number exceeds its critical value. The theory of Rayleigh Bénard convection for a fluid of single component under varying assumptions of hydrodynamics and hydromagnetics is given in the monograph by Chandrasekhar [3]. When the convective motions in the fluid are produced due to two different components, the convection is known as binary convention or double diffusive convection. Stern [4] first established that the interesting effects in binary convection are due to sharp contrast between thermal and solutal diffusivities.

Citation: Jyoti Sharma, Urvashi Gupta, and R. K. Wanchoo, “Magneto Binary Nanofluid Convection in Porous Medium,”International Journal of Chemical Engineering, vol. 2016, Article ID 9424036, 8 pages, 2016. doi:10.1155/2016/9424036.

Modern nanotechnology deals with the manufacturing of objects with dimensions of less than hundred nanometers. As a consequence, the idea of suspending these nanoparticles in a base liquid (known as nanofluid) for improving thermal conductivity has been proposed by Choi [5]. He showed a substantial augmentation of heat transport in suspensions of copper or aluminum nanoparticles in water and other liquids. Pioneer works of Masuda et al. [6] and Eastman et al. [7] introduced the thermal conductivity enhancement of nanofluids to the scientific community. Since then, a large number of experimental and theoretical studies have been carried out by numerous researchers. Keblinski et al. [8] and Wang et al. [9] suggested new mechanisms for the transport of thermal energy in nanofluids. An analytical investigation to study the thermal instability of a nanofluid layer using Buongiorno's model [10] was made by Tzou [11] and found that regular fluids exhibit higher stability than nanofluids. Nield and Kuznetsov [12, 13] derived the expression for thermal Rayleigh number for instability of a nanofluid layer in porous medium and conditions for oscillatory motions. Gupta et al. [14] investigated the onset of convection in a horizontal nanofluid layer under vertical magnetic field and found that stability rises with the rise in magnetic field parameter. Rudraiah et al. [15] studied the effect of magnetic field on the convection in a rectangular enclosure and found that the magnetic field reduces the rate of heat transfer. Hayat et al. [16] described the heat transfer analysis with thermal radiation on the two-dimensional magnetohydrodynamic flow in a channel with porous walls. Recently, Sheikholeslami et al. [17,18] investigated MFD viscosity effect and thermal radiation effect on magnetohydrodynamic nanofluid flow numerically using control volume based finite element method, respectively. The effect of rotation on nanofluid convection using Darcy-Brinkman model was considered by Chand and Rana [19]. An expression for thermal Rayleigh number for stationary convection was obtained. A nonlinear study on the nanofluid convection in a rotating porous layer in terms of Nusselt number was made by Bhadauria and Agarwal [20].

For the preparation of nanofluid, when a binary liquid is used, it is termed as binary nanofluid. Instability of a binary nanofluid layer in porous medium was investigated analytically by Kuznetsov and Nield [21]. The results were approximated by limiting to the case of large nanoparticle Lewis number and large Prandtl number. Further, Gupta et al. [22] and Yadav et al. [23] studied the same problem analytically as well as numerically for alumina-water nanofluid independent of the restrictions on parameters. The influence of magnetic field on binary nanofluid convection was considered by Gupta et al. [24]. They have made valid approximations in the complex expressions for analytical study.

Convection in binary nanofluids with magnetic field has its importance in enhanced oil recovery from underground reservoirs which motivated us to study the influence of magnetic field on the stability of a binary nanofluid layer in porous medium. Darcy revealed proportionality between flow rate and applied pressure difference for a fluid flowing through porous medium which is used in the model under investigation. The nanoscale effects, Brownian motion (movement of nanoparticles from high concentration to low concentration) and thermophoresis (movement of nanoparticles from high temperature site to low temperature site), play an important role in enhancement of effective thermal conductivity of the fluid. Thus the partial differential equations incorporating both the effects of Brownian motion and thermophoresis are considered. Also, the presence of solute (with much lower diffusivity than that of nanoparticles and heat) in nanofluid layer makes the convection problem a triple diffusion process resulting in the introduction of three Rayleigh numbers due to heat, solute, and nanoparticles. The problem is solved using normal mode analysis and residuals method to get an eigenvalue equation which is further analyzed for stationary and oscillatory mode of convection in the fluid layer. Complex expressions are simplified using approximations on the parameters involved. Mathematica software is used to plot the stability curves for the impact of magnetic field, porosity, solute concentration difference, nanoparticle volume fraction difference, Dufour parameter, and Soret parameter on metallic (Cu, Ag) and semiconducting (TiO_2, SiO_2) nanoparticles.

PROBLEM FORMULATION

Let us consider a horizontal binary nanofluid layer of thickness d in the presence of uniform magnetic field $(0, 0, h_0)$. The temperature, solute concentration, and nanoparticle volume fraction at the lower and upper layers are taken to be T_1 and T_0, C_1 and C_0, and ϕ_0 and ϕ_1, respectively, with $T_1 > T_0$, $C_1 > C_0$, and $\phi_1 > \phi_0$. The basic equations which express the interaction between the fluid motion and the magnetic field as given by Buongiorno [10], Kuznetsov and Nield [21], and Gupta et al. [24] are

$$\nabla \cdot \mathbf{u}_D = 0, \tag{1}$$

$$\frac{\partial \phi}{\partial t} + \frac{\mathbf{u}_D}{\varepsilon} \cdot \nabla \phi = \nabla \cdot \left[D_B \nabla \phi + D_T \frac{\nabla T}{T} \right], \tag{2}$$

$$0 = -\nabla p - \frac{\mu}{k} \mathbf{u}_D + \left[\phi \rho_p + (1 - \phi) \left\{ \rho \left(1 - \beta_T (T - T_0) - \beta_C (C - C_0) \right) \right\} \right] \mathbf{g} + \frac{\mu_e}{4\pi} (\mathbf{h} \cdot \nabla) \mathbf{h}, \tag{3}$$

$$(\rho c)_m \frac{\partial T}{\partial t} + (\rho c)_f \mathbf{u}_D \cdot \nabla T = \left(k_m \nabla^2 T \right) + \varepsilon (\rho c)_p \cdot \left[D_B \nabla \phi \cdot \nabla T + D_T \frac{\nabla T \cdot \nabla T}{T} \right] + \rho c D_{TC} \nabla^2 C, \tag{4}$$

$$\frac{\partial C}{\partial t} + \frac{\mathbf{u}_D}{\varepsilon} \cdot \nabla C = D_S \nabla^2 C + D_{CT} \nabla^2 T, \tag{5}$$

where u_D, C, T, ϕ, D_B, D_T, D_S, DT_C, DCT, μ, h, μ_e, $(\rho c)_f$, $(\rho c)_p$, $(\rho c)_m$, k_m, ε, k, ρ_f, t, g, β_T, and β_C denote, respectively, the Darcy velocity, concentration of solute, temperature, volume fraction of nanoparticles, Brownian diffusion coefficient, thermophoretic diffusion coefficient, solute diffusivity coefficient, Dufour diffusivity, Soret diffusivity, fluid viscosity, magnetic field, magnetic permeability, fluid heat capacity, nanoparticles heat capacity, medium heat capacity, thermal conductivity of the medium, porosity, permeability of the medium, base-fluid's density, time, acceleration due to gravity, thermal volumetric coefficient, and solutal volumetric coefficient. The Maxwell's equations are

$$\frac{\partial \mathbf{h}}{\partial t} + \left(\frac{\mathbf{u}_D}{\varepsilon} \cdot \nabla\right)\mathbf{h} = (\mathbf{h} \cdot \nabla)\frac{\mathbf{u}_D}{\varepsilon} + \eta\nabla^2\mathbf{h}, \tag{6}$$

$$\nabla \cdot \mathbf{h} = 0, \tag{7}$$

where η is the resistivity of the fluid. Let us consider the nondimensional variables as

$$(x^*, y^*, z^*) = \frac{(x, y, z)}{d},$$

$$t^* = \frac{t\alpha_m}{\sigma d^2},$$

$$\mathbf{u}^* = \mathbf{u}_D\frac{d}{\alpha_m},$$

$$p^* = \frac{pk}{\mu\alpha_m},$$

$$\mathbf{h}^* = \frac{\mathbf{h}}{h_0},$$

$$\phi^* = \frac{\phi - \phi_1}{\phi_1 - \phi_0},$$

$$C^* = \frac{C - C_0}{C_1 - C_0},$$

$$T^* = \frac{T - T_0}{T_1 - T_0},$$

$$\text{with } \alpha_m = \frac{k_m}{(\rho c)_f},\ \sigma = \frac{(\rho c)_m}{(\rho c)_f}. \tag{8}$$

Using nondimensional variables, (1)–(7) (after removing the asterisks) now become

$$\nabla \cdot \mathbf{u} = 0,$$

$$0 = -\nabla p - \mathbf{u} - \mathrm{Rm}\,\hat{k} - \mathrm{Rn}\,\phi\hat{k} + R_a T\hat{k} + \frac{\mathrm{Rs}}{\mathrm{Le}}C\hat{k} + Q\frac{\mathrm{Pr}_1}{\mathrm{Pr}_2}(\mathbf{h} \cdot \nabla)\mathbf{h},$$

$$\frac{\partial T}{\partial t} + \mathbf{u} \cdot \nabla T = \nabla^2 T + \frac{N_B}{\mathrm{Ln}}\nabla\phi \cdot \nabla T + \frac{N_A N_B}{\mathrm{Ln}}\nabla T \cdot \nabla T + S_{TC}\nabla^2 C,$$

$$\frac{1}{\sigma}\frac{\partial \phi}{\partial t} + \frac{1}{\varepsilon}\mathbf{u} \cdot \nabla\phi = \frac{1}{\mathrm{Ln}}\nabla^2\phi + \frac{N_A}{\mathrm{Ln}}\nabla^2 T,$$

$$\frac{1}{\sigma}\frac{\partial C}{\partial t} + \frac{1}{\varepsilon}\mathbf{u} \cdot \nabla C = \frac{1}{\mathrm{Le}}\nabla^2 C + S_{CT}\nabla^2 T,$$

$$\nabla \cdot \mathbf{h} = 0,$$

$$\frac{\partial \mathbf{h}}{\partial t} + \frac{1}{\varepsilon}(\mathbf{u} \cdot \nabla)\mathbf{h} = (\mathbf{h} \cdot \nabla)\frac{1}{\varepsilon}\mathbf{v} + \frac{\mathrm{Pr}_1}{\mathrm{Pr}_2}\nabla^2\mathbf{h}, \tag{9}$$

where $\mathrm{Pr}_1 = \mu/\rho f\alpha_m$, $\mathrm{Ln} = \alpha_m/D_B$, $R_a = \rho g\beta_T\kappa d(T_1 - T_0)/\mu\alpha_m$, $\mathrm{R_m} = [\rho p\phi_0 + \rho(1-\phi_0)]g\,\kappa d/\mu\alpha m$, $\mathrm{Le} = \alpha_m/D_S$, $\mathrm{R_s} = \rho g\beta_C\kappa d(C_1 - C_0)/\mu D_S$, $\mathrm{R_n} = (\rho_p - \rho)(\phi_1 - \phi_0)g\,\kappa d/\mu\alpha_m$, $N_A = D_T(T_1 - T_0)/D_B T_0(\phi_1 - \phi_0)$, $N_B = (\varepsilon(\rho_c)P/\rho_c)(\phi_1 - \phi_0)$, $S_{TC} = D_{TC}(C_1 - C_0)/\alpha_m(T_1 - T_0)$, and $S_{CT} = D_{CT}(T_1 - T_0)/\alpha_m(C_1 - C_0)$ are the Prandtl number, the nanofluid Lewis number, the thermal Rayleigh number, the basic density Rayleigh number, the solute Lewis number, the solute Rayleigh number, the concentration Rayleigh number, the diffusivity ratio, the particle-density increment, regular Dufour parameter, and regular Soret parameter, respectively. Due to the inclusion of magnetic field, we have two additional parameters: the magnetic Prandtl number and the Chandrasekhar number defined as $\mathrm{Pr}_2 = \mu/\rho\eta$ and $Q = \mu_e h_o^2 k/4\pi\eta\mu$, respectively.

NORMAL MODE TECHNIQUE AND GALERKIN METHOD

Initially, the fluid layer is assumed to be in a state of rest. Therefore the basic solution as derived from (9) becomes

$$T_i = 1 - z,$$

$$C_i = 1 - z,$$

$$\phi_i = z, \tag{10}$$

where the subscript "i" represents the initial variable. Let us introduce small perturbations on the initial state as

$$\mathbf{u} = 0 + \tilde{\mathbf{u}},$$

$$p = p_i + \tilde{p},$$

$$T = T_i + \tilde{T},$$

$$C = C_i + \tilde{C},$$

$$\phi = \phi_i + \tilde{\phi},$$

$$\mathbf{h} = (0, 0, 1) + \tilde{\mathbf{h}}, \tag{11}$$

where the superscript "~" denotes the perturbed variable. Using (11) on the set of (9), we get a new set of equations known as perturbation equations. To solve perturbation equations, normal mode analysis is used as follows:

$$\left(\tilde{w}, \tilde{T}, \tilde{C}, \tilde{\phi}\right) = (W, T, \Psi, \Phi)(z)\,e^{ilx+iky+nt}, \tag{12}$$

where lx and ky are the wave numbers along x and y directions, respectively, and n is the growth rate. For neutral stability, let us write $n = i\omega$; ω is a real number and represents the frequency of oscillation. The conditions for two free boundaries are

$$W = 0,$$

$$D^2W = 0,$$

$$T = 0,$$

$$\Psi = 0,$$

$$\Phi = 0 \tag{13}$$

$$\text{at } z = 0,\ z = 1,\ \text{with } D \equiv \frac{d}{dz}.$$

The Galerkin weighted residual method is employed to determine the eigenvalue equation which gives

$$(W, T, \Psi, \Phi) = (A_1, B_1, C_1, D_1)\sin\pi z, \tag{14}$$

satisfying the boundary conditions given by (13). Using (14) in perturbation equations produces the eigenvalue equation for the system as

$$\mathrm{Rn}\,\alpha^2\left(\varepsilon J+\frac{\varepsilon\,\mathrm{Pr}_2}{\sigma\,\mathrm{Pr}_1}n\right)\left(\frac{J^2+Jn\,\mathrm{Le}+Jn-J^2S_{CT}S_{TC}\mathrm{Le}}{\varepsilon}\right.$$
$$\left.+\frac{n^2\mathrm{Le}}{\sigma}+\frac{N_A}{\mathrm{Ln}}J\left(J+\frac{n\mathrm{Le}}{\sigma}-\frac{JS_{TC}\mathrm{Le}}{\varepsilon}\right)\right)-\left(\frac{n}{\sigma}\right.$$
$$\left.+\frac{J}{\mathrm{Ln}}\right)\left(J\left(\varepsilon J+\frac{\varepsilon\,\mathrm{Pr}_2}{\sigma\,\mathrm{Pr}_1}n\right)+Q\pi^2J\right)\left(J^2+\frac{Jn}{\sigma}+Jn\right.$$
$$\left.+\frac{n^2}{\sigma}-J^2S_{CT}S_{TC}\mathrm{Le}\right)+\mathrm{Rs}\alpha^2\left(\frac{n}{\sigma}+\frac{J}{\mathrm{Ln}}\right)\left(\varepsilon J\right.$$
$$\left.+\frac{\varepsilon\,\mathrm{Pr}_2}{\sigma\,\mathrm{Pr}_1}n\right)\left(\frac{J+n}{\varepsilon}-JS_{CT}\right)+R_a\alpha^2\left(\varepsilon J+\frac{\varepsilon\,\mathrm{Pr}_2}{\sigma\,\mathrm{Pr}_1}n\right)$$
$$\cdot\left(J+\frac{n}{\sigma}-\frac{JS_{TC}\mathrm{Le}}{\varepsilon}\right)\left(\frac{n}{\sigma}+\frac{J}{\mathrm{Ln}}\right)=0, \tag{15}$$

where $\alpha=(lx^2+ky^2)^{1/2}$, $J=\pi^2+\alpha^2$.

RESULTS AND DISCUSSION

Nonoscillatory Convection

For stationary convection, put ω=0 in (15). This gives the Rayleigh number as

$$R_a=\frac{1}{(\varepsilon-\mathrm{Le}\,S_{TC})}\left[\left(\frac{(\pi^2+\alpha^2)^2\varepsilon}{\alpha^2}+\frac{(\pi^2+\alpha^2)Q\pi^2}{\alpha^2}\right.\right.$$
$$\left.\left.-\mathrm{Rn}\,\mathrm{Ln}\right)(1-S_{CT}S_{TC}\mathrm{Le})-\mathrm{Rs}\,(1-\varepsilon S_{CT})\right]$$
$$-\mathrm{Rn}\,N_A. \tag{16}$$

Note that (16) attains its minimum when $x^2 = Q/\varepsilon + 1$ with $x=\alpha^2/\pi^2$. Thus the critical wave number increases with increase in Chandrasekhar number and decreases with increase in porosity while it is independent of the presence of solute and nanoparticles. When the base fluid is one component in nonporous medium (S_{TC}, S_{CT}, and Rs are zeros; ε is unity), (16) reduces to

$$R_a=\frac{(\pi^2+\alpha^2)}{\alpha^2}\left[(\pi^2+\alpha^2)^2+Q\pi^2\right]$$
$$-\mathrm{Rn}\,(\mathrm{Ln}+N_A), \tag{17}$$

which agrees with the results given by Gupta et al. [14] for the instability in nanofluids with magnetic field.

The presence of magnetic field delays the onset of convection as coefficient of Q in (16) is positive. Also, for the large value of Rn, Rayleigh number R_a takes negative value meaning thereby that the top heavy arrangement of nanoparticles makes the system very unstable. Also N_B does not appear in the expression for R_a and N_A appears only with Rn. This is due to the fact that the regular cross-diffusion terms dominate the nanofluid cross-diffusion terms. It is worthwhile to mention that effect of N_A on R_a is much less than Rn meaning thereby that the influence of density of nanoparticles on the stability is more than their conductivity.

Oscillatory Convection

For oscillatory convection ω = 0./ Let us substitute $n = i\omega$ in (15) to find the Rayleigh number for oscillatory motions. As the binary nanofluid convection is a triple diffusion process, the expression for thermal Rayleigh number for oscillatory convection is complex. Thus in order to analyze the problem analytically, let us make valid approximations as Lewis number and Prandtl number approach to infinity, Dufour and Soret parameters are negligible, and heat capacity ratio is unity. Thus the real and imaginary parts of (15) reduce to

$$\frac{2}{\alpha^2}\left(J^2+\frac{J\pi^2Q}{\varepsilon}\right)=R_a-\frac{J^2\mathrm{Rn}}{\varepsilon\omega^2}+\frac{\mathrm{Rn}}{\varepsilon}+\frac{\mathrm{Rs}}{\varepsilon}, \tag{18}$$

$$\frac{1}{\alpha^2}\left(1-\frac{\omega^2}{J^2}\right)\left(J^2+\frac{J\pi^2Q}{\varepsilon}\right)=R_a+\frac{2\mathrm{Rn}}{\varepsilon}+\frac{\mathrm{Rs}}{\varepsilon}. \tag{19}$$

Equations (18) and (19) give

$$\omega^2=-a^2\mathrm{Rn}\left(\varepsilon\left(1+\frac{\pi^2Q}{\varepsilon J}\right)\right)^{-1}. \tag{20}$$

Note that ω^2 is negative when Rn is positive meaning thereby that the oscillatory mode of heat transfer is not possible for top heavy nanoparticle distribution in the fluid layer. This was expected as for oscillations to occur two of the buoyancy forces must act in opposite directions. Also, the value of ω^2 diminishes with the rise in Chandrasekhar number and porosity while it is independent of presence of solute in the fluid layer. Using (20) in (18) gives

$$R_a\ (\text{Oscillatory})=\frac{(\pi^2+\alpha^2)^2}{\alpha^2}+\frac{(\pi^2+\alpha^2)Q\pi^2}{\alpha^2\varepsilon}$$
$$-\frac{\mathrm{Rn}}{\varepsilon}-\frac{\mathrm{Rs}}{\varepsilon}. \tag{21}$$

The expression for R_a on neglecting Dufour and Soret parameters in (16) becomes

$$R_a\ (\text{Stationary})$$
$$=\left[\left(\frac{(\pi^2+\alpha^2)^2}{\alpha^2}+\frac{(\pi^2+\alpha^2)Q\pi^2}{\alpha^2\varepsilon}-\frac{\mathrm{Rn\,Ln}}{\varepsilon}\right)\right.$$
$$\left.-\frac{\mathrm{Rs}}{\varepsilon}\right]-\mathrm{Rn}\,N_A. \tag{22}$$

For Rn negative, the value of R_a for oscillatory convection is less than that for stationary convection. Thus mode of convection must be oscillatory for bottom heavy distribution of nanoparticles in the fluid layer as variations in the density caused by heat compete with the density gradient caused by nanoparticles and solute concentration at the bottom of the layer which makes bottom heavy nanofluids more stable than regular fluids. Also, coefficient of Rn is very small in (21) which establishes the result that oscillatory mode of convections is not much influenced by the presence of nanoparticles at the bottom of the layer.

It is interesting to note that while both the nanoscale effects (thermophoresis and Brownian motion) are responsible for the convection in a nanofluid layer, thermophoresis at a higher value of thermophoretic diffusivity is found to have more influence on the value of Rayleigh number and hence on the stability of the nanofluids. Also, note that the radius of nanoparticles appears in the expression of Brownian diffusion coefficient (refer Gupta et al. [14]) and the effects of Lewis number and diffusivity ratio are stabilizing/destabilizing for bottom/top heavy configuration of nanoparticles and hence increase in the radius of nanoparticles makes the system more stable/unstable for bottom/top heavy nanofluids.

NUMERICAL RESULTS AND DISCUSSION

Keeping in mind the fact that oscillatory motions are not much influenced by the properties of different nanoparticles, the stability curves are drawn for stationary mode of heat transfer (see (16)) with the help of Mathematica software. Figures 1–12 show the role played by magnetic field, porosity, solute concentration difference, nanoparticle volume fraction difference, Dufour parameter, and Soret parameter on Rayleigh number for metallic (Cu, Ag) and nonmetallic (TiO_2, SiO_2) nanoparticles in water based nanofluids. Table 1shows the physical properties of metallic and nonmetallic nanoparticles considered in the study.

Table 1: Physical properties of water and nanoparticles under consideration [25]

Physical properties	Water	Cu	Ag	TiO_2	SiO_2
ρ (Kg/m^3)	997.1	8933	10500	4250	2600
k (W/mK)	0.613	401	429	8.9	10.4

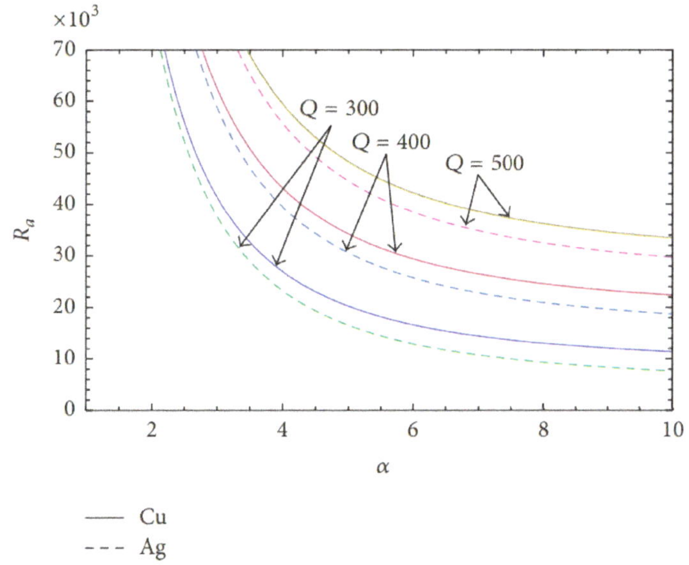

Figure 1: Effect of Q on metallic nanoparticles.

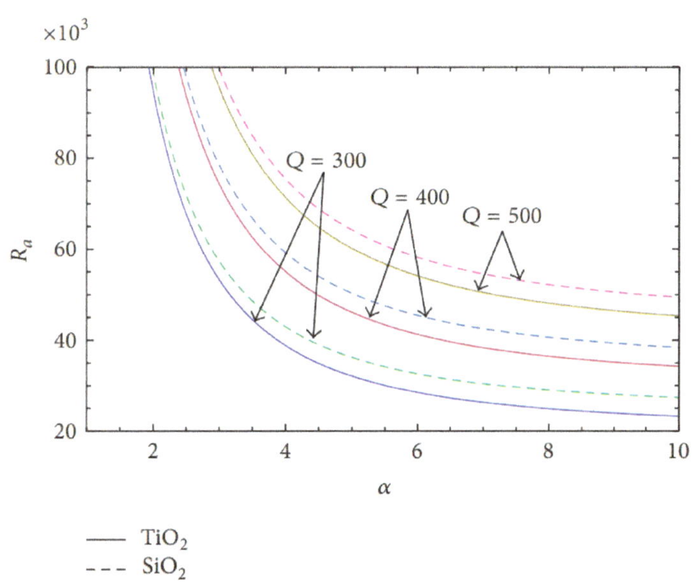

Figure 2: Effect of Q on semiconducting nanoparticles.

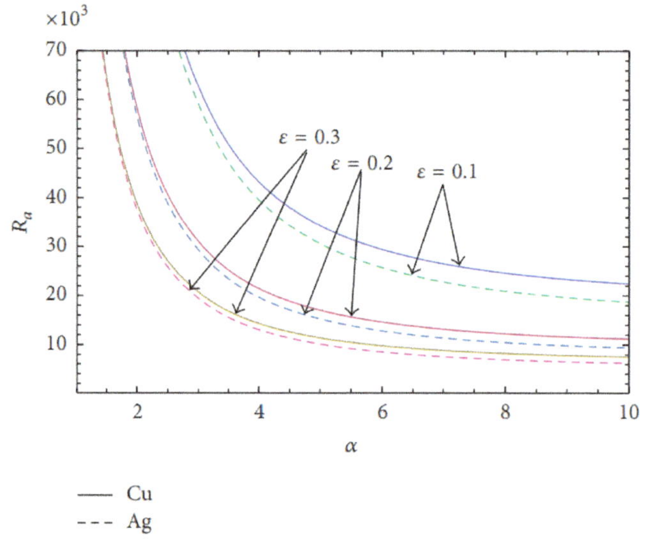

Figure 3: Effect of ε on metallic nanoparticles.

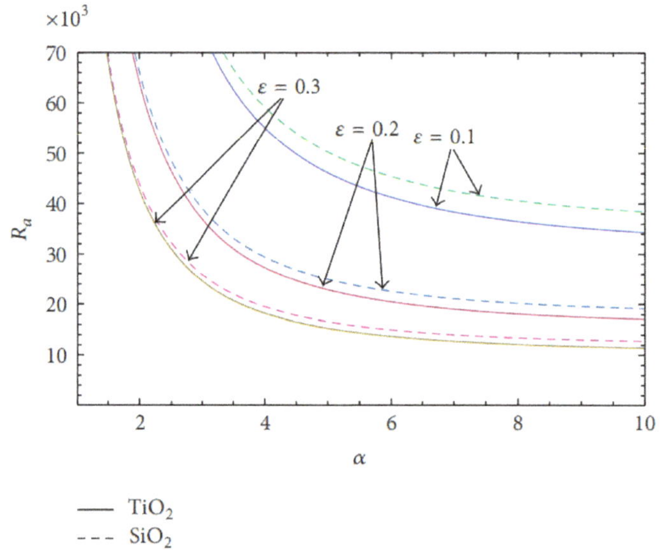

Figure 4: Effect of ε on semiconducting nanoparticles.

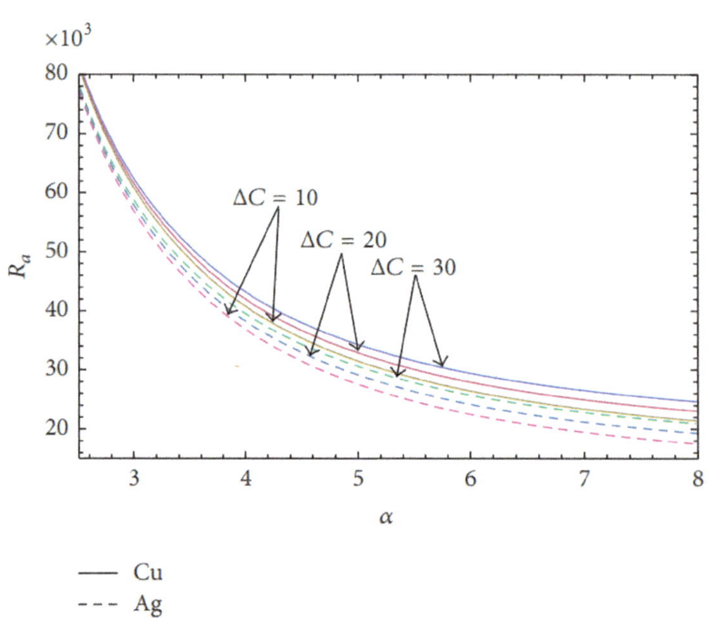

Figure 5: Effect of ΔC on metallic nanoparticles.

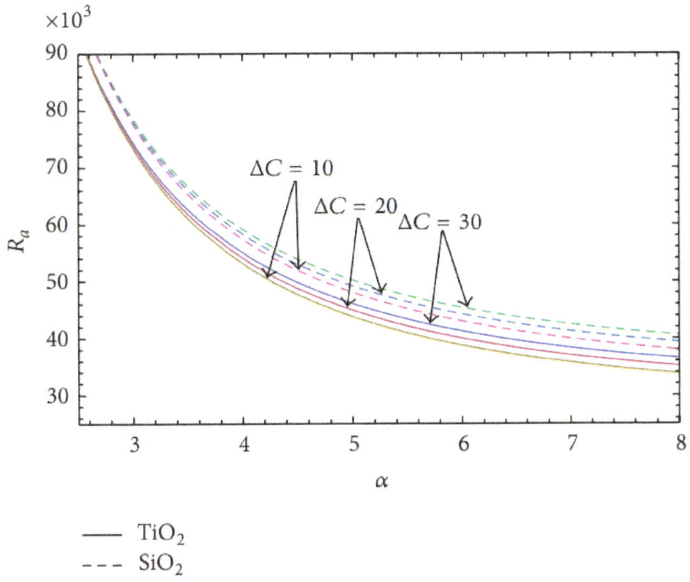

Figure 6: Effect of ΔC on semiconducting nanoparticles.

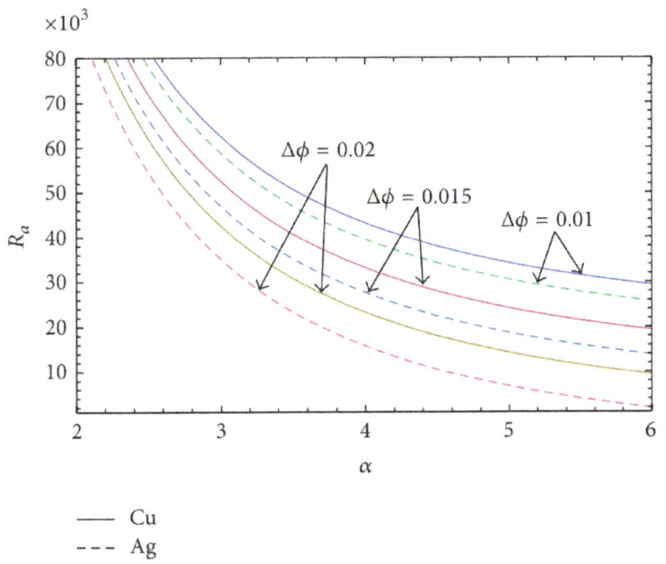

Figure 7: Effect of $\Delta\phi$ on metallic nanoparticles.

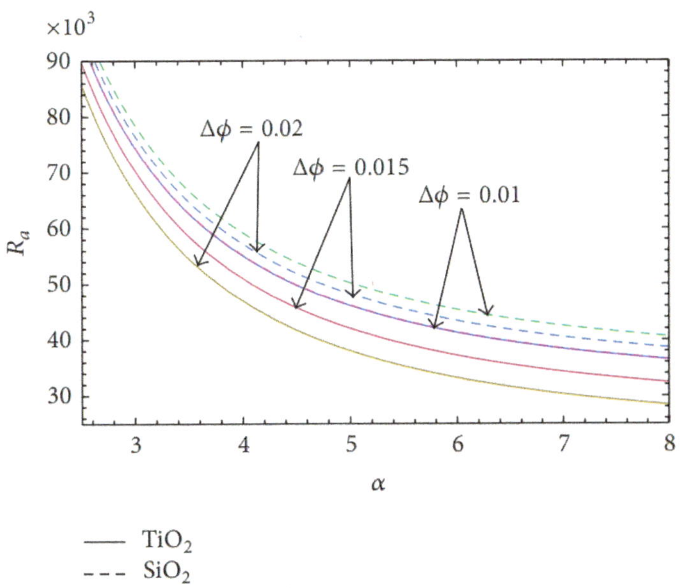

Figure 8: Effect of $\Delta\phi$ on semiconducting nanoparticles.

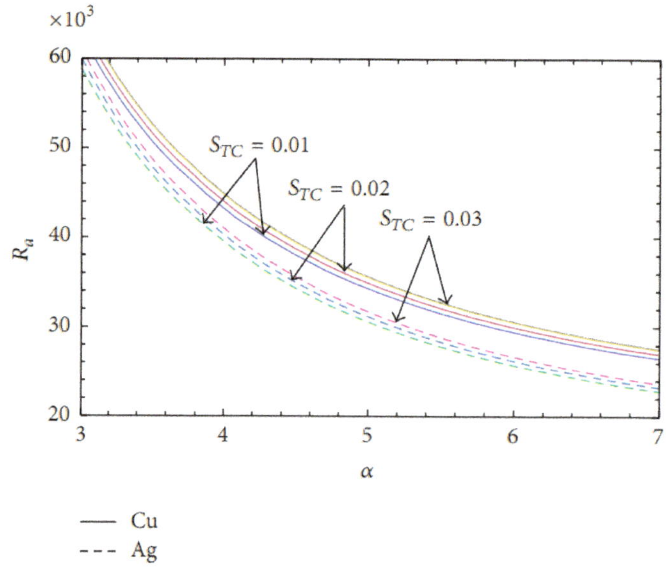

Figure 9: Effect of S_{TC} on metallic nanoparticles.

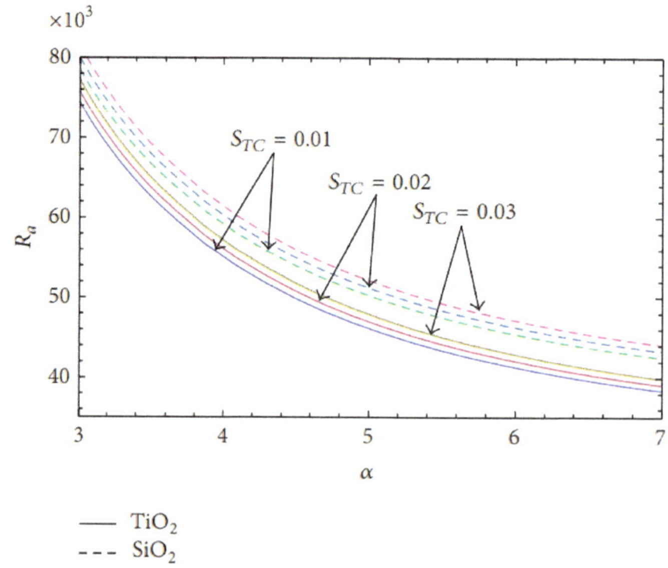

Figure 10: Effect of S_{TC} on semiconducting nanoparticles.

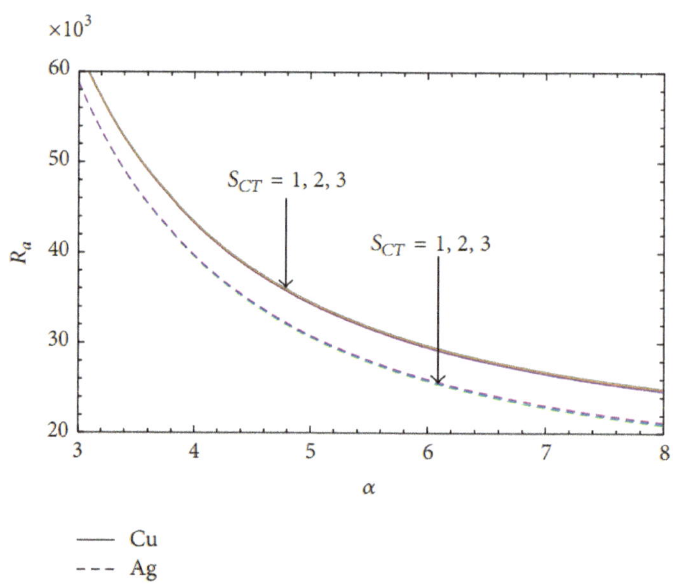

Figure 11: Effect of Sc_T on metallic nanoparticles.

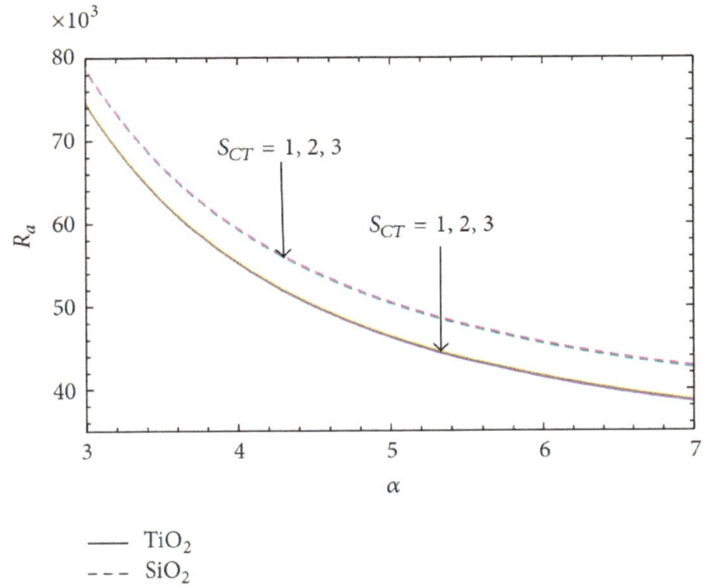

Figure 12: Effect of Sc_T on semiconducting nanoparticles.

In (16), we have three parameters Ln, Rn, and N_A which depend on nanofluid properties. According to Buongiorno [10], for water based nanofluid, Ln = 5000. Using Table 1, values of nanofluid parameters Rn and N_A are found as shown in Table 2. Note that the higher values of nanoparticle Rayleigh number lead to negative values of thermal Rayleigh number and hence system becomes unstable. Thus the values of nanoparticle Rayleigh number are taken to be small as these values can be controlled with the depth of the layer and diffusivity ratio for typical nanofluid is fixed. Further, metallic and nonmetallic (semiconducting) nanofluids have been considered with the existing data available in the literature to investigate the stability problem. We fix other parameters as Le = 2, $\varepsilon = 0.1$, $R_s = 200$, $S_{CT} = 1$, $S_{TC} = 0.001$, and $Q = 400$.

Table 2: Nanofluidparameters for metallic/nonmetallic nanoparticles in water based nanofluids under consideration

Nanofluid parameters	Cu-water nanofluid	Ag-water nanofluid	SiO_2-water nanofluid	TiO_2-water nanofluid
Rn (nanoparticle Rayleigh number)	0.392	0.465	0.0785	0.159
N_A (diffusivity ratio)	0.5	0.5	17.5	20

Figures 1 and 2 illustrate the impact of Chandrasekhar number on the Rayleigh number for metals (Cu, Ag) and nonmetals (TiO_2, SiO_2) in water based nanofluids. The value of R_a increases appreciably with the increase in Q. Thus magnetic field delays the onset of convection in a binary nanofluid layer system and this stabilizing effect is more pronounced for the case of nonmetallic nanofluids.

Figures 3 and 4 show the effect of porosity on the system. The value of Ra decreases with the increase in ε and hence porosity hastens the onset of instability. The Rayleigh number for nonmetallic nanofluids is higher than that for metallic nanofluids. To assess the influence of solute concentration difference on the stability of the fluid layer system, it is necessary to note that the solute concentration difference is entering in the expression for Rayleigh number through solute parameters (Rs, S_{CT}, S_{TC}). With an increase in ΔC ($C_1 - C_0$), R_a increases as shown in Figures 5 and 6. Thus solute concentration difference on the boundaries stabilizes the binary nanofluid layer system.

Further, the destabilizing impact of nanoparticle volume fraction difference $\Delta\phi$ ($\phi_1 - \phi_0$) (appears through Rn and N_A) on the fluid layer is depicted in Figures 7 and 8. This destabilizing influence is more pronounced for metallic nanofluids due to their higher densities as compared to semiconducting nanoparticles.

Let us study the influence of Dufour and Soret parameters arising due to the presence of solute in the fluid layer which otherwise were assumed to be negligible in Section 4 to get analytical results. Dufour parameter stabilizes the system for both metals and nonmetals almost at the same rate (Figures 9 and 10) while destabilizing influence of Soret parameter on the stability is so small that it is not depicted in Figures 11 and 12. Interestingly, the stabilizing influence of nonmetals in binary nanofluids is more than that of metals. This may be due to the fact that density of nonmetals is much lower than metals. Also, It is noteworthy that copper-water nanofluid is more stable than silver-water nanofluid (refer to Figures 1, 3, 5, 7, 9, and 11) and silicon oxide-water nanofluid is more stable than titanium oxide-water nanofluid (refer to Figures 2, 4, 6, 8, 10, and 12).

CONCLUSIONS

The influence of magnetic field on the stability of binary nanofluid layer in porous medium is investigated using normal mode technique and Galerkin method. It is assumed that the temperature, nanoparticle volume fraction, and solute concentration are constant along the walls. It is found that the regular cross-diffusion terms dominate the nanofluid terms. Out of the two nanoscale effects, thermophoresis is more responsible to initiate instability as compared to Brownian motion. The critical wave number rises with the rise in Chandrasekhar number and decreases with the rise in porosity while it is independent of the presence of solute and nanoparticles in the fluid layer. Influence of density of nanoparticles on the stability is more than their conductivity. The stability of the system increases/decreases with the increase in size of nanoparticles for bottom/top heavy distribution of nanoparticles in the fluid layer. The presence of nanoparticles on the top of the layer makes it less stable and oscillatory motions are not possible. Also, distribution of nanoparticles at the top of the layer makes system so unstable that the applied magnetic field must be increased or temperature at the bottom of the layer must be decreased relative to the top for neutral stability. Nonmetals stabilize the water based fluid layer more than metals. Copper-nanofluid is more stable than silver-water while silicon oxide-water nanofluid is more stable than titanium oxide-water nanofluid. Porosity destabilizes the system and solute concentration difference on the boundaries stabilizes it slightly. The destabilizing effect of nanoparticle volume fraction difference is more for metals than semiconductors due to their higher densities. Dufour parameter stabilizes the system while destabilizing influence of Soret parameter is almost negligible. Magnetic field is found to have a strong stabilizing impact on the fluid layer.

REFERENCES

1. H. Bénard, "Les Tourbillions Cellulaires dans une Nappe Liquide," Revue Générale des Sciences, vol. 11, pp. 1261–1271, 1900.
2. L. Rayleigh, "On convection currents in a horizontal layer of fluid, when the higher temperature is on the underside," Philosophical Magazine, vol. 6, no. 32, pp. 529–546, 1916.
3. S. Chandrasekhar, Hydrodynamic and Hydromagnetic Stability, Dover Publications, New York, NY, USA, 1981.
4. M. E. Stern, "The 'salt-fountain' and thermohaline convection," Tellus, vol. 12, no. 2, pp. 172–175, 1960. ·
5. S. Choi, "Enhancing thermal conductivity of fluids with nanoparticles," in Development and Applications of Non-Newtonian Flows, D. A. Siginer and H. P. Wang, Eds., vol. 66, ASME FED-231/MD, pp. 99–105, ASME, 1995.
6. H. Masuda, A. Ebata, K. Teramae, and N. Hishinuma, "Alteration of thermal conductivity and viscosity of liquid by dispersing ultra-fine particles. dispersion of Al_2O_3, SiO_2 and TiO_2 ultra-fine particles," Netsu Bussei, vol. 7, no. 4, pp. 227–233, 1993. ·
7. J. A. Eastman, S. U. S. Choi, S. Li, W. Yu, and L. J. Thompson, "Anomalously increased effective thermal conductivities of ethylene glycol-based nanofluids containing copper nanoparticles," Applied Physics Letters, vol. 78, no. 6, pp. 718–720, 2001.
8. P. Keblinski, S. R. Phillpot, S. U. S. Choi, and J. A. Eastman, "Mechanisms of heat flow in suspensions of nano-sized particles (nanofluids)," International Journal of Heat and Mass Transfer, vol. 45, no. 4, pp. 855–863, 2001.
9. W. Wang, L. Lin, Z. X. Feng, and S. Y. Wang, "A comprehensive model for the enhanced thermal conductivity of nanofluids," Journal of Advanced Research in Physics, vol. 3, no. 2, Article ID 021209, 2012.
10. J. Buongiorno, "Convective transport in nanofluids," ASME Journal of Heat Transfer, vol. 128, no. 3, pp. 240–250, 2006.
11. D. Y. Tzou, "Thermal instability of nanofluids in natural convection," International Journal of Heat and Mass Transfer, vol. 51, no. 11-12, pp. 2967–2979, 2008.
12. D. A. Nield and A. V. Kuznetsov, "Thermal instability in a porous medium layer saturated by a nanofluid," International Journal of Heat and Mass Transfer, vol. 52, no. 25-26, pp. 5796–5801, 2009.
13. A. V. Kuznetsov and D. A. Nield, "Thermal instability in a porous medium layer saturated by a nanofluid: Brinkman model," Transport in Porous Media, vol. 81, no. 3, pp. 409–422, 2010. ·
14. U. Gupta, J. Ahuja, and R. K. Wanchoo, "Magneto convection in a nanofluid layer," International Journal of Heat and Mass Transfer, vol. 64, pp. 1163–1171, 2013. · ·
15. N. Rudraiah, R. M. Barron, M. Venkatachalappa, and C. K. Subbaraya, "Effect of a magnetic field on free convection in a rectangular enclosure," International Journal of Engineering Science, vol. 33, no. 8, pp. 1075–1084, 1995. ·
16. T. Hayat, R. Sajjad, Z. Abbas, M. Sajid, and A. A. Hendi, "Radiation effects on MHD flow of Maxwell fluid in a channel with porous medium," International Journal of Heat and Mass Transfer, vol. 54, no. 4, pp. 854–862, 2011. ·
17. M. Sheikholeslami, M. M. Rashidi, T. Hayat, and D. D. Ganji, "Free convection of magnetic nanofluid considering MFD viscosity effect," Journal of Molecular Liquids, vol. 218, pp. 393–399, 2016. ·
18. M. Sheikholeslami, T. Hayat, and A. Alsaedi, "MHD free convection of Al2O3–water nanofluid considering thermal radiation: a numerical study," International Journal of Heat and Mass Transfer, vol. 96, pp. 513–524, 2016. ·
19. R. Chand and G. C. Rana, "On the onset of thermal convection in rotating nanofluid layer saturating a Darcy-Brinkman porous medium," International Journal of Heat and Mass Transfer, vol. 55, no. 21-22, pp. 5417–5424, 2012.
20. B. S. Bhadauria and S. Agarwal, "Natural convection in a nanofluid saturated rotating porous layer: a nonlinear study," Transport in Porous Media, vol. 87, no. 2, pp. 585–602, 2011. ·
21. A. V. Kuznetsov and D. A. Nield, "The onset of double-diffusive nanofluid convection in a layer of a saturated porous medium," Transport in Porous Media, vol. 85, no. 3, pp. 941–951, 2010. ·
22. U. Gupta, J. Sharma, and R. K. Wanchoo, "Thermosolutal convection in a horizontal nanofluid layer: introduction of oscillatory motions," in Proceedings of the Recent Advances in Engineering and Computational Sciences (RAECS ‹14), pp. 1–6, IEEE, Chandigarh, India, March 2014. · ·
23. D. Yadav, G. S. Agrawal, and R. Bhargava, "The onset of convection in a binary nanofluid saturated porous

layer," International Journal of Theoretical and Applied Multiscale Mechanics, vol. 2, no. 3, pp. 198–224, 2012.

24. U. Gupta, J. Sharma, and V. Sharma, "Instability of binary nanofluids with magnetic field," Applied Mathematics and Mechanics, vol. 36, no. 6, pp. 693–706, 2015.

25. M. Turkyilmazoglu, "Exact analytical solutions for heat and mass transfer of MHD slip flow in nanofluids," Chemical Engineering Science, vol. 84, pp. 182–187, 2012.

INDEX